INTRODUCTION TO THE FINITE ELEMENT METHOD

INTRODUCTION TO THE FINITE ELEMENT METHOD

A NUMERICAL METHOD FOR ENGINEERING ANALYSIS

CHANDRAKANT S. DESAI

Department of Civil Engineering
Virginia Polytechnic Institute and State University
Blacksburg, Virginia

JOHN F. ABEL

School of Civil and Environmental Engineering
Cornell University
Ithaca, New York

VNR VAN NOSTRAND REINHOLD COMPANY
NEW YORK CINCINNATI TORONTO LONDON MELBOURNE

Van Nostrand Reinhold Company Regional Offices:
New York Cincinnati

Van Nostrand Reinhold Company International Offices:
London Toronto Melbourne

Copyright ©1972 by **Van Nostrand Reinhold Company Inc.**

Library of Congress Catalog Card Number: 70-153193

ISBN: 0-442-22083-9

All rights reserved. No part of this work covered by the copyright hereon may be reproduced or used in any form or by any means—graphic, electronic, or mechanical, including photocopying, recording, taping, or information storage and retrieval systems—without written permission of the publisher.

Manufactured in the United States of America

Published by Van Nostrand Reinhold Company Inc.
135 West 50th Street, New York, N.Y. 10020

Published simultaneously in Canada by Van Nostrand Reinhold, Ltd.
15 14 13 12 11 10 9 8

PREFACE

High-speed electronic digital computers have enabled engineers to employ various numerical discretization techniques for approximate solutions of complex problems. The finite element method is one such technique. It was originally developed as a tool for structural analysis, but the theory and formulation have been progressively so refined and generalized that the method has been applied successfully to such other fields as heat flow, seepage, hydrodynamics, and rock mechanics. As a result of this broad applicability and the systematic generality of the associated computer codes, the method has gained wide acceptance by designers and research engineers. It is now being taught to both students and practicing engineers at many universities and will soon be a regular part of the curriculum at almost all colleges of engineering.

In order to apply the finite element method to complex modern problems, engineers must be familiar with the method's fundamental theory, assumptions, and limitations. However, a beginner may encounter difficulties in acquiring this familiarity, because there is a scarcity of unified introductory treatments that are sufficiently elementary. This text, therefore, is designed to prepare undergraduates and practicing engineers with bachelor's degrees both to solve their specific problems and to read further in the current finite element literature. It is assumed that the reader has some background in the diverse mathematical fields associated with the finite element method: matrix algebra, mechanics, variational methods, and computer skills.

The book is divided into three parts: introduction and background material, a description of the method, and applications. A relatively simple computer code is included in Appendix I. At the end of each chapter bibliographic information is provided in two sections, References and Further Reading. A brief descriptive commentary is included with most of the references cited. Abbreviations used in the references are summarized in Appendix II.

Part A is incorported both to guide the reader to useful texts on the background material and to summarize without detailed derivations the methods and equations that are employed in Parts B and C. Part B is a detailed description of the theory of the finite element method. To provide a physical insight, the formulation of the theory as well as the examples in the first three chapters of this section are based largely on the displacement method of analysis, which is common in structural and soils engineering. However, the last chapter in this section is devoted to the generalization of the theory and terminology, so that the method can be applied to other fields of engineering. Part C, Applications, consists of an extended series of examples from several engineering fields. These solved problems involve various specializations of the material covered in Parts A and B, as well as some new ideas and techniques not presented elsewhere in the text. Each chapter of Part C deals with a particular topic, so the reader may concentrate on the subject most relevant to his fields of interest. To demonstrate how aspects of the theory are incorporated in a code, Appendix I presents a simplified finite element program complete with a flow diagram, an explanation of the symbols and arrays, a user's guide, a FORTRAN IV listing, and the input and output for a sample problem.

A number of practice problems, many with hints and solutions, are included at the ends of some of the chapters in Part B.

The main goal of the authors has been a systematic, simplified summary of the available knowledge on the finite element method.

C. S. DESAI
J. F. ABEL

ACKNOWLEDGMENTS

This book is a synthesis of the work of many investigators. Among the individuals whose contributions have greatly influenced the authors are Professors J. H. Argyris, E. B. Becker, R. W. Clough, R. H. Gallagher, H. C. Martin, J. T. Oden, T. H. H. Pian, E. P. Popov, L. C. Reese, B. Fraeijs de Veubeke, E. L. Wilson, O. C. Zienkiewicz, and Dr. C. A. Felippa.

Many persons at the U.S. Army Engineer Waterways Experiment Station have provided inspiration to the authors. We wish to express gratitude to J. P. Sale, S. J. Johnson, W. C. Sherman, Dr. G. H. Keulegan, Col. L. A. Brown, F. R. Brown, and W. R. Martin for encouragement and sustained interest during the preparation of this book.

Thanks are due to Dr. G. S. Orenstein for carefully reading the manuscript and for offering valuable suggestions. We also thank Professors E. P. Popov, D. W. Murray, and R. S. Sandhu; and J. Crawford, D. C. Banks, C. J. Huval, J. B. Palmerton, and J. E. Ahlberg for helpful comments.

Finally, we are deeply grateful to our wives, P. L. Desai and V. L. S. Abel, for reading the entire manuscript, for making a number of suggestions toward improving the presentation of the material, and for assisting in the preparation of the book.

Authorship of any sort is a fantastic indulgence of the ego. It is well, no doubt, to reflect on how much one owes to others.

J. K. Galbraith, *The Affluent Society*,
Houghton-Mifflin Co., Boston, Mass.

LIST OF SYMBOLS

The following list defines the principal symbols used in this book and gives the sections where they are explained further. Other symbols are defined in context. Rectangular matrices are indicated by brackets [ ], and column vectors by braces { }. (A thorough description of the matrix notation employed is given in Section 2-1.) Overdots indicate differentiation with respect to time, and primes usually denote differentiation with respect to the space variable. For forcing parameters used in variational principles, an overbar indicates a prescribed or known value (Section 4-3), whereas within the finite element equations this notation indicates a condensed form (Sections 2-3 and 5-9).

a_i, b_i, c_i	Parameters for natural coordinate systems (5-4).
A	Cross-sectional area (1-3), general functional which is the integral of another functional (4-1), area of a two-dimensional element (5-4).
$[A]$	General rectangular matrix (2-1), coefficient matrix in eigenvalue problem of standard form (2-3), transformation relating generalized and nodal displacements (5-3).
B	Semiband width of coefficient matrix (2-2, 6-1).
$[B]$, $[B_\alpha]$	Transformation relating strains and displacements (5-5).
c	Cohesive strength (3-5).
$[C]$	Viscous damping matrix (2-4, 11-3), constitutive (stress-strain) matrix (3-3).
$[C^e]$, $[C^p]$, $[C^{ep}]$	Constitutive (stress-strain) matrices (3-5, 7-5).
D	The maximum value for an assemblage of the largest difference between the node numbers of an element (6-1).
$[D]$	Constitutive (strain-stress) matrix (3-3), general diagonal matrix (2-2, 11-3, 11-4).
e	Superscript indicating "elastic" (3-5), subscript or superscript indicating "element" (6-3).
ep	Superscript indicating "elastic-plastic" (3-5).
E	Young's modulus of elasticity (1-3, 3-3).
E_s, E_t	Secant and tangent moduli (3-5).
f	Yield surface (3-5), number of degrees of freedom at a node (6-1).
F	General functional (4-1).

$[F]$	Flexibility matrix (2-3, 9-2).
g	Acceleration due to gravity, subscript indicating global coordinate system (6-5).
$\{g\}$	Vector of gradients of field variable (12-1).
G	Shear modulus (3-3).
h	Thickness of two-dimensional element (5-1).
H	Hydraulic head (8-2), hydrostatic stress (9-3).
I_1, I_2, I_3	Strain invariants (3-2).
$[I]$	Identity matrix.
J_1, J_2, J_3	Stress invariants (3-1).
$[J]$	Jacobian matrix (5-4).
$[k], [k_\alpha]$	Element elastic stiffness matrix (1-3, 5-6, 5-7), general property matrix (8-1).
$[k_G]$	Element geometric stiffness matrix (7-6).
K	Bulk modulus (3-3).
$[K]$	Assemblage elastic stiffness matrix (1-3, 6-5).
ℓ	Length of one-dimensional element (1-3, 5-1), subscript indicating local coordinate system (6-5).
L	Lagrangian functional (4-2), natural coordinate for one-dimensional element (5-4), general differential operator (8-4).
L_i	Natural coordinates of triangle or tetrahedron (5-4).
$[L_1]$	Unit lower triangular matrix (2-2).
$[m], [m_\alpha]$	Element mass matrix (11-2).
$[M]$	Assemblage mass matrix (2-4, 11-2).
N	Number of equations for assemblage (2-2).
$[N]$	Coefficient matrix for interpolation field variable (displacement) model (5-3, 5-4).
$\{O\}, [O]$	Null vector and matrix.
P, p	Arbitrary load (1-2), superscript indicating "plastic" (3-5).
$\{q\}$	Vector of nodal field variables (displacements) for element (1-3, 5-3).
$\{Q\}, \{Q_\alpha\}$	Vector of nodal actions (forces) for element (1-3, 5-6, 5-7).
$\{Q_o\}$	Vector of nodal actions (forces) due to initial effects (initial strains) (7-1).
r	Radial coordinate (3-4), natural coordinate for hexahedral element (5-4).
$\{r\}$	Vector of nodal field variables (displacements) for assemblage (1-3, 6-5).
$\{R\}$	Vector of nodal actions (forces) for assemblage (1-3, 6-5).
s	Natural coordinate for quadrilateral or hexahedral elements (5-4).

$[s_i]$	Transformation for skewed coordinates at i^{th} node (6-6).
S_1, S_2	Surface areas over which tractions and displacements are prescribed, respectively (4-2).
$[S]$	Overall transformation for skewed coordinates (6-6), general symmetric matrix (2-2).
t	Time, natural coordinate for quadrilateral or hexahedral element (5-4).
$[t]$	Transformation from local to global coordinates (6-5).
T	Superscript indicating transpose (2-1), kinetic energy functional (4-2), temperature (7-1).
T_x, T_y, T_z	Surface traction components (4-2).
$\{T\}$	Surface traction vector (4-2).
$[T]$	Overall transformation from local to global coordinates (6-5, 7-6).
u, v, w	Displacements in cartesian coordinates (3-2), general field variables (8-1).
U, U_c	Strain energy and complementary strain energy (4-2).
V	General volume (4-2), volume of a tetrahedral element (5-4).
W, W_c	Work and complementary work (4-2).
W_p, W_{pc}	Potential and complementary potential of loads (4-2).
x, y, z	Cartesian coordinates.
X, Y, Z	Components of body force intensity (4-2).
$\{X\}$	Body force intensity vector (4-2).
$[X]$	Modal matrix (2-3, 11-4).
α	Coefficient of thermal expansion (7-1).
α_i	Generalized coordinates (5-3).
$\{\alpha\}$	Vector of element generalized coordinates (5-3).
γ	Shear strain (3-2).
δ	Variational operator (4-1).
Δ	Prefix indicating a finite increment (7-2).
ε	Normal strain (3-2).
$\{\varepsilon\}$	Vector of strains (3-2, 5-5).
θ	Circumferential coordinate (3-4), slope of a beam (5-3), angle of twist per unit length in torsion (12-4).
λ	Eigenvalue (2-3, 7-6), Lamé constant (3-3).
$[\Lambda]$	Spectral matrix (2-3, 11-4).
μ	Lamé constant (3-3), viscosity (14-4).
ν	Poisson ratio (3-3).
ν_t	Tangent Poisson ratio (3-5).
Π, Π_c	Total potential and complementary potential functionals (4-2).

Π_R	Reissner's functional (4-2).
ρ	Mass density (4-2, 11-2).
σ	Normal stress (3-1).
$\{\sigma\}$	Stress vector (3-1, 5-5, 9-2).
τ	Shear stress (3-1).
ϕ	Angle of internal friction (3-5), potential function (14-3).
$\{\phi\}$, $[\phi]$	Coefficient matrix for generalized coordinate model (5-3).
ψ	General field variable (8-3).
$\{\psi\}$	Vector of nodal field variables (12-1).
ω	Frequency (2-3, 11-4).

CONTENTS

PART A

INTRODUCTION AND BACKGROUND MATERIAL

1
INTRODUCTION

The finite element method has developed simultaneously with the increasing use of high-speed electronic digital computers and with the growing emphasis on numerical methods for engineering analysis. Although the method was originally developed for structural analysis, the general nature of the theory on which it is based has also made possible its successful application for solutions of problems in other fields of engineering.

1-1 BACKGROUND AND APPLICATIONS

It is not possible to obtain analytical mathematical solutions for many engineering problems. An analytical solution is a mathematical expression that gives the values of the desired unknown quantity at any location in a body, and as a consequence it is valid for an infinite number of locations in the body. Analytical solutions can be obtained only for certain simplified situations. For problems involving complex material properties and boundary conditions, the engineer resorts to numerical methods that provide approximate, but acceptable, solutions. In most of the numerical methods, the solutions yield approximate values of the unknown quantities only at a discrete number of points in the body. The process of selecting only a certain number of discrete points in the body can be termed *discretization*. One of the ways to discretize a body or a structure is to divide it into an equivalent system of smaller bodies, or units. The assemblage of such units then represents the original body. Instead of solving the problem for the entire body in one operation, the solutions are formulated for each constituent unit and combined to obtain the solution for the original body or structure. This approach is known as *going from part to whole*. Although the analysis procedure is thereby considerably simplified, the amount of data to be handled is dependent upon the number of smaller bodies into which the original body is divided. For a large number of subdivisions it is a formidable task to handle the volume of data manually, and recourse must be made to automatic electronic computation.

Many of the numerical methods developed before the era of electronic computers are now adapted for use with these machines. Perhaps the best known is the finite difference method.[1†] Other types of classical methods that have been adapted to modern computation are such residual methods as the method of least squares and such variational methods as the Ritz method.[2]

In contrast to the techniques mentioned above, the finite element method is essentially a product of the electronic digital computer age. Therefore, although the approach shares many of the features common to the previous numerical approximations, it possesses certain characteristics that take advantage of the special facilities offered by the high-speed computers. In particular, the method can be systematically programmed to accommodate such complex and difficult problems as nonhomogeneous materials, nonlinear stress-strain behavior, and complicated boundary conditions. It is difficult to accommodate these complexities in the methods mentioned above. Another favorable aspect of the finite element method is the variety of levels at which we may develop an understanding of the technique. One may take a very physical or intuitive approach to the learning and using of the method. This approach is similar to the extension of the familiar concepts of the analysis of framed structures as one-dimensional bodies to problems involving two- and three-dimensional structures. On the other hand, one may develop a rigorous mathematical interpretation of the method. In this volume we shall steer a middle course with a slant toward the intuitive or physical approach.

A large number of publications, most of which have appeared since 1960, discuss the finite element method, particularly its application to structural mechanics. A few of the formative works are given as References 3 through 9, while many others will appear as annotated references at the ends of later chapters.

The finite element method is applicable to a wide range of boundary value problems in engineering. In a *boundary value problem*,[2] a solution is sought in the region of the body, while on the boundaries (or edges) of the region the

† The superscript numbers refer to references at the end of the chapter. Details of abbreviations for references used in this text are given in Appendix II.

values of the dependent variables (or their derivatives) are prescribed. In Table 1-1, examples of specific applications of the method are given for the three major categories of boundary value problems: equilibrium or steady state problems, eigenvalue problems, and propagation or transient problems.[2]

Since the majority of the applications of the method is in the realm of solid mechanics (including structural, soil, and rock mechanics), the descriptions in this volume are presented primarily in terms of these fields of study. Problems in these fields are usually tackled by one of three approaches: the *displacement method*, the *equilibrium method*, or the *mixed method*. Displacements are assumed as primary unknown quantities in the displacement method; stresses are assumed as primary unknown quantities in the equilibrium method; and some displacements and some stresses are assumed as unknown quantities in the mixed method.

For the benefit of the beginner, the explanation of the theory of the method is further simplified by restricting the descriptions in the early part of this book to the displacement method, with examples drawn from the fields of structural and soil mechanics. Since most engineers are familiar with displacement analyses involving such terms as stresses, strains, and equilibrium, our emphasis on the displacement approach should help the reader grasp the fundamentals of the method. Indeed, the great majority of literature on the finite element method has been written in terms of the displacement method. Once the basic concepts of the finite element method are understood in these simple terms, one can study the advanced literature and apply the method to specialties other than structures or soils. In Chapter 8, therefore, the theory and terminology will be generalized so that the method can be readily extended to the other fields shown in Table 1-1. Moreover, further discussion of the various applications will be given in Part C, Applications.

1-2 GENERAL DESCRIPTION OF THE METHOD

In brief, the basis of the finite element method is the representation of a body or a structure by an assemblage of subdivisions called *finite elements*, Figure 1-1. These elements are considered interconnected at joints which are called *nodes* or *nodal points*. Simple functions are chosen to approximate the distribution or variation of the actual displacements over each finite element. Such assumed functions are called *displacement functions* or *displacement models*. The unknown magnitudes or amplitudes of the displacement functions are the displacements (or the derivatives of the displacements) at the nodal points, Figure 1-2. Hence, the final solution will yield the approximate displacements at discrete locations in the body, the nodal points. A displacement model can be expressed in various simple forms, such as polynomials and

TABLE 1-1 ENGINEERING APPLICATIONS OF THE FINITE ELEMENT METHOD

	Fields of Study	*Equilibrium Problems*	*Eigenvalue Problems*	*Propagation Problems*
1	Structural engineering, structural mechanics, and aerospace engineering.	Analysis of beam, plate, and shell structures. Analysis of complex or hybrid structures. Two- and three-dimensional stress analysis. Torsion of prismatic sections.	Stability of structures. Natural frequencies and modes of vibration of structures. Linear viscoelastic damping.	Propagation of stress waves. Dynamic response of structures to aperiodic loads. Coupled thermoelasticity and thermoviscoelasticity. Viscoelastic problems.
2	Soil mechanics, foundation engineering, and rock mechanics.	Two- and three-dimensional stress analyses. Construction and excavation problems. Slope stability problems. Soil-structure interaction. Analysis of dams, tunnels, boreholes, culverts, locks, etc. Steady-state seepage of fluids in soils and rocks.	Natural frequencies and modes of vibration of soil-structure combinations.	Transient seepage in soils and rocks. Flow-consolidation in deformable porous media. Propagation of stress-waves through soils and rocks. Dynamic soil-structure interaction.
3	Heat conduction.	Steady-state temperature distribution in solids and fluids.		Transient heat flow in solids and fluids.

TABLE 1-1 (Continued)

Fields of Study	Equilibrium Problems	Eigenvalue Problems	Propagation Problems
4 Hydrodynamics, hydraulic engineering, and water resources.	Solutions for potential flow of fluids. Solutions for viscous flow of fluids. Steady-state seepage in aquifers and porous media. Analysis of hydraulic structures and dams.	Seiche of lakes and harbors (natural periods and modes of oscillation). Sloshing of fluids in rigid and flexible containers.	Salinity and pollution studies of estuaries (diffusion). Sediment transport. Unsteady fluid flow. Wave propagation. Transient seepage in porous media and aquifers.
5 Nuclear engineering.	Analysis of reactor containment structures. Steady state temperature distributions in reactors and reactor structures.		Dynamic analysis of reactor containment structures. Thermoviscoelastic analysis of reactor structures. Unsteady temperature distribution in reactors and reactor structures.

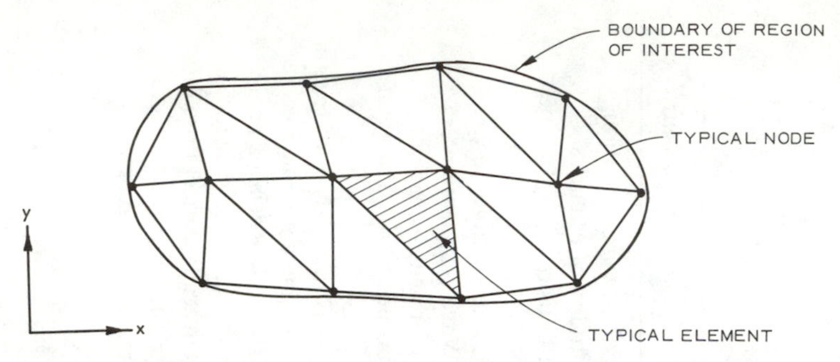

Figure 1-1 Two-dimensional region represented as an assemblage of triangular elements.

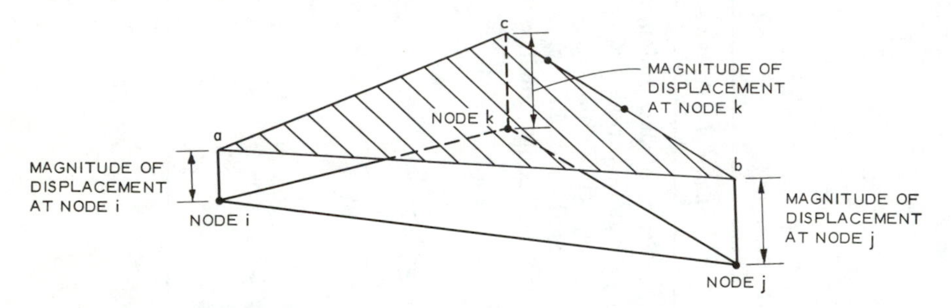

Figure 1-2 Isometric view of triangular element with linear displacement model plotted in the third dimension.

trigonometric functions. Since polynomials offer ease in mathematical manipulations, they have been employed commonly in finite element applications.

A variational principle of mechanics, such as the principle of minimum potential energy, is usually employed to obtain the set of equilibrium equations for each element. The potential energy of a loaded elastic body or structure is represented by the sum of the internal energy stored as a result of the deformations and the potential energy of the external loads. If the body is in a state of equilibrium, this energy is a minimum. This is a simple statement of the principle of minimum potential energy. In Chapter 4, we shall present a more general statement of this principle.

To illustrate the concept of the potential energy, we consider an elementary example. Figure 1-3 shows a linear spring, with spring constant s lb/in., under a load of P lbs. The spring deflects by the amount u inches under the load. The internal strain energy stored in the spring is given by

$$\begin{aligned} U &= \tfrac{1}{2}(\text{Force in the spring})\,(\text{Displacement}) \\ &= \tfrac{1}{2}(su)(u) \\ &= \tfrac{1}{2}su^2 \end{aligned} \tag{1.1}$$

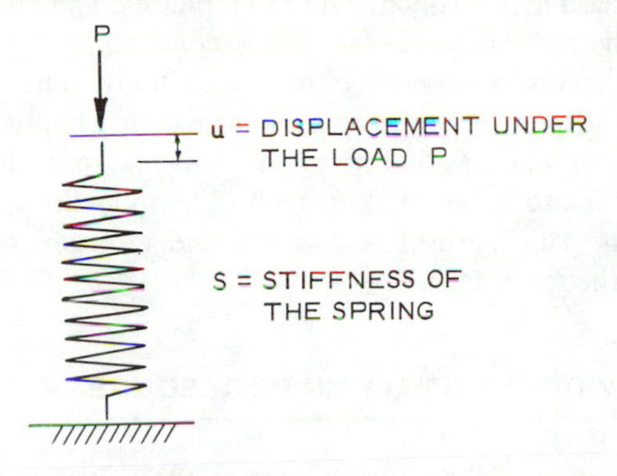

Figure 1-3 Linear spring.

The potential energy of the external load P is

$$W_p = \text{(Load)(Displacement from zero potential state)} \qquad (1.2)$$
$$= -Pu$$

where the zero potential state corresponds to $u = 0$. Hence the total potential energy Π is

$$\Pi = U + W_p = \tfrac{1}{2}su^2 - Pu \qquad (1.3)$$

The minimum of Π can be obtained by differentiating it with respect to u, and equating the result to zero:

$$\frac{\partial \Pi}{\partial u} = su - P = 0 \qquad (1.4)$$

or

$$su = P$$

Equation (1.4) is the equilibrium equation or load-displacement relation for the loaded spring.

The equilibrium equations for the entire body are then obtained by combining the equations for the individual elements in such a way that continuity of displacements is preserved at the interconnecting nodes. These equations are modified for the given displacement boundary conditions and then solved to obtain the unknown displacements. In many types of problems, the desired solution is in terms of the strains or stresses rather than the displacements, so additional manipulations or calculations may be necessary.

The general description given above is recapitulated in a step-by-step procedure in Section 1-3. Nevertheless, it is apparent that the theory of the finite element method may be divided into two phases. The first phase consists of the study of the *individual element*. The second phase is the study of the *assemblage* of elements representing the entire body. The theoretical and practical aspects of these two phases are covered in Chapters 5 and 6, respectively. The reader who has had no previous exposure to the finite element method may find it profitable to skim Part C, Applications, before reading Chapters 5 and 6. This perusal will contribute a broader perspective to our study of the theory in Part B.

1-3 SUMMARY OF THE ANALYSIS PROCEDURE

The general description of the finite element method given in Section 1-2 can be detailed in a step-by-step procedure. This sequence of steps describes the actual solution process that is followed in setting up and solving any equilibrium problem. Although the present summary is based upon a procedure developed for structural mechanics applications,[10] it can be generalized for applications to other fields as well (Chapter 8).

The following six steps summarize the finite element analysis procedure:

1. *Discretization of the continuum.* The *continuum* is the physical body, structure, or solid being analyzed. Discretization may be simply described as the process in which the given body is subdivided into an equivalent system of finite elements. The finite elements may be triangles, groups of triangles, or quadrilaterals for a two-dimensional continuum. Figure 1-1 shows an arbitrary two-dimensional body discretized into a system of triangular finite elements. For three-dimensional analysis, the finite elements may be tetrahedra, rectangular prisms, or hexahedra. A detailed description of various types of finite elements is given in Chapter 5. Although some efforts have been made to automate the process of subdivision, it remains essentially a judgmental process on the part of the engineer. He must decide what number, size, and arrangement of finite elements will give an effective representation of the given continuum for the particular problem considered. Some of the criteria that influence this decision are discussed in Chapter 6.

The notion of a continuum is usually clear cut. For example, in elasticity the continuum to be subdivided is merely the deformable body or the structure. However, in some cases the extent of the continuum to be modeled may not be clearly defined. In a seepage problem, for instance, one may consider a porous medium of infinite or very large extent in one or more dimensions. Only a significant portion of such a continuum need be considered and discretized. Indeed, practical limitations require that one include only the significant portion of any large continuum in the finite element analysis. This aspect will be clarified further in Parts B and C.

2. *Selection of the displacement models.* The assumed displacement functions or models represent only approximately the actual or exact distribution of the displacements. For example, a displacement function is commonly assumed in a polynomial form, and practical considerations limit the number of terms that can be retained in the polynomial. The simplest displacement model that is commonly employed is a linear polynomial. The shaded portion *abc* over the triangular finite element *ijk* in Figure 1-2 is a typical example of linear variation of displacement given by a linear polynomial. Obviously, it is generally not possible to select a displacement function that can represent exactly the actual variation of displacement in the element. Hence, the basic approximation of the finite element method is introduced at this stage.

There are three interrelated factors which influence the selection of a displacement model. First, the type and the degree of the displacement model must be chosen. (Usually, since a polynomial is chosen, only the degree of the polynomial is open to decision.) Second, the particular displacement magnitudes that describe the model must be selected. These are usually the displacements of the nodal points, but they may also include derivatives of the displacements at some or all of the nodes. Third, the model should satisfy certain requirements which ensure that the numerical results approach the correct solution. All of these factors are detailed in Chapter 5.

3. *Derivation of the element stiffness matrix using a variational principle.* The stiffness matrix consists of the coefficients of the equilibrium equations derived from the material and geometric properties of an element and obtained by use of the principle of minimum potential energy. The stiffness relates the displacements at the nodal points (the *nodal displacements*) to the applied forces at the nodal points (the *nodal forces*). The distributed forces applied to the structure are converted into equivalent concentrated forces at the nodes. The equilibrium relation between the stiffness matrix $[k]$, nodal force vector $\{Q\}$, and nodal displacement vector $\{q\}$ is expressed as a set of simultaneous linear algebraic equations,†

$$[k]\{q\} = \{Q\} \tag{1.5}$$

The elements of the stiffness matrix are the *influence coefficients*. Recall that a stiffness of a structure is an influence coefficient that gives the force at one point on a structure associated with a unit displacement of the same or a different point.

† The notation for matrices and vectors used in this book is

Rectangular matrix: $[A]$
Column matrix (vector): $\{A\}$
Row matrix (vector): $\{A\}^T$

where the superscript T indicates the transpose.

We consider a simple, rather trivial, illustration to clarify the meaning of equation (1.5). Figure 1-4 shows a column (length 20 inches, elastic modulus E), loaded axially and having cross sectional areas A_1 and A_2. The column is divided into two one-dimensional finite elements with three nodes. This structure can also be represented by two linear springs with stiffness s_1 and s_2. Figure 1-4(d) shows a typical spring element corresponding to a typical finite element for the column. The stiffness $[k]$ of the typical simple element of this example can be evaluated without recourse to an assumed displacement model by directly computing exact influence coefficients. For example, in Figure 1-4(c) if we apply a unit downward displacement at point 1 and restrain point 2, we induce a force (influence coefficient) equal to AE/ℓ at point 1, and a force (influence coefficient) equal to $-AE/\ell$ at point 2. Similarly, we can evaluate the influence coefficients caused by a unit movement at node 2 (node 1 restrained) and obtain the stiffness matrix $[k]$ for the element as

$$[k] = \begin{bmatrix} \dfrac{AE}{\ell} & -\dfrac{AE}{\ell} \\[2ex] -\dfrac{AE}{\ell} & \dfrac{AE}{\ell} \end{bmatrix} = \frac{AE}{\ell}\begin{bmatrix} 1 & -1 \\ -1 & 1 \end{bmatrix}$$

We can now write equation (1.5) for our column element as

$$\frac{AE}{\ell}\begin{bmatrix} 1 & -1 \\ -1 & 1 \end{bmatrix}\begin{Bmatrix} u_1 \\ u_2 \end{Bmatrix} = \begin{Bmatrix} Q_1 \\ Q_2 \end{Bmatrix} \tag{1.6}$$

where u_1 and u_2 are the displacements of nodes 1 and 2, and Q_1 and Q_2 are the known externally applied loads at points 1 and 2.

The stiffness matrix for an element depends upon (1) the displacement model, (2) the geometry of the element, and (3) the local material properties or constitutive relations. For an elastic isotropic body a pair of parameters such as the Young's modulus E and the Poisson's ratio ν define the local material properties. Since material properties are assigned to a particular finite element, it is possible to account for nonhomogeneity by assigning different material properties to different finite elements in the assemblage. A summary of constitutive relations is given in Chapter 3. Complete details regarding general methods of deriving element stiffness matrices and element load vectors are given in Chapter 5.

In this text we emphasize the utility of variational methods for the derivation of stiffness matrices and load vectors for finite elements. However, the use of variational principles, though always convenient for finite element formulations, is not essential.[11]

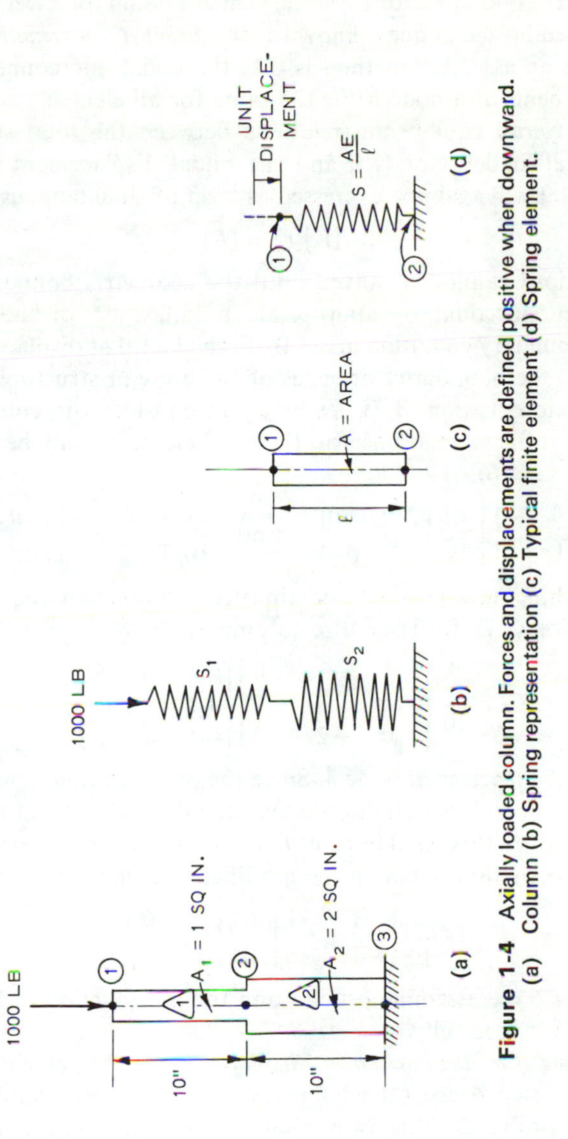

Figure 1-4 Axially loaded column. Forces and displacements are defined positive when downward. (a) Column (b) Spring representation (c) Typical finite element (d) Spring element.

4. *Assembly of the algebraic equations for the overall discretized continuum.* This process includes the assembly of the overall or global stiffness matrix for the entire body from the individual element stiffness matrices, and the overall or global force or load vector from the element nodal force vectors. The most common assembly technique is known as the *direct stiffness method.* In general, the basis for an assembly method is that the nodal interconnections require the displacements at a node to be the same for all elements adjacent to that node. The overall equilibrium relations between the total stiffness matrix $[K]$, the total load vector $\{R\}$, and the nodal displacement vector for the entire body $\{r\}$ will again be expressed as a set of simultaneous equations.

$$[K]\{r\} = \{R\} \tag{1.7}$$

These equations cannot be solved until the geometric boundary conditions are taken into account by appropriate modification of the equations. A geometric boundary condition arises from the fact that displacements may be prescribed at the boundaries or edges of the body or structure.

To illustrate equation (1.7), let us again consider the column in Figure 1-4(a). The stiffness relations for the two elements can be computed by using equation (1.6).

$$\frac{E}{10}\begin{bmatrix} 1 & -1 \\ -1 & 1 \end{bmatrix}\begin{Bmatrix} u_1 \\ u_2 \end{Bmatrix} = \begin{Bmatrix} 1000 \\ 0 \end{Bmatrix}, \quad \text{and} \quad \frac{2E}{10}\begin{bmatrix} 1 & -1 \\ -1 & 1 \end{bmatrix}\begin{Bmatrix} u_2 \\ u_3 \end{Bmatrix} = \begin{Bmatrix} 0 \\ R_3 \end{Bmatrix}$$

We can combine these results to obtain three simultaneous equilibrium equations, equation (1.7), for the entire column as

$$\frac{E}{10}\begin{bmatrix} 1 & -1 & 0 \\ -1 & 3 & -2 \\ 0 & -2 & 2 \end{bmatrix}\begin{Bmatrix} u_1 \\ u_2 \\ u_3 \end{Bmatrix} = \begin{Bmatrix} 1000 \\ 0 \\ R_3 \end{Bmatrix} \tag{1.8}$$

where R_3 is the reaction at node 3. Since the base of the column is restrained, the geometric boundary condition is that the displacement at node 3 is zero. We can introduce this condition into equation (1.8) by deleting the last row and the last column and obtain the modified equilibrium relation as

$$\frac{E}{10}\begin{bmatrix} 1 & -1 \\ -1 & 3 \end{bmatrix}\begin{Bmatrix} u_1 \\ u_2 \end{Bmatrix} = \begin{Bmatrix} 1000 \\ 0 \end{Bmatrix} \tag{1.9}$$

In Chapter 6 the assembly process and the introduction of boundary conditions are discussed fully.

5. *Solutions for the unknown displacements.* The algebraic equations assembled in step 4 are solved for the unknown displacements. In linear equilibrium problems, this is a relatively straightforward application of matrix algebra techniques. However, for nonlinear problems the desired solutions are obtained by a sequence of steps, each step involving the modification of the stiffness matrix and/or load vector. Some methods for the

solution of large sets of simultaneous equations for equilibrium, eigenvalue, and propagation problems will be stated in Chapter 2.

The solution of equation (1.9) is elementary:

$$u_1 = \frac{15{,}000}{E} \text{ inches, and } u_2 = \frac{5{,}000}{E} \text{ inches}$$

6. *Computation of the element strains and stresses from the nodal displacements.* In certain cases the magnitudes of the primary unknowns, that is the nodal displacements, will be all that are required for an engineering solution. More often, however, other quantities derived from the primary unknowns, such as strains and/or stresses, must be computed.

By using the computed displacements, the strains for the two elements in our column example may be computed as follows:

$$\varepsilon_1 = \frac{(\Delta\ell)_1}{\ell_1} = \frac{u_2 - u_1}{10} = -\frac{1000}{E}$$

$$\varepsilon_2 = \frac{(\Delta\ell)_2}{\ell_2} = \frac{u_3 - u_2}{10} = -\frac{500}{E}$$

Hence the axial stresses are

$$\sigma_1 = E\varepsilon_1 = -1000 \text{ psi}$$
$$\sigma_2 = E\varepsilon_2 = -500 \text{ psi}$$

where the negative signs indicate compression.

In general, the stresses and strains are proportional to the derivatives of the displacements; and in the domain of each element meaningful values of the required quantities are calculated. These "meaningful values" are usually taken as some average value of the stress or strain at the center of the element. The procedure for obtaining element stresses and strains is discussed in Chapters 5 and 6.

1-4 FUNDAMENTALS FOR THE UNDERSTANDING OF THE METHOD

From the background and descriptions given in Sections 1-1 through 1-3, it is apparent that the reader must have a working capability in several fundamental areas before he can thoroughly understand the theory and applications of the finite element method. These fields include matrix algebra, solid mechanics, variational methods, and computer skills. The reader is assumed to have some background in these four areas. However, to make the text as self-contained as possible, a short chapter on each of the first three topics is included, wherein the methods and equations relevant to the finite element method are stated. Annotated reference lists are provided for those readers desiring further study of these three topics.

Matrix techniques are definitely the most efficient and systematic way to handle the algebra of the finite element method. Basically, matrix algebra provides a scheme by which a large number of equations can be stored and manipulated. Chapter 2 is devoted to this topic and touches upon the solution of equations for equilibrium, eigenvalue, and propagation problems.

Because the majority of applications of the finite element method is in structural and soil mechanics, the relevant equations of solid mechanics are summarized in Chapter 3. Here some emphasis is placed upon material characterization.

An introduction to minimum principles and the calculus of variations is presented in Chapter 4. Chapter 4 also includes statements of some specific variational principles of solid mechanics.

Finally, the description of the finite element method is not complete without a discussion of its advantages and limitations. However, the reader will better appreciate this discussion after being exposed to the details of the method. This aspect, therefore, shall be postponed until Chapter 15, which concludes the book.

REFERENCES

(1) Forsythe, G. E., and Wasow, W. R., *Finite Difference Methods for Partial Differential Equations*, New York, John Wiley and Sons, Inc., 1960. This rigorous treatment includes sections on computers, hyperbolic equations, parabolic equations, elliptic equations, and initial-value problems.

(2) Crandall, S. H., *Engineering Analysis*, New York, McGraw-Hill Book Company, 1956. A survey of numerical procedures for engineering problems, this book gives applications of extremum principles to equilibrium, eigenvalue, and propagation problems. It also includes the Ritz method.

(3) Argyris, J. H., *Energy Theorems and Structural Analysis*, London, Butterworth's, 1960. (Reprinted from *Aircraft Engineering*, Oct. 1954–May 1955.) This is one of the classics of matrix methods of structural analysis. It unifies and generalizes the fundamental energy principles for elastic structures and develops practical methods of analysis.

(4) Turner, M. J., Clough, R. W., Martin, H. C., and Topp, L. C., "Stiffness and Deflection Analysis of Complex Structures," *J. Aero. Sci.*, Vol. 23, No. 9 (Sept. 1956). A method is developed for calculating stiffness influence coefficients of complex structures. It is shown how these stiffnesses can be used in analysis procedures that can be carried out conveniently on automatic digital computers.

(5) Turner, M. J., "The Direct Stiffness Method of Structural Analysis," Structures and Materials Panel Paper, AGARD Meeting, Aachen, Germany, Sept.

17, 1959. This paper contains the first formal development of the direct stiffness method for assembling the overall equations of the finite element method.

(6) Clough, R. W., "The Finite Element Method in Plane Stress Analysis," *Proc. 2nd Conf. Electronic Computation*, ASCE, Pittsburg, Pa., Sept. 8–9, 1960. This paper marks the first use of the term *finite element method.*

(7) Melosh, R. J., "Basis for Derivation of Matrices for the Direct Stiffness Method," *AIAA J.*, Vol. 1, No. 7 (July 1963). The advantages of using the minimum variational theorems are reviewed, and requirements on displacements for the convergence of the potential energy approach are developed.

(8) Felippa, C. A., and Clough, R. W., "The Finite Element Method in Solid Mechanics," *Numerical Solution of Field Problems in Continuum Physics, SIAM-AMS* Proceedings, Vol. II, American Mathematical Society, Providence, R. I., 1970. This paper contains a mathematical interpretation of the method, a survey of both its applications and techniques, and an extensive bibliography.

(9) Zienkiewicz, O. C., Irons, B. M., Ergatoudis, J., Ahmad, S., and Scott, F. C., "Isoparametric and Associated Element Families for Two- and Three-Dimensional Analysis," *FEM Tapir*, 1969. This paper summarizes the isoparametric finite element concept, a powerful generalized tool for formulating complete and conforming elements with geometry and models of any polynomial order. Examples of the application of various elements are given. Superparametric and subparametric elements are also considered.

(10) Clough, R. W., "The Finite Element Method in Structural Mechanics," In *Stress Analysis: Recent Developments in Numerical and Experimental Methods*, edited by O. C. Zienkiewicz and G. S. Holister, New York, John Wiley and Sons, 1954, Chap. 7. This article is one of three chapters dealing with the finite element method in this book. It presents the philosophy and basic concepts of the method as applied to two-dimensional elasticity.

(11) Oden, J. T., "A General Theory of Finite Elements: Part I. Topological Considerations," *IJNME*, Vol. 1, No. 2 (1969). This paper presents a rigorous general theory of the finite element method valuable for wider applications to problems in mathematical physics.

FURTHER READING

Argyris, J. H., "Continua and Discontinua," *First Conf.* This is a broad survey of developments in the field of finite element analysis, including static and dynamic analysis, plate and shell structures, elastic-plastic analysis, plastic collapse, finite displacements, and structural optimization.

Fraeijs de Veubeke, B. (ed), *Matrix Methods of Structural Analysis*, AGARDograph 72, New York, MacMillan Company, 1964.

Przemieniecki, J. S., *Theory of Matrix Structural Analysis*, New York, McGraw-Hill Book Co., 1968. The text includes basic principles of structural idealization, matrix methods, assembly processes, and solution of static, dynamic, and nonlinear problems. An extensive reference list is appended.

2
MATRIX TECHNIQUES

Finite element methods of analysis usually generate a relatively large number of linear simultaneous equations. It is impractical or impossible to solve these equations without the use of an electronic digital computer. To ensure systematic and efficient solutions, it is imperative to employ matrix methods of analysis, which represent a union of linear algebra and numerical analysis. Therefore, the understanding of the basic principles of matrix algebra is essential to the understanding of the finite element method.

Many references cover the topic of matrix algebra, and for this information the reader may consult References 1 through 4. We shall define the notation used in this book and briefly introduce a few aspects of matrix analysis that are important in the application of the finite element method.

2-1 MATRIX NOTATION

A two-dimensional matrix is denoted herein by square brackets, [ ]. Individual elements of such a matrix are represented by doubly subscripted letters, where the first subscript gives the row number and the second gives the column number. A column vector is denoted by braces, { }, and its elements by singly subscripted letters. In this book the superscript T always designates a transpose of a matrix or vector. Therefore, a row vector is indicated as the transpose of a column vector, that is $\{\ \}^T$. Finally, the inverse of a square matrix is signified by the superscript -1, that is $[\ ]^{-1}$.

By using the above notation, an nth order system of linear algebraic equations can be written in the following general form:

$$[A]\{x\} = \{b\} \tag{2.1}$$

where $[A]$ is the $n \times n$ square matrix of coefficients A_{ij}, $\{x\}$ is the $n \times 1$ column vector of unknowns, and $\{b\}$ is the $n \times 1$ column vector of constants or "right-hand sides."

2-2 SOLUTION OF LARGE SYSTEMS OF ALGEBRAIC EQUATIONS

The two basic approaches for the solution of large systems of equations are *elimination* and *iteration*.[1,4] The former, also known as the *direct* approach and typified by Gaussian elimination, is a procedure wherein the matrix $[A]$ is transformed to a triangular form which can be solved directly for the unknowns. The latter is a series of successive corrections to an original estimate for the unknowns, the process being carried out repetitively until the size of the necessary corrections becomes negligible.[4]

There has been some utilization of iterative methods for the large systems of equations generated in applications of the finite element method to three-dimensional problems. However, with the improvement in digital computer hardware, elimination techniques have proved to be more versatile and reliable.[5,6,7]

Fortunately, in most finite element applications, the equations are amenable to solution by techniques that take advantage of the special character of such systems of equations. In particular, the coefficient (stiffness) matrices generated in the analyses are usually symmetric and are thinly populated or sparse.

A general form for a coefficient matrix obtained in a two-dimensional finite element analysis is shown schematically in Figure 2-1. Only the shaded areas of the $N \times N$ matrix contain nonzero elements. We denote $2B-1$ as the *band width* of the matrix where B is the *semiband* or *half-band width*. We can achieve economy of core storage by taking into account the fact that the coefficient matrix is usually symmetric. Thus we need to store only the $N \times B$ portion of the matrix as indicated in Figure 2-2. For extremely large systems of equations even this method of storage may be inadequate, and we must partition the matrix as in Figure 2-3. Then only a few of the triangular submatrices need be stored in the computer core *at a given time*, while the remaining portions are kept in peripheral storage, for example on disk or tape.[5,7]

Because the symmetric coefficient matrices are typically positive definite, we know that a solution is possible by elimination methods without pivoting. This is advantageous, because pivoting would destroy the banded nature

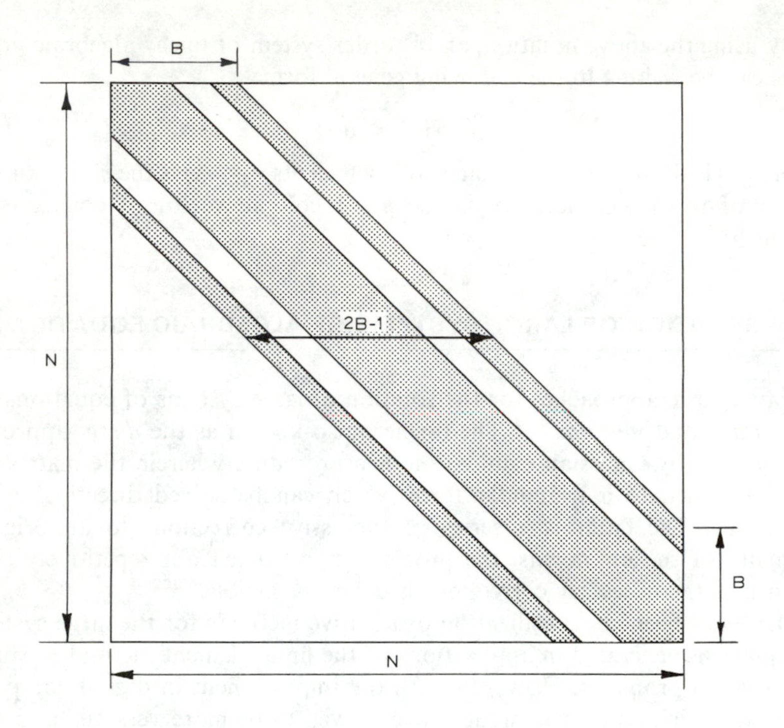

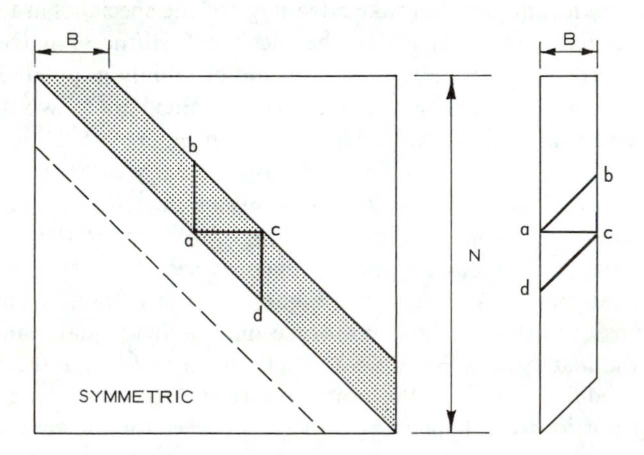

Figure 2-1 (Top) General form of banded coefficient matrix.

Figure 2-2 (Bottom) Two-dimensional storage of upper band of symmetric coefficient matrix.

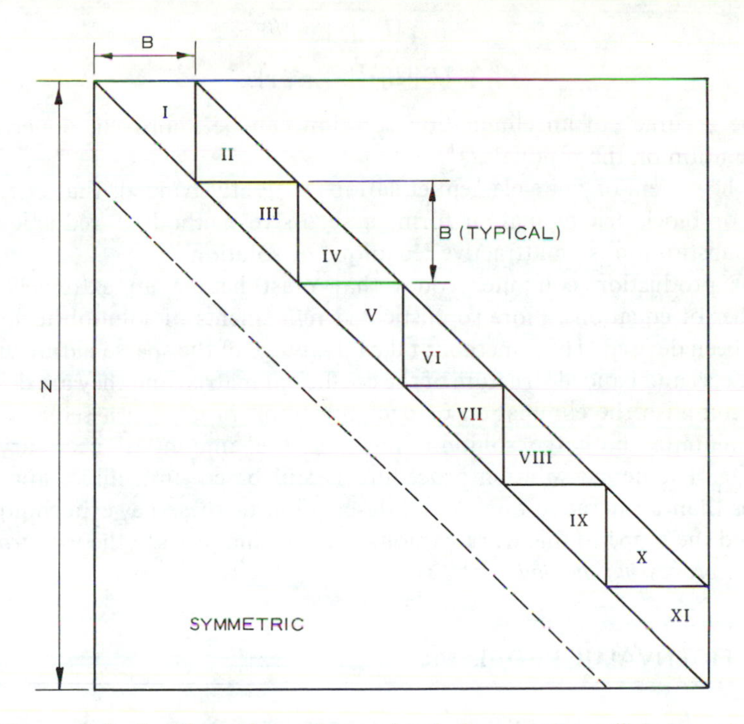

Figure 2-3 Partitioned form of upper band of symmetric coefficient matrix.

of the matrix. In addition, without pivoting, elimination performed using one row affects only the triangle of elements within the band below that row. For example, in Figure 2-2, reduction involving row *ac* modifies only the triangle *acd*. This permits us to carry out the elimination with only a few of the submatrices of Figure 2-3 in core.

The technique of direct solution generally favored is a simple symmetric Gauss-Doolittle decomposition of the symmetric matrix.[1] In effect, this elimination decomposes the symmetric matrix $[S]$ as follows:

$$[S] = [L_1][D][L_1]^{\mathrm{T}} \tag{2.2}$$

where $[L_1]$ is a unit lower triangular matrix and $[D]$ is a diagonal matrix. In the compact storage scheme indicated in Figures 2-2 and 2-3, we need to retain only the product $[D][L_1]^{\mathrm{T}}$ and we can overwrite this upper triangular band matrix on the upper band of $[S]$ as we compute the decomposition. By using the decomposed coefficient matrix, we can solve for any given right-hand side $\{b\}$ by the sequential operations of forward reduction and back solution:[1, 8]

$$[L_1]\{y\} = \{b\} \tag{2.3a}$$

$$[D][L_1]^T\{x\} = \{y\} \tag{2.3b}$$

The accuracy of an elimination solution can be improved, if necessary, by iteration on the residuals.[5,8,9]

If the system of finite element equations is tightly banded, that is tridiagonal or block tridiagonal in form, a recursive method of reduction and backsubstitution is an attractive technique of solution.[10]

For production computer codes that must handle an extremely large number of equations, more sophisticated refinements of solution techniques have been devised. These methods take advantage of the sparse nature and the symmetry and banded structure of the coefficient matrix, and they are designed to accomplish the elimination or decomposition process in a sequence that will minimize both the solution time and the amount of necessary core storage. The actual solution procedure is still based on a direct approach, rather than an iterative one. A full description of these newer techniques is beyond the scope of this work. Among these techniques are the *wavefront*, or *frontal processing methods.*[11,12]

2-3 EIGENVALUE PROBLEMS

In many engineering applications of the finite element method, we obtain an algebraic eigenvalue problem written in *standard form* as

$$[A]\{x_i\} = \lambda_i\{x_i\} \qquad \text{for} \quad i = 1, 2, 3, \ldots, n \tag{2.4}$$

or

$$[A][X] = [X][\Lambda] \tag{2.5}$$

where $[X]$ is the *modal matrix* and is the square matrix whose columns are the eigenvectors $\{x_i\}$ of $[A]$, $[\Lambda]$ is the *spectral matrix* and is the diagonal matrix of the eigenvalues λ_i of $[A]$, and n is the order of the system of equations.

In finite element applications, we usually obtain symmetric matrices. If the matrix is both symmetric and positive definite, all the eigenvalues will be real, positive numbers. Moreover, the eigenvectors of a symmetric matrix are independent; therefore, the matrix $[X]$ is nonsingular.[1,3] Another useful property of a symmetric matrix is that if the eigenvectors are normalized in such a way that $\{x_i\}^T\{x_i\} = 1$, the inverse of the modal matrix is equal to its transpose, that is the modal matrix is *orthogonal.*

We can solve the eigenvalue problem for large systems by numerical schemes that are either direct or iterative. The direct methods are more general and are commonly employed, although the iterative schemes are suitable for computations when only one or a few of the eigenvalues are needed.[1] Among various direct approaches to be found in literature are the Jacobi, Givens, Householder, and QR methods. Among the iterative techniques are the power (or Stodola-Vianello) method and inverse iteration.[1,3,9] A discussion of specific methods is beyond the scope of this work.

Most subroutines available from computer program libraries are written for the solution of the problem in the standard form. However, in displacement finite element method formulations, we usually obtain an eigenvalue problem in the form

$$[K]\{r\} = \omega^2[M]\{r\} \tag{2.6}$$

This equation must be converted to the standard form if available library subroutines are to be used.

In equation (2.6) we have used the notation of a free vibration problem in which ω is the natural frequency of the system, $[M]$ is the mass matrix, and $[K]$ is the structural stiffness matrix. Equations similar to equation (2.6) are also obtained for such other types of eigenvalue problems as structural stability problems.

We shall limit our attention to the situation in which $[A]$ is symmetric and $[K]$ and $[M]$ are symmetric and banded. Furthermore, we shall consider three common cases: (1) $[M]$ is of the same band width as $[K]$; (2) $[M]$ is diagonal with no zero values on the principal diagonal; and (3) $[M]$ is diagonal with some zero values on the principal diagonal. The reductions to standard form given here follow Felippa.[13]

For the first case, we decompose $[M]$ in Cholesky form

$$[M] = [L][L]^{\mathrm{T}} \tag{2.7a}$$

where $[L]$ is a banded lower triangular matrix. Then the standard form is given by

$$[A] = [L]^{-1}[K][L^{\mathrm{T}}]^{-1} \tag{2.7b}$$

$$\{x\} = [L]^{\mathrm{T}}\{r\} \tag{2.7c}$$

$$\lambda = \omega^2 \tag{2.7d}$$

The results in equation (2.7b) can be obtained efficiently by solving the following two sets of linear triangular equations:

$$[L][B] = [K] \tag{2.7e}$$

$$[L][A] = [B]^{\mathrm{T}} \tag{2.7f}$$

In general, $[A]$ and $[B]$ are full matrices, and $[A]$ may be overwritten on $[B]$.

For the second case, the standard form is given by

$$[A] = [M]^{-1/2}[K][M]^{-1/2} \tag{2.8a}$$

$$\{x\} = [M]^{1/2}\{r\} \tag{2.8b}$$

$$\lambda = \omega^2 \tag{2.8c}$$

where $[A]$ is now symmetric and banded, and therefore can be stored and solved efficiently. For example, it can be stored by overwriting on $[K]$.

The third case is not quite so straightforward. A technique used to transform the equations to standard form involves removal of the equations corresponding to zero values on the diagonal of $[M]$. If the total number of equations is N, and the number of nonzero elements on the diagonal of $[M]$ is $N_r < N$, the following sets of equations are solved:

$$[K]\{f_i\} = \{e_i\}, i = 1, 2, 3, \ldots, N_r \tag{2.9a}$$

This can be done efficiently by using equations (2.2) and (2.3). The vector $\{e_i\}$ is a unit vector with zeros in all locations except the one corresponding to the ith nonzero diagonal of $[M]$. As a result, $\{f_i\}$ is a column of the *flexibility matrix*, defined as

$$[F] = [K]^{-1} \tag{2.9b}$$

It is possible to select the N_r elements of each of the $\{f_i\}$ vectors which correspond only to the nonzero diagonal elements of $[M]$ and thus to construct the condensed $(N_r \times N_r)$ flexibility matrix $[\bar{F}]$. The standard form of the condensed eigenvalue problem is now given by

$$[A] = [\bar{M}]^{1/2}[\bar{F}][\bar{M}]^{1/2} \tag{2.9c}$$

$$\{x\} = [\bar{M}]^{1/2}\{r\} \tag{2.9d}$$

$$\lambda = 1/\omega^2 \tag{2.9e}$$

Here $[A]$ is symmetric and full, and the overbars indicate the condensed forms.

2-4 SOLUTION OF PROPAGATION PROBLEMS

The third class of problems solved by the finite element method consists of propagation or initial value problems. Typically, these problems involve time as one of the independent variables, and initial values of the dependent variables are given in addition to boundary conditions. Let us consider the matrix form of a problem of this type which is frequently encountered in dynamics for forced vibration problems

$$[M]\{\ddot{r}\} + [C]\{\dot{r}\} + [K]\{r\} = \{R(t)\} \tag{2.10}$$

Here the overdots indicate derivatives with respect to time. Hence, if $\{r\}$ is the vector of unknown displacements of a structure, $[M]$ is the mass matrix, $[C]$ is the viscous damping matrix, $[K]$ is the structural stiffness matrix, and the vector $\{R(t)\}$ is the known time history of the loads. We assume that the boundary conditions have already been incorporated into equation (2.10) and that the initial displacements $\{r(0)\}$ and initial velocities $\{\dot{r}(0)\}$ are also known. Equation (2.10) is a set of second order ordinary differential equations in time, which is usually the highest order encountered in engineering applications.

There are several different approaches available for the solution of propagation problems, and they fall into three general categories:[8] (1) the *derivative methods*, especially the Taylor series method, (2) the *finite difference methods*, and (3) the *Lagrangian methods*, including Runge-Kutta integration and step-by-step methods. A primary concern in choosing and employing a solution procedure is the *stability* of the method.[8] If either the method or the problem is *inherently unstable*, we will fail to obtain meaningful results. Moreover, even with potentially stable methods, if our choice of time increment is too large, the accumulation of rounding errors will be unacceptable. This divergence from the correct solution is called *induced instability*. A detailed study of stability considerations and of various solution methods is beyond the scope of this work and the reader is referred to texts on numerical analysis.[8]

However, we will consider one method for the solution of propagation problems which has been employed successfully in conjunction with the finite element method. This is a modified version of a linear-acceleration, step-by-step method.[14] The solution is based on the assumption that the acceleration during any small time interval Δt can be approximated by a linear variation; this corresponds to parabolic velocity and cubic displacement distributions over the interval. Using a subscript to denote the time level of the calculation, we can therefore write

$$\{\ddot{r}_t\} = \frac{6}{(\Delta t)^2}\{r_t\} - \{A_t\} \tag{2.11a}$$

$$\{\dot{r}_t\} = \frac{3}{\Delta t}\{r_t\} - \{B_t\} \tag{2.11b}$$

$$\{A_t\} = \frac{6}{(\Delta t)^2}\{r_{t-\Delta t}\} + \frac{6}{\Delta t}\{\dot{r}_{t-\Delta t}\} + 2\{\ddot{r}_{t-\Delta t}\} \tag{2.11c}$$

$$\{B_t\} = \frac{3}{\Delta t}\{r_{t-\Delta t}\} + 2\{\dot{r}_{t-\Delta t}\} + \frac{\Delta t}{2}\{\ddot{r}_{t-\Delta t}\} \tag{2.11d}$$

We see that the acceleration and velocity at a particular time step can be computed from the current displacements and from the acceleration, velocity, and displacement at the previous time step. Substituting equations (2.11) into equation (2.10), we obtain

$$[\bar{K}]\{r_t\} = \{\bar{R}_t\} \tag{2.12a}$$

$$[\bar{K}] = [K] + \frac{3}{\Delta t}[C] + \frac{6}{(\Delta t)^2}[M] \tag{2.12b}$$

$$\{\bar{R}_t\} = \{R_t\} + [C]\{B_t\} + [M]\{A_t\} \tag{2.12c}$$

Equation (2.12a) is the basic equation to be solved by one of the methods outlined in Section 2-2.

The technique given by equations (2.11) and (2.12) is stable if the time increment is small in comparison to the shortest natural period of vibration represented in the algebraic discretization of the dynamic problem. When instabilities do occur in the method, they begin as an oscillation about the true solution. The modification of the method used to minimize this oscillatory instability is the use of a time interval of $2\Delta t$. The accelerations at the end of the interval, $\{\ddot{r}_{t+\Delta t}\}$, are computed by the above technique. We then drop back to the middle of the time increment to obtain more stable values of the accelerations, velocities, and displacements to be used at the beginning of the next time increment. These are obtained as follows:

$$\{\ddot{r}_t\} = \tfrac{1}{2}(\{\ddot{r}_{t+\Delta t}\} + \{\ddot{r}_{t-\Delta t}\}) \tag{2.13a}$$

$$\{\dot{r}_t\} = \{\dot{r}_{t-\Delta t}\} + \frac{\Delta t}{2}(\{\ddot{r}_{t-\Delta t}\} + \{\ddot{r}_t\}) \tag{2.13b}$$

$$\{r_t\} = \{r_{t-\Delta t}\} + \Delta t\{\dot{r}_{t-\Delta t}\} + \frac{(\Delta t)^2}{6}(2\{\ddot{r}_{t-\Delta t}\} + \{\ddot{r}_t\}) \tag{2.13c}$$

This modified linear-acceleration, step-by-step method was formulated by Wilson.[14]

A solution method different from any of the approaches mentioned above has been proposed for time dependent problems. This proposal is that time be considered directly as a dimension or variable in the finite element method.[15] This extension of the finite element concept to include the time domain has not yet been completely developed and tested. We shall discuss this possibility further in Chapter 8.

REFERENCES

(1) Fox, L., *An Introduction to Numerical Linear Algebra*, New York, Oxford University Press, 1965. An excellent introduction to both linear algebra and its associated numerical analysis, including discussion of various practical algorithms for the solution of linear equations and eigenvalue problems.

(2) Gere, J. M., and Weaver, W., Jr., *Matrix Algebra for Engineers*, New York, Van Nostrand Reinhold Co., 1965. This paperback covers basics of matrix and vector algebra. It is quite readable and gives many examples and exercises.
(3) Faddeeva, V. N., *Computational Methods of Linear Algebra*, New York, Dover Publications, Inc., 1959. This paperback contains a rigorous treatment of many theorems and calculation techniques of linear algebra.
(4) Varga, R. S., *Matrix Iterative Analysis*, Englewood Cliffs, Prentice-Hall, Inc., 1962. The theory and practice of iterative methods of solutions of equations is covered and an extensive bibliography is included.
(5) See Reference 8, Chapter 1.
(6) White, R. N., "Optimum Solution Techniques for Finite-Difference Equations," *Proc. ASCE, J.ST Dn*, Vol. 89, ST4 (Aug. 1963). This paper presents a comparison of Guassian elimination with iterative and gradient methods.
(7) Schkade, A. F., "Solution Techniques for Large Systems of Stiffness Equations," M. S. Thesis, Department of Aerospace Engineering, The University of Texas, Austin, Jan. 1969. A comparison of banded Gaussian elimination and block iteration methods.
(8) Fox, L., and Mayers, D. F., *Computing Methods for Scientists and Engineers*, London, Oxford University Press, 1968. A medium-level treatment of numerical analysis presented in terminology and mathematics readily understandable by engineers.
(9) Wilkinson, J. H., *The Algebraic Eigenvalue Problem*, London, Oxford University Press, 1965. A detailed and complete discourse on both the theory and practicalities of eigenvalue problems and the solution of linear equations.
(10) Acton, F. S., *Numerical Methods That Work*, New York, Harper & Row, 1970. Efficient numerical methods for the solution of algebraic, transcendental and differential equations. The discussion is geared to engineering applications.
(11) Melosh, R. J., and Bamford, R. M., "Efficient Solution of Load Deflection Equations," *Proc. ASCE, J.ST Dn*, Vol. 95, ST4 (Apr. 1969). Discussion ST 12 (Dec. 1969), Vol. 96, ST 1 (Jan. 1970), ST2 (Feb. 1970), and ST 5 (May 1970). Presents a wavefront processing method and a modified elimination solution method. The wavefront method involves the allocation of data according to structural connectivity rather than the bandwidth.
(12) Irons, B. M., "A Frontal Solution Program for Finite Element Analysis," *IJNME*, Vol. 2, No. 1 (1970). Gives program and FORTRAN code to assemble and solve symmetric, positive-definite equations.
(13) Felippa, C. A., "Refined Finite Element Analysis of Linear and Nonlinear Two-Dimensional Structures," Ph. D. Dissertation, Department of Civil Engineering, University of California, Berkeley, 1966. Rigorous treatment of the systematic development of refined two-dimensional elements for plane strain, plane stress and plate bending. A version of this without plate bending but including computer codes is published as SESM Rept. 66–22, Department of Civil Engineering, University of California, Berkeley, Oct. 1966.
(14) Wilson, E. L., "A Computer Program for the Dynamic Stress Analysis of Underground Structures," USAEWES, Contract Report No. 1–175 (Jan. 1968).

Finite element method and a linear-acceleration, step-by-step integration procedure is used for the dynamic response of linearly elastic two-dimensional structures.

(15) See Reference 10, Chapter 1.

FURTHER READING

Cakiroglu, A., and Ozmen, G., "Numerical-Integration of Forced Vibration Equations," *Proc. ASCE, J.EM Dn*, Vol. 94, EM3 (June 1968) Discussion, Vol. 95, EM1 (Feb. 1969) and EM6 (Dec. 1969).

Forysthe, G. E., and Moler, C. B., *Computer Solutions of Linear Algebraic Systems*, Englewood Cliffs, Prentice-Hall, Inc., 1967.

Franklin, J. N., *Matrix Theory*, Englewood Cliffs, Prentice-Hall, Inc., 1968.

Fried, I., "A Gradient Computational Procedure for the Solution of Large Problems Arising from the Finite Element Discretization Method," *IJNME*, Vol. 2, No. 4 (1970).

Jensen, H. G., and Parks, G. A., "Efficient Solutions for Linear Matrix Equations," *Proc. ASCE, J.ST Dn*, Vol. 96, ST1 (Jan. 1970).

Melosh, R. J., "Manipulation Errors in Finite Element Analysis," *Recent Advances.*

Roy, J. R., "Numerical Error in Structural Solutions," *Proc. ASCE, J. ST Dn*, Vol. 97, ST4, (Apr. 1971).

See Reference 1, Chapter 11.

See Reference 46, Chapter 9.

3

BASIC EQUATIONS FROM SOLID MECHANICS

Because the vast majority of the literature on the finite element method treats problems in structural and continuum mechanics, including soil and rock mechanics, the descriptions in this volume are presented primarily in terms of these fields of study. These fields are founded upon the theory of solid mechanics, and a sufficient background in the basic principles of solid mechanics is essential for the study of the finite element method. Among the many texts that deal with this topic are References 1 through 9.

We shall consider only those results and equations that will be used in later chapters. This summary emphasizes results from the theory of elasticity.

Although tensor notation would permit a more compact presentation of the necessary equations, we shall employ a vector notation that is consistent with the matrix formulation of finite element theory treated in Part B.

3-1 STRESS

The state of stress in an elemental volume of a loaded body is defined in terms of six components of stress, expressed in a vector form as

$$\{\sigma\}^{\mathrm{T}} = [\sigma_x \ \ \sigma_y \ \ \sigma_z \ \ \tau_{xy} \ \ \tau_{yz} \ \ \tau_{zx}] \tag{3.1}$$

where σ_x, σ_y, and σ_z are the normal components of stress, and τ_{xy}, τ_{yz}, and τ_{zx} are the components of shear stress. Stresses acting on a positive face of the elemental volume in a positive coordinate direction are positive; those acting on a negative face in a negative direction are positive; all others are negative. A positive face is one on which a normal vector directed outward from the element points in a positive direction.

If the coordinate axes are principal axes, the stress vector becomes

$$\{\sigma\}^{T} = [\sigma_1 \ \sigma_2 \ \sigma_3 \ 0 \ 0 \ 0] \tag{3.2}$$

The invariants of the stress are

$$\begin{aligned} J_1 &= \sigma_x + \sigma_y + \sigma_z = \sigma_1 + \sigma_2 + \sigma_3 \\ J_2 &= \sigma_x\sigma_y + \sigma_y\sigma_z + \sigma_z\sigma_x - \tau_{xy}^2 - \tau_{yz}^2 - \tau_{zx}^2 \\ &= \sigma_1\sigma_2 + \sigma_2\sigma_3 + \sigma_3\sigma_1 \\ J_3 &= \sigma_x\sigma_y\sigma_z + 2\tau_{xy}\tau_{yz}\tau_{zx} - \sigma_x\tau_{yz}^2 - \sigma_y\tau_{zx}^2 - \sigma_z\tau_{xy}^2 \\ &= \sigma_1\sigma_2\sigma_3 \end{aligned} \tag{3.3}$$

Among the alternative representations of the state of stress is the sum of the volumetric and deviatoric stresses.[6]

$$\begin{aligned} \{\sigma\}^{T} &= \{\sigma_v\}^{T} + \{\sigma_D\}^{T} \\ \{\sigma_v\}^{T} &= [J_1/3 \ J_1/3 \ J_1/3 \ 0 \ 0 \ 0] \\ \{\sigma_D\}^{T} &= [\sigma_x - J_1/3, \ \sigma_y - J_1/3, \ \sigma_z - J_1/3, \ \tau_{xy}, \ \tau_{yz}, \ \tau_{zx}] \\ &= [\sigma_{Dx} \ \sigma_{Dy} \ \sigma_{Dz} \ \tau_{xy} \ \tau_{yz} \ \tau_{zx}] \end{aligned} \tag{3.4}$$

Still another form of expressing the stresses is in reference to the octahedral plane, which makes equal angles with each of the principal directions. The octahedral normal stress is

$$\sigma_{\text{oct}} = J_1/3 \tag{3.5a}$$

and the octahedral shear stress is

$$\tau_{\text{oct}} = \tfrac{1}{3}[(\sigma_1 - \sigma_2)^2 + (\sigma_2 - \sigma_3)^2 + (\sigma_3 - \sigma_1)^2]^{1/2} \tag{3.5b}$$

3-2 STRAIN AND KINEMATICS

Corresponding to the six stress components, equation (3.1), the state of strain at a point can be divided into six strain components given by the following strain vector:[2,3]

$$\{\varepsilon\}^{T} = [\varepsilon_x \ \varepsilon_y \ \varepsilon_z \ \gamma_{xy} \ \gamma_{yz} \ \gamma_{zx}] \tag{3.6}$$

The principal strains correspond to the principal stresses, so for these coordinate directions

$$\{\varepsilon\}^{T} = [\varepsilon_1 \ \varepsilon_2 \ \varepsilon_3 \ 0 \ 0 \ 0] \tag{3.7}$$

Moreover, the following strain invariants are analogous to the stress invariants:[2]

$$
\begin{aligned}
I_1 &= \varepsilon_x + \varepsilon_y + \varepsilon_z = \varepsilon_1 + \varepsilon_2 + \varepsilon_3 \\
I_2 &= \varepsilon_x\varepsilon_y + \varepsilon_y\varepsilon_z + \varepsilon_z\varepsilon_x - \tfrac{1}{4}(\gamma_{xy}^2 + \gamma_{yz}^2 + \gamma_{zx}^2) \\
&= \varepsilon_1\varepsilon_2 + \varepsilon_2\varepsilon_3 + \varepsilon_3\varepsilon_1 \\
I_3 &= \varepsilon_x\varepsilon_y\varepsilon_z + \tfrac{1}{4}(\gamma_{xy}\gamma_{yz}\gamma_{zx} - \varepsilon_x\gamma_{yz}^2 - \varepsilon_y\gamma_{zx}^2 - \varepsilon_z\gamma_{xy}^2) \\
&= \varepsilon_1\varepsilon_2\varepsilon_3
\end{aligned}
\tag{3.8}
$$

Here the factors of $\frac{1}{4}$ arise from the difference in the definitions of engineering strain and of the strain tensor, for example $\gamma_{xy} = 2\varepsilon_{xy}$.

The strain at a point can be decomposed into its volumetric and deviatoric components in the same manner as the stress

$$
\begin{aligned}
\{\varepsilon\}^{\mathrm{T}} &= \{\varepsilon_v\} + \{\varepsilon_D\} \\
\{\varepsilon_v\}^{\mathrm{T}} &= [I_1/3 \;\; I_1/3 \;\; I_1/3 \;\; 0 \;\; 0 \;\; 0] \\
\{\varepsilon_D\}^{\mathrm{T}} &= [\varepsilon_x - I_1/3,\; \varepsilon_y - I_1/3,\; \varepsilon_z - I_1/3,\; \gamma_{xy},\; \gamma_{yz},\; \gamma_{zx}] \\
&= [\varepsilon_{Dx} \;\; \varepsilon_{Dy} \;\; \varepsilon_{Dz} \;\; \gamma_{xy} \;\; \gamma_{yz} \;\; \gamma_{zx}]
\end{aligned}
\tag{3.9}
$$

Finally, the octahedral normal and shear strains corresponding to the octahedral stresses are

$$\varepsilon_{\text{oct}} = I_1/3 \tag{3.10a}$$

$$\gamma_{\text{oct}} = \tfrac{2}{3}[(\varepsilon_1 - \varepsilon_2)^2 + (\varepsilon_2 - \varepsilon_3)^2 + (\varepsilon_3 - \varepsilon_1)^2]^{1/2} \tag{3.10b}$$

Strain-Displacement Equations

The relations between the components of strain and the displacement components u, v, and w at a point are[4]

$$
\begin{aligned}
\varepsilon_x &= \frac{\partial u}{\partial x} + \frac{1}{2}\left[\left(\frac{\partial u}{\partial x}\right)^2 + \left(\frac{\partial v}{\partial x}\right)^2 + \left(\frac{\partial w}{\partial x}\right)^2\right] \\
\varepsilon_y &= \frac{\partial v}{\partial y} + \frac{1}{2}\left[\left(\frac{\partial u}{\partial y}\right)^2 + \left(\frac{\partial v}{\partial y}\right)^2 + \left(\frac{\partial w}{\partial y}\right)^2\right] \\
\varepsilon_z &= \frac{\partial w}{\partial z} + \frac{1}{2}\left[\left(\frac{\partial u}{\partial z}\right)^2 + \left(\frac{\partial v}{\partial z}\right)^2 + \left(\frac{\partial w}{\partial z}\right)^2\right] \\
\gamma_{xy} &= \frac{\partial v}{\partial x} + \frac{\partial u}{\partial y} + \frac{\partial u}{\partial x}\frac{\partial u}{\partial y} + \frac{\partial v}{\partial x}\frac{\partial v}{\partial y} + \frac{\partial w}{\partial x}\frac{\partial w}{\partial y} \\
\gamma_{yz} &= \frac{\partial w}{\partial y} + \frac{\partial v}{\partial z} + \frac{\partial u}{\partial y}\frac{\partial u}{\partial z} + \frac{\partial v}{\partial y}\frac{\partial v}{\partial z} + \frac{\partial w}{\partial y}\frac{\partial w}{\partial z} \\
\gamma_{zx} &= \frac{\partial u}{\partial z} + \frac{\partial w}{\partial x} + \frac{\partial u}{\partial z}\frac{\partial u}{\partial x} + \frac{\partial v}{\partial z}\frac{\partial v}{\partial x} + \frac{\partial w}{\partial z}\frac{\partial w}{\partial x}
\end{aligned}
\tag{3.11a}
$$

Equations (3.11a) are one version of the strain-displacement equations, in which the strain components are expressed in terms of only the first (linear) and the second order changes in the displacement components while the higher order terms are neglected. The expressions for the strain components can be further simplified by retaining only the first order or linear terms and neglecting the second order terms, that is

$$\begin{aligned}
\varepsilon_x &= \frac{\partial u}{\partial x} & \gamma_{xy} &= \frac{\partial v}{\partial x} + \frac{\partial u}{\partial y} \\
\varepsilon_y &= \frac{\partial v}{\partial y} & \gamma_{yz} &= \frac{\partial w}{\partial y} + \frac{\partial v}{\partial z} \\
\varepsilon_z &= \frac{\partial w}{\partial z} & \gamma_{zx} &= \frac{\partial u}{\partial z} + \frac{\partial w}{\partial x}
\end{aligned} \tag{3.11b}$$

The relations in equations (3.11b) are considered valid if the body experiences only small deformations, that is, each derivative in equations (3.11a) is much smaller than unity. If the body experiences large or finite deformations or strains, higher order terms must be retained, as in equations (3.11a). These terms represent significant changes in the geometry of the body and thus are called geometric nonlinearities.[5]

3-3 LINEAR CONSTITUTIVE EQUATIONS

The reader is probably familiar with the simplest stress-strain equation, Hooke's law. The Hooke's law for uniaxial deformation states that the deformation (strain) is proportional to the force (stress). In the more general case of three-dimensional bodies, six components of stress and strain will be present. As a natural extension of the Hooke's law, each of the six stress components may be expressed as a linear function of the six components of strain,[5] and vice versa. Such relations constitute the generalized Hooke's law:

$$\begin{Bmatrix} \sigma_x \\ \sigma_y \\ \sigma_z \\ \tau_{xy} \\ \tau_{yz} \\ \tau_{zx} \end{Bmatrix} = \begin{bmatrix} C_{11} & C_{12} & \dots & C_{16} \\ C_{21} & C_{22} & \dots & C_{26} \\ \cdot & & & \cdot \\ \cdot & & & \cdot \\ \cdot & & & \cdot \\ C_{61} & C_{62} & \dots & C_{66} \end{bmatrix} \begin{Bmatrix} \varepsilon_x \\ \varepsilon_y \\ \varepsilon_z \\ \gamma_{xy} \\ \gamma_{yz} \\ \gamma_{zx} \end{Bmatrix} \tag{3.12a}$$

or

$$\{\sigma\} = [C]\{\varepsilon\} \tag{3.12b}$$

and

$$\begin{Bmatrix} \varepsilon_x \\ \varepsilon_y \\ \varepsilon_z \\ \gamma_{xy} \\ \gamma_{yz} \\ \gamma_{zx} \end{Bmatrix} = \begin{bmatrix} D_{11} & D_{12} & \cdots & D_{16} \\ D_{21} & D_{22} & \cdots & D_{26} \\ \cdot & & & \cdot \\ \cdot & & & \cdot \\ \cdot & & & \cdot \\ D_{61} & D_{62} & \cdots & D_{66} \end{bmatrix} \begin{Bmatrix} \sigma_x \\ \sigma_y \\ \sigma_z \\ \tau_{xy} \\ \tau_{yz} \\ \tau_{zx} \end{Bmatrix} \tag{3.12c}$$

or

$$\{\varepsilon\} = [D]\{\sigma\} \tag{3.12d}$$

where $\{\sigma\}$ and $\{\varepsilon\}$ are the stress and strain vectors, respectively.

Equations (3.12a) or (3.12c) represent the constitutive law for a linear, elastic, anistropic, and homogeneous material. The matrices $[C]$ and $[D]$ are symmetric;[2,3] hence, a complete constitutive description of a general anistropic solid necessitates experimental evaluation of 21 elastic constants.[10,11]

Certain materials exhibit symmetry with respect to planes within the body, so the number of material constants will be reduced from the 21 required in the anistropic case.[10,11,12] For example, equation (3.12a) for an orthotropic material is expressed in terms of nine constants:

$$\begin{Bmatrix} \sigma_x \\ \sigma_y \\ \sigma_z \\ \tau_{xy} \\ \tau_{yz} \\ \tau_{zx} \end{Bmatrix} = \begin{bmatrix} C_{11} & C_{12} & C_{13} & 0 & 0 & 0 \\ & C_{22} & C_{23} & 0 & 0 & 0 \\ & & C_{33} & 0 & 0 & 0 \\ & & & C_{44} & 0 & 0 \\ & \text{Symmetrical} & & & C_{55} & 0 \\ & & & & & C_{66} \end{bmatrix} \begin{Bmatrix} \varepsilon_x \\ \varepsilon_y \\ \varepsilon_z \\ \gamma_{xy} \\ \gamma_{yz} \\ \gamma_{zx} \end{Bmatrix} \tag{3.13a}$$

The strain-stress equations for orthotropic materials may be written in terms of the Young's moduli and Poisson's ratios as follows:[10,11]

$$\begin{aligned} \varepsilon_x &= \frac{\sigma_x}{E_x} - \frac{\nu_{yx}}{E_y}\sigma_y - \frac{\nu_{zx}}{E_z}\sigma_z \\ \varepsilon_y &= -\frac{\nu_{xy}}{E_x}\sigma_x + \frac{\sigma_y}{E_y} - \frac{\nu_{zy}}{E_z}\sigma_z \\ \varepsilon_z &= -\frac{\nu_{xz}}{E_x}\sigma_x - \frac{\nu_{yz}}{E_y}\sigma_y + \frac{\sigma_z}{E_z} \\ \gamma_{xy} &= \frac{\tau_{xy}}{G_{xy}}, \quad \gamma_{yz} = \frac{\tau_{yz}}{G_{yz}}, \quad \gamma_{zx} = \frac{\tau_{zx}}{G_{zx}} \end{aligned} \tag{3.13b}$$

These equations may be inverted to obtain the values of the elements of $[C]$ in equations (3.13a). Note that there are twelve material parameters in equations (3.13b); however, only nine of these are independent because the following must be true:

$$\frac{E_x}{\nu_{xy}} = \frac{E_y}{\nu_{yx}}, \quad \frac{E_y}{\nu_{yz}} = \frac{E_z}{\nu_{zy}}, \quad \frac{E_z}{\nu_{zx}} = \frac{E_x}{\nu_{xz}} \tag{3.13c}$$

Linear Isotropic Elasticity

The simplest specialization of the generalized Hooke's law is the case in which the material is assumed to be linear, isotropic, and elastic. An isotropic material is one that has point symmetry; that is, every plane in the body is a plane of symmetry of material behavior. It can be shown that only two independent elastic constants are necessary to represent the behavior in the case of such symmetry.[4] Hence, equation (3.12c) in terms of E and ν becomes

$$\begin{Bmatrix} \varepsilon_x \\ \varepsilon_y \\ \varepsilon_z \\ \gamma_{xy} \\ \gamma_{yz} \\ \gamma_{zx} \end{Bmatrix} = \begin{bmatrix} 1/E & -\nu/E & -\nu/E & 0 & 0 & 0 \\ & 1/E & -\nu/E & 0 & 0 & 0 \\ & & 1/E & 0 & 0 & 0 \\ & & & \dfrac{2(1+\nu)}{E} & 0 & 0 \\ & \text{Symmetrical} & & & \dfrac{2(1+\nu)}{E} & 0 \\ & & & & & \dfrac{2(1+\nu)}{E} \end{bmatrix} \begin{Bmatrix} \sigma_x \\ \sigma_y \\ \sigma_z \\ \tau_{xy} \\ \tau_{yz} \\ \tau_{zx} \end{Bmatrix} \tag{3.12e}$$

or in terms of stress components, equation (3.12a) becomes

$$\begin{Bmatrix} \sigma_x \\ \sigma_y \\ \sigma_z \\ \tau_{xy} \\ \tau_{yz} \\ \tau_{zx} \end{Bmatrix} = \frac{E}{(1+\nu)(1-2\nu)} \begin{bmatrix} 1-\nu & \nu & \nu & 0 & 0 & 0 \\ & 1-\nu & \nu & 0 & 0 & 0 \\ & & 1-\nu & 0 & 0 & 0 \\ \text{Symmetrical} & & & \dfrac{1-2\nu}{2} & 0 & 0 \\ & & & & \dfrac{1-2\nu}{2} & 0 \\ & & & & & \dfrac{1-2\nu}{2} \end{bmatrix} \begin{Bmatrix} \varepsilon_x \\ \varepsilon_y \\ \varepsilon_z \\ \gamma_{xy} \\ \gamma_{yz} \\ \gamma_{zx} \end{Bmatrix} \tag{3.12f}$$

Hooke's law for isotropic materials is sometimes expressed in terms of Lamé's constants,[5] λ and μ,

$$\begin{aligned} C_{11} &= C_{22} = C_{33} = \lambda + 2\mu \\ C_{12} &= C_{13} = C_{23} = \lambda \\ C_{44} &= C_{55} = C_{66} = \mu = G \end{aligned} \tag{3.14a}$$

where

$$\lambda = \frac{E\nu}{(1+\nu)(1-2\nu)}, \quad \mu = G = \frac{E}{2(1+\nu)} \tag{3.14b}$$

The inverse relationships for E and ν in terms of λ and μ are

$$E = \frac{\mu(3\lambda + 2\mu)}{\lambda + \mu}, \quad \nu = \frac{\lambda}{2(\lambda + \mu)} \tag{3.14c}$$

The shear modulus, G, above represents the behavior of a material under pure shearing stresses. To characterize the behavior of the material as a result of volumetric (or hydrostatic) stresses, we use the bulk modulus (or the modulus of volume deformation), K, which is expressed in terms of E and ν or of λ and μ as

$$K = \frac{E}{3(1-2\nu)} = \frac{3\lambda + 2\mu}{3} \tag{3.15}$$

Sometimes it is convenient to express the stress-strain relations in terms of the bulk and shear moduli

$$\begin{Bmatrix} \sigma_x \\ \sigma_y \\ \sigma_z \\ \tau_{xy} \\ \tau_{yz} \\ \tau_{zx} \end{Bmatrix} = \begin{bmatrix} K + \frac{4}{3}G & K - \frac{2}{3}G & K - \frac{2}{3}G & 0 & 0 & 0 \\ & K + \frac{4}{3}G & K - \frac{2}{3}G & 0 & 0 & 0 \\ & & K + \frac{4}{3}G & 0 & 0 & 0 \\ \text{Symmetrical} & & & G & 0 & 0 \\ & & & & G & 0 \\ & & & & & G \end{bmatrix} \begin{Bmatrix} \varepsilon_x \\ \varepsilon_y \\ \varepsilon_z \\ \gamma_{xy} \\ \gamma_{yz} \\ \gamma_{zx} \end{Bmatrix} \tag{3.12g}$$

or

$$\{\sigma\} = \{\sigma_v\} + \{\sigma_D\} = 3K\{\varepsilon_v\} + 2G\{\varepsilon_D\} \tag{3.12h}$$

3-4 TWO-DIMENSIONAL SPECIALIZATIONS OF ELASTICITY

It is costly and time-consuming to perform finite element analyses of three-dimensional problems in solid mechanics. Practical situations, however, may have geometry and loading configurations that reduce these three-dimensional problems to problems in one or two dimensions. As is the case for analytical approaches, we exploit these simplifications whenever possible in applying the finite element method.

Plane Strain

Problems involving a long body whose geometry and loading do not vary significantly in the longitudinal direction are referred to as plane strain problems. Some examples of this configuration, shown in Figure 3-1, are a loaded semi-infinite half-space such as a strip footing on a soil mass; a long cylinder such as a tunnel, culvert, or buried pipe; a laterally loaded retaining wall; and a long earth dam. In these problems the dependent variables can be assumed to be functions of only the x and y coordinates, provided we consider a cross section some distance away from the ends. If we further assume that w, the displacement component in the z direction, is zero at every cross section, the strain components ε_z, γ_{yz}, and γ_{zx} will vanish and the remaining nonzero strain components will be

$$\varepsilon_x = \frac{\partial u}{\partial x}, \quad \varepsilon_y = \frac{\partial v}{\partial y} \quad \text{and} \quad \gamma_{xy} = \frac{\partial u}{\partial y} + \frac{\partial v}{\partial x} \tag{3.16a}$$

Moreover, from the vanishing of ε_z, the stress σ_z can be expressed in terms of σ_x and σ_y as

$$\sigma_z = \nu(\sigma_x + \sigma_y) \tag{3.16b}$$

and σ_x, σ_y, and τ_{xy} are thus the only dependent stress variables. The constitutive law for elastic isotropic material, equation (3.12f), reduces to

$$\begin{Bmatrix} \sigma_x \\ \sigma_y \\ \tau_{xy} \end{Bmatrix} = \frac{E}{(1+\nu)(1-2\nu)} \begin{bmatrix} 1-\nu & \nu & 0 \\ \nu & 1-\nu & 0 \\ 0 & 0 & \dfrac{1-2\nu}{2} \end{bmatrix} \begin{Bmatrix} \varepsilon_x \\ \varepsilon_y \\ \gamma_{xy} \end{Bmatrix} \tag{3.16c}$$

In plane strain problems we usually consider only a slice of unit thickness.

Plane Stress

In contrast to the plane strain condition, in which the longitudinal dimension in the z direction is large compared to the x and y dimensions, the *plane stress* condition is characterized by very small dimensions in the z direction. A thin plate loaded in its plane is the well-known example of the plane stress approximation, Figure 3-2. We consider the case where no loadings are applied on the surfaces of the plate. Then the stress components τ_{yz} and τ_{zx} vanish on the surfaces, and σ_z is zero throughout the thickness. The nonzero components

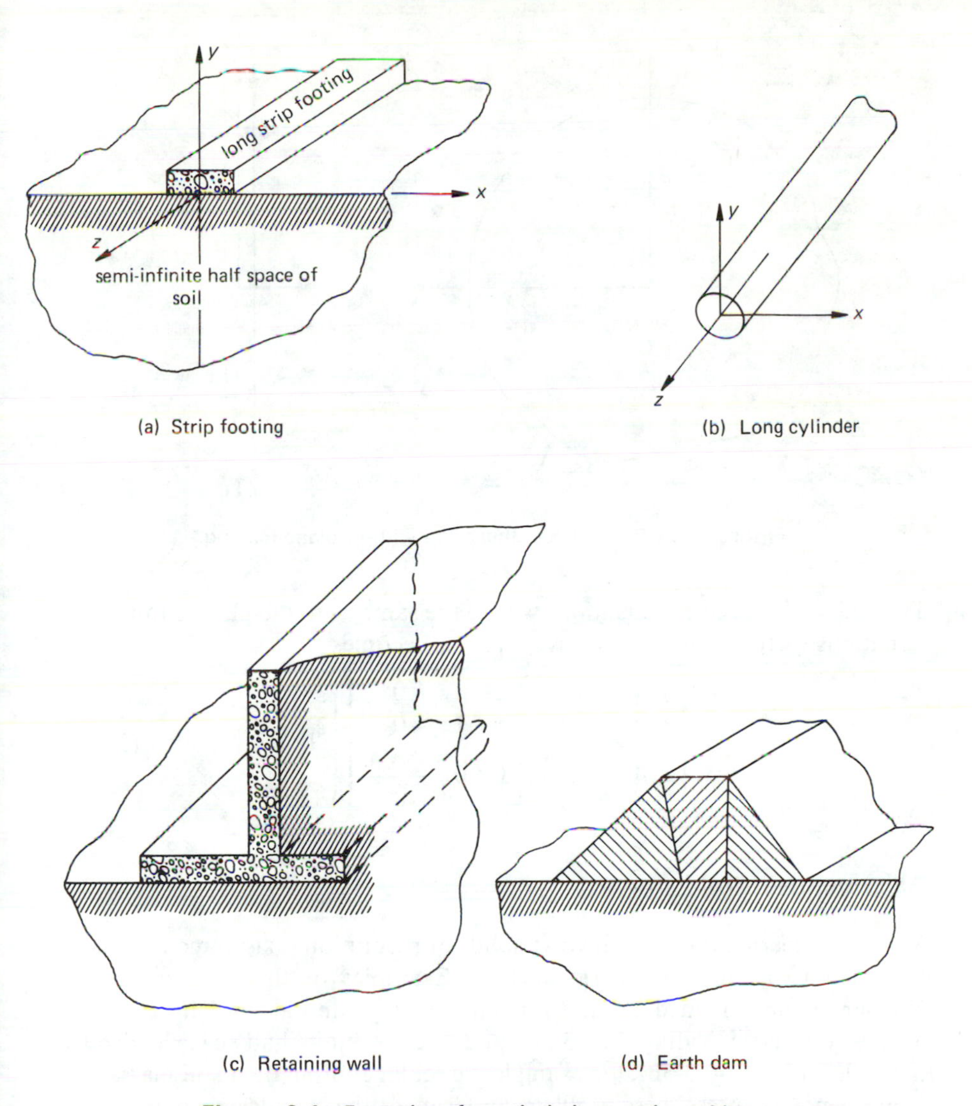

Figure 3-1 Examples of practical plane strain problems.

σ_x, σ_y, and τ_{xy} may be averaged over the thickness and assumed to be independent of z. The state of stress characterized by the above description is referred to as *generalized plane stress*. The strain components γ_{yz} and γ_{zx} vanish on the surfaces, while the component ε_z is given by

$$\varepsilon_z = -\frac{\nu}{1-\nu}(\varepsilon_x + \varepsilon_y) \tag{3.17a}$$

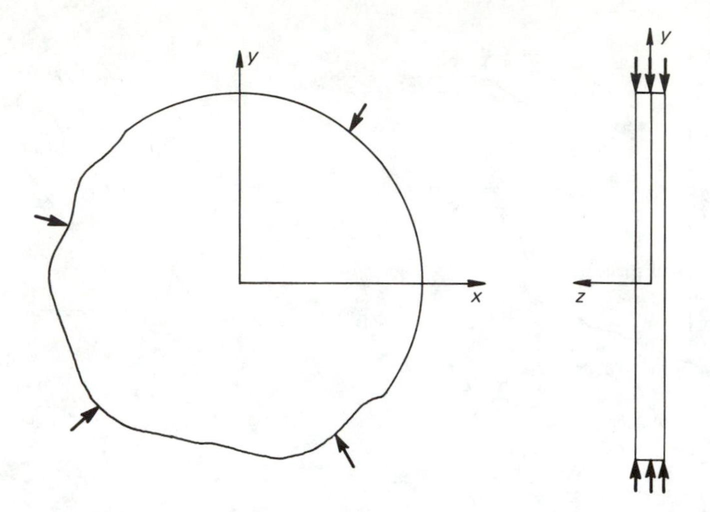

Figure 3-2 Plane stress: thin plate with in-plane loading.

The strain-displacement equations will be the same as in the plane strain case, equations (3.16a). The constitutive relation becomes

$$\begin{Bmatrix} \sigma_x \\ \sigma_y \\ \tau_{xy} \end{Bmatrix} = \frac{E}{1-\nu^2} \begin{bmatrix} 1 & \nu & 0 \\ \nu & 1 & 0 \\ 0 & 0 & \dfrac{1-\nu}{2} \end{bmatrix} \begin{Bmatrix} \varepsilon_x \\ \varepsilon_y \\ \gamma_{xy} \end{Bmatrix} \tag{3.17b}$$

Axisymmetric Problems

Many engineering problems involve solids of revolution (axisymmetric solids) subjected to axially symmetric loading. Examples of this situation are a circular cylinder loaded by uniform internal or external pressure or other axially symmetric loading (Fig. 3-3a), and a semi-infinite half space loaded by a circular area (Fig. 3-3b), for example a circular footing on a soil mass. It is convenient to express these problems in terms of the cylindrical coordinates[2] shown in Figure 3-3. Because of symmetry, the stress components are independent of the angular (θ) coordinate; hence, all derivatives with respect to θ vanish and the components v, $\gamma_{r\theta}$, $\gamma_{\theta z}$, $\tau_{r\theta}$, and $\tau_{\theta z}$ are zero. The nonzero stress components are σ_r, σ_θ, σ_z, and τ_{rz}. The strain-displacement relations for the nonzero strains become

$$\varepsilon_r = \frac{\partial u}{\partial r}, \; \varepsilon_\theta = \frac{u}{r}, \; \varepsilon_z = \frac{\partial w}{\partial z}, \; \gamma_{rz} = \frac{\partial u}{\partial z} + \frac{\partial w}{\partial r} \tag{3.18a}$$

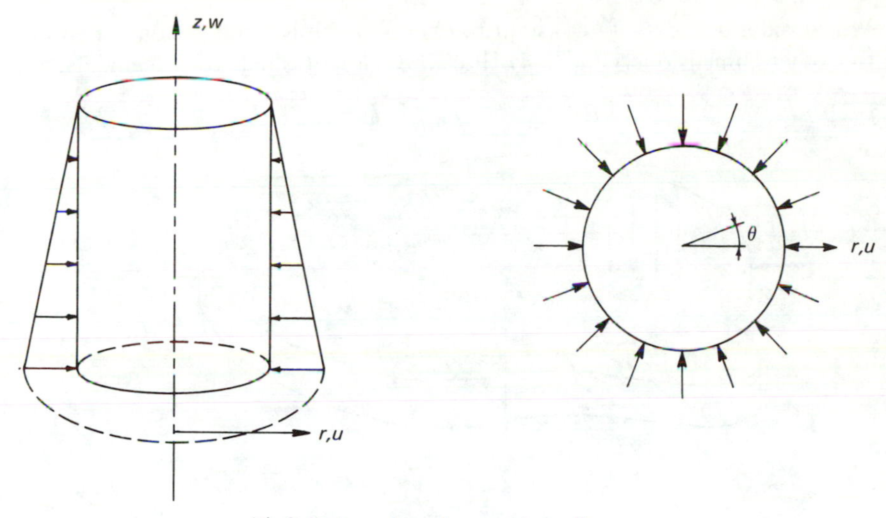

(a) Cylinder under axisymmetric loading

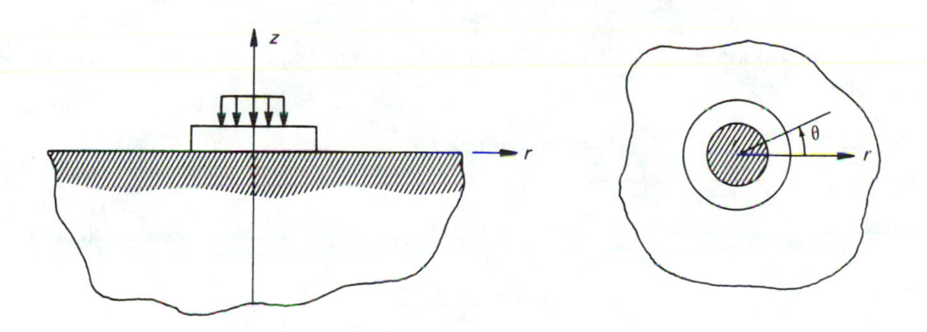

(b) Circular footing on soil mass

Figure 3-3 Axisymmetric problems.

The constitutive relation is

$$\begin{Bmatrix} \sigma_r \\ \sigma_z \\ \sigma_\theta \\ \tau_{rz} \end{Bmatrix} = \frac{E}{(1+\nu)(1-2\nu)} \begin{bmatrix} 1-\nu & \nu & \nu & 0 \\ & 1-\nu & \nu & 0 \\ & & 1-\nu & 0 \\ \text{Symmetrical} & & & \dfrac{1-2\nu}{2} \end{bmatrix} \begin{Bmatrix} \varepsilon_r \\ \varepsilon_z \\ \varepsilon_\theta \\ \gamma_{rz} \end{Bmatrix} \quad (3.18b)$$

Elementary Theory of Plates

We consider a body with constant thickness which is small in comparison to the other dimensions (Fig. 3-4). For the case of small displacements, we

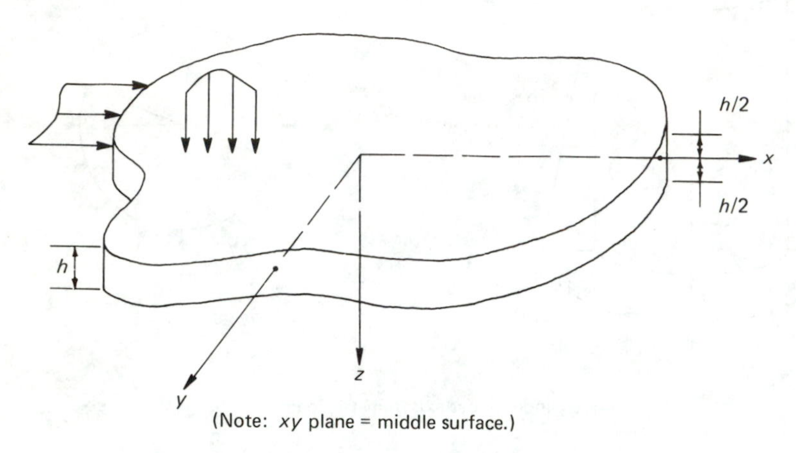

Figure 3-4 Thin plate under transverse and in-plane loadings.

follow Poisson-Kirchhoff theory and assume that the displacements take the form

$$\begin{aligned} u &= u^o(x,y) - z\frac{\partial w}{\partial x} \\ v &= v^o(x,y) - z\frac{\partial w}{\partial y} \\ w &= w^o(x,y) \end{aligned} \tag{3.19a}$$

where the superscript o indicates the displacements of the middle plane. This linear variation of u and v displacements across the thickness is analogous to the assumption of beam theory that plane sections remain plane. Applying the strain-displacement equations, equations (3.6b), we obtain

$$\begin{aligned} \varepsilon_x &= \frac{\partial u^o}{\partial x} - z\,\frac{\partial^2 w}{\partial x^2} = \varepsilon_x^o - z\,\frac{\partial^2 w}{\partial x^2} \\ \varepsilon_y &= \frac{\partial v^o}{\partial y} - z\,\frac{\partial^2 w}{\partial y^2} = \varepsilon_y^o - z\,\frac{\partial^2 w}{\partial y^2} \\ \gamma_{xy} &= \frac{\partial u^o}{\partial y} + \frac{\partial v^o}{\partial x} - 2z\,\frac{\partial^2 w}{\partial x\,\partial y} = \gamma_{xy}^o - 2z\,\frac{\partial^2 w}{\partial x\,\partial y} \end{aligned} \tag{3.19b}$$

The stress-strain relations are the same as for generalized plane stress, equations (3.17b). However, instead of working with the stresses directly, the usual engineering approach is to formulate the plate problem in terms of stress resultants which are the integrals of the stresses over the thicknesses of the plate, h. These are defined by

$$N_x = \int_{-h/2}^{h/2} \sigma_x \, dz, \quad N_y = \int_{-h/2}^{h/2} \sigma_y \, dz, \quad N_{xy} = \int_{-h/2}^{h/2} \tau_{xy} \, dz$$

$$M_x = -\int_{-h/2}^{h/2} z\sigma_x \, dz, \quad M_y = -\int_{-h/2}^{h/2} z\sigma_y \, dz, \quad M_{xy} = \int_{-h/2}^{h/2} \tau_{xy} z \, dz \qquad (3.19c)$$

$$Q_x = \int_{-h/2}^{h/2} \tau_{xz} \, dz, \quad Q_y = \int_{-h/2}^{h/2} \tau_{yz} \, dz$$

As with elementary beam theory, we find that the stretching and bending are independent; that is, we can express the in-plane behavior by using only the terms with superscripts o and the N_x, N_y, and N_{xy} stress resultants, and the bending behavior by using the remaining terms. We neglect the effects of the strains ε_z, γ_{xz}, and γ_{yz} and can exclude them from our formulation. Values of the stresses σ_z, τ_{xz}, and τ_{yz} at the surfaces enter the loading terms. For example, the intensity of transverse loading per unit surface area is given by

$$q = \sigma_z \Big|_{z=h/2} - \sigma_z \Big|_{z=-h/2} + \int_{-h/2}^{h/2} Z \, dz \qquad (3.19d)$$

where Z is the body force intensity in the z direction.

Theory of Shells

One specialization of general theory that is beyond the scope of this chapter is the *theory of shells*. However, in Part C some applications of the finite element method to shell structures will be presented.

3-5 NONLINEAR MATERIAL BEHAVIOR

Application of the finite element method to problems involving materials that obey linear constitutive laws is straightforward, because the material parameters are constant and because only one application of the solution process is required to obtain results for a particular loading case. Nevertheless, the method can be applied to nonlinear elastic and elastic-plastic problems. In this section we shall summarize some aspects of nonlinear material behavior used in finite element formulations.

Nonlinear Elastic Behavior

For nonlinear elastic behavior the material parameters depend upon the state of stress. An incremental approach often used for nonlinear analysis requires a separate solution process for each of the several increments of the load. Essentially the incremental technique approximates the behavior as *piecewise linear*. In other words, during the application of each load increment, the material is considered to be linear and elastic, but different material properties are used for different increments. Hence, the principles developed for linear elastic behavior become applicable in the range of each small increment. In Chapter 7 we will go further into this incremental procedure.

Figure 3-5(a) shows a typical nonlinear elastic behavior. For a uniaxial test $\bar{\sigma}$ and $\bar{\varepsilon}$ represent the axial stress and strain. In the piecewise linear approximation, suitable elastic constants, such as E and ν for the isotropic case, may be obtained in one of two ways. The tangent modulus at any point P is defined as

$$E_t = \left.\frac{d\bar{\sigma}}{d\bar{\varepsilon}}\right|_P \tag{3.20a}$$

that is, the slope of the $\bar{\sigma}$ vs $\bar{\varepsilon}$ curve at point P. E_t can be approximately evaluated by an expression such as

$$E_t \cong \frac{\Delta\bar{\sigma}}{\Delta\bar{\varepsilon}} \tag{3.20b}$$

where Δ denotes a finite increment. Equation (3.20b) is shown as the dashed line in Figure 3-5(a). Alternatively, a secant modulus† can also be defined in terms of the total stress and strain at a given stage such as

$$E_s = \left.\frac{\bar{\sigma}}{\bar{\varepsilon}}\right|_P \tag{3.21}$$

If volumetric measurements are available, Figure 3-5(a), the Poisson's ratio for the increment can be similarly defined. For instance, the tangent Poisson's ratio is

$$\nu_t = -\frac{d\bar{\varepsilon}_v}{d\bar{\varepsilon}} \cong -\frac{\Delta\bar{\varepsilon}_v}{\Delta\bar{\varepsilon}} \tag{3.22}$$

where $\bar{\varepsilon}_v$ denotes some measure of lateral strain.

If the experimental behavior is obtained from other tests, the elastic parameters can be appropriately defined. For instance, a form of triaxial test[13] behavior is shown in Figure 3-5(b), where σ_1 and σ_3 are the axial or the major

† A secant modulus computed between two points is known as a *chord modulus* when neither of the points is the origin of the stress-strain curve.

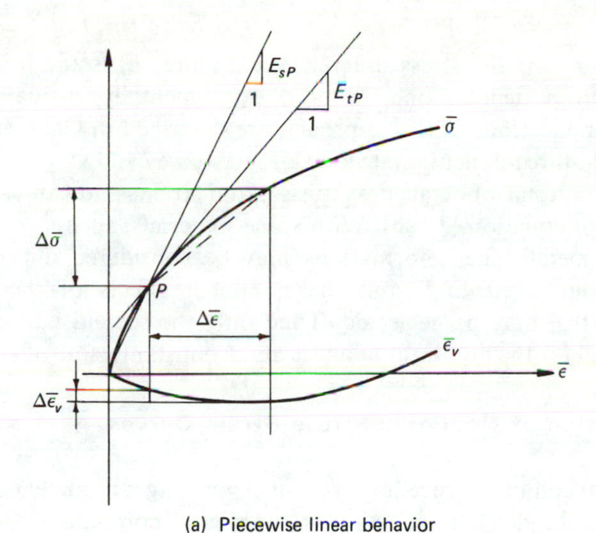

(a) Piecewise linear behavior

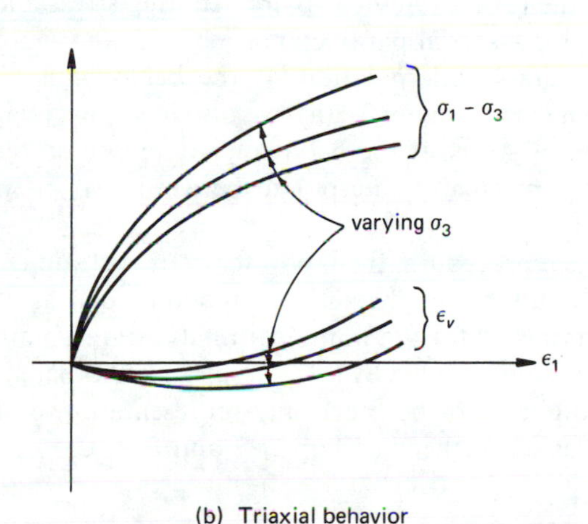

(b) Triaxial behavior

Figure 3-5 Nonlinear elastic behavior.

principal stress and the confining or the minor principal stress, respectively. The intermediate principal stress σ_2 is the same as σ_3. ε_v and ε_1 are the measured volume change and the axial strain, respectively. The elastic constants, E and ν, are now dependent on the confining pressure σ_3 in addition to the state of stress, $\sigma_1 - \sigma_3$. One way of defining them is by mathematical expressions fit to the experimental data. These take the form

$$\begin{aligned} E_t &= E_t[(\sigma_1 - \sigma_3), (\sigma_1 - \sigma_3)_f, \sigma_3, E_i, A] \\ \nu_t &= \nu_t[(\sigma_1 - \sigma_3), (\sigma_1 - \sigma_3)_f, \sigma_3, E_i, A] \end{aligned} \tag{3.23}$$

where $(\sigma_1 - \sigma_3)_f$ is the stress difference at failure, E_i is the initial tangent modulus, and A denotes one or more experimentally evaluated material constants. Applications of this approach are described in Chapter 10.

For general three-dimensional states of stress, we can express the constitutive relation in terms of octahedral stresses and strains and can use equations of the form of equation (3.12h). With some materials, such as cohesive soils and certain metals, the deformations may be considered due only to the deviatoric component $\{\sigma_D\}$, and the nonlinear effects of the volumetric components $\{\sigma_v\}$ may be neglected. Then only the second part of equation (3.12h) is used for the nonlinear analysis and a constant value of K is adopted.

Representation of Nonlinear Stress-Strain Curves

There are two common procedures for incorporating a nonlinear stress-strain law into a finite element formulation for digital computation. The stress-strain law derived from a laboratory test can be used directly in a tabular or digital form. Several points on the curve, such as P in Figure 3-5(a), are selected and are input in the form of number pairs denoting stress and strain at those points. The variable material parameters such as E and ν are obtained from such curves by suitable interpolation. If the behavior is represented by a single stress-strain curve, Figure 3-5(a), we obtain strains by interpolation for a calculated state of stress. If the behavior is represented by several curves, Figure 3-5(b), we must also interpolate between two curves for different confining pressures.

In the alternative procedure, the laboratory stress-strain curve is expressed in the form of a suitable mathematical function, such as equations (3.23). The material parameters for the nonlinear analysis are obtained on the basis of the state of stress represented by $(\sigma_1 - \sigma_3)$, σ_3, and other stress parameters. The tabular or digital and the functional representation of the constitutive laws will be discussed further in Chapter 7 and Part C.

Plastic or Inelastic Behavior

In an elastic-plastic material, we can use the linear elastic formulations for the behavior prior to the proportional limit. In the zone between the proportional limit and the yield point, the material exhibits nonlinear elastic behavior; and we can employ the piecewise linear approximation. However, the plastic or inelastic behavior beyond the yield point may require a different approach; and in the remainder of this section we shall present some useful methods of representing such plastic behavior.

It is necessary to have a yield criterion to ascertain the state of stress at which yielding is considered to begin. We also need a flow rule to explain the post-yielding behavior. Correspondingly, there are two major theories of plastic behavior, the deformation theory and the incremental or flow theory.[7,14–18] The latter is more general than the former.[14] In the deformation theory, the plastic strains are uniquely defined by the state of stress, whereas in the incremental theory the plastic strains depend upon a combination of factors, such as the increments of stress and strain and the state of stress.

In both theories the total strain vector $\{\varepsilon\}$ is decomposed into its elastic $\{\varepsilon^e\}$ and plastic $\{\varepsilon^P\}$ components as follows:

$$\{\varepsilon\} = \{\varepsilon^e\} + \{\varepsilon^P\} \tag{3.24}$$

Notice that the vector $\{\varepsilon^e\}$ now corresponds to $\{\varepsilon\}$ in equations (3.12).

Yield Criteria

We shall now consider some of the common yield conditions used to specify the critical or yield stress at which the plastic deformations initiate. Such a criterion is expressed through an equation of a yield surface of the form

$$f(\{\sigma\}) = f(J_1, J_2, J_3) = 0 \tag{3.25}$$

This is the equation of a surface in the three-dimensional space with coordinates σ_1, σ_2, σ_3, the principal stresses. If the state of stress is such that $f < 0$, the material is still in the elastic range, that is $\{\varepsilon^P\} = 0$. When $f = 0$, a plastic state is attained and one of the theories of plasticity must be used to determine subsequent plastic behavior under increasing stress or strain. No significance is attached to the case $f > 0$.

The *von Mises yield criterion* is one form of equation (3.25) and presumes that yielding is caused by the maximum distortion energy.[1] The yield surface is

$$f = J_{D2} - \bar{\sigma}_y = 0 \tag{3.26a}$$

where $\bar{\sigma}_y$ indicates an experimentally determined yield stress in simple shear, and J_{D2} is the second deviatoric stress invariant. In terms of stresses, the von Mises criterion is

$$\begin{aligned} f &= [\tfrac{1}{2}(\sigma_x - \sigma_y)^2 + \tfrac{1}{2}(\sigma_y - \sigma_z)^2 + \tfrac{1}{2}(\sigma_z - \sigma_x)^2 + 3\tau_{xy}^2 + 3\tau_{yz}^2 + 3\tau_{zx}^2]^{1/2} - \bar{\sigma}_y = 0 \\ &= (\sigma_1 - \sigma_2)^2 + (\sigma_2 - \sigma_3)^2 + (\sigma_3 - \sigma_1)^2 - 2\bar{\sigma}_y^2 = 0 \end{aligned} \tag{3.26b}$$

This yield rule is applicable to the behavior of some metals and rocks and to saturated clays.

The *Tresca criterion* is often used for soils and rocks that possess only cohesive strength, c. The yield surface is determined by the maximum shearing stress.[1]

$$f = (\sigma_1 - \sigma_3) - 2c = 0 \tag{3.27a}$$

Recall that σ_1 and σ_3 are the maximum and minimum principal stresses, respectively. The Tresca condition can also be represented as follows:[7,14]

$$f = 4J_{D2}^3 - 27J_{D3}^2 - 36c^2J_{D2}^2 + 96c^4J_{D2} - 64c^6 = 0 \tag{3.27b}$$

The above two criteria presume that yielding is caused by a deviatoric state of stress and they neglect the effects of the volumetric component of stress, which for some materials like sands, concrete, and certain rocks do play a significant role in yielding. The von Mises and Tresca criteria may be modified to account for this effect, which is associated with the so-called frictional strength of the material. The *modified von Mises criterion* is

$$f = J_2 - F(J_1/3) \tag{3.28a}$$

If we take F as

$$F(J_1/3) = \bar{\sigma}_y J_1^2/9 \tag{3.28b}$$

we obtain the *extended von Mises criterion.* The *modified Tresca condition* is expressed as

$$f = [(\sigma_1 - \sigma_2)^2 - F(J_1/3)][(\sigma_2 - \sigma_3)^2 - F(J_1/3)][(\sigma_3 - \sigma_1)^2 - F(J_1/3)] = 0 \tag{3.28c}$$

If equation (3.28b) applies, we obtain the *extended Tresca rule.*

The *Coulomb criterion* for the two-dimensional case is[19]

$$R = c \cos \phi - \frac{\sigma_x + \sigma_y}{2} \sin \phi \tag{3.29a}$$

where c is the cohesion, ϕ denotes the internal friction, and R is the radius of the Mohr's circle. In terms of principal stresses, equations (3.29a) can be written as

$$f = (1 + \sin \phi)\sigma_3 - (1 - \sin \phi)\sigma_1 - 2c \cos \phi = 0 \tag{3.29b}$$

or

$$\sigma_1 = \sigma_3 N_\phi + 2c\sqrt{N_\phi}$$

where

$$N_\phi = \tan^2 (\pi/4 + \phi/2)$$

A general form of the Coulomb rule known as *Mohr-Coulomb criterion* is

$$f = \tfrac{1}{2}(\sigma_3 - \sigma_1) - F_1[\tfrac{1}{2}(\sigma_3 + \sigma_1)] = 0 \tag{3.29c}$$

where F_1 is a function of σ_1, σ_3, and ϕ. The Mohr-Coulomb condition accounts for both the cohesive strength c and the frictional strength ϕ of the material, but does not include the effects of the intermediate principal stress, σ_2. A generalization of the Mohr-Coulomb rule, proposed by Drucker and Prager, is[18,19]

$$f = \alpha J_1 + J_{D2}^{1/2} - k = 0 \tag{3.29d}$$

where α and k are positive constants. In expanded form, this law is

$$\begin{aligned} f = \alpha(\sigma_x + \sigma_y + \sigma_z) + \{\tfrac{1}{6}[(\sigma_x - \sigma_y)^2 + (\sigma_y - \sigma_z)^2 + (\sigma_z - \sigma_x)^2] \\ + \tau_{xy}^2 + \tau_{yz}^2 + \tau_{zx}^2\}^{1/2} - k = 0 \end{aligned} \tag{3.29e}$$

Deformation Theory of Plasticity

In deformation (or Hencky) theory, it is assumed that the plastic strain is incompressible, that is, the volumetric strain is negligible, $\{\varepsilon_v^P\} = 0$. The strain-stress relations may then be written as[11,45]

$$\{\varepsilon_v^e\} = \{\varepsilon_v\} = \{\sigma_v\}/3K = (1 - 2v)\{\sigma_v\}/E \tag{3.30a}$$

$$\{\varepsilon_D^e\} = \{\sigma_D\}/G \tag{3.30b}$$

$$\{\varepsilon^P\} = \{\varepsilon_D^P\} = P\{\sigma_D\} \tag{3.30c}$$

For an isotropic material, P is a function of the state of stress, in particular of the stress invariants J_2 and J_3. Equation (3.30c) is the flow rule uniquely relating the plastic strain to the deviatoric stress which characterizes the deformation theory. For elastic-perfectly plastic materials P is indeterminate.[14] Often, an approximate bilinear representation for a nonlinear material is obtained by adopting a constant value of P for the plastic region.

Incremental or Flow Theory of Plasticity

In the incremental theory we express the increments of plastic strain $\{d\varepsilon^P\}$ as functions of the current stress, the strain increments, and the stress increments:

$$\{d\varepsilon^P\} = \{d\varepsilon^P\}(\{\sigma\},\{d\varepsilon\},\{d\sigma\}) \tag{3.31a}$$

Here d denotes an increment. Equation (3.24) can be written in the incremental form as

$$\{d\varepsilon\} = \{d\varepsilon^e\} + \{d\varepsilon^P\} \tag{3.31b}$$

in which

$$\{d\varepsilon^e\} = [D]\{d\sigma\} = [C^e]^{-1}\{d\sigma\} \tag{3.31c}$$

The last two equations can be combined to give

$$\{d\sigma\} = [C^e](\{d\varepsilon\} - \{d\varepsilon^P\}) \tag{3.32a}$$

which may be rewritten as

$$\{d\sigma\} = [C^{ep}]\{d\varepsilon\} \tag{3.32b}$$

where $[C^{ep}]$ is called the *elastic-plastic matrix*, and is expressed as

$$[C^{ep}] = [C^e] - [C^p] \tag{3.32c}$$

Equation (3.32c) indicates that the plastic action reduces the strength of the material by reducing the magnitudes of the parameters in the elasticity matrix $[C^e]$.

To obtain the elastic-plastic matrix $[C^{ep}]$ we need to compute $[C^p]$, which in turn requires computations of the plastic strain increments $\{d\varepsilon^P\}$. In many finite element applications, $\{d\varepsilon^P\}$ is evaluated on the basis of the von Mises criterion and the Prandtl-Reuss equations.[14,20,21,22,23] Without going into the details of the derivation of $[C^{ep}]$, we present here the final result for the plasticity matrix $[C^p]$ in equation (3.32c).

$$[C^p] = \frac{2G}{\frac{2}{3}\bar{\sigma}^2\left(1 + \dfrac{H'}{3G}\right)} \begin{bmatrix} \sigma_{Dx}^2 & \sigma_{Dx}\sigma_{Dy} & \sigma_{Dx}\sigma_{Dz} & \sigma_{Dx}\tau_{xy} & \sigma_{Dx}\tau_{yz} & \sigma_{Dx}\tau_{zx} \\ & \sigma_{Dy}^2 & \sigma_{Dy}\sigma_{Dz} & \sigma_{Dy}\tau_{xy} & \sigma_{Dy}\tau_{yz} & \sigma_{Dy}\tau_{zx} \\ & & \sigma_{Dz}^2 & \sigma_{Dz}\tau_{xy} & \sigma_{Dz}\tau_{yz} & \sigma_{Dz}\tau_{zx} \\ & & & \tau_{xy}^2 & \tau_{xy}\tau_{yz} & \tau_{xy}\tau_{zx} \\ & \text{Symmetrical} & & & \tau_{yz}^2 & \tau_{yz}\tau_{zx} \\ & & & & & \tau_{zx}^2 \end{bmatrix} \tag{3.33}$$

σ_{Dx} etc. are given in equation (3.4). $\bar{\sigma}$ is called the *equivalent* or *effective stress* which is related to the equivalent plastic strain $\bar{\varepsilon}^P$ by

$$H' = d\bar{\sigma}/d\bar{\varepsilon}^P \tag{3.34}$$

H' is the slope of the curve relating effective stress, $\bar{\sigma}$, and effective strain, $\int d\bar{\varepsilon}^P$.

With the von Mises law the effective stress and strain are computed from the following expressions:

$$\begin{aligned} \bar{\sigma} &= \frac{1}{\sqrt{2}}\{(\sigma_1 - \sigma_2)^2 + (\sigma_2 - \sigma_3)^2 + (\sigma_3 - \sigma_1)^2\}^{1/2} \\ \bar{\varepsilon} &= \frac{\sqrt{2}}{3}\{(\varepsilon_1 - \varepsilon_2)^2 + (\varepsilon_2 - \varepsilon_3)^2 + (\varepsilon_3 - \varepsilon_1)^2\}^{1/2} \end{aligned} \tag{3.35}$$

Equation (3.32b) with equation (3.33) is valid for isotropic materials. Recently, Yamada[21] obtained a general expression for the matrix $[C^{ep}]$ for elastic-plastic work hardening materials which permits anistropic properties in both the elastic and plastic zones. Similar expressions were also obtained by Zienkiewicz *et al.*[23] and Marcal and King.[22]

The above derivation is based on von Mises law with the assumption that the yielding of materials is generally not affected by the volumetric component of stresses, $\{\sigma_v\}$. This assumption is valid for materials such as steel, some metals, and saturated cohesive soils under certain loading conditions. The behavior of materials such as concrete, rocks, and other soils is significantly affected by the volumetric stresses. For such materials, the generalized Mohr-Coulomb rule, equation (3.29d), proposed by Drucker and Prager[19] is relevant. Formulations based on this criterion can account for both deviatoric and volumetric effects and can also permit changes in anistropy in the elastic and plastic zones. Finite element formulations based on the Drucker-Prager rule have been developed and used for two-dimensional problems by Reyes and Deere,[24] and Shieh and Sandhu.[25] The symmetric plastic matrix, $[C^p]$, for plane strain is given by[24,25]

$$
\begin{aligned}
C^p_{11} &= 2G(1 - h_2 - 2h_1\sigma_x - h_3\sigma_x^2) \\
C^p_{22} &= 2G(1 - h_2 - 2h_1\sigma_y - h_3\sigma_y^2) \\
C^p_{33} &= 2G(\tfrac{1}{2} - h_3\tau_{xy}^2) \\
C^p_{12} &= C^p_{21} = -2G[h_2 + h_1(\sigma_x + \sigma_y) + h_3\sigma_x\sigma_y] \\
C^p_{13} &= C^p_{31} = -2G(h_1\tau_{xy} + h_3\sigma_x\tau_{xy}) \\
C^p_{23} &= C^p_{32} = -2G(h_1\tau_{xy} + h_3\sigma_y\tau_{xy})
\end{aligned}
\tag{3.36}
$$

where $G = E/2(1 + \nu)$ and

$$h_1 = (3K\alpha/2G - J_1/6J_{D2}^{1/2})/\{J_{D2}^{1/2}(1 + 9\alpha^2K/G)\}$$

$$h_2 = \frac{(\alpha - J_1/6J_{D2}^{1/2})(3K\alpha/G - J_1/3J_{D2}^{1/2})}{(1 + 9\alpha^2K/G)} - \frac{3\nu Kk}{EJ_{D2}^{1/2}(1 + 9\alpha^2K/G)}$$

$$h_3 = 1/\{2J_{D2}(1 + 9\alpha^2K/G)\}$$

3-6 MATERIAL CHARACTERIZATION

The nature of this introductory volume does not permit us to go further into the details of material characterization. Usually the behavior of real meterials is highly complex and is influenced by such factors as the physical properties, the magnitude and nature of the loads, the temperature, the time, the rate of loading, and the previous history of the material. A constitutive relation is generally derived from field and/or laboratory experiments on the material. Ideally, these tests should simulate all significant factors and conditions existing in the actual body or prototype. Only then will the constitutive law be accurate.

The results from any numerical or analytical technique are valid only to the extent that the constitutive model is accurate. Hence for realistic results from finite element analyses, it is imperative that the constitutive relations be carefully determined from proper tests.

REFERENCES

(1) Popov, E. P., *Introduction to Mechanics of Solids*, Englewood Cliffs, Prentice-Hall, 1968. A fundamental text for undergraduate study.

(2) Timoshenko, S., and Goodier, J. N., *Theory of Elasticity*, New York, McGraw-Hill Book Co., 1951. A treatment directed to the application of the theory to engineering problems.

(3) Wang, C. T., *Applied Elasticity*, New York, McGraw-Hill Book Co., 1953. Gives a number of applications, numerical methods and variational principles as they relate to elasticity.

(4) Love, A. E. H., *A Treatise on the Mathematical Theory of Elasticity*, New York, Dover Publications, 1944. Classical treatise on the subject.

(5) Sokolnikoff, I. S., *Mathematical Theory of Elasticity*, New York, McGraw-Hill Book Co., 1956. An excellent book for mathematical aspects; uses tensor notation. Last chapter includes applications of variational principles to elasticity.

(6) Drucker, D. C., *Introduction to Mechanics of Deformable Solids*, New York, McGraw-Hill Book Co., 1967.

(7) Fung, Y. C., *Foundations of Solid Mechanics*, Englewood Cliffs, Prentice-Hall, 1965. An excellent, diverse text for the elementary graduate level.

(8) Eringen, A. C., *Nonlinear Theory of Continuous Media*, New York, McGraw-Hill Book Co., 1962.

(9) Flügge, W., *Viscoelasticity*, Boston, Blaisdell Publishing Co., 1967. Concise treatment of linear viscoelasticity.

(10) Lekhnitskii, S. G., *Theory of Elasticity of an Anistropic Elastic Body*, San Francisco, Holden-Day, 1963.

(11) Hearman, R. F. S., *An Introduction to Applied Anistropic Elasticity*, New York, Oxford University Press, 1961.

(12) Zienkiewicz, O. C., Cheung, Y. K., and Stagg, K. G., "Stresses in Anistropic Media with Particular Reference to Problems of Rock Mechanics," *J. of Strain Analysis*, Vol. 1, No. 2 (1966). Finite element method is used for stratified anistropic rock masses and the results compared with analytical solutions.

(13) Bishop, A. W., and Henkel, D. J., *The Measurement of Soil Properties in the Triaxial Test*, London, Edward Arnold Publishers, 1964. Deals with the theory and use of the triaxial test apparatus.

(14) Hill, R., *The Mathematical Theory of Plasticity*, Oxford, Clarendon Press, 1950. A pioneering work on plasticity; deals with theory and applications.
(15) Hoffmann, O., and Sachs, G., *Introduction to the Theory of Plasticity for Engineers*, New York, McGraw-Hill Book Co., 1953.
(16) Drucker, D. C., "A more Fundamental Approach to Plastic Stress-Strain Relations," *Proc. 1st US Nat. Congress Appl. Mech.*, Chicago, 1951. An important paper in the work-hardening theory of plasticity. Establishes convexity of the yield surface and the normality rule.
(17) Prager, W., "The Theory of Plasticity: A Survey of Recent Achievements," *Proc. Inst. of Mech. Engrs.*, Vol. 109 (1955).
(18) Naghdi, P. M., "On Stress-Strain Relationships in Plasticity and Thermoelasticity," *Proc. 2nd Symp. on Naval Structural Mech.*, Stanford, 1960.
(19) Drucker, D. C., and Prager, W., "Soil Mechanics and Plastic Analysis on Limit Design," *Quart. of Appl. Math.*, Vol. 10, No. 2 (1952). Proposes plasticity theory for behavior of soil. Uses extended Mohr-Coulomb law. Shows applications to problems such as bearing capacities, slope stability, and lateral pressures on retaining structures.
(20) Yamada, Y., Kawai, T., and Yoshimura, N., "Analysis of the Elastic-Plastic Problems by the Matrix Displacement Method," *Second Conf.* Uses incremental theory in the finite element method. On the basis of von Mises criterion, derives general elastic-plastic matrix and specializes for axisymmetric, plane stress, and torsion problems.
(21) Yamada, Y., "Recent Japanese Developments in Matrix Displacement Method for Elastic-Plastic Problems," *Recent Advances*. Develops general elastic-plastic stress-strain relations on the basis of von Mises rule. Allows for anisotropy. Applications include notched or perforated plates under tension, contact or indentation; block compressed between rough plates; and thermal stress due to rapid heating.
(22) Marcal, P. V., and King, I. P., "Elastic-Plastic Analysis of Two-Dimensional Stress Analysis Systems by the Finite Element Method," *Int. J. Mech. Sci.*, Vol. 9 (1967).
(23) Zienkiewicz, O. C., Valliapan, S., and King, I. P., "Elasto-plastic Solutions of Engineering Problems—'Initial Stress' Finite Element Method," *IJNME*, Vol. 1, No. 1 (1969). Details in Chapter 7.
(24) Reyes, S. F., and Deere, D. U., "Elasto-plastic Analysis of Underground Openings by the Finite Element Method," *Proc. 1st Int. Cong. Rock Mech.*, Vol. II, Lisbon, 1966.
(25) Shieh, W. Y. J., and Sandhu, R. S., "Application of Elasto-plastic Analysis in Earth Structures," *Proc. Nat. Meeting on Water Resources Eng.*, ASCE, Memphis, Tenn. (Jan. 1970). Uses extended Mohr-Coulomb law and the incremental theory of plasticity for analyses of tunnels, slopes, and riverbanks.

FURTHER READING

Newmark, N. M., "Failure Hypotheses for Soils," *Proc. Res. Conf. on Shear Strength of Soils*, ASCE—University of Colorado, Boulder, June 1960. Proposes use of octahedral stress approach for soil and rock behavior.

4
VARIATIONAL METHODS

It is useful to consider the finite element procedure basically as a variational approach. This conception has contributed significantly to the convenience in formulating the method and to its generality. Therefore, in order to understand fully the material presented in Parts B and C of this text, the reader should be familiar with basics of variational methods. Only a summary of relevant results is given in this chapter. For detailed study, the reader may consult References 1 through 4 and other works listed under *Further Reading* at the end of this chapter.

4-1 CALCULUS OF VARIATIONS

The two principal goals of this brief section are to review the basic procedure of the calculus of variations and to distinguish the two essentially different types of boundary conditions, *natural* and *geometric*, encountered in variational formulations of boundary value problems.

Let us consider a functional expressed as

$$A = \int_{x_1}^{x_2} F(x, u, u', u'')\, dx \tag{4.1}$$

where the variable u and its first and second derivatives with respect to x, u' and u'', are functions of x. F and A are called *functionals* because they are functions of other functions. In mechanics, the functional usually possesses a

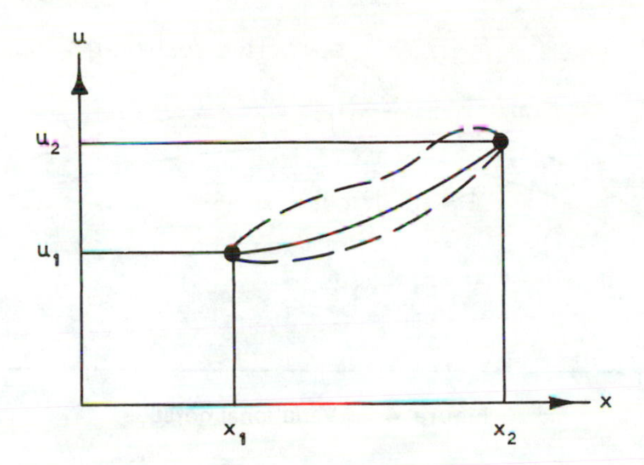

Figure 4-1 Domain $[x_1, x_2]$.
——Actual solutions.
- - - - -Tentative solutions.

clear physical meaning, for example the potential energy of a deformable solid. We shall encounter some additional examples of functionals later in this chapter. The integral in equation (4.1) is defined in the region or domain $[x_1, x_2]$ (Fig. 4-1). In this figure, let the values of u be prescribed on the boundaries

$$\begin{aligned} u(x_1) &= u_1 \\ u(x_2) &= u_2 \end{aligned} \tag{4.2}$$

In variational procedures a tentative solution is tried for a given problem, and the functional is expressed in terms of the tentative solution. From all such possible solutions satisfying the boundary conditions, the solution which satisfies the variational principle governing the behavior will be the one which makes the functional, A, stationary. We shall consider some specific variational principles for solid mechanics in the next section. The mathematical procedure used to select the correct solution from a number of tentative solutions is the *calculus of variations.*

Any tentative solution, $\bar{u}$, in the neighborhood of the exact solution, may be represented by the sum of the exact solution, u, and a *variation of* u, δu (Fig. 4-2).

$$\bar{u} = u + \delta u \tag{4.3}$$

The variation in $u = u(x)$ is defined as an infinitesimal arbitrary change in u for a fixed value of the independent variable x, that is, for $\delta x = 0$. The notation δu used in equation (4.3) is called variational notation. The δ of variational

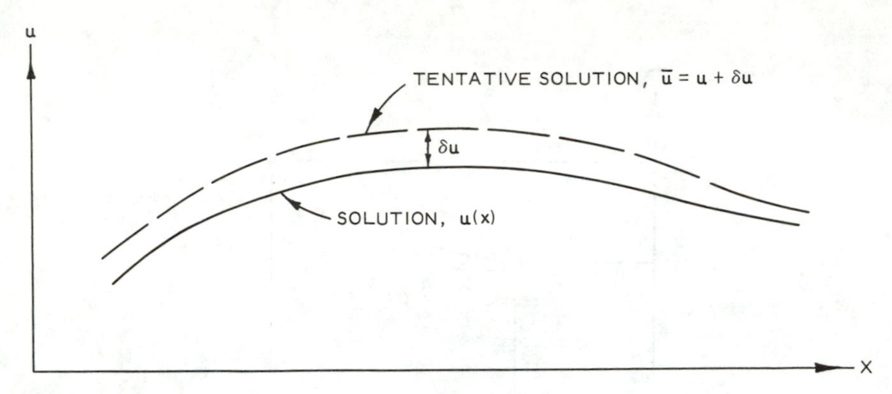

Figure 4-2 Variational notation.

notation can be treated as an operator, similar to the differential operator, *d*. The operation of variation is commutative with both integration and differentiation, i.e.

$$\delta(\int F\,dx) = \int(\delta F)\,dx \tag{4.4a}$$

and

$$\delta(du/dx) = d(\delta u)/dx \tag{4.4b}$$

We define the variation of a function of several variables or of a functional in a manner similar to the calculus definition of a total differential.

$$\delta F = \frac{\partial F}{\partial u}\,\delta u + \frac{\partial F}{\partial u'}\,\delta u' + \frac{\partial F}{\partial u''}\,\delta u'' \tag{4.4c}$$

Let us now consider the functional A in equation (4.1). We consider small changes or variations in this functional to correspond to variations in the solution. By analogy to the maximization or minimization of simple functions in ordinary calculus, the vanishing of the first variation of the functional A will give the condition for A to be stationary. We shall denote the variations in A by the variational notation δA.

$$\delta A = \int_{x_1}^{x_2}\left(\frac{\partial F}{\partial u}\,\delta u + \frac{\partial F}{\partial u'}\,\delta u' + \frac{\partial F}{\partial u''}\,\delta u''\right)dx = 0 \tag{4.5a}$$

or

$$\delta A = \int_{x_1}^{x_2} \delta F\,dx = 0 \tag{4.5b}$$

Integration by parts for the respective terms in equation (4.5a) gives:

$$\int_{x_1}^{x_2} \frac{\partial F}{\partial u}\, \delta u\, dx \tag{4.6a}$$

$$\int_{x_1}^{x_2} \frac{\partial F}{\partial u'}\, \delta u'\, dx = \left[\frac{\partial F}{\partial u'}\, \delta u\right]\Bigg|_{x_1}^{x_2} - \int_{x_1}^{x_2} \frac{d}{dx}\left(\frac{\partial F}{\partial u'}\right) \delta u\, dx \tag{4.6b}$$

and

$$\int_{x_1}^{x_2} \frac{\partial F}{\partial u''}\, \delta u''\, dx = \left[\frac{\partial F}{\partial u''}\, \delta u'\right]\Bigg|_{x_1}^{x_2} - \left[\frac{d}{dx}\left(\frac{\partial F}{\partial u''}\right) \delta u\right]\Bigg|_{x_1}^{x_2} + \int_{x_1}^{x_2} \frac{d^2}{dx^2}\left(\frac{\partial F}{\partial u''}\right) \delta u\, dx \tag{4.6c}$$

Adding equations (4.6a), (4.6b), and (4.6c), we obtain

$$\begin{aligned} \delta A = & \int_{x_1}^{x_2} \left[\frac{\partial F}{\partial u} - \frac{d}{dx}\left(\frac{\partial F}{\partial u'}\right) + \frac{d^2}{dx^2}\left(\frac{\partial F}{\partial u''}\right)\right] \delta u\, dx \\ & + \left[\frac{\partial F}{\partial u'} - \frac{d}{dx}\left(\frac{\partial F}{\partial u''}\right)\right] \delta u \Bigg|_{x_1}^{x_2} \\ & + \left[\left(\frac{\partial F}{\partial u''}\right) \delta u'\right]\Bigg|_{x_1}^{x_2} = 0 \end{aligned} \tag{4.7}$$

To satisfy the condition in equation (4.7), recall that the variation δu is arbitrary; hence, the terms must vanish individually:

$$\frac{\partial F}{\partial u} - \frac{d}{dx}\left(\frac{\partial F}{\partial u'}\right) + \frac{d^2}{dx^2}\left(\frac{\partial F}{\partial u''}\right) = 0 \tag{4.8a}$$

$$\left[\frac{\partial F}{\partial u'} - \frac{d}{dx}\left(\frac{\partial F}{\partial u''}\right)\right] \delta u \Bigg|_{x_1}^{x_2} = 0 \tag{4.8b}$$

$$\left[\left(\frac{\partial F}{\partial u''}\right) \delta u'\right]\Bigg|_{x_1}^{x_2} = 0 \tag{4.8c}$$

Equation (4.8a) is the governing differential equation for the problem and is called the *Euler equation* or the *Euler-Lagrange equation*. Equations (4.8b) and (4.8c) give the boundary conditions. The conditions

$$\left[\frac{\partial F}{\partial u'} - \frac{d}{dx}\left(\frac{\partial F}{\partial u'}\right)\right]\Bigg|_{x_1}^{x_2} = 0 \qquad \text{and} \qquad \frac{\partial F}{\partial u''}\Bigg|_{x_1}^{x_2} = 0$$

are called the *natural boundary conditions*; if they are satisfied, they are called *free boundary conditions*. If one of the natural boundary conditions is not satisfied, the corresponding portions of equations (4.8b) and (4.8c) must be satisfied.

$$\delta u(x_1) = 0, \ \delta u(x_2) = 0$$

$$\delta u'(x_1) = 0, \ \delta u'(x_2) = 0$$

The last conditions are called *geometric boundary conditions*, or *forced boundary conditions*. In summary, for this type of problem the calculus of variations has produced one governing differential equation and four boundary conditions. The latter may be fulfilled by any combination of forced and free conditions so that equations (4.8b) and (4.8c) are satisfied.

It is apparent that the calculus of variations approach, when applied to a quantity which is the integral of a functional, provides a complete description of the problem. It gives not only the governing differential equation of the problem, but also the inherent boundary conditions. The above formulation is known as the *Euler-Lagrange formulation*.

When the finite element equations are derived on the basis of the variational principles, the natural boundary conditions are automatically incorporated in the formulation. Only the geometric boundary conditions need special treatment. These aspects will become clear in Chapter 6.

Example: Beam on Elastic Foundation. Consider a beam on an elastic foundation, with one end rigidly supported and the other end propped on an elastic spring support as shown in Figure 4-3. The geometric boundary conditions are given by

$$u(x = 0) = u'(x = 0) = 0$$

The total potential of the beam is

$$\Pi = \int_0^\ell \frac{1}{2} EI(u'')^2 \, dx + \int_0^\ell \frac{k_f u^2}{2} \, dx + \frac{K_s}{2} u^2(\ell) - \int_0^\ell pu \, dx$$

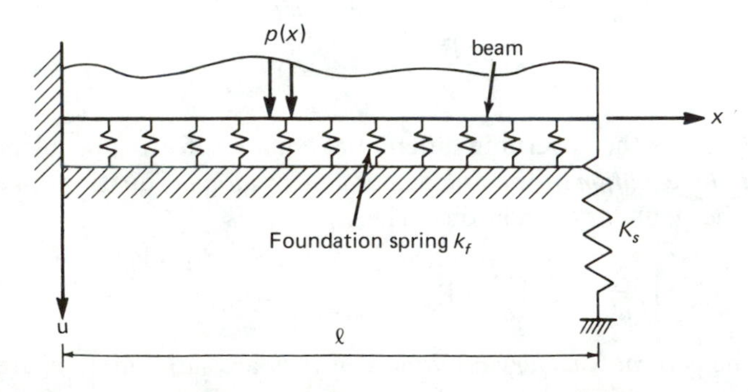

Figure 4-3 Beam on elastic foundation.

where k_f is the subgrade reaction spring constant per unit length and K_s is the spring constant for flexible end support. Note that Π here corresponds to A in equation (4.1). Using the variational notation and extremizing Π, we obtain

$$\delta\Pi = \int_0^\ell (EIu'')\,\delta u''\,dx + \int_0^\ell k_f u\,\delta u\,dx$$

$$+ K_s u(\ell)\,\delta u(\ell) - \int_0^\ell p\,\delta u\,dx = 0$$

Integration by parts and application of the geometric boundary conditions

$$\delta u\Big|_{x=0} = \delta u'\Big|_{x=0} = 0$$

gives

$$\int_0^\ell [(EIu'')'' - p + k_f u]\,\delta u\,dx + (EIu''\,\delta u')\Big|_{x=\ell}$$

$$- [(EIu'')' - K_s u]\,\delta u\Big|_{x=\ell} = 0$$

Since δu is arbitrary, each term in the last equation must vanish individually. The first term gives the governing differential equation

$$(EIu'')'' = p - k_f u$$

and the remaining terms give the natural boundary conditions

$$(EIu'')|_{x=\ell} = 0$$

and

$$EI(u'')'|_{x=\ell} + K_s u(\ell) = 0$$

4-2 VARIATIONAL PRINCIPLES OF SOLID MECHANICS

In this section we shall consider statements of the variational theorems of solid mechanics that are commonly utilized to derive finite element equations. These theorems are stated without proof, but they can be derived from the well known energy principles of mechanics; specifically, the principle of virtual work can be considered the basis of these variational formulations.[5] The majority of the theorems presented below are also called *minimum principles* because the stationary value of the functional can be shown to be a minimum.

Principle of Minimum Potential Energy

The total potential, or the potential energy, of an elastic body is defined as[1,2,3]

$$\Pi = U + W_p \tag{4.9}$$

where U is the strain energy, and W_p is the potential of the applied loads. Because the forces are assumed to remain constant during a variation of the displacements, we can relate the variations of the work done by the loads, W, and of the potential of the loads as follows:[5]

$$\delta W = -\delta W_p \tag{4.10}$$

The *principle of minimum potential energy* is

$$\delta\Pi = \delta U + \delta W_p = \delta U - \delta W = 0 \tag{4.11}$$

The principle and its accompanying conditions can be stated as follows: *Of all possible displacement configurations a body can assume which satisfy compatibility and the constraints or kinematic boundary conditions, the configuration satisfying equilibrium makes the potential energy assume a minimum value.* Here it is important to note that variations of displacements are taken while forces and stresses are assumed constant. Moreover, the resulting equations are equilibrium equations.

The potential energy for a linearly elastic body can be expressed as the sum of the internal work (the strain energy due to internal stresses) and the potential of the body forces and surface tractions. Following the terminology of Chapter 3, the potential energy functional is[6,7]

$$\begin{aligned}\Pi = \iiint_V dU(u, v, w) - \iiint_V (\bar{X}u + \bar{Y}v + \bar{Z}w)\, dV \\ - \iint_{S_1} (\bar{T}_x u + \bar{T}_y v + \bar{T}_z w)\, dS_1\end{aligned} \tag{4.12}$$

where S_1 is the surface of the body on which surface tractions are prescribed and $dU(u, v, w)$ denotes the *strain energy per unit volume*, or the *strain energy density*. The last two integrals in equation (4.12) represent the work done by the constant external forces, that is, the body forces, $\bar{X}, \bar{Y}, \bar{Z}$, and surface tractions, $\bar{T}_x$, $\bar{T}_y$, $\bar{T}_z$. A bar at the top of a letter indicates that the quantity is specified.

Using the notation of Chapters 2 and 3, we can write the functional in equation (4.12) in matrix form. The strain energy density for a linear elastic body is defined as

$$dU = \tfrac{1}{2}\{\varepsilon\}^{\mathrm{T}}\{\sigma\}\, dV = \tfrac{1}{2}\{\varepsilon\}^{\mathrm{T}}[C]\{\varepsilon\}\, dV.$$

Hence the total potential becomes

$$\Pi = \tfrac{1}{2}\iiint_V (\{\varepsilon\}^{\mathrm{T}}[C]\{\varepsilon\} - 2\{u\}^{\mathrm{T}}\{\bar{X}\})\, dV - \iint_{S_1} \{u\}^{\mathrm{T}}\{\bar{T}\}\, dS_1 \tag{4.13}$$

where

$$\{u\}^{\mathrm{T}} = [u\ v\ w]$$
$$\{\bar{X}\}^{\mathrm{T}} = [\bar{X}\ \bar{Y}\ \bar{Z}]$$
$$\{\bar{T}\}^{\mathrm{T}} = [\bar{T}_x\ \bar{T}_y\ \bar{T}_z]$$

Principle of Minimum Complementary Energy

The complementary energy of an elastic body without initial strains is defined as

$$\Pi_c = U_c + W_{pc} \tag{4.14}$$

where U_c is the complementary strain energy, and W_{pc} is the complementary potential of the applied loads. Because displacements are assumed constant during a variation of forces, we can relate the variations of the complementary work of the loads, W_c, and of the complementary potential of the loads as follows:[5]

$$\delta W_c = -\delta W_{pc} \tag{4.15}$$

The *principle of minimum complementary energy* is

$$\delta \Pi_c = \delta U_c + \delta W_{pc} = \delta U_c - \delta W_c = 0 \tag{4.16}$$

The principle and its accompanying conditions can be stated as follows: *Of all possible force and stress states which satisfy equilibrium and the stress boundary conditions, the state satisfying compatibility makes the complementary energy assume a minimum value.* In contrast with the minimum potential energy principle, here variations of force and stress are taken while displacements are assumed constant. Moreover, the resulting equations are compatibility equations.

The total complementary energy is expressed as[3,6,7]

$$\Pi_c = \iiint_V dU_c(\sigma_x, \sigma_y, \sigma_z, \ldots, \tau_{zx})\, dV - \iint_{S_2} (T_x\bar{u} + T_y\bar{v} + T_z\bar{w})\, dS_2 \tag{4.17}$$

where S_2 is that part of the surface of the body on which displacements are prescribed, and dU_c is the *complementary strain energy density*. Note that the barred quantities here denote prescribed displacement components.

In matrix notation, the complementary strain energy density for a linear elastic body is defined as

$$dU_c = \tfrac{1}{2}\{\sigma\}^{\mathrm{T}}\{\varepsilon\}\,dV = \tfrac{1}{2}\{\sigma\}^{\mathrm{T}}[D]\{\sigma\}\,dV$$

so equation (4.17) can be expressed as

$$\Pi_c = \frac{1}{2}\iiint_V \{\sigma\}^{\mathrm{T}}[D]\{\sigma\}\,dV - \iint_{S_2} \{T\}^{\mathrm{T}}\{\bar{u}\}\,dS_2 \tag{4.18}$$

Hamilton's Principle

The variational principle useful for problems in dynamics is Hamilton's principle. The functional for this principle is the *Lagrangian*, L, defined as

$$L = T - U - W_p \tag{4.19}$$

where T is the kinetic energy. In matrix notation, the kinetic energy density is defined as

$$dT = \tfrac{1}{2}\rho\{\dot{u}\}^{\mathrm{T}}\{\dot{u}\}\,dV$$

where ρ is the density of the material, and the overdot denotes the derivative with respect to time. The functional in equation (4.19) is expressed in matrix form for a linear elastic body as

$$\begin{aligned} L = \tfrac{1}{2}\iiint_V (\rho\{\dot{u}\}^{\mathrm{T}}\{\dot{u}\} - \{\varepsilon\}^{\mathrm{T}}[C]\{\varepsilon\} + 2\{u\}^{\mathrm{T}}\{\bar{X}\})\,dV \\ + \iint_{S_1} \{u\}^{\mathrm{T}}\{\bar{T}\}\,dS_1 \end{aligned} \tag{4.20}$$

The statement of *Hamilton's principle* is as follows:

$$\delta \int_{t_1}^{t_2} L\,dt = 0 \tag{4.21}$$

Among all possible time histories of displacement configurations which satisfy compatibility and the constraints or kinematic boundary conditions and which also satisfy conditions at times t_1 and t_2, the history which is the actual solution makes the Lagrangian functional a minimum.

Hellinger-Reissner Principle

In the principle of minimum potential energy the functional Π was expressed in terms of displacements (u, v, w), while the functional Π_c for the complementary energy was expressed in terms of stresses (σ_x, σ_y, ... τ_{zx}). A mixed principle, wherein a functional is expressed in terms of both displacements and stresses, is called the *Hellinger-Reissner principle*. This is a derived principle in the sense that it can be obtained from the potential energy principle or the complementary energy principle by applying suitable constraint conditions.[2,8] The functional for this principle is

$$\begin{aligned}\Pi_R = \iiint_V \Big\{ & \sigma_x \frac{\partial u}{\partial x} + \sigma_y \frac{\partial v}{\partial y} + \cdots + \tau_{zx}\left(\frac{\partial w}{\partial x} + \frac{\partial u}{\partial z}\right) - dU_c(\sigma_x, \sigma_y, \ldots \tau_{zx}) \\ & - (\bar{X}u + \bar{Y}v + \bar{Z}w)\Big\}\, dV - \iint_{S_1} (\bar{T}_x u + \bar{T}_y v + \bar{T}_z w)\, dS_1 \\ & - \iint_{S_2} \{(u - \bar{u})T_x + (v - \bar{v})T_y + (w - \bar{w})T_z\}\, dS_2 \end{aligned} \tag{4.22}$$

In matrix notation, the functional for a linear elastic material is

$$\begin{aligned}\Pi_R = & \iiint_V (\{\sigma\}^T\{\varepsilon\} - \tfrac{1}{2}\{\sigma\}^T[D]\{\sigma\} - \{u\}^T\{\bar{X}\})\, dV \\ & - \iint_{S_1} \{u\}^T\{\bar{T}\}\, dS_1 - \iint_{S_2} \{(u - \bar{u})\}^T\{T\}\, dS_2 \end{aligned} \tag{4.23}$$

The functional Π_R can now be varied by considering variations in both displacements and stresses simultaneously. A set of stress-displacement equations, boundary condition equations, and equilibrium equations is derived by obtaining a stationary value of the functional.

$$\delta \Pi_R = 0 \tag{4.24}$$

Strictly speaking, the Hellinger-Reissner principle is a stationary principle not a minimum principle because the stationary value cannot be shown to be a minimum.

4-3 APPLICATIONS TO THE FINITE ELEMENT METHOD

Most problems are too complex to use the energy principles to obtain the exact solution directly. The usual technique is to guess, *a priori*, a trial family of solutions for the unknown quantity in order to construct a variational functional. It is generally impossible to guess the correct family of trial solutions. Therefore, we choose an approximate pattern for the trial solution. If we apply the energy principles to the approximate functional, we obtain the equations necessary to find the approximate solution.

In the theory of the finite element method for structural analysis, we first select the proper variational principle for the given problem. Next, we express the functional involved in terms of approximate assumed displacement (and/or stress) functions which satisfy the geometric boundary conditions. We then minimize the approximate functional to obtain a set of governing equations. We shall see the details of this approach in Chapter 5.

It may be useful to reemphasize the basis of the application of the energy principles to problems in solid mechanics. When we apply the principle of minimum potential energy, we assume approximate *displacement* functions (or models) and express the functional Π in terms of the approximate models. The governing equations that we generate are *approximate equilibrium equations*. Similarly, when we consider the principle of minimum complementary energy, we assume approximate *stress* functions and establish the approximate functional Π_c. This formulation results in governing relations in the form of *approximate compatibility equations.*

REFERENCES

(1) Lanczos, C., *The Variational Principles of Mechanics,* Toronto, University of Toronto Press, 1970. Gives fundamental philosophy of analytical mechanics and the calculus of variations. Includes lucid presentation of Lagrangian equations, virtual work principles and Hamilton-Jacobi equation.
(2) Washizu, K., *Variational Methods in Elasticity and Plasticity*, Oxford, Pergamon Press, 1968. A recent treatment of variational formulations in elasticity and plasticity including small displacement and finite displacement theories.
(3) Langhaar, H. L., *Energy Methods in Applied Mechanics*, New York, John Wiley & Sons, 1962. A textbook on application of energy principles to engineering problems.
(4) See Reference 2, Chapter 1.
(5) See Reference 3, Chapter 1.
(6) See Reference 5, Chapter 3.
(7) See Reference 3, Chapter 3.
(8) Pian, T. H. H., and Tong, P., "Basis of Finite Element Methods for Solid Continua," *IJNME*, Vol. 1, No. 1 (1969). Reviews the various variational methods that have been and could be applied to the finite element method. Classifies the methods according to variational principles.

FURTHER READING

Courant, R., and Hilbert, D., *Methods of Mathematical Physics*, Vol. 1, New York, Inter-Science Publishers, 1965.
Forray, M. J., *Variational Calculus in Science and Engineering*, New York, McGraw-Hill Book Co., 1968.

Hildebrand, F. B., *Methods of Applied Mathematics*, Englewood Cliffs, Prentice-Hall, 1965.

Kantorovich, L. V., and Krylov, V. I., *Approximate Methods of Higher Analysis*, Groningen, P. Noordhoff, 1958.

Mikhlin, S. G., *Variational Methods in Mathematical Physics*, New York, Macmillan, 1964.

Volterra, V., *Theory of Functionals and of Integral and Integro-Differential Equations*, New York, Dover Publications, 1959.

PART B

THE THEORY OF THE FINITE ELEMENT METHOD

5

THE BASIC COMPONENT—A SINGLE ELEMENT

In Chapter 1 we saw that the finite element method is based upon the general principle known as *going from part to whole*. This principle is familiar not only in engineering analysis but in other spheres of human endeavor as well. The humanities, social sciences, and physical sciences all utilize inductive reasoning based on such principles as "particular to general" and "individual to universal." In any field the application of the principle is an attempt to obtain information about the "whole" by understanding the "parts." In engineering we are faced with problems which cannot be solved in closed form, that is, as a "whole." Therefore, we consider the physical medium as an assemblage of many small parts. Analysis of the basic part forms the first step toward a solution. In this chapter we present the necessary theory regarding this basic component—the finite element.

5-1 THE CONCEPT OF AN ELEMENT

The predominant interpretation of the finite element is a physical visualization of a body or structure as an assemblage of building block-like elements, interconnected at the nodal points. In addition to the engineer's natural preference for such a physical interpretation, this viewpoint also stems from the traditional idealization techniques for the analysis of framed structures such as those shown in Figure 5-1. The chord and web members of the truss and the

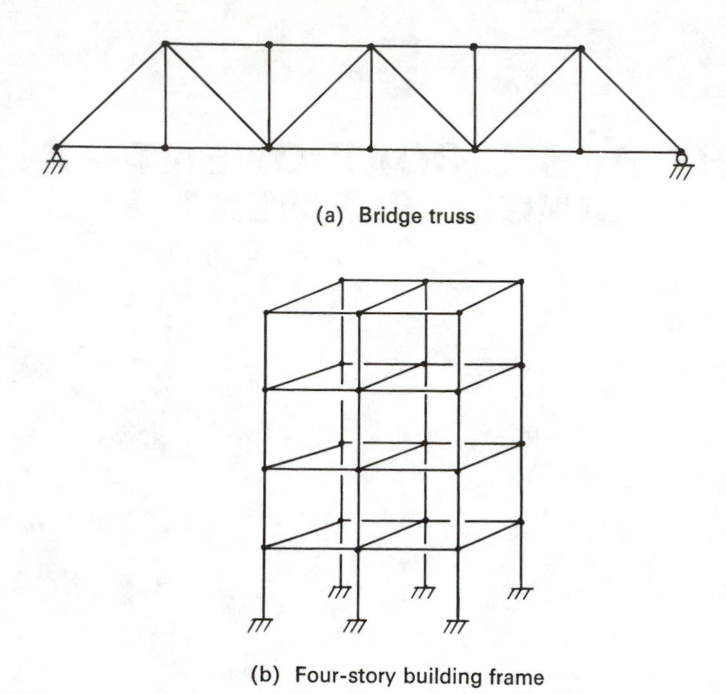

(a) Bridge truss

(b) Four-story building frame

Figure 5-1 Idealized frame structures.

beams and columns of the building are physically distinct members or elements which interconnect only at the joints. The finite element method is an offspring of framed-structure analysis. This fact will be emphasized further in the next chapter, where we shall see that the solution of an assemblage of elements is based directly upon the standard methods of frame analysis.

A significant intermediary step in the evolution of the physical concept of finite elements was the representation of two- and three-dimensional bodies and structures as assemblages of small grids or lattices.[1] This approach still enjoys substantial application in structural mechanics and is sometimes called the *discrete element method.* One example of its use is shown in Figure 5-2, in which a discrete element model of a plate or slab is represented as a combination of elastic blocks, rigid bars, and torsion bars.[2] It has been shown that the discrete element approach is mathematically equivalent to the finite difference method, and thus it may be considered as a physical interpretation of the finite difference method.

Physically, the finite element concept differs from the lattice analogy in that the elements themselves are two- or three-dimensional bodies.[3] Hence

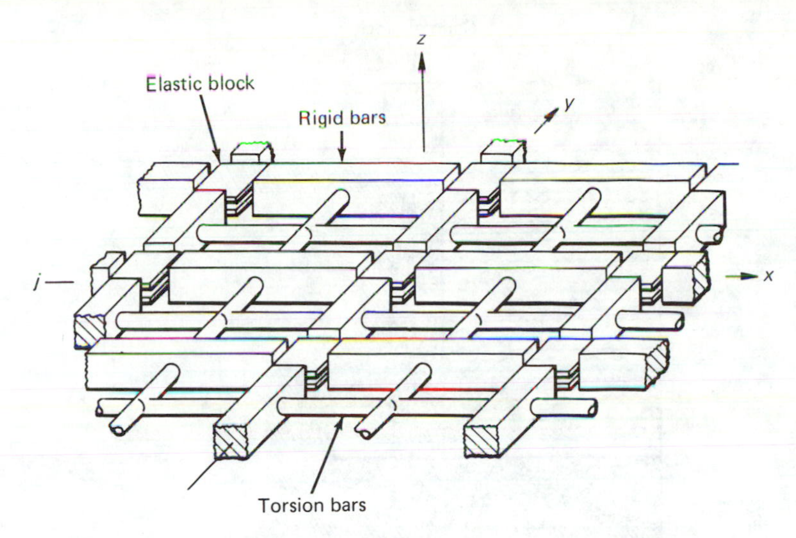

Figure 5-2 Discrete element model of a plate or slab (from Reference 2).

the body or structure is more easily subdivided because each part is essentially a piece of the whole. In addition, continuous elements provide a more natural representation of the properties of the original continuum. Moreover, the concept of two- or three-dimensional elements has permitted the finite element method to be generalized and applied to nonstructural problems, such as those tabulated in Table 1-1. An example of a finite element idealization appears in Figure 5-3. The concrete lock structure and the surrounding soil are idealized by an assemblage of two-dimensional quadrilateral and triangular finite elements. Because the lock is long compared with its width, the cross-section shown may be analyzed as a plane strain problem.

An alternative but equivalent interpretation of the finite element method derives from the physical continuity of elements in a two- or three-dimensional body. This notion, which is mathematical rather than physical, does not consider the body or structure to be subdivided into separate parts that are reassembled in the analysis procedure. Instead, the continuum is zoned into regions by imaginary lines (two-dimensional bodies) or imaginary planes (three-dimensional bodies) inscribed on the body. Note that no physical separation is envisaged at these lines or planes. Using this concept, we may apply variational procedures to the analysis by assuming a patchwork of solutions or displacement models, each of which applies to a single region.

Whether we regard a finite element as a "piece" of the continuum or as a "region" of the body, we are first interested in the behavior of a single

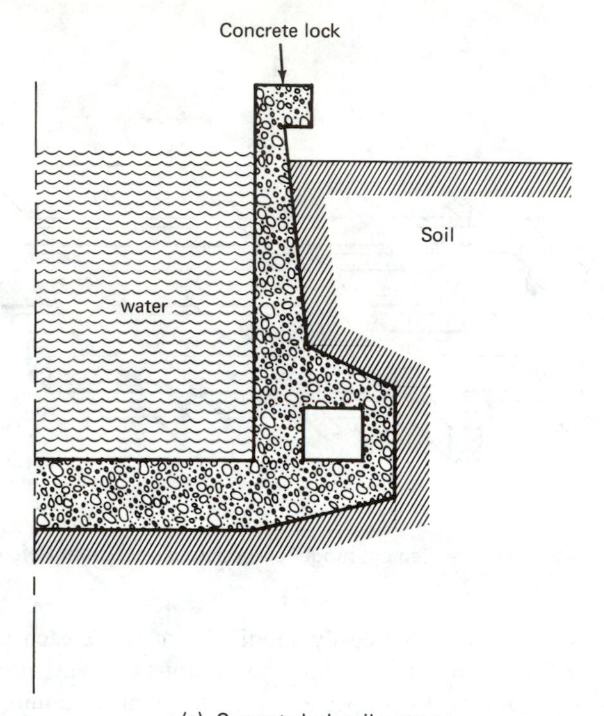

(a) Concrete lock-soil system

Figure 5-3 Finite element idealization of lock-backfill-foundation system.

element. We need to be able to define this behavior in terms of the element's geometry and of its material properties so that the general element formulation can be applied to any element of the assemblage. However, while we examine the individual element in this chapter, we should also recognize that we plan eventually to consider an assemblage or collection of elements.

Sample Problems

A number of problems are presented at the end of this chapter to illustrate the application of the theory. In addition, two sample problems will be worked out step by step throughout the chapter. The ideas covered in each major section will be applied to these two problems at the end of the section as an aid to the clear understanding of the material as it is presented. The two types of elements to be considered will be a simple one-dimensional element for the bending of a beam, Figure 5-4, and a triangular element for (two-dimensional) plane strain analysis, Figure 5-5.

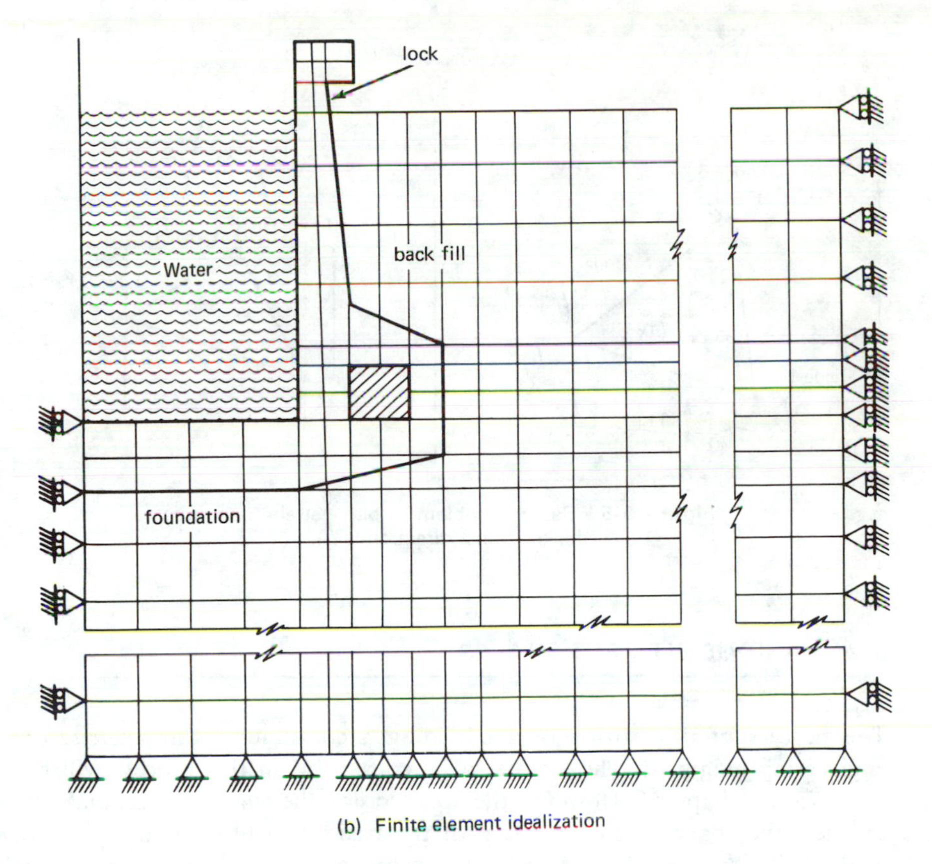

(b) Finite element idealization

Figure 5.3 *continued*

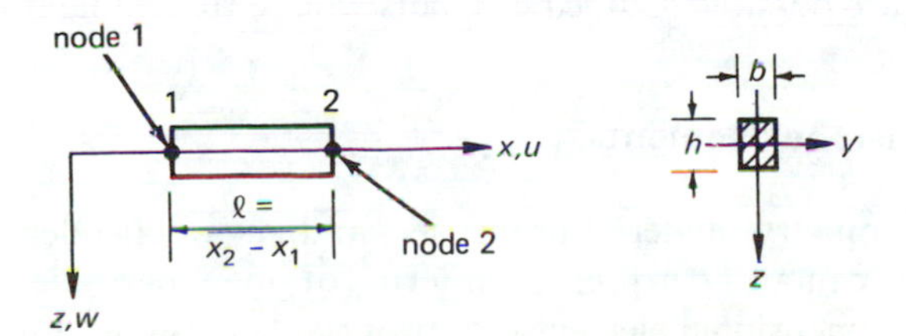

Figure 5-4 Sample Problem 1 : beam element.

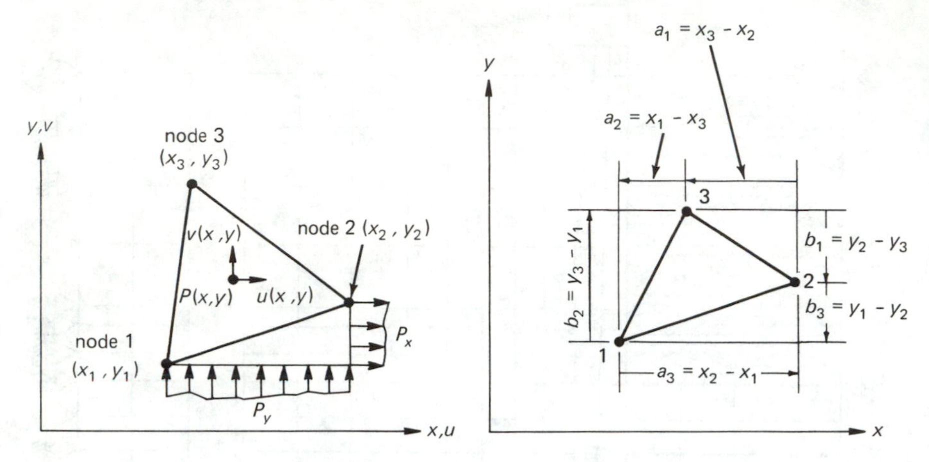

Figure 5-5 Sample Problem 2: plane strain triangle (thickness in z direction is h).

5-2 VARIOUS ELEMENT SHAPES

The process of discretizing or subdividing a continuum is an exercise of engineering judgment. The factors to be considered in this process will be discussed in Chapter 6. However, the first decision the engineer must make is to select the shape or configuration of the basic element to be used in the analysis. This choice depends upon the geometry of the body or structure and upon the number of independent space coordinates (e.g. x, y, or z) necessary to describe the problem. As we shall see, a finite element usually has a simple one-, two-, or three-dimensional configuration. The boundaries of elements are often straight lines, although for problems that can be best represented in curvilinear coordinates, it is advantageous for the element shapes to be similarly defined. In this section we shall emphasize elements with straight-line boundaries; in all cases, however, analogous elements with curved-line boundaries are possible. Curved elements will be treated later in this chapter.

One-Dimensional Elements

When the geometry, material properties, and such dependent variables as a displacement can all be expressed in terms of one independent space coordinate, a one-dimensional element is appropriate. This coordinate is measured along the axis of the element. This type of element is used for structures that

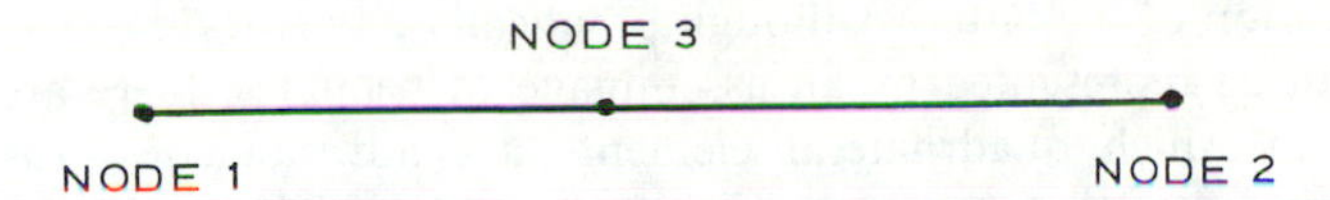

Figure 5-6 One-dimensional element.

can be idealized by line drawings, such as the truss and frame shown in Figure 5-1. In a strict sense, therefore, we are speaking of frame analysis when we speak of one-dimensional elements, because a distinguishing characteristic of the finite element method is that the elements themselves are continuous two- or three-dimensional bodies. However, a finite element approach to the analysis of beams and framed structures produces some useful insights, particularly when the geometry and properties of the structure vary continuously or discontinuously with the axial coordinates. In addition, one-dimensional problems are a simple but useful tool for developing our understanding of the finite element method. This utility will be implemented in Sample Problem 1 (Fig. 5-4).

A one-dimensional element may be represented by a straight line whose ends are nodal points, Figure 5-6. These nodal points, numbered 1 and 2 in the figure, are called *external nodes* because they represent connecting points to the adjacent elements. Some applications may require additional nodal points, such as node 3 in Figure 5-6. Because no connection to other elements occurs at this intermediate node, it is called an *internal node*.

Two-Dimensional Elements

In Chapter 3 we saw that many problems in solid mechanics could be approximated by a two-dimensional formulation. Among these specializations are plane strain, plane stress, and plate bending. The simplest element for two-dimensional problems is such a triangle as shown in Figures 5-5 and 5-7(a). There are two possible types of external nodes for triangular elements. The corner nodes indicated by 1, 2, and 3 in Figure 5-7(a) are called *primary external nodes*.

When additional nodes occur on the sides of the element, like nodes 4, 5, and 6 in Figure 5-7(a), we shall refer to them as *secondary external nodes*. This distinction is necessary because the secondary nodes may have fewer displacements of interest than the corner nodes. Finally, *internal nodes*, such as node 7 in Figure 5-7(a), are sometimes also used in triangular elements.

Other common types of two-dimensional elements are the rectangular and quadrilateral shapes, Figures 5-7(b) and 5-7(c). The former can be considered a specialization of the latter. Although practically any two-dimensional continuum can be represented by an assemblage of triangles, there are certain problems in which quadrilateral elements are advantageous. Instead of directly using quadrilateral elements, it is possible to construct such shapes from two or four triangular elements, as shown in Figures 5-7(d) and 5-7(e). In addition to the primary external nodes shown in Figures 5-7(b) through 5-7(e), each of the quadrilateral elements may also have secondary external nodes and one or more internal nodes.

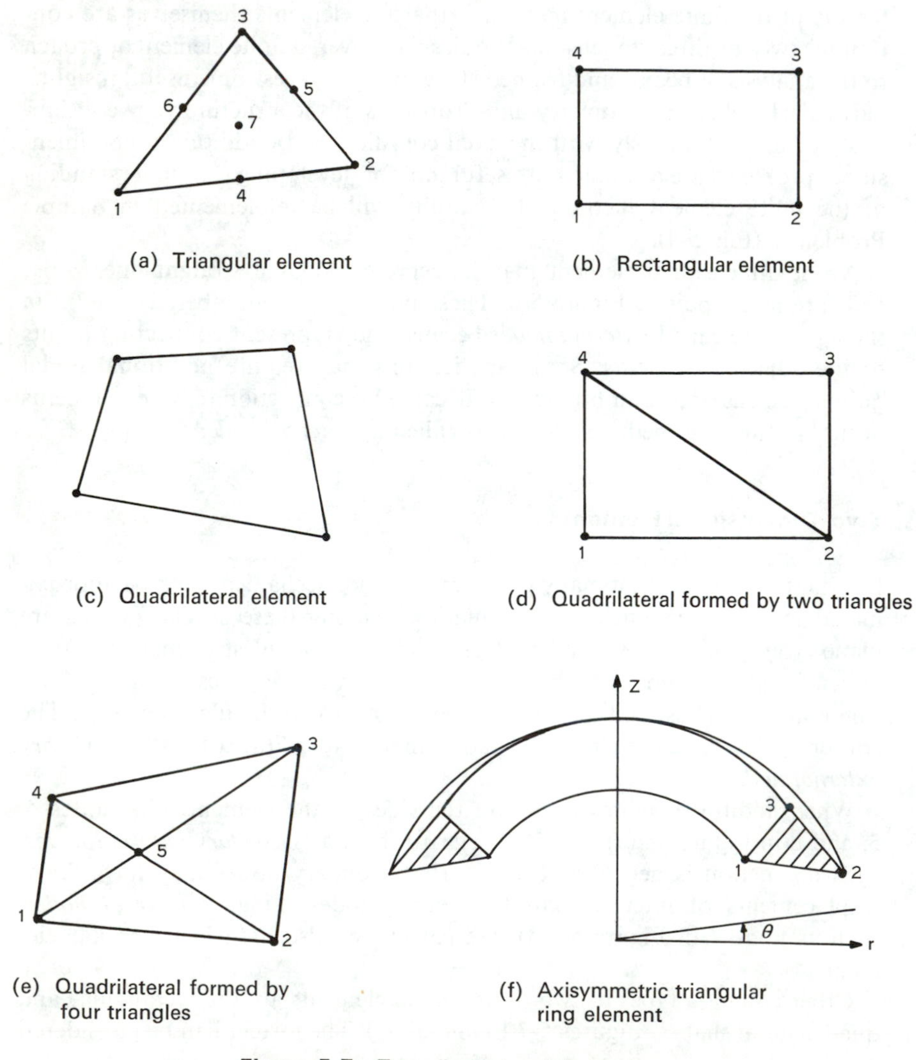

Figure 5-7 Two-dimensional elements.

Axisymmetric Elements

Another type of three-dimensional problem which can be represented completely in terms of one or two independent variables is the axisymmetric problem. Here a cylindrical coordinate system, r, z, and θ, is used; and all properties and variables are independent of the θ coordinate. The finite elements used for these problems are circular rings, called toroidal elements, with cross-sections in the r-z plane similar to the shapes in Figures 5-6 and 5-7. Thus a one-dimensional axisymmetric element is a conical frustum, and a two-dimensional axisymmetric element is a ring with a triangular [Fig. 5-7(f)] or quadrilateral cross-section.

Three-Dimensional Elements

Corresponding to the triangle, the tetrahedron shown in Figure 5-8(a) is the basic finite element for three-dimensional problems. A tetrahedron has four primary external nodes. Three-dimensional elements with eight primary external nodes are either in the form of a general hexahedron, Figure 5-8(c), or a rectangular prism, Figure 5-8(b), which is the specialization of a hexahedron. If necessary, we can introduce secondary external nodes or internal

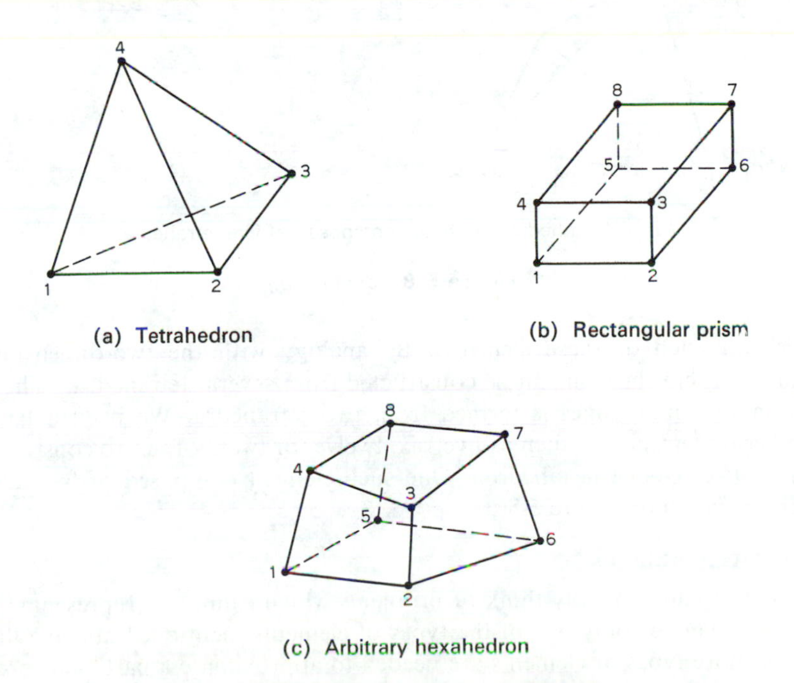

Figure 5-8 Three-dimensional elements.

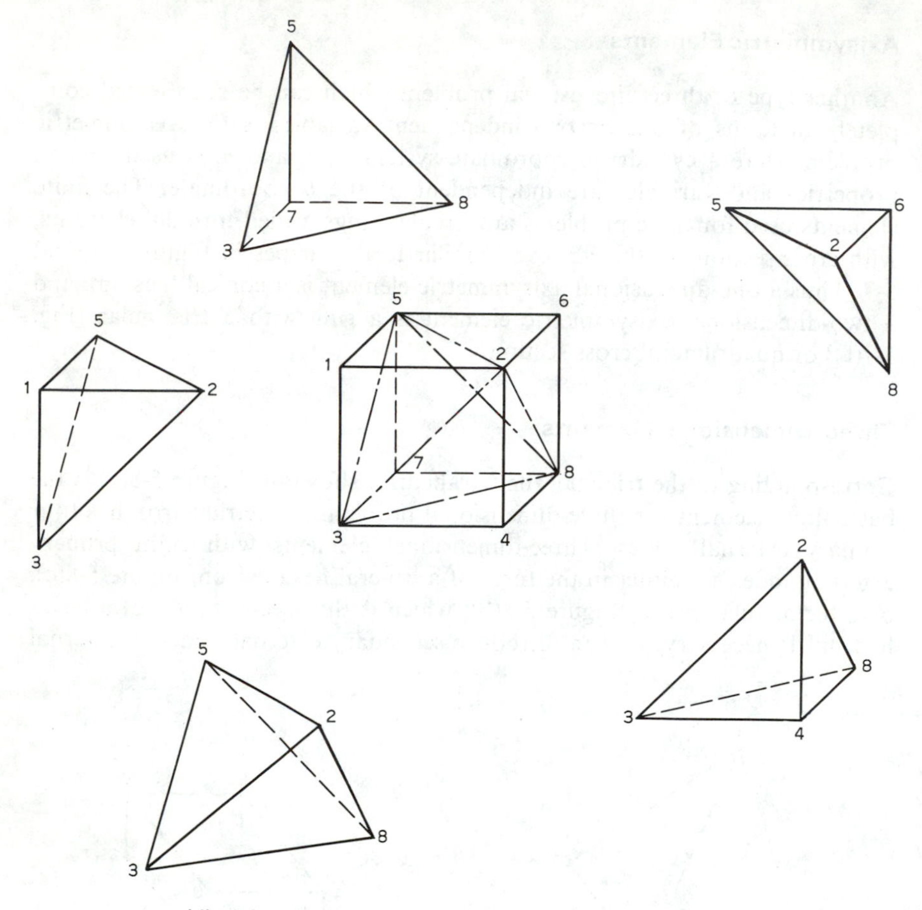

(d) A hexahedron (cuboid) composed of five tetrahedra

Figure 5-8 *continued*

nodes for each of these elements. By analogy with the two-dimensional elements, a hexahedron can be constructed from several tetrahedra. A hexahedron with nine edges is formed from two tetrahedra. We need a larger number of tetrahedra, such as five, six, twelve, or twenty-four, to construct a general twelve-edge hexahedron. One such element composed of five tetrahedra is shown in Figure 5-8(d).

Mixed Assemblages

The reader can probably think of problems which cannot be represented by an assemblage of only one of the types of elements mentioned above. Often two or more types of elements are needed to approximate a particular problem. One example of this would be the bending of a plate which is stiffened by edge beams. Here the plate itself would be represented by two-dimensional elements such as rectangles, while the edge beams would be approximated by one-dimensional elements appropriately connected to the plate elements.

5-3 DISPLACEMENT MODELS

The basic philosophy of the finite element method is piecewise approximation. That is, we approximate a solution to a complicated problem by subdividing the region of interest and representing the solution within each subdivision by a relatively simple function. Specifically, we have spoken of the displacement method of structural analysis, in which the structure or body is divided into finite elements. We now consider the simple functions which are assumed to approximate the displacements for each element. These functions are called *displacement models*, *displacement functions*, *displacement fields*, or *displacement patterns*.

Generalized Coordinate Form of Displacement Models

In Chapter 1 we stated that a polynomial is the most common form of displacement model. There are two principal reasons why this form is so widely used. First, it is easy to handle the mathematics of polynomials in formulating the desired equations for various elements and in performing digital computation. In particular, the use of polynomials permits us to differentiate and integrate with relative ease. Second, a polynomial of arbitrary order permits a recognizable approximation to the true solution. We can see that a polynomial of infinite order corresponds to an exact solution; however, for practical purposes we are limited to one of finite order. By truncating an infinite polynomial at different orders, we clearly vary the degree of approximation, as shown for the simple one-dimensional case in Figure 5-9. In this figure, an "exact" solution for the displacement $u(x)$ is approximated by various degree polynomials of the general form

$$u(x) = \alpha_1 + \alpha_2 x + \alpha_3 x^2 + \cdots + \alpha_{n+1} x^n \tag{5.1a}$$

The greater the number of terms included in the approximation, the more closely the exact solution is represented. (Obviously, this last statement does not apply to the case wherein the exact solution is a polynomial of some finite order m. Here terms in excess of m do not improve the representation.)

In equation (5.1a), the coefficients of the polynomial, the α's, are known as *generalized coordinates* or *generalized displacement amplitudes*. The number

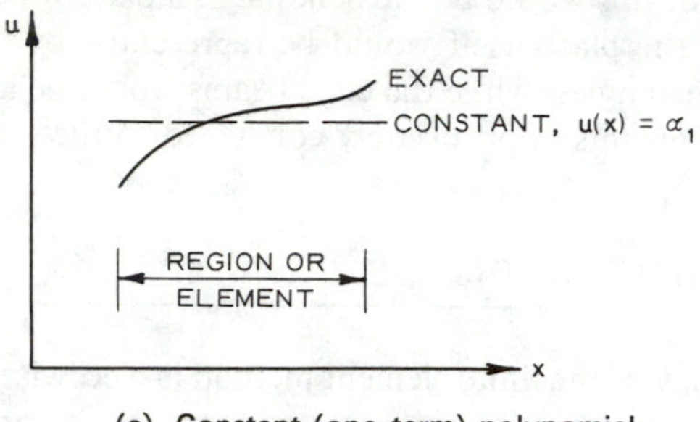

(a) Constant (one-term) polynomial

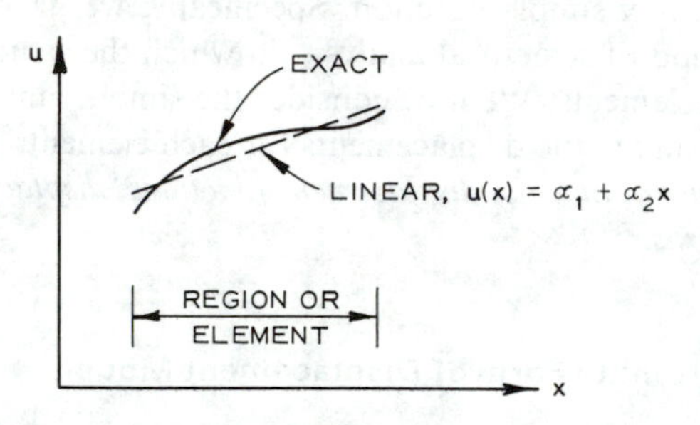

(b) Linear (two-term) polynomial

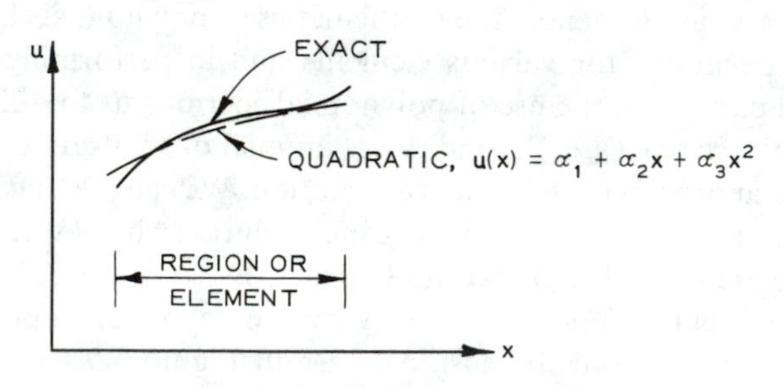

(c) Quadratic (three-term) polynomial

Figure 5-9 Polynomial approximation in one dimension.

of terms retained in the polynomial determines the shape of the displacement model, whereas the magnitudes of the generalized coordinates govern the amplitude. These amplitudes are called "generalized" because they are not necessarily identified with the physical displacements of the element on a one-to-one basis; rather, they are linear combinations of some of the nodal displacements and perhaps of some of the derivatives of displacements at the nodes as well. The generalized coordinates represent the minimum number of parameters necessary to specify the polynomial amplitude. Their physical significance will become more apparent in the next section.

Equation (5.1a) can be expressed in matrix form as

$$u(x) = \{\phi\}^{\mathrm{T}}\{\alpha\}$$

$$\{\phi\}^{\mathrm{T}} = [1 \;\; x \;\; x^2 \;\; \cdots \;\; x^n] \tag{5.1b}$$

$$\{\alpha\}^{\mathrm{T}} = [\alpha_1 \;\; \alpha_2 \;\; \alpha_3 \;\; \cdots \;\; \alpha_{n+1}]$$

The general polynomial form of a two-dimensional displacement model is

$$\begin{aligned} u(x, y) &= \alpha_1 + \alpha_2 x + \alpha_3 y + \alpha_4 x^2 + \alpha_5 xy + \alpha_6 y^2 + \cdots + \alpha_m y^n \\ v(x, y) &= \alpha_{m+1} + \alpha_{m+2} x + \alpha_{m+3} y + \alpha_{m+4} x^2 + \alpha_{m+5} xy + \alpha_{m+6} y^2 \\ &\qquad + \cdots + \alpha_{2m} y^n \end{aligned} \tag{5.2a}$$

where

$$m = \sum_{i=1}^{n+1} i$$

and where u and v denote the components of displacements in the x and y directions, respectively (Fig. 5-5). Equation (5.2a) can also be expressed in matrix form

$$\{u(x, y)\} = \begin{Bmatrix} u(x, y) \\ v(x, y) \end{Bmatrix} = [\phi]\{\alpha\} = \begin{bmatrix} \{\phi_1\}^{\mathrm{T}}\{0\}^{\mathrm{T}} \\ \{0\}^{\mathrm{T}}\{\phi_1\}^{\mathrm{T}} \end{bmatrix}\{\alpha\}$$

$$\{\phi_1\}^{\mathrm{T}} = [1 \;\; x \;\; y \;\; x^2 \;\; xy \;\; y^2 \;\; \cdots \;\; y^n] \tag{5.2b}$$

$$\{\alpha\}^{\mathrm{T}} = [\alpha_1 \;\; \alpha_2 \;\; \alpha_3 \;\; \cdots \;\; \alpha_{2m}]$$

Finally, a three-dimensional displacement model of nth order is given by

$$\begin{aligned} u(x, y, z) &= \alpha_1 + \alpha_2 x + \alpha_3 y + \alpha_4 z + \alpha_5 zx + \cdots + \alpha_m z^n \\ v(x, y, z) &= \alpha_{m+1} + \alpha_{m+2} x + \alpha_{m+3} y + \alpha_{m+4} z + \alpha_{m+5} zx + \cdots + \alpha_{2m} z^n \\ w(x, y, z) &= \alpha_{2m+1} + \alpha_{2m+2} x + \alpha_{2m+3} y + \alpha_{2m+4} z + \alpha_{2m+5} zx + \cdots + \alpha_{3m} z^n \end{aligned} \tag{5.3a}$$

where

$$m = \sum_{i=1}^{n+1} i(n + 2 - i)$$

and where u, v, and w are the components of displacement. Equation (5.3a) can be written in matrix form as

$$\{u(x, y, z)\} = \begin{Bmatrix} u(x, y, z) \\ v(x, y, z) \\ w(x, y, z) \end{Bmatrix} = [\phi]\{\alpha\} = \begin{bmatrix} \{\phi_2\}^T & \{0\}^T & \{0\}^T \\ \{0\}^T & \{\phi_2\}^T & \{0\}^T \\ \{0\}^T & \{0\}^T & \{\phi_2\}^T \end{bmatrix} \{\alpha\}$$

$$\{\phi_2\}^T = [1 \;\; x \;\; y \;\; z \;\; zx \;\; x^2 \;\; xy \;\; y^2 \;\; yz \;\; z^2 \;\; \cdots \;\; z^n] \tag{5.3b}$$

$$\{\alpha\}^T = [\alpha_1 \;\; \alpha_2 \;\; \alpha_3 \;\; \cdots \;\; \alpha_{3m}]$$

Each of these general polynomial forms of displacement models in equations (5.1), (5.2), and (5.3) can be truncated at any desired degree to give constant, linear, quadratic, cubic, or higher order patterns. For example, in the two-dimensional case, a linear model is

$$\begin{aligned} u &= \alpha_1 + \alpha_2 x + \alpha_3 y \\ v &= \alpha_4 + \alpha_5 x + \alpha_6 y \end{aligned} \tag{5.2c}$$

and a quadratic model is

$$\begin{aligned} u &= \alpha_1 + \alpha_2 x + \alpha_3 y + \alpha_4 x^2 + \alpha_5 xy + \alpha_6 y^2 \\ v &= \alpha_7 + \alpha_8 x + \alpha_9 y + \alpha_{10} x^2 + \alpha_{11} xy + \alpha_{12} y^2 \end{aligned} \tag{5.2d}$$

Because the displacement models in equations (5.1) through (5.3) are expressed in terms of generalized coordinates, $\{\alpha\}$, they are referred to as *generalized coordinate displacement models.* This is the elementary form of models for the finite element method. In Section 5-4 we shall consider an alternative representation of polynomial displacement fields that facilitates the formulation of the basic equations for the element. However, for the remainder of this section we shall continue to discuss the characteristics of displacement models in the elementary form.

Convergence Requirements

In any acceptable numerical formulation, the numerical solution must converge or tend to the exact solution of the problem. For the finite element method, it has been shown that under certain circumstances the displacement formulation provides an upper bound to the true stiffness of the structure.[4] In other words, the stiffness coefficients for a given displacement model have magnitudes higher than those for the exact solution. Thus, under a given load, the simulated structure deforms less than the actual structure. It follows that, as the finite element subdivision of the structure is made finer, the approximate displacement solution will bound the exact solution from below, that is, we obtain lower bounds to the true solution. In order for this boundedness to be rigorously assured, three conditions must be met.[5]

1. *The displacement models must be continuous within the elements, and the displacements must be compatible between adjacent elements.* The first part of this requirement is readily met by choosing polynomial models, which are inherently continuous. The second part implies that adjacent elements must deform without causing openings, overlaps, or discontinuities between the elements. This will be discussed further in the next chapter.† Inter-element compatibility cannot be satisfied unless the displacements along the side of an element depend only upon the displacements of the nodes occurring on that side.

2. *The displacement models must include the rigid body displacements of the element.* A rigid body displacement is the most elementary deformation that an element may undergo. Basically, this condition states that there should exist combinations of values of the generalized coordinates that cause all points on the element to experience the same displacement. One such combination should occur for each of the possible rigid body translations and rotations. For instance, in the displacement model of equation (5.1) the constant term α_1 provides for a rigid body displacement.

3. *The displacement models must include the constant strain states of the element.* This requirement can be stated in terms similar to those of the second condition. There should exist combinations of values of the generalized coordinates that cause all points on the element to experience the same strain. One such combination should occur for each possible strain. The necessity of this requirement can be understood physically if we imagine the subdivision of a body into smaller and smaller elements. As these elements approach infinitesimal size, the strains in each element approach constant values. Unless our approximation includes these constant strains, we cannot hope to converge to a correct solution. The terms associated with α_2 and α_{m+3} in equation (5.2a) provide for uniform strains ε_x and ε_y in an element.

In the finite element literature, formulations which satisfy the first condition above are called *compatible* or *conforming*. Elements which meet both the second and third requirements are known as *complete*. Later we shall see how variational principles are used in the finite element formulation. To use such principles meaningfully, we must ensure that the displacement model will allow continuous nonzero derivatives of the highest order appearing in the potential energy functional. This is a more general statement of the completeness criteria.

Displacement models of the types given in equation (5.1), (5.2), and (5.3), when applied to axial tension, plane strain, and three-dimensional elasticity,

† In Section 6-4 we shall see that interelement compatibility must be enforced for displacements and their derivatives up to the order $n - 1$, where n is the highest order derivative in the energy functional.

respectively, satisfy the three conditions given above if at least the constant and linear terms are retained in the polynomials. In other applications, such as the analysis of beam, plate, and shell structures, it is more difficult to satisfy the first two requirements. However, although all three conditions are sufficient to prove convergence in the general case, practical results for elements that satisfy only the third requirement appear to converge acceptably.[6] In particular, elements which relax the first criterion, that is, which are complete but nonconforming, have been widely and successfully used. The principal disadvantage of nonconforming elements is that we no longer know in advance that the stiffness will be an upper bound. On the other hand, nonconforming formulations are usually less stiff (in other words, more flexible) than compatible ones, so the former *may* converge more quickly than the latter. Some of the possible convergence curves for incompatible elements are qualitatively compared to a compatible element in Figure 5-10.

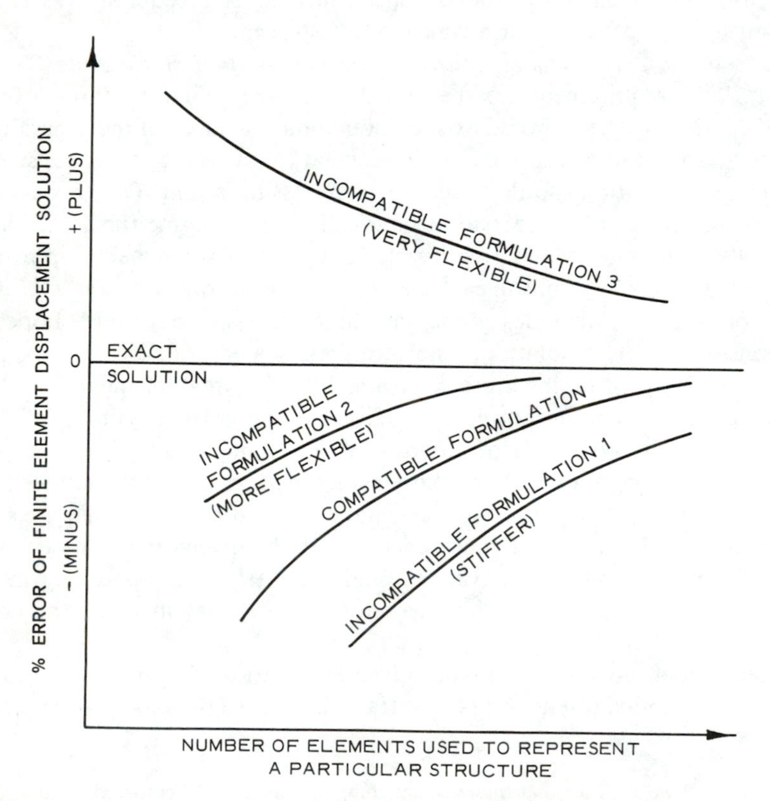

Figure 5-10 Schematic convergence of three hypothetical incompatible elements in comparison to a compatible element.[6]

Selection of the Order of the Polynomial

In selecting the order of a polynomial displacement model, the completeness and compatibility requirements are important considerations; but they are not the only factors we must take into account. An additional consideration in the selection of the order of the model is that the pattern should be independent of the orientation of the local coordinate system.[7] This property of the model is known as *geometric isotropy*, *spatial isotropy*, or *geometric invariance*. For polynomials of linear order, the isotropy requirement is usually equivalent to the necessity of including constant strain states. For higher order patterns, we can see intuitively that it is undesirable to have a preferential coordinate direction, in other words to have displacement shapes which change with a change in local coordinate system. Experience confirms this intuition. One way to envision the achievement of isotropy is to consider the Pascal triangle for the variable terms of the two-dimensional polynomial, equations (5.2).[6]

$$
\begin{array}{ccccccccccccl}
 & & & & & & 1 & & & & & & \text{constant terms} \\
 & & & & & x & | & y & & & & & \text{linear terms} \\
 & & & & x^2 & & xy & & y^2 & & & & \text{quadratic terms} \\
 & & & x^3 & & x^2y & | & xy^2 & & y^3 & & & \text{cubic terms} \\
 & & x^4 & & x^3y & & x^2y^2 & & xy^3 & & y^4 & & \text{quartic terms} \\
 & x^5 & & x^4y & & x^3y^2 & | & x^2y^3 & & xy^4 & & y^5 & \text{quintic terms} \\
 & & & & & & \text{axis of symmetry} & & & & & &
\end{array}
\tag{5.4}
$$

For the two-dimensional approach, we should not include any term from one side of the axis of symmetry of the triangle without including its counterpart from the other side. For example, if we wish to construct a cubic model with eight terms, the following are geometrically isotropic: (1) all the constant, linear, and quadratic terms plus the x^3 and y^3 terms; or (2) all the constant, linear, and quadratic terms plus the x^2y and xy^2 terms.

Immediately below, we shall define element degrees of freedom and discuss their selection. Meanwhile, the final consideration in the selection of the displacement polynomial is that the total number of generalized coordinates for an element must be equal to or greater than the number of joint or external degrees of freedom of the element. The usual procedure employs the same number of generalized coordinates as the degrees of freedom. It is possible to utilize an excess of generalized coordinates to improve the element stiffness matrix, that is, to make it less stiff or more flexible.[9] These excess coordinates are generally associated with internal nodes and improve the approximation of equilibrium within the element. However, they do not improve interelement equilibrium. Therefore, more than a few of such extra coordinates are rarely justifiable; in fact, their inclusion may be detrimental in some cases.[8]

Nodal Degrees of Freedom

The nodal displacements, rotations, and/or strains necessary to specify completely the deformation of the finite element are the *degrees of freedom* of the element. The degrees of freedom differ from the generalized coordinates in that each is specifically identified with a single nodal point and represents a displacement (or rotation or strain) having a clear physical interpretation. We shall distinguish between degrees of freedom occurring at external and internal nodes by referring to them as *joint* or *nodal degrees of freedom* and *internal degrees of freedom*, respectively. In the formulation of individual element properties, however, we need not make this distinction; it is only during the assembly process that the difference becomes important.

The minimum number of degrees of freedom (or generalized coordinates) necessary for a given element is determined by the completeness requirements for convergence, the requirements of geometric isotropy, and the necessity of an adequate representation of the terms in the potential energy functional. Additional degrees of freedom beyond the minimum number may be included for any element by adding secondary external nodes or by specifying as degrees of freedom higher order derivatives of displacements at the primary nodes. The latter approach is preferred because it leads to a more compact numerical formulation in the assembly process. Elements with additional degrees of freedom are called *higher order elements*. The more such additional degrees of freedom are added, the more flexible becomes the element stiffness for an element of given size. However, this improvement in the upper bound to the stiffness is purchased at the price of an increasingly complex formulation of the individual element properties. The nature of this tradeoff will be discussed more fully in the next chapter.

We may now relate the degrees of freedom and the generalized coordinates by employing the displacement model. We can evaluate the displacements at the nodes by substituting the nodal coordinates into the model. For example, using a model of the form given by the first equations in equations (5.1b), (5.2b), and (5.3b), we may write:

$$\{u\} = [\phi]\{\alpha\} \tag{5.5}$$

$$\{q\} = \begin{Bmatrix} \{u \text{ (node } 1)\} \\ \{u \text{ (node } 2)\} \\ \cdots \\ \{u \text{ (node } M)\} \end{Bmatrix} = \begin{bmatrix} [\phi \text{ (node } 1)] \\ [\phi \text{ (node } 2)] \\ \cdots \\ [\phi \text{ (node } M)] \end{bmatrix} \{\alpha\} = [A]\{\alpha\} \tag{5.6}$$

Here M is the total number of nodes for the element being considered, $\{q\}$ is the vector of nodal displacements, and the notation in parentheses indicates that the dependent variables are assigned their values at the particular node. We may invert equation (5.6) to give

$$\{\alpha\}=[A^{-1}]\{q\} \tag{5.7}$$

where $[A^{-1}]$ is a *displacement transformation matrix*. Note that $[A]$ is a square matrix; hence, the total number of generalized coordinates equals the total number of joint and internal degrees of freedom. When equation (5.7) is substituted into equation (5.5), we can eliminate the generalized coordinates and obtain

$$\{u\}=[\phi][A^{-1}]\{q\}=[N]\{q\} \tag{5.8}$$

Equation (5.8) expresses the displacements $\{u\}$ at any point P (Fig. 5-5) within the element in terms of the displacements of the nodes, $\{q\}$.

One limitation of generalized coordinate displacement models is that it is not always possible to obtain the inverse of matrix $[A]$.[7] The interpolation function representation of displacement models discussed in the next section avoids this difficulty.

We shall now illustrate the material covered in this section by applying it to the two sample problems.

Sample Problem 1

The standard Navier assumption of beam theory that plane cross sections of a beam remain plane in bending allows us to represent the axial displacements in terms of the beam slopes (Fig. 5-4),

$$u(x, z) = -z\frac{dw}{dx} \tag{5.9}$$

Here the x axis is coincident with the neutral axis. The axial strains are given by

$$\varepsilon(x, z) = \frac{du}{dx} = -z\frac{d^2w}{dx^2} \tag{5.10}$$

For a beam problem we are primarily interested in the transverse displacements w; moreover, the axial displacements u are directly related to the transverse displacements by equation (5.9). Therefore, we need only select a displacement model for w. Equation (5.10) indicates that the strains are proportional to the second derivative of w. Hence, the minimum order of the polynomial of the type in equation (5.1) for a beam problem is quadratic. Such a polynomial with three generalized coordinates is expressed as

$$w(x) = \alpha_1 + \alpha_2 x + \alpha_3 x^2$$

A rigid body translation is possible since nonzero α_1 is permitted; a rigid body rotation is possible because nonzero α_2 is allowed; and a constant bending strain state (that is, a constant curvature) is possible as nonzero α_3 is allowed. Because interelement compatibility is no problem with one-dimensional elements (there are no element "sides," only nodes), the model proposed satisfies all the completeness and compatibility requirements. In addition, geometric isotropy is identically satisfied in the one-dimensional case. The joint degrees of freedom will be the transverse displacements at the end nodes, designated w_1 and w_2. An additional (internal) degree of freedom is necessary in order to employ the suggested model because there are three generalized coordinates. Instead, let us employ two additional joint degrees of freedom, in order to preserve interelement compatibility for the slope (see Section 6-4). We define the slope as

$$\theta = \frac{dw}{dx} \tag{5.11}$$

which is positive clockwise. Hence, the two additional joint degrees of freedom are θ_1 and θ_2. Correspondingly, we must increase the number of generalized coordinates from three to four. For convenience, let us transform our coordinate system as follows:

$$\bar{x} = x - x_1 \tag{5.12}$$

Hence, $\bar{x}_1 = 0$ and $\bar{x}_2 = \ell$, and our displacement model becomes

$$\begin{aligned} w(x) &= \underset{1\times 4}{\{\phi\}^{\mathrm{T}}} \ \underset{4\times 1}{\{\alpha\}} \\ \{\phi\}^{\mathrm{T}} &= [1 \ \ \bar{x} \ \ \bar{x}^2 \ \ \bar{x}^3] \\ \{\alpha\}^{\mathrm{T}} &= [\alpha_1 \ \ \alpha_2 \ \ \alpha_3 \ \ \alpha_4] \end{aligned} \tag{5.13}$$

Also

$$\theta(x) = [0 \ \ 1 \ \ 2\bar{x} \ \ 3\bar{x}^2]\{\alpha\}$$

We now express the nodal displacement in terms of the generalized coordinates using equations (5.6) and (5.11).

$$\{q\} = \begin{Bmatrix} w(x=0) \\ \theta(x=0) \\ w(x=\ell) \\ \theta(x=\ell) \end{Bmatrix} = \begin{Bmatrix} w_1 \\ \theta_1 \\ w_2 \\ \theta_2 \end{Bmatrix} = \begin{bmatrix} 1 & 0 & 0 & 0 \\ 0 & 1 & 0 & 0 \\ 1 & \ell & \ell^2 & \ell^3 \\ 0 & 1 & 2\ell & 3\ell^2 \end{bmatrix} \{\alpha\} = [A]\{\alpha\} \tag{5.14a}$$

Inverting this equation, we obtain

$$\{\alpha\} = [A^{-1}]\{q\} = \begin{bmatrix} 1 & 0 & 0 & 0 \\ 0 & 1 & 0 & 0 \\ -3/\ell^2 & -2/\ell & 3/\ell^2 & -1/\ell \\ 2/\ell^3 & 1/\ell^2 & -2/\ell^3 & 1/\ell^2 \end{bmatrix} \{q\} \qquad (5.14b)$$

This equation gives the generalized coordinates as a linear combination of the nodal displacements. The formulation of equation (5.8) for this problem can be accomplished by the matrix multiplication arising from the substitution of equation (5.14b) into equation (5.13). This will be deferred until later in the chapter.

Sample Problem 2

In plane strain the only possible displacements are in the x and y directions. Thus a displacement model of the type given in equations (5.2) is necessary. Consider the linear model which is a special case of equation (5.5).

$$\{u(x, y)\} = \begin{Bmatrix} u(x, y) \\ v(x, y) \end{Bmatrix} = \begin{bmatrix} 1 & x & y & 0 & 0 & 0 \\ 0 & 0 & 0 & 1 & x & y \end{bmatrix} \underset{6\times 1}{\{\alpha\}} = [\phi]\{\alpha\} \qquad (5.15)$$

We note that a rigid body displacement in the x direction is possible if α_1 is a nonzero generalized coordinate; similarly, a rigid body v displacement may occur if α_4 is nonzero. A rigid body rotation may occur if $\alpha_3 = -\alpha_5 \neq 0$. Finally, the constant states of the three nonvanishing plane strains, ε_x, ε_y, and γ_{xy}, equation (3.16a), are possible if $\alpha_2 \neq 0$, $\alpha_6 \neq 0$, and $\alpha_3 = \alpha_5 \neq 0$ are nonzero coordinates, respectively. Hence the completeness, compatibility, and isotropy requirements are satisfied by the model given in equation (5.15). The joint degrees of freedom will be the u and v displacements at each of the three corner nodes (Fig. 5-5). Thus there are a total of six nodal displacements. Because this is the same as the number of generalized coordinates, the model proposed is the simplest one that is sufficient for a plane strain analysis. Expressing the nodal displacements in terms of the generalized coordinates using equation (5.6), we obtain

$$\{q\} = \begin{Bmatrix} u_1 \\ u_2 \\ u_3 \\ v_1 \\ v_2 \\ v_3 \end{Bmatrix} = \begin{bmatrix} 1 & x_1 & y_1 & 0 & 0 & 0 \\ 1 & x_2 & y_2 & 0 & 0 & 0 \\ 1 & x_3 & y_3 & 0 & 0 & 0 \\ 0 & 0 & 0 & 1 & x_1 & y_1 \\ 0 & 0 & 0 & 1 & x_2 & y_2 \\ 0 & 0 & 0 & 1 & x_3 & y_3 \end{bmatrix} \{\alpha\} = [A]\{\alpha\} \qquad (5.16)$$

Note that we have arranged the nodal displacements to obtain a compact form of the $[A]$ matrix.

5-4 ISOPARAMETRIC ELEMENTS

The formulation of displacement models and the calculation of element stiffnesses have been both simplified and generalized by the concept known as *isoparametric elements.*[10] Before we can define and discuss this concept, we must consider the two essential ingredients of this approach: natural coordinate systems and interpolation displacement models.

Natural Coordinate Systems

A *local coordinate system* is one that is defined for a particular element and not necessarily for the entire body or structure; the coordinate system for the entire body is called the *global system.* One example of local coordinates is the barred system defined in equation (5.12), although that is a rather trivial case. A *natural coordinate system* is a local system which permits the specification of a point within the element by a set of dimensionless numbers whose magnitudes never exceed unity. Moreover, these systems usually are arranged so that some of the natural coordinates have unit magnitude at primary external nodal points. Not only does such a coordinate system generalize and simplify the formulation, but we shall see that it also often facilitates the integration which is required to obtain the element stiffness. Figure 5-11 illustrates natural coordinate systems for one-, two-, and three-dimensional elements.

A natural coordinate system for a one-dimensional line element is shown in Figure 5-11(a). The relationship between the natural coordinates, L_1 and L_2, of any point P and the cartesian coordinate x can be written as

$$\begin{Bmatrix} 1 \\ x \end{Bmatrix} = \begin{bmatrix} 1 & 1 \\ x_1 & x_2 \end{bmatrix} \begin{Bmatrix} L_1 \\ L_2 \end{Bmatrix} \tag{5.17a}$$

where

$$L_1 = \ell_1/\ell \quad \text{and} \quad L_2 = \ell_2/\ell \tag{5.17b}$$

Moreover, since $L_1 + L_2 = 1$, only one of the two natural coordinates is independent. The inverse of equation (5.17a) is

$$\begin{Bmatrix} L_1 \\ L_2 \end{Bmatrix} = \frac{1}{\ell} \begin{bmatrix} x_2 & -1 \\ -x_1 & 1 \end{bmatrix} \begin{Bmatrix} 1 \\ x \end{Bmatrix} \tag{5.17c}$$

Differentiation with respect to x is given by the formula

$$\frac{d}{dx} = \sum_{i=1}^{2} \frac{\partial L_i}{\partial x} \frac{\partial}{\partial L_i} = \frac{1}{\ell}\left(\frac{\partial}{\partial L_2} - \frac{\partial}{\partial L_1}\right) \tag{5.17d}$$

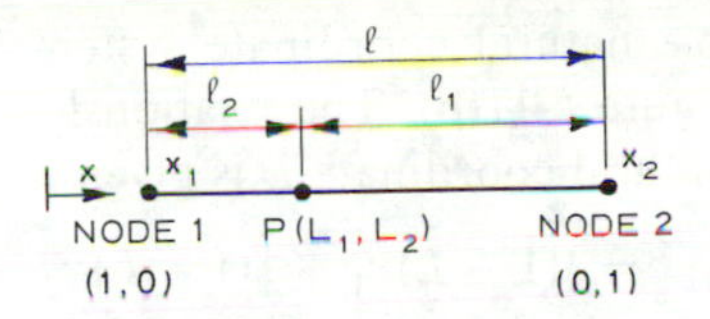

(a) Natural coordinate for line element

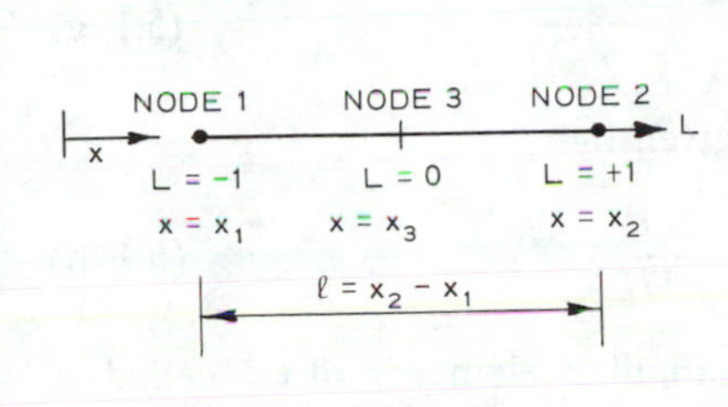

(b) A simple natural coordinate for line element

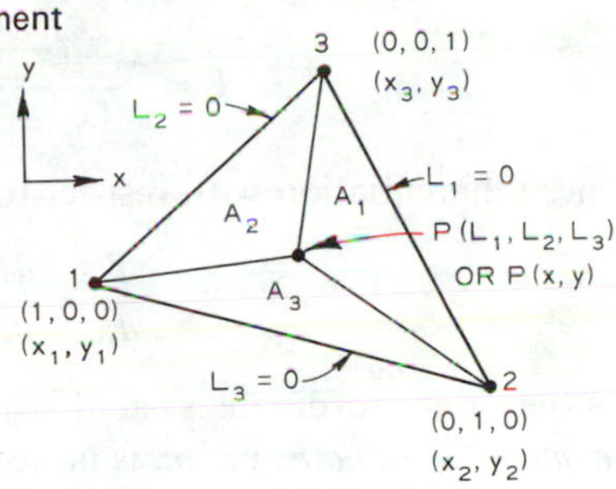

(c) Triangular coordinates

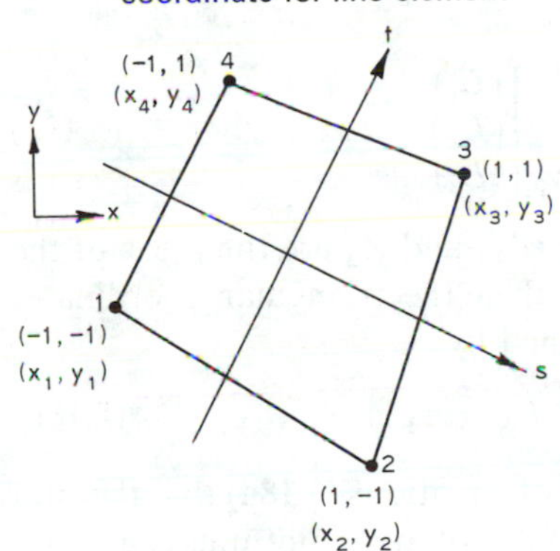

(d) Quadrilateral coordinates

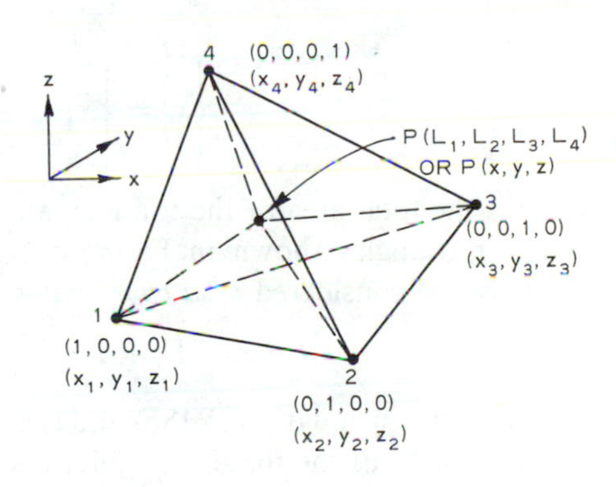

(e) Tetrahedral coordinates

Figure 5-11 Natural coordinate systems.

Integration of polynomial terms in the natural coordinate system is conveniently expressed by the formula

$$\int_{\ell} L_1^p L_2^q \, d\ell = \frac{p!\,q!}{(p+q+1)!}\ell \tag{5.17e}$$

where $p!$ is the factorial product $p(p-1)(p-2)\cdots(1)$, and $0!$ is defined as unity. As an example, consider $p = q = 1$. Then

$$\int_{\ell} L_1^1 L_2^1 \, d\ell = \frac{1!\,1!}{3!}\ell = \frac{\ell}{6}$$

An alternative, simple natural coordinate system for a one-dimensional element is shown in Figure 5-11(b). The relationship between the cartesian coordinate x and the natural coordinate L is given by

$$x = \tfrac{1}{2}(1 - L)x_1 + \tfrac{1}{2}(1 + L)x_2 \tag{5.17f}$$

This can be inverted to give

$$L = \frac{x - (x_1 + x_2)/2}{(x_2 - x_1)/2} = \frac{x - x_3}{\ell/2} \tag{5.17g}$$

Hence differentiation with respect to x is given by

$$\frac{d}{dx} = \frac{dL}{dx}\frac{d}{dL} = \frac{2}{\ell}\frac{d}{dL} \tag{5.17h}$$

A natural coordinate system for a triangular element, often called a *triangular coordinate system*, is indicated in Figure 5-11(c). Here three coordinates, L_1, L_2, and L_3, are used to define the location of a point, but only two of these are independent. Their relation to cartesian coordinates, x, y, is given by

$$\begin{Bmatrix} 1 \\ x \\ y \end{Bmatrix} = \begin{bmatrix} 1 & 1 & 1 \\ x_1 & x_2 & x_3 \\ y_1 & y_2 & y_3 \end{bmatrix} \begin{Bmatrix} L_1 \\ L_2 \\ L_3 \end{Bmatrix} \tag{5.18a}$$

If A is the total area of the triangle, and A_1, A_2, and A_3 are the areas of the smaller triangles shown in Figure 5-11(c), then the triangular coordinates can also be considered *area coordinates* defined by

$$L_1 = A_1/A, L_2 = A_2/A, L_3 = A_3/A \tag{5.18b}$$

It is clear from equation (5.18b) and the first of equations (5.18a) that the sum $L_1 + L_2 + L_3$ of the three triangular coordinates of any point must be unity. This is the condition that makes only two of the coordinates independent. We define A_{ij} to be the area of the triangle for which the nodes i and j and the origin of the cartesian (x–y) coordinate system are vertices. Furthermore, we define the following (Fig. 5-5):

$$\begin{aligned} a_1 &= x_3 - x_2 & b_1 &= y_2 - y_3 \\ a_2 &= x_1 - x_3 & b_2 &= y_3 - y_1 \\ a_3 &= x_2 - x_1 & b_3 &= y_1 - y_2 \end{aligned} \tag{5.18c}$$

By using these definitions, the inverse of equation (5.18a) may be written

$$\begin{Bmatrix} L_1 \\ L_2 \\ L_3 \end{Bmatrix} = \frac{1}{2A} \begin{bmatrix} 2A_{23} & b_1 & a_1 \\ 2A_{31} & b_2 & a_2 \\ 2A_{12} & b_3 & a_3 \end{bmatrix} \begin{Bmatrix} 1 \\ x \\ y \end{Bmatrix} \tag{5.18d}$$

where the area of the triangle is given by $2A = a_3 b_2 - a_2 b_3 = a_1 b_3 - a_3 b_1 = a_2 b_1 - a_1 b_2$. From equation (5.18d), we can derive the following useful differentiation formulae

$$\frac{\partial}{\partial x} = \sum_{i=1}^{3} \frac{\partial L_i}{\partial x} \frac{\partial}{\partial L_i} = \sum_{i=1}^{3} \frac{b_i}{2A} \frac{\partial}{\partial L_i}$$

$$\frac{\partial}{\partial y} = \sum_{i=1}^{3} \frac{\partial L_i}{\partial y} \frac{\partial}{\partial L_i} = \sum_{i=1}^{3} \frac{a_i}{2A} \frac{\partial}{\partial L_i} \tag{5.18e}$$

Although the above formulation may seem cumbersome, one of its principal advantages is the ease with which polynomial terms may be integrated according to the following formula

$$\int_A L_1^p L_2^q L_3^r \, dA = \frac{p!\,q!\,r!}{(p+q+r+2)!} 2A \tag{5.18f}$$

The triangular coordinate formulation given above follows Felippa.[8] To illustrate use of equation (5.18f), consider

$$\int_A L_1 L_2 L_3 \, dA = \frac{(1)!(1)!(1)!}{(1+1+1+2)!} 2A = \frac{2A}{5!} = \frac{A}{60}$$

$$\int_A L_1^3 \, dA = \frac{3!0!0!}{(3+0+0+2)!} 2A = \frac{3!}{5!} 2A = \frac{A}{10}$$

Cartesian coordinates and natural coordinates, s and t, for a quadrilateral element, Figure 5-11(d), may be related as follows

$$\begin{Bmatrix} x \\ y \end{Bmatrix} = \begin{bmatrix} \{z\}^T & \{0\}^T \\ \{0\}^T & \{z\}^T \end{bmatrix} \begin{Bmatrix} \{x_n\} \\ \{y_n\} \end{Bmatrix} \tag{5.19a}$$

where

$$\{z\}^T = \tfrac{1}{4}[(1-s)(1-t),\ (1+s)(1-t),\ (1+s)(1+t),\ (1-s)(1+t)]$$

$$\{x_n\}^T = [x_1 \ x_2 \ x_3 \ x_4]$$

$$\{y_n\}^T = [y_1 \ y_2 \ y_3 \ y_4]$$

where x_1 and y_1 are the cartesian coordinates of node 1, and so on. Because equation (5.19a) cannot be inverted specifically, we must use a numerical process to obtain differentiation formulae.[10] Without covering the detailed derivation, we write

$$\begin{Bmatrix} \dfrac{\partial}{\partial x} \\ \dfrac{\partial}{\partial y} \end{Bmatrix} = [J]^{-1} \begin{Bmatrix} \dfrac{\partial}{\partial s} \\ \dfrac{\partial}{\partial t} \end{Bmatrix} \tag{5.19b}$$

where $[J]$ is the 2×2 Jacobian matrix[11] which is evaluated numerically from

$$\underset{2\times 2}{[J]} = \underset{2\times 4}{\begin{bmatrix} \dfrac{\partial}{\partial s}\{z\}^{\mathrm{T}} \\ \\ \dfrac{\partial}{\partial t}\{z\}^{\mathrm{T}} \end{bmatrix}} \underset{4\times 2}{[\{x_n\}\,\{y_n\}]} \tag{5.19c}$$

In addition, integrations must be performed numerically with the limits of both s and t being -1 and $+1$ and with[10]

$$dx\,dy = \det([J])\,ds\,dt \tag{5.19d}$$

The subject of numerical integration will be discussed later.

An alternative type of natural coordinates is possible for quadrilateral elements. These coordinates are more similar to triangular coordinates than the type shown in Figure 5-11(d). The alternative type utilizes four coordinates, two of which are independent. Along each side of the element, one of the four coordinates is zero and another is unity.[12] We will not discuss these in detail here.

Tetrahedral coordinates, Figure 5-11(e), are analogous to triangular coordinates. They are related to the cartesian coordinates as follows:

$$\begin{Bmatrix} 1 \\ x \\ y \\ z \end{Bmatrix} = \begin{bmatrix} 1 & 1 & 1 & 1 \\ x_1 & x_2 & x_3 & x_4 \\ y_1 & y_2 & y_3 & y_4 \\ z_1 & z_2 & z_3 & z_4 \end{bmatrix} \begin{Bmatrix} L_1 \\ L_2 \\ L_3 \\ L_4 \end{Bmatrix} \tag{5.20a}$$

The total volume of a tetrahedron element, V, is given by one sixth of the determinant of the 4×4 matrix in equation (5.20a). If V_i is the volume of the smaller tetrahedron which has vertices P and the three nodes other than the node i, then the tetrahedral coordinates can also be considered *volume coordinates* defined as

$$L_i = V_i/V \qquad i = 1, 2, 3, 4 \tag{5.20b}$$

From equation (5.20b) and the first of equations (5.20a), we see that the sum $L_1 + L_2 + L_3 + L_4$ of the four tetrahedral coordinates of any point within the element is unity; therefore, only three of the four natural coordinates are independent. The inverse of equation (5.20a) is of the form

$$\begin{Bmatrix} L_1 \\ L_2 \\ L_3 \\ L_4 \end{Bmatrix} = \frac{1}{6V} \begin{bmatrix} V_{234} & a_1 & b_1 & c_1 \\ V_{341} & a_2 & b_2 & c_2 \\ V_{412} & a_3 & b_3 & c_3 \\ V_{123} & a_4 & b_4 & c_4 \end{bmatrix} \begin{Bmatrix} 1 \\ x \\ y \\ z \end{Bmatrix} \tag{5.20c}$$

where V_{ijk} is the volume of the tetrahedron with the vertices i, j, k, and the origin of the cartesian system; and where a_i, b_i, and c_i are the projected areas of face i onto the x, y, and z coordinate planes, respectively. The latter terms are actually cofactors of terms in the determinant of the 4 × 4 matrix in equation (5.20a). For example, a_2 is the cofactor of x_2, and c_3 is the cofactor of z_3 in this determinant. The differentiation formulae follow from equation (5.20c):

$$\frac{\partial}{\partial x} = \sum_{i=1}^{4} \frac{\partial L_i}{\partial x} \frac{\partial}{\partial L_i} = \sum_{i=1}^{4} \frac{a_i}{6V} \frac{\partial}{\partial L_i}$$

$$\frac{\partial}{\partial y} = \sum_{i=1}^{4} \frac{\partial L_i}{\partial y} \frac{\partial}{\partial L_i} = \sum_{i=1}^{4} \frac{b_i}{6V} \frac{\partial}{\partial L_i} \tag{5.20d}$$

$$\frac{\partial}{\partial z} = \sum_{i=1}^{4} \frac{\partial L_i}{\partial z} \frac{\partial}{\partial L_i} = \sum_{i=1}^{4} \frac{c_i}{6V} \frac{\partial}{\partial L_i}$$

Finally, the polynomial terms may be integrated by the following formula

$$\int_V L_1^p L_2^q L_3^r L_4^s \, dV = \frac{p!\,q!\,r!\,s!}{(p+q+r+s+3)!} 6V \tag{5.20e}$$

For $p = q = r = s = 1$,

$$\int_V L_1 L_2 L_3 L_4 \, dV = \frac{6V}{7!} = \frac{V}{840}$$

The tetrahedral coordinate notation used here follows Clough.[13]

Natural coordinates are also used for hexahedral elements.[13] These are analogous to quadrilateral coordinates and are considered in Example 5-8.

The applications of the differentiation and integration formulae for the one-, two-, and three-dimensional cases above will be clarified by their subsequent use in the sample and example problems.

Interpolation Displacement Models

The basic process of constructing a displacement model was described in Section 5-3 and summarized in equations (5.5) through (5.8). From these equations, it is apparent that if we can directly construct the matrix $[N]$ defined in equation (5-8), we can avoid the necessity of computing and inverting the matrix $[A]$. Indeed, in many cases it is possible to avoid this inversion by selecting interpolation functions as the basis for the displacement model. An *interpolation function*, also known as a *shape function*, is a function which has unit value at one nodal point, and zero value at all other nodal points.[13] We use interpolation functions which are polynomials and select the order of the polynomial to satisfy the requirements discussed in Section 5-3. For each displacement in the vector $\{u\}$, we need an interpolation function corresponding to every nodal degree of freedom represented in the vector $\{q\}$. This concept can be demonstrated by considering various order interpolation models for one-dimensional and triangular elements.

Interpolation functions are expressed most conveniently in terms of natural coordinate systems. Figure 5-12 illustrates this for the simple case of a one-dimensional element. By substituting the natural coordinates of the nodes into the given shape functions, we can easily see that the definition of an interpolation function stated above is satisfied by the linear and quadratic shapes indicated in Figures 5-12(b) and 5-12(c), respectively. For example, the first

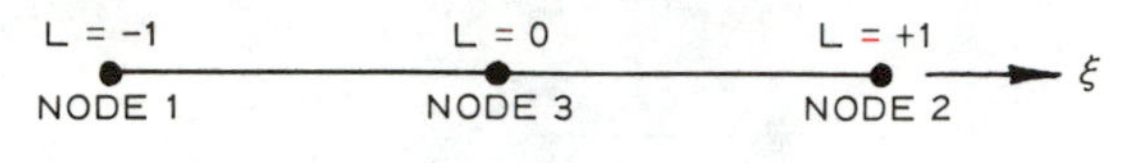

(a) One-dimensional element

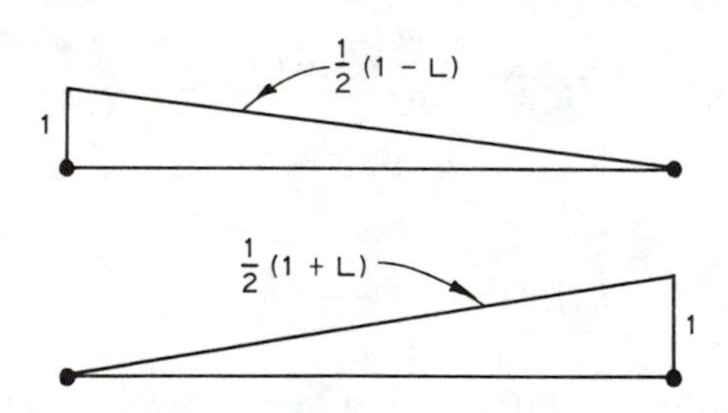

(b) Linear interpolation functions (two nodes)

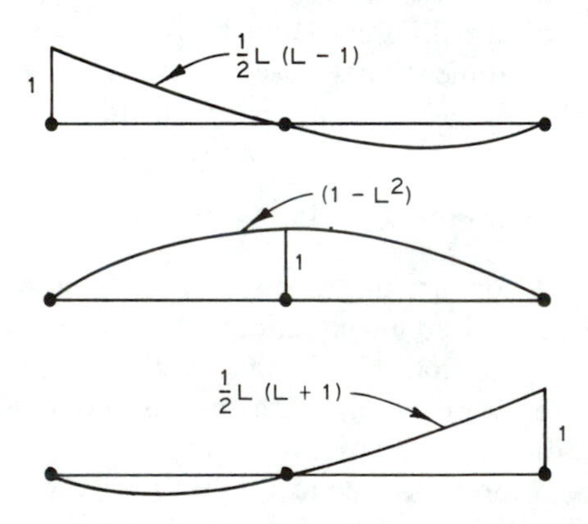

(c) Quadratic interpolation functions (three nodes)

Figure 5-12 Interpolation functions for one-dimensional element.

linear interpolation function shown, when evaluated at the two nodes, gives

$$\tfrac{1}{2}(1-L)\Big|_{\text{node }1} = \tfrac{1}{2}[1-(-1)] = 1$$

$$\tfrac{1}{2}(1-L)\Big|_{\text{node }2} = \tfrac{1}{2}(1-1) = 0$$

We can thus write an equation of the type (5.8) for linear interpolation as

$$u = [\tfrac{1}{2}(1-L) \;\; \tfrac{1}{2}(1+L)]\begin{Bmatrix} u_1 \\ u_2 \end{Bmatrix} = \{N\}^{\mathrm{T}}\{q\} \tag{5.21}$$

Here each of the two terms in $\{N\}^{\mathrm{T}}$ is an *interpolation function* or *shape function*. The entire vector is a (linear) *interpolation model*. For instructional purposes only, let us carry out the multiplication in equation (5.21) and regroup the terms to obtain

$$u = \tfrac{1}{2}(u_1 + u_2) + \tfrac{1}{2}(u_2 - u_1)L$$

Substituting for L from equation (5.17g) and rearranging, we find

$$u = \left[\frac{1}{2}(u_1 + u_2) - \frac{x_3}{\ell}(u_2 - u_1)\right] + \left[\frac{1}{\ell}(u_2 - u_1)\right]x$$

Comparing this result with the linear form of equation (5.1a), we see that the terms in brackets are the generalized coordinates α_1 and α_2, respectively. This gives an insight into the physical nature of the generalized coordinates for the simplest finite element and the lowest order displacement model.

For the quadratic variation, we can write the interpolation model for a three-node element as

$$u = [\tfrac{1}{2}L(L-1) \;\; \tfrac{1}{2}L(L+1) \;\; (1-L^2)]\begin{Bmatrix} u_1 \\ u_2 \\ u_3 \end{Bmatrix} = \{N\}^{\mathrm{T}}\{q\} \tag{5.22}$$

Quadratic interpolation functions for a triangular element are shown in Figure 5-13. We can apply tests and detailed analysis to this case similar to those performed above for the one-dimensional element. The resulting quadratic interpolation model of the form of equation (5.8) is given by

$$\{u\} = \begin{Bmatrix} u \\ v \end{Bmatrix} = \begin{bmatrix} \{N_1\}^{\mathrm{T}} & \{0\}^{\mathrm{T}} \\ \{0\}^{\mathrm{T}} & \{N_1\}^{\mathrm{T}} \end{bmatrix}\begin{Bmatrix} \{q_u\} \\ \{q_v\} \end{Bmatrix} = [N]\{q\} \tag{5.23}$$

where

$$\{N_1\}^{\mathrm{T}} = [L_1(2L_1 - 1), L_2(2L_2 - 1), L_3(2L_3 - 1), 4L_1L_2, 4L_2L_3, 4L_3L_1]$$

$$\{q_u\}^{\mathrm{T}} = [u_1 \;\; u_2 \;\; u_3 \;\; u_4 \;\; u_5 \;\; u_6]$$

$$\{q_v\}^{\mathrm{T}} = [v_1 \;\; v_2 \;\; v_3 \;\; v_4 \;\; v_5 \;\; v_6]$$

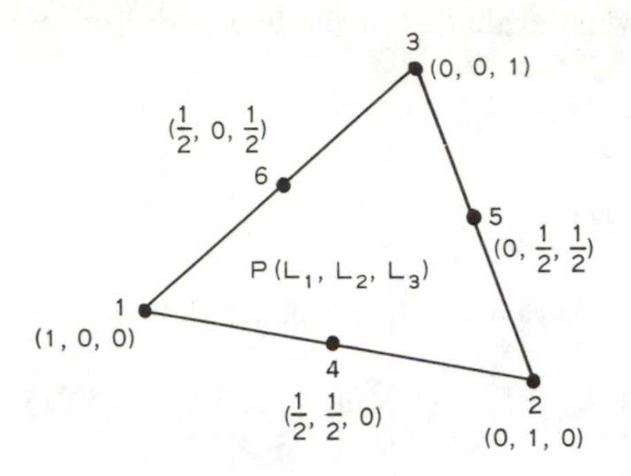

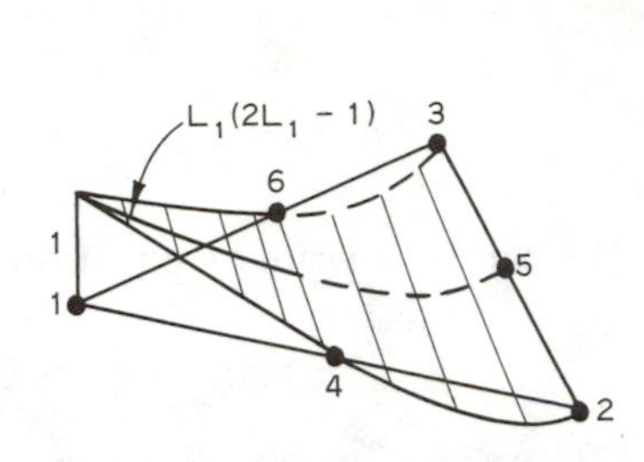

(a) Triangular coordinates of six-node element

(b) Typical quadratic shape function for primary node

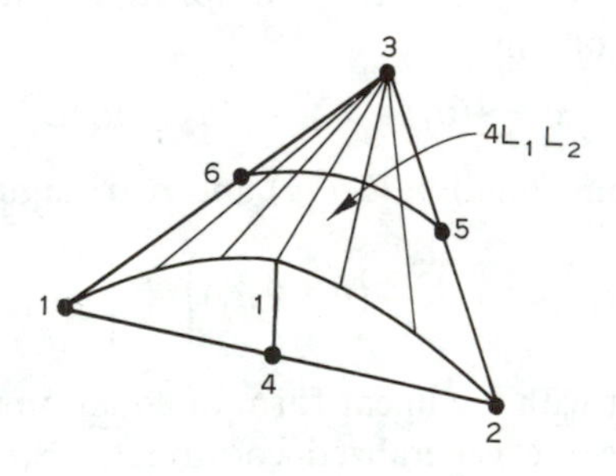

(c) Typical quadratic shape function for a secondary node

Figure 5-13 Isometric views of graphs of quadratic shape functions.[8] Note that the magnitudes of the functions are plotted normal to the planar element; however, the functions may represent displacements that occur either in the plane or normal to it.

Models of different order are possible for each of the two element types used in the examples above.[8] In addition, a similar approach can be applied to elements of other configurations.[6,10,13] More examples will appear at the end of this section and in applications to be discussed later in the book.

In general, interpolation functions can be derived from the standard polynomial interpolation methods of numerical analysis, such as Lagrangian and Hermitian interpolation (Example 5-9, Section 5-10). Alternatively, shape functions can be constructed by judicious combination of natural coordinate polynomial terms.

The Isoparametric Element Concept

If we compare equations (5.17f) and (5.21), we see that these equations are of exactly the same form:

$$x = \{N\}^{\mathrm{T}}\{x_n\} \qquad \text{and} \qquad u = \{N\}^{\mathrm{T}}\{q\}$$

In other words, the geometry and displacements of the element are described in terms of the same parameters and are of the same order. Elements such as this are called *isoparametric elements.*[10] The isoparametric concept enables us to formulate elements of any order which satisfy the completeness and compatibility requirements given in Section 5-3 and which have isotropic displacement models.

If the shape functions in natural coordinates fulfill continuity of geometry and displacements both within the element and between adjacent elements, it can be shown that the compatibility requirement is satisfied in global coordinates.[10] It is apparent that the polynomial interpolation models discussed above are continuous within the element. Also, we can see that the displacements along any side of the element depend only upon nodal displacements occurring on that side when interpolation in natural coordinates is employed. For example, consider the shape functions shown in Figure 5-13. Displacements along side 1-4-2 of the triangle will be affected only by displacements at nodes 1, 4, and 2.

Moreover, if the interpolation displacement model provides rigid body displacements in the local natural coordinate system, the conditions of both rigid body displacements and constant strain states are satisfied in global coordinates.[10] We can see that it is easily possible to select some combinations of nodal displacements to cause all points on the element to experience the same displacement. For the interpolation model case, this occurs if all nodal displacements are the same (Exercise 5-5).

Typical shape functions which satisfy the sufficient convergence conditions and which are geometrically isotropic are given in Table 5-1 for various two- and three-dimensional elements. Functions of higher order can also be derived, but an increasing number of external and internal nodal degrees of freedom is required to define these functions. In practice, functions higher than cubic order are seldom employed.

It is not necessary that the geometry and displacements of an element be expressed by the same order model. To include this possibility in our terminology, we define *subparametric elements* as those elements for which the geometry is determined by a lower order model than the displacements. Elements for which the converse is true are called *superparametric elements.* For all subparametric elements and for some forms of superparametric elements used in the analysis of shells, the convergence criteria are satisfied.[10]

Referring to Figure 1-1, we see that there are obvious drawbacks in attempting to represent a body with curved boundaries by straight-sided elements. A large number of elements may be necessary to obtain a reasonable resemblance between the original body and the assemblage. One way of approaching this problem is to use elements with curved boundaries, and the isoparametric representation is most efficient in this respect.

TABLE 5-1 TYPICAL SHAPE FUNCTIONS FOR ISOPARAMETRIC ELEMENTS[8,10,12,13,20]

Order of Model	Node i	Shape Functions for Various Elements		
		Triangle or Tetrahedron (note 1)	Quadrilateral	Hexahedron
Linear	Primary External (corner)	L_i	$\frac{1}{4}(1+ss_i)(1+tt_i)$	$\frac{1}{8}(1+rr_i)(1+ss_i)(1+tt_i)$
Quadratic	Primary External (corner)	$L_i(2L_i-1)$	$\frac{1}{4}(1+ss_i)(1+tt_i)(ss_i+tt_i-1)$	$\frac{1}{8}(1+rr_i)(1+ss_i)(1+tt_i)\cdot(rr_i+ss_i+tt_i-2)$
	Secondary External (mid-edge)	$4L_jL_k$ (note 2)	$\frac{1}{2}(1-s^2)(1+tt_i)$ for $s_i=0$; $\frac{1}{2}(1+ss_i)(1-t^2)$ for $t_i=0$	$\frac{1}{4}(1-r^2)(1+ss_i)(1+tt_i)$ for $r_i=0$ etc.
Cubic (note 5)	Primary External (corner)	$\frac{L_i}{2}(3L_i-1)(3L_i-2)$	$\frac{1}{32}(1+ss_i)(1+tt_i)[9(s^2+t^2)-10]$	$\frac{1}{64}(1+rr_i)(1+ss_i)(1+tt_i)\cdot[9(r^2+s^2+t^2)-19]$
	Secondary External ($\frac{1}{3}$ points of edge)	$\frac{9}{2}L_jL_k(3L_j-1)$ (note 3)	$\frac{9}{32}(1-s^2)(1+9ss_i)(1+tt_i)$ for $s_i=\pm\frac{1}{3}$, $t_i=\pm1$ etc.	$\frac{9}{64}(1-r^2)(1+9rr_i)(1+ss_i)\cdot(1+tt_i)$ for $r=\pm\frac{1}{3}$, $s=\pm1$, $t=\pm1$ etc.
	(note 4)	$27\,L_jL_kL_\ell$	—	—

1. The corner nodes of triangles and tetrahedra are numbered as shown in Figures 5-7(a) and 5-8(a).
2. j and k are node numbers of corners adjacent to edge containing node i. For example see the model used in eq. (5-23).
3. For triangle, coordinates of node i are $L_j=2/3$, $L_k=1/3$, $L_\ell=0$. For tetrahedron, $L_j=2/3$, $L_k=1/3$, $L_\ell=L_m=0$.
4. For triangle, this is an internal node at the centroid and $j=1$, $k=2$, $\ell=3$. For tetrahedron, this is a secondary external node at the centroid of the face $L_m=0$.
5. For models of quartic order and higher, internal nodes are necessary to maintain geometric isotropy.[10]

Equations of the form

$$\{x\} = [N]\{x_n\} \tag{5.24a}$$

may be considered a transformation between cartesian coordinates and curvilinear coordinates if $[N]$ is a matrix of shape functions of order higher than linear. Here the curvilinear coordinates are the element natural coordinates, such as L_1, L_2, L_3 or s, t, embedded in the interpolation functions. Equation (5.24a) may also be interpreted as the mapping of a straight-sided element in local coordinates into a curved-sided element in the global cartesian coordinate system. This approach permits us to represent complex geometries with fewer elements than if we used only elements with straight sides. The principal limitation on this technique is that we must be certain that the transformation is unique in that there is a one-to-one correspondence between points in the two coordinate systems.[10] In other words, the mapping must not cause distortions violent enough to fold back the element upon itself.

To illustrate the isoparametric approach, consider the "triangular" element with curved sides shown in Figure 5-14. We shall apply a quadratic representation of the curved sides and quadratic interpolation models, Figure 5-13. The geometry of the element is given by the natural coordinates analogous to those defined in equations (5.18).

$$\begin{Bmatrix} 1 \\ x \\ y \end{Bmatrix} = \begin{bmatrix} 1 & 1 & 1 & 1 & 1 & 1 \\ x_1 & x_2 & x_3 & x_4 & x_5 & x_6 \\ y_1 & y_2 & y_3 & y_4 & y_5 & y_6 \end{bmatrix} \{N_1\} \tag{5.25a}$$

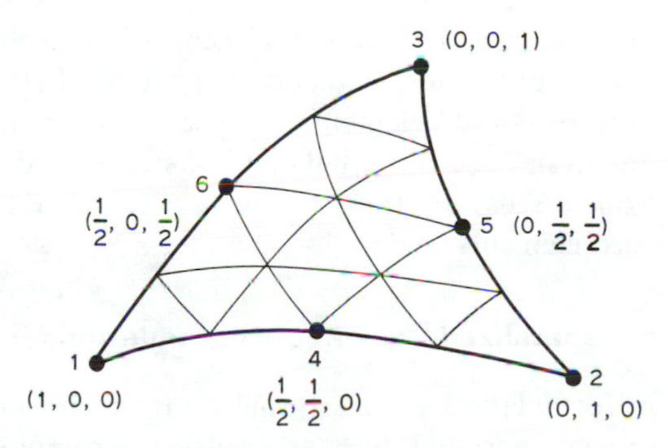

Figure 5-14 Plane element with three quadratically curved sides, shown in global cartesian coordinates.[12] Inscribed on the element are lines on which one of the natural coordinates (L_1, L_2, or L_3) has a constant value.

$\{N_1\}$ is defined in equation (5.23). Similarly, the displacement model given in equation (5.23) can be rewritten as follows:

$$\begin{Bmatrix} u \\ v \end{Bmatrix} = \begin{bmatrix} u_1 & u_2 & u_3 & u_4 & u_5 & u_6 \\ v_1 & v_2 & v_3 & v_4 & v_5 & v_6 \end{bmatrix} \{N_1\} \tag{5.25b}$$

Equation (5.25a), the relationship between the cartesian coordinates and local coordinates, cannot be inverted in literal form as was equation (5.18a). Therefore, a scheme analogous to equations (5.19) must be used to perform differentiation and integration numerically.

In some problems, such as the analyses of shell structures, we map an element which has only two independent natural coordinates into three-dimensional global coordinate space, that is, we use an element that is plane in terms of local coordinates to represent a curved surface. Similarly, we may need elements which can be described in terms of a single natural coordinate but which are curved lines in two- or three-dimensional global coordinates. In each of these cases the Jacobian matrix, $[J]$, which appears in such equations as equations (5.19), is not square and cannot be inverted. However, we can circumvent this difficulty because the differentiations and integrations necessary for the formulation of the finite element equations are performed in only as many coordinate directions as are necessary to describe the element in natural coordinates. We can therefore use a local orthogonal coordinate system in which the directions normal to the element do not affect the differentiations and integrations of the formulation.[10] Hence the transformation of equation (5.24a) may be modified as follows:

$$\{x'\} = [T]\{x\} = [T][N]\{x_n\} \tag{5.24b}$$

Here the prime indicates the local orthogonal system, and the number of local coordinates in $\{x'\}$ is the same as the number of natural coordinates for the element.

The isoparametric concept is a powerful generalized technique for constructing complete and conforming elements of any order. In the isoparametric formulation, the use of curved elements becomes a systematic extension of the finite element analysis procedure that employs straight-sided elements. For the remainder of our discussion, therefore, we shall continue to concentrate on the straight-sided elements.

Interpolation vs. Generalized Coordinate Formulations

The interpolation formulation is clearly more direct and elegant than the generalized coordinate representation of displacement models. One of the drawbacks of generalized coordinates is that they do not have a readily evident physical significance. With interpolation models we work directly with nodal displacements rather than generalized displacements. Hence, not only is it much easier to see whether a particular interpolation displacement model satisfies the convergence criteria, but also the possibility of a singular $[A]$ matrix is no longer a concern. Moreover, computational economy is attained by eliminating the matrix $[A]$ from the formulation, equations (5.6) to (5.8). A typical finite element analysis may involve hundreds of elements. Thus we can avoid the necessity of computing and inverting $[A]$ hundreds of times.

Nevertheless, it is important to note that the isoparametric concept and interpolation models are not universally employed. A significant body of literature exists which utilizes the generalized coordinate approach, and some work of this nature continues to appear. We shall therefore consider both approaches.

Sample Problem 1 (Continued)

By using the natural coordinate system defined in Figure 5-11(a), we can express the cubic interpolation model for the beam element as

$$w = [L_1^2(3 - 2L_1), L_1^2 L_2 \ell, L_2^2(3 - 2L_2), - L_1 L_2^2 \ell]\{q\} \tag{5.26}$$

where $\{q\}$ is defined in equation (5.14a) and L_1 and L_2 are shown in Figure 5-15.

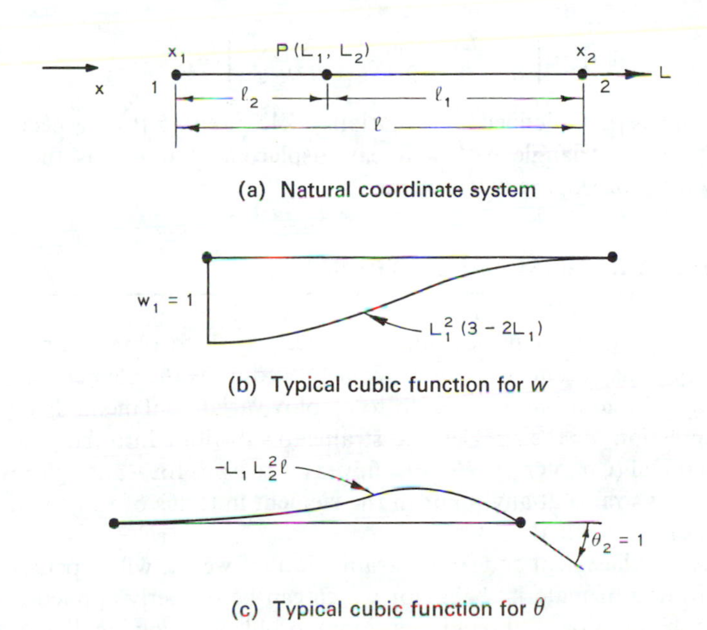

(a) Natural coordinate system

(b) Typical cubic function for w

(c) Typical cubic function for θ

Figure 5-15 Interpolation model for Sample Problem 1.

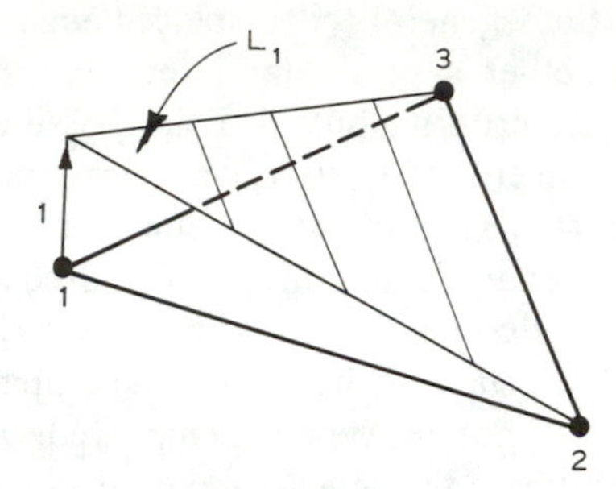

Figure 5-16 Isometric view of graph of typical linear shape function for a triangular element.

We observe that, since the geometry of the element can be expressed by a linear model, equation (5.17a), the element is subparametric.

Sample Problem 2 (Continued)

The triangular coordinates shown in Figure 5-11(c) and described by equations (5.18) are directly applicable to the plane strain element of this problem. Figure 5-16 shows a typical shape function which is included in the following model:

$$\{u\} = \begin{bmatrix} L_1 & L_2 & L_3 & 0 & 0 & 0 \\ 0 & 0 & 0 & L_1 & L_2 & L_3 \end{bmatrix} \{q\} \tag{5.27}$$

where $\{u\}$ and $\{q\}$ are defined in equations (5.15) and (5.16), respectively. This straight-sided triangle with a linear displacement model is the most elementary form of isoparametric element.

5-5 ELEMENT STRESSES AND STRAINS

One of the important quantities that we usually wish to obtain for design analysis from a displacement finite element procedure is the element stresses and/or strains. In addition, if we wish to employ variational methods we will need an expression for the stresses and strains to substitute into the potential energy functional (Chapter 4). We thus have two reasons for wanting to write the stresses and strains at any point in the element in terms of the nodal displacements of the element.

The strain-displacement and stress-strain relations we use will depend upon the specialization of material behavior which can be properly applied to the problem at hand. For a three-dimensional problem involving linear and elastic behavior, we would use the strain-displacement relations given by equation (3.11b) and the generalized Hooke's Law given by equation (3.12). On the other hand, many of our applications in structures and soils are concerned with nonlinear behavior or with plane strain or stress, plate bending, and similar problems, for which appropriate strain-displacement and stress-strain relations are given in Chapter 3.

So far, we have formulated a displacement model as given in equations (5.8) or (5.5) by using either an interpolation model or a simple polynominal model in terms of generalized coordinates. The strains are expressed in terms of some combination of the derivatives of the displacements, $\{u\}$. Since neither the nodal displacements, $\{q\}$, nor the generalized coordinates, $\{\alpha\}$, are functions of the spatial coordinates, these derivatives must be formed in terms of one of the matrices $[N]$ or $[\phi]$.

If $\{\varepsilon\}$ is the vector of the relevant strain components at an arbitrary point within the finite element, we use the strain-displacement equations and the displacement model to write

$$\{\varepsilon\} = [B_\alpha]\{\alpha\} \tag{5.28a}$$

or

$$\{\varepsilon\} = [B]\{q\} \tag{5.28b}$$

It is important to note that $\{\varepsilon\}$ and $[B_\alpha]$ or $[B]$ are functions of the independent space coordinates, whether they be cartesian coordinates, x, y, z for $[B_\alpha]$, or some convenient set of natural coordinates (e.g. L_1, L_2, L_3) for $[B]$. However, if $[B_\alpha]$ and $[B]$ are in the same coordinate system they may be related by

$$[B] = [B_\alpha][A^{-1}]$$

If $\{\sigma\}$ is the vector of stresses corresponding to the strains $\{\varepsilon\}$, we may use an appropriate matrix form of the stress-strain equations to write the element stresses as follows:

$$\{\sigma\} = [C][B_\alpha]\{\alpha\} \tag{5.29a}$$

or

$$\{\sigma\} = [C][B]\{q\} \tag{5.29b}$$

where $[C]$ is the matrix of material constants. Usually we assume that $[C]$ is constant in each element, but later we shall see how we can accommodate varying material properties in an element.

Sample Problem 1 (Continued)

Earlier, we obtained the following strain-displacement result:

$$\varepsilon(x, z) = -z\frac{d^2w}{dx^2} \tag{5.10}$$

Differentiating the expression of the displacement model

$$w(x) = \alpha_1 + \alpha_2 \bar{x} + \alpha_3 \bar{x}^2 + \alpha_4 \bar{x}^3 \tag{5.13}$$

we obtain

$$\varepsilon(x, z) = -z[0 \;\; 0 \;\; 2 \;\; 6\bar{x}]\{\alpha\} \tag{5.30a}$$

Assuming a linearly elastic material with modulus E, the stresses are given by

$$\sigma(x, z) = -z\, E[0 \;\; 0 \;\; 2 \;\; 6\bar{x}]\{\alpha\} \tag{5.30b}$$

However, for engineering applications of beam theory we generally do not work in terms of stresses and strains; but we use moments and curvatures instead. Defining the curvature and moment as

$$w''(x) = \frac{d^2w}{dx^2}, \qquad M(x) = -\int_{-h/2}^{h/2} z\sigma(x, z)\, dz \tag{5.31}$$

we can substitute the following for equations (5.30)

$$w''(x) = [0 \;\; 0 \;\; 2 \;\; 6\bar{x}]\{\alpha\}$$

$$M(x) = EI(x)[0 \;\; 0 \;\; 2 \;\; 6\bar{x}]\{\alpha\} \tag{5.32}$$

If we consider the interpolation displacement model given by equation (5.26), we can write the curvatures and moments as

$$w''(x) = \frac{1}{\ell^2}\,[(6 - 12L_1), \ell(2L_2 - 4L_1), (6 - 12L_2), \ell(4L_2 - 2L_1)]\{q\} \tag{5.33a}$$

$$M(x) = \frac{EI(L_1, L_2)}{\ell^2}\,[(6 - 12L_1), \ell(2L_2 - 4L_1), (6 - 12L_2), \ell(4L_2 - 2L_1)]\{q\} \tag{5.33b}$$

Sample Problem 2 (Continued)

In Chapter 3, we stated the strain-displacement relations

$$\varepsilon_x = \frac{\partial u}{\partial x}, \qquad \varepsilon_y = \frac{\partial v}{\partial y}, \qquad \text{and} \qquad \gamma_{xy} = \frac{\partial u}{\partial y} + \frac{\partial v}{\partial x} \tag{3.17a}$$

Now, by using the displacement model

$$u = \alpha_1 + \alpha_2 x + \alpha_3 y \qquad \text{and} \qquad v = \alpha_4 + \alpha_5 x + \alpha_6 y \tag{5.15}$$

we obtain

$$\{\varepsilon\} = \begin{Bmatrix} \dfrac{\partial u}{\partial x} \\ \dfrac{\partial v}{\partial y} \\ \dfrac{\partial u}{\partial y} + \dfrac{\partial v}{\partial x} \end{Bmatrix} = \begin{bmatrix} 0 & 1 & 0 & 0 & 0 & 0 \\ 0 & 0 & 0 & 0 & 0 & 1 \\ 0 & 0 & 1 & 0 & 1 & 0 \end{bmatrix} \{\alpha\} = [B_\alpha]\{\alpha\} \tag{5.34a}$$

and the corresponding stresses are given by

$$\{\sigma\} = [C][B_\alpha]\{\alpha\}$$
$$\{\sigma\}^T = [\sigma_x \ \sigma_y \ \tau_{xy}] \tag{5.34b}$$

where the constitutive matrix, $[C]$, is as follows:

$$[C] = \frac{E}{(1+v)(1-2v)} \begin{bmatrix} 1-v & v & 0 \\ v & 1-v & 0 \\ 0 & 0 & \dfrac{1-2v}{2} \end{bmatrix} \tag{5.34c}$$

On the other hand, we can use the interpolation model,

$$\{u\} = \begin{bmatrix} L_1 & L_2 & L_3 & 0 & 0 & 0 \\ 0 & 0 & 0 & L_1 & L_2 & L_3 \end{bmatrix} \{q\}$$
$$\{q\}^T = [u_1 \ u_2 \ u_3 \ v_1 \ v_2 \ v_3] \tag{5.27}$$

to obtain the strains. Using the differentiation formulae, equation (5.18e), we obtain

$$\frac{\partial u}{\partial x} = \sum_{i=1}^{3} \frac{\partial L_i}{\partial x}\frac{\partial u}{\partial L_i} = \frac{\partial L_1}{\partial x}\frac{\partial u}{\partial L_1} + \frac{\partial L_2}{\partial x}\frac{\partial u}{\partial L_2} + \frac{\partial L_3}{\partial x}\frac{\partial u}{\partial L_3}$$

$$= \frac{1}{2A}\left[\frac{\partial}{\partial x}(2A_{23} + b_1 x + a_1 y)u_1 + \frac{\partial}{\partial x}(2A_{31} + b_2 x + a_2 y)u_2 + \frac{\partial}{\partial x}(2A_{12} + b_3 x + a_3 y)u_3\right]$$

$$= \frac{1}{2A}(b_1 u_1 + b_2 u_2 + b_3 u_3)$$

and so on. Hence,

$$\{\varepsilon\} = \frac{1}{2A}\begin{bmatrix} b_1 & b_2 & b_3 & 0 & 0 & 0 \\ 0 & 0 & 0 & a_1 & a_2 & a_3 \\ a_1 & a_2 & a_3 & b_1 & b_2 & b_3 \end{bmatrix}\{q\} = [B]\{q\} \tag{5.35a}$$

and

$$\{\sigma\} = [C][B]\{q\} \tag{5.35b}$$

The definition of $[C]$ remains unchanged. We observe from equation (5.34a) or equation (5.35a) that the strains for this element are *constant*, that is, they are not functions of the spatial coordinates. Therefore, the element is referred to as a *constant strain triangle* (CST).[8]

5-6 DIRECT FORMULATION OF ELEMENT STIFFNESS AND LOADS

The next step in the finite element procedure is the computation of the stiffness matrix of the element and the vector of forces to be applied at the nodes of the element. Before discussing the general method of calculating these element properties by the application of variational principles, we shall review briefly the elementary *direct method* which derives from the basic techniques of structural analysis. In the area where frame analysis and the finite element method overlap, that is, where the structure may be represented by an assemblage of one-dimensional elements only, the direct method is as powerful as the variational method. For more complex structures, the variational method is often demonstrably superior. Nonetheless, the direct technique enhances our understanding of the physical interpretation of the finite element method.

Direct Method for Stiffnesses

In Chapter 1 we saw that an element of a stiffness matrix is a *stiffness influence coefficient* which gives the force at one point on the structure arising from a unit displacement at the same or a different point of the structure. This notion is the basis of the direct method.

For simple, one-dimensional structures, we can consider an element restrained against all nodal displacements save one, which is given a unit value. The nodal forces (or reactions) necessary to maintain this displacement are elements of the stiffness matrix, specifically, the column of the matrix corresponding to the nodal displacement that is given the unit value.[3] For instance, if an element has four nodal degrees of freedom and the third is given a unit value while all others are kept zero, the resulting reaction forces are the elements of the third column of the matrix. In the stiffness equation

$$\{Q\} = \begin{Bmatrix} Q_1 \\ Q_2 \\ Q_3 \\ Q_4 \end{Bmatrix} = \begin{bmatrix} k_{11} & k_{12} & k_{13} & k_{14} \\ k_{21} & k_{22} & k_{23} & k_{24} \\ k_{31} & k_{32} & k_{33} & k_{34} \\ k_{41} & k_{42} & k_{43} & k_{44} \end{bmatrix} \begin{Bmatrix} q_1 \\ q_2 \\ q_3 \\ q_4 \end{Bmatrix} = [k]\{q\}$$

where $\{Q\}$ is the vector of nodal loads and $[k]$ is the stiffness matrix, the above idea is symbolically stated as follows: if $q_1 = q_2 = q_4 = 0$ and $q_3 = 1$, then k_{13} is the reaction at node 1 of the type and direction associated with q_1, etc. This is illustrated in a subsequent example.

For more complex, multi-dimensional structures, the direct method is not so simple, because we are using an assembly of multi-dimensional elements to represent a continuous body. The above direct technique based on influence coefficients, if used for such multi-dimensional bodies, would result in a highly nonconforming assemblage. In other words, gaps or overlaps would develop at the interfaces between the elements, and compatibility will be maintained only at the nodes. Moreover, such analysis would lead to unrealistic stress concentrations at the nodes.[3,14] To compensate for these factors, we must assume that all elements deform in a similar manner and that these deformations resemble or approximate those of the continuous medium.[3] The application of the direct method to a two-dimensional element will be illustrated by Sample Problem 2 at the end of this section.

Example: Stiffness of a Truss Member. We shall apply the direct method to the calculation of the stiffness matrix for the truss member shown in Figure 5-17. This member has four degrees of freedom, as indicated in Figure 5-17(a). Using the method of influence coefficients, we shall develop the first and fourth columns of the 4 × 4 stiffness matrix. We assume that the unit nodal displacements are small in comparison to the length of the member so that linear theory applies, and the change in angle of inclination of the member is negligible. In Figure 5-17(b), the axial force necessary to sustain a change in length, ΔL_1, is given by

$$P_1 = -\frac{AE}{L}(\Delta L_1) = -\frac{AE}{L}\cos\theta$$

where the negative sign indicates compression. Hence, the components of the reactions at the node 1 are given by

$$k_{11} = \frac{AE}{L}\cos^2\theta$$

$$k_{21} = \frac{AE}{L}\cos\theta\sin\theta$$

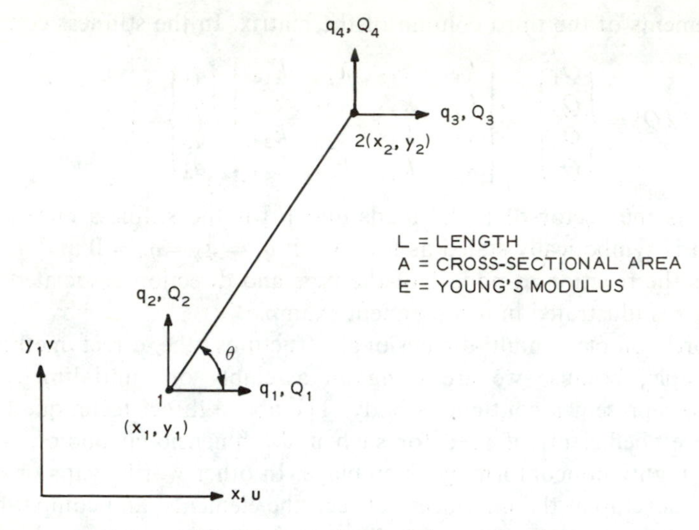

(a) Truss member and sign convention

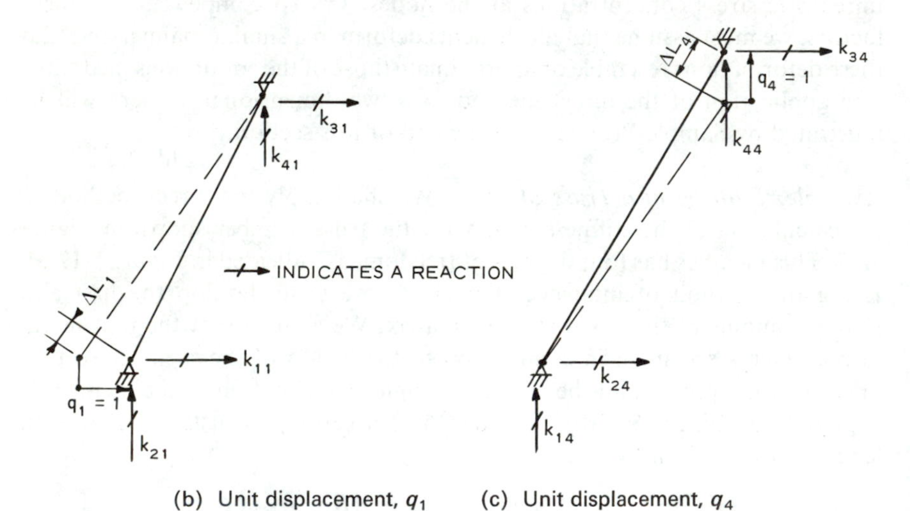

(b) Unit displacement, q_1 (c) Unit displacement, q_4

Figure 5-17 Direct method applied to a truss member.

From equilibrium of the member, we obtain the remaining reactions

$$k_{31} = -k_{11} = -\frac{AE}{L}\cos^2\theta$$

$$k_{41} = -k_{21} = -\frac{AE}{L}\sin\theta\cos\theta$$

A similar calculation is used to obtain column 4 of the stiffness after the axial force of Figure 5-17(c) is determined:

$$P_4 = \frac{AE}{L}(\Delta L_4) = \frac{AE}{L}\sin\theta$$

The overall stiffness matrix of the truss member is given by

$$[k] = \frac{AE}{L}\begin{bmatrix} \lambda^2 & \lambda\mu & -\lambda^2 & -\lambda\mu \\ \lambda\mu & \mu^2 & -\lambda\mu & -\mu^2 \\ -\lambda^2 & -\lambda\mu & \lambda^2 & \lambda\mu \\ -\lambda\mu & -\mu^2 & \lambda\mu & \mu^2 \end{bmatrix} \tag{5.36}$$

where $\lambda = \cos\theta$ and $\mu = \sin\theta$. Note that the stiffness matrix is symmetric, which is consistent with the Maxwell-Betti reciprocal theorem. In effect, this theorem states that all stiffness matrices for linear structures must be symmetric.

Lumped Loads

Generally, when subdividing a structure we select nodal locations that coincide with the locations of concentrated external forces. However, we usually encounter distributed loading of some kind. Therefore, we must have some technique of expressing the element load vector to account for all such loads. Both the direct procedure and the variational approach described subsequently offer methods of constructing these load characteristics for an element.

In the direct method of formulation, the usual approach is to consider the distributed loads on an element as divided among the nodes of the element. The loads at a particular node are then taken as the static equivalent of that portion of the distributed loading corresponding to a region tributary to the node and are called *lumped loads*. For example, in Figure 5-18(a) the tributary region (length) for a node of the element is the half-length of the element. The type and sign convention of these loads correspond to nodal degrees of freedom on a one-to-one basis.

Sample Problem 1 (Continued)

Using the notation and sign conventions previously defined for this problem, Figure 5-18(a), we can easily compute the element stiffness by the direct method. The reactions associated with typical end displacements and rotations are shown in Figures 5-18(b) and 5-18(c). These are well known results from elementary strength of materials.[15] The stiffness equation from the beam element can be assembled from these basic results as follows:

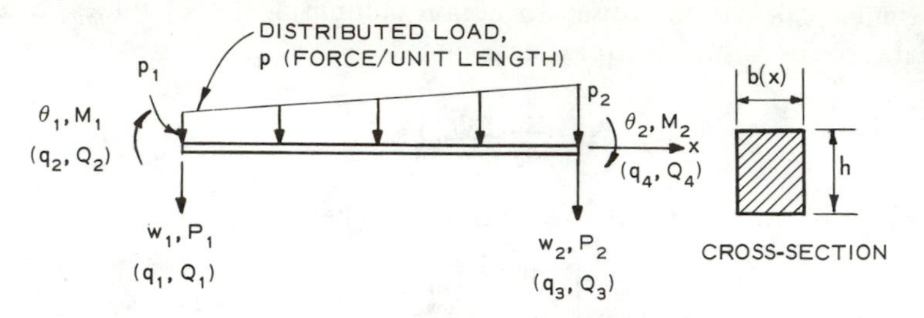

(a) Notation and sign convention

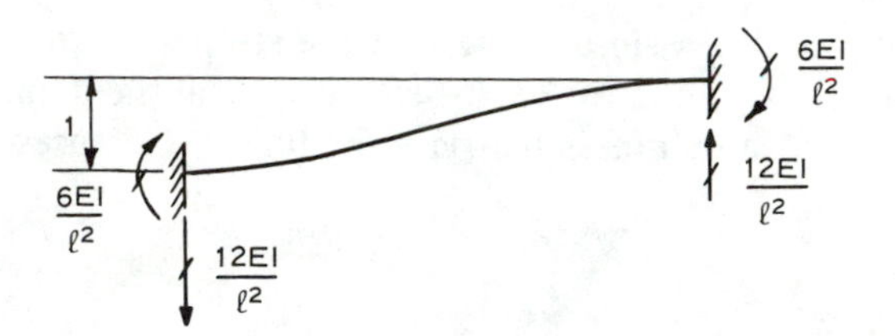

(b) Beam reactions under unit translation of one end

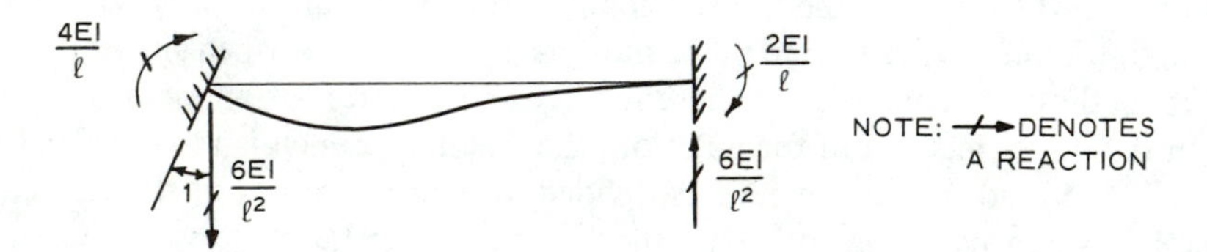

(c) Beam reactions under unit rotation of one end

Figure 5-18 Beam element stiffness by the direct method.

$$\frac{EI}{\ell^3}\begin{bmatrix} 12 & 6\ell & -12 & 6\ell \\ 6\ell & 4\ell^2 & -6\ell & 2\ell^2 \\ -12 & -6\ell & 12 & -6\ell \\ 6\ell & 2\ell^2 & -6\ell & 4\ell^2 \end{bmatrix}\begin{Bmatrix} w_1 \\ \theta_1 \\ w_2 \\ \theta_2 \end{Bmatrix} = \begin{Bmatrix} P_1 \\ M_1 \\ P_2 \\ M_2 \end{Bmatrix} \tag{5.37}$$

or

$$[k]\{q\} = \{Q\} \tag{1.5}$$

The load vector for uniformly distributed loading p can be assembled by noting that the tributary length for each node is $\ell/2$. The resultant force in each area is thus $p\ell/2$ acting at a distance $\ell/4$ from the node. The static equivalent of this loading is a transverse load of $p\ell/2$ and a moment of $p\ell^2/8$ at each node. Therefore, we can write

$$\{Q\}^{\mathrm{T}} = \frac{p\ell}{8}[4 \;\; \ell \;\; 4 \;\; -\ell] \tag{5.38}$$

Sample Problem 2 (Continued)

A direct method of obtaining the stiffness for a plane strain triangle was given by Turner *et al.*[16] We assume constant strains given by

$$\{\varepsilon\}^{\mathrm{T}} = \left[\frac{\partial u}{\partial x} \;\; \frac{\partial v}{\partial y} \;\; \frac{\partial u}{\partial y} + \frac{\partial v}{\partial x}\right] = [a \;\; b \;\; c]$$

where a, b, and c are constants. By integrating these equations, we obtain

$$\begin{Bmatrix} u \\ v \end{Bmatrix} = \begin{bmatrix} a & (c+d)/2 & e \\ (c-d)/2 & b & f \end{bmatrix} \begin{Bmatrix} x \\ y \\ 1 \end{Bmatrix}$$

where d, e, and f are additional constants resulting from the integration. We can express the constants $a, b, \ldots, f$ in terms of the nodal displacement by substituting the nodal coordinates into the last equations.

$$\{q\} = \begin{Bmatrix} u_1 \\ u_2 \\ u_3 \\ v_1 \\ v_2 \\ v_3 \end{Bmatrix} = \begin{bmatrix} x_1 & 0 & y_1/2 & y_1/2 & 1 & 0 \\ x_2 & 0 & y_2/2 & y_2/2 & 1 & 0 \\ x_3 & 0 & y_3/2 & y_3/2 & 1 & 0 \\ 0 & y_1 & x_1/2 & -x_1/2 & 0 & 1 \\ 0 & y_2 & x_2/2 & -x_2/2 & 0 & 1 \\ 0 & y_3 & x_3/2 & -x_3/2 & 0 & 1 \end{bmatrix} \begin{Bmatrix} a \\ b \\ c \\ d \\ e \\ f \end{Bmatrix} = [\bar{B}]^{-1}\{\bar{\varepsilon}\} = [\bar{B}]^{-1}\begin{Bmatrix} \{\varepsilon\} \\ \{\varepsilon'\} \end{Bmatrix}$$

This can be inverted to give

$$\{\bar{\varepsilon}\} = \begin{Bmatrix} \{\varepsilon\} \\ \{\varepsilon'\} \end{Bmatrix} = \begin{bmatrix} [B_d] \\ [B'] \end{bmatrix}\{q\} = [\bar{B}]\{q\}$$

By using the first of these matrix equations, it is now possible to express the stresses in terms of the nodal displacements

$$\{\sigma\} = [C][B_d]\{q\} \tag{5.39a}$$

where $[C]$ is given in equation (5.34c) and the subscript d is added to $[B]$ to distinguish it from the matrix in equations (5.35). We now wish to relate the nodal force vector to the constant stress distribution. The constant stress state is represented in Figure 5-19. These stresses have been replaced by statically equivalent forces at the midsides of the triangles by using the notation of equations (5.18c). (The triangle is assumed to have thickness h in the z-direction.) Each of the midside forces is divided equally between the two adjacent

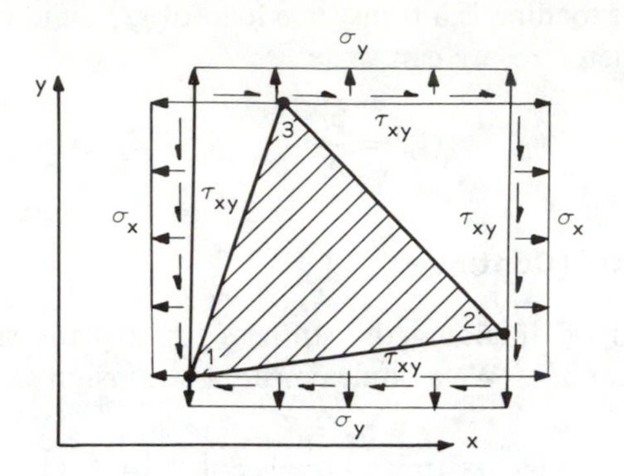

(a) Constant stress distribution

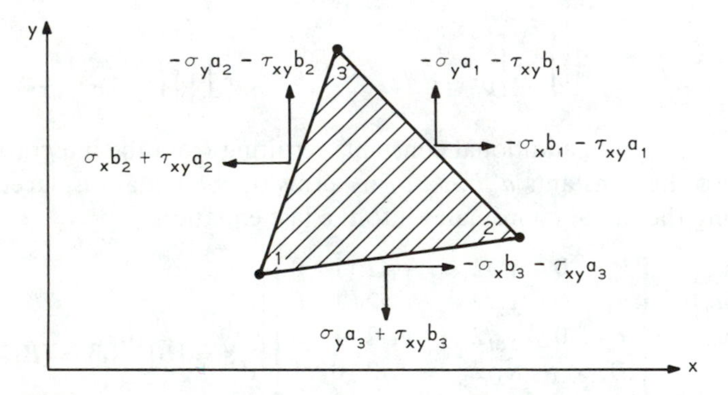

(b) Statically equivalent forces at mid-sides (per unit thickness)

Figure 5-19 Plane strain triangle force vector by direct method.[16]

nodes. We can therefore relate the nodal forces to the stresses as follows:

$$\{Q\} = \begin{Bmatrix} P_{u_1} \\ P_{u_2} \\ P_{u_3} \\ P_{v_1} \\ P_{v_2} \\ P_{v_3} \end{Bmatrix} = -\frac{h}{2}\begin{bmatrix} b_3 + b_2 & 0 & a_3 + a_2 \\ b_1 + b_3 & 0 & a_1 + a_3 \\ b_2 + b_1 & 0 & a_2 + a_1 \\ 0 & a_3 + a_2 & b_3 + b_2 \\ 0 & a_1 + a_3 & b_1 + b_3 \\ 0 & a_2 + a_1 & b_2 + b_1 \end{bmatrix}\begin{Bmatrix} \sigma_x \\ \sigma_y \\ \tau_{xy} \end{Bmatrix} = [F]\{\sigma\} = hA[B]^{\mathrm{T}}\{\sigma\} \tag{5.39b}$$

Substituting this equation into equation (5.39a), we obtain

$$\{Q\} = [F][C][B_d]\{q\} = [k]\{q\} \tag{5.40}$$

Equation (5.40) gives the stiffness of the element in terms of the product $[F][C][B_d]$. We will not write this matrix explicitly because the inversion of $[\bar{B}]^{-1}$ necessary to obtain $[B_d]$ is laborious, except in numerical form.

5-7 VARIATIONAL FORMULATION OF ELEMENT STIFFNESS AND LOADS

A generalized method of calculating the element stiffness matrix and load vector is the application of one of the variational principles of solid mechanics (Chapter 4). Specifically, for the displacement method of analysis we use the principle of minimum potential energy for static problems. The load vector obtained by this approach is usually termed *consistent* to distinguish it from the *lumped* loads. We call it consistent because the energy or work which it effects is more consistently correlated to the energy of the actual distributed loads than it is in the direct method. Consistent loads are also called *kinematically equivalent forces*. The pioneering work in the consistent formulation was performed by Melosh.[17]

Before going into detail on the variational formulation, we should note that we really apply the minimum principle to the structure or body as a whole, that is, to the assemblage of elements. However, because our displacement models are separately assumed for each element or subregion of the continuum with interelement compatibility maintained to the necessary degree, we can think of any integral in the variational principle as operating on the sum of the various displacement subregions. Because the integral of the summation is the same as the sum of the individual integrals, we can apply the principle to the elements separately. For example, in the principle of minimum potential energy, we must perform an integral to obtain the strain energy, U, for a linear elastic body.

$$2U = \iiint\limits_{V_{\text{body}}} \{\varepsilon\}^{\mathrm{T}}\{\sigma\}\, dV_{\text{body}} = \iiint\limits_{V_{\text{body}}} \sum_{e=1}^{N} \{\varepsilon_e\}^{\mathrm{T}}\{\sigma_e\}\, dV_{\text{body}} = \sum_{e=1}^{N} \iiint\limits_{V_e} \{\varepsilon_e\}^{\mathrm{T}}\{\sigma_e\}\, dV_e \tag{5.41}$$

where e is an index denoting an element, and N is the total number of elements used to represent the body. In effect, we are performing integrals over the individual elements before carrying out the summation inherent to the assembly process (Chapter 6).

Element Characteristics

If we formulate our displacement model in terms of *interpolation functions*, our element displacements, strains, and stresses are given in terms of equations (5.8), (5.28b), and (5.29b), respectively. Substituting these equations into equation (4.13), we can write the functional in terms of the displacement model,

$$\Pi = \frac{1}{2}\iiint\limits_V (\{q\}^T[B]^T[C][B]\{q\} - 2\{q\}^T[N]^T\{\bar{X}\})\,dV - \iint\limits_{S_1} \{q\}^T[N]^T\{\bar{T}\}\,dS_1 \tag{5.42}$$

where V is the volume of the element, and S_1 is that portion of the element over which surface tractions are specified. Applying the variational principle, equation (4.11), we obtain

$$\{\delta q\}^T\left(\iiint\limits_V [B]^T[C][B]\,dV\{q\} - \iiint\limits_V [N]^T\{\bar{X}\}\,dV - \iint\limits_{S_1} [N]^T\{\bar{T}\}\,dS_1\right) = 0 \tag{5.43}$$

Since the variations of the nodal displacements $\{\delta q\}$ are arbitrary, the expression in the parentheses must vanish. This gives the equilibrium equations for the element

$$[k]\{q\} = \{Q\} \tag{5.44}$$

where the stiffness matrix and the nodal load vector† are respectively defined as

$$[k] = \iiint\limits_V [B]^T[C][B]\,dV \tag{5.45a}$$

$$\{Q\} = \iiint\limits_V [N]^T\{\bar{X}\}\,dV + \iint\limits_{S_1} [N]^T\{\bar{T}\}\,dS_1 \tag{5.45b}$$

We can see that the application of the variational principle has produced general formulae for the essential element characteristics. The examples at the end of this section will show how these formulae can readily be applied to specific elements.

When our displacement models are formulated in terms of *generalized coordinates* $\{\alpha\}$, then the displacements, strains, and stresses are given respectively by equations (5.5), (5.28a), and (5.29a). By using these equations, the element characteristics derived from equation (5.43) are

$$[k_\alpha] = \iiint\limits_V [B_\alpha]^T[C][B_\alpha]\,dV \tag{5.46a}$$

† Section 7–1 indicates how the load vector may be modified for thermal strain and other initial strains.

$$\{Q_\alpha\} = \iiint\limits_V [\phi]^{\mathrm{T}}\{\bar{X}\}\,dV + \iint\limits_{S_1} [\phi]^{\mathrm{T}}\{\bar{T}\}\,dS_1 \tag{5.46b}$$

Equation (5.7) in conjunction with equation (5.42) gives the transformation necessary to express the characteristics in terms of the nodal displacements as

$$[k] = [A^{-1}]^{\mathrm{T}}[k_\alpha][A^{-1}] \tag{5.47a}$$

$$\{Q\} = [A^{-1}]^{\mathrm{T}}\{Q_\alpha\} \tag{5.47b}$$

Hence, equation (5.44) applies for both the generalized coordinate and the interpolation model cases.

The formulae for the element stiffness matrix and the load vector given by equations (5.45) and (5.46) require integrations. For constant strain elements in cartesian coordinates, the matrices $[B]$ are constant (Sample Problem 2), and the integrations necessary to obtain the stiffness are trivial. However, the integrals for the loads, and for the stiffness and load characteristics with higher order models, will involve polynomials. The integration formulae such as equations (5.17e), (5.18f), and (5.20e) can be used for these integrations. However, it is often convenient to perform the more difficult integrations numerically. This technique will be discussed further below.

The general formulae for the element stiffness and load vector in equations (5.45a), (5.45b), and (5.46a), (5.46b) remain the same irrespective of the type of the element. For different types, however, the order of the matrix and vector will change. For instance, in the case of a plane triangular element with a linear displacement model, $[k]$ is 6×6 and $\{Q\}$ is 6×1; but for the same element shape with a quadratic model, the corresponding orders are 12×12 and 12×1. For a tetrahedron with a linear displacement model, $[k]$ is 12×12 and $\{Q\}$ is 12×1.

An additional point of importance is that the stiffness matrix in equations (5.45) and (5.46) is always symmetric. Not only is the constitutive matrix $[C]$ symmetric, but any multiplication of the type $[B]^{\mathrm{T}}[C][B]$ gives symmetric matrix products.

Alternative Stiffness Formulation for Refined Elements

For elements with displacement models of quadratic or higher order, often called *refined elements*, the method for computing the stiffness matrix given by equation (5.45a) is laborious. An alternative, more compact procedure is possible. It simplifies the necessary integrations.[8] This approach expresses the stresses and strains in terms of their nodal values. Because the order of the stress and strain polynomials is less than the order of the displacement model by at least one, a smaller number of nodal values of the stresses and strains is usually sufficient to define the element stresses and strains. For example, to specify a quadratic displacement model for a triangle, Figure 5-13 and equation (5.23), we need six nodal values of each displacement. However, the strain variation in this case is linear, so only three nodal values of each strain, say at the corner nodes, will suffice to define the element strains.

The stresses and strains may be expressed in terms of their nodal values by interpolation models

$$\{\varepsilon\} = [N_\varepsilon]\{\varepsilon_n\} \qquad \{\sigma\} = [N_\sigma]\{\sigma_n\} \tag{5.48}$$

where the subscript n indicates a vector of nodal values. If the material properties are uniform for the element in question, the interpolation models $[N_\varepsilon]$ and $[N_\sigma]$ are identical. Furthermore, we may write the nodal values of the strains in terms of the nodal displacements by evaluating the matrix $[B]$ of equation (5.28b) at the appropriate nodes

$$\{\varepsilon_n\} = [B_n]\{q\} \tag{5.49}$$

We relate the stresses at the nodes to the strains at the nodes by a stress-strain matrix which has the appropriate nodal values of the material properties

$$\{\sigma_n\} = [C_n]\{\varepsilon_n\} \tag{5.50}$$

As a result of the above, we may formulate the strain energy of the element as follows:

$$U = \frac{1}{2}\iiint\limits_V \{\varepsilon\}^T\{\sigma\}\, dV = \frac{1}{2}\{\varepsilon_n\}^T \iiint\limits_V [N_\varepsilon]^T[N_\sigma]\, dV\{\sigma_n\}$$

$$= \frac{1}{2}\{q\}^T[B_n]^T \iiint\limits_V [N_\varepsilon]^T[N_\sigma]\, dV\, [C_n][B_n]\{q\}$$

Therefore, we can define the stiffness matrix as

$$[k] = [B_n]^T[D][C_n][B_n] \tag{5.51a}$$

where

$$[D] = \iiint\limits_V [N_\varepsilon]^T[N_\sigma]\, dV \tag{5.51b}$$

Usually the integration in equation (5.51b) is much easier than that of equation (5.45a), because we can take advantage of repetitive patterns, which occur in the former. Moreover, we can save the product $[C_n][B_n]$ for later use in calculating the nodal stresses, once the nodal displacements are determined.

Elements with Variable Properties

So far we have tacitly assumed that, within the element, the material properties do not vary, and that the element geometry is constant. However, the variational approach can account for variation of properties in the element. Indeed, we have already seen how the shape of the element boundaries can be varied by using the isoparametric concept (Section 5-4). Similarly, we can represent any property of the element as an interpolation polynomial in terms of its known values at the nodes. For simple variation of properties, this representation can be exact. However, we often adopt a linear or quadratic variation to approximate a complicated variation.

We routinely assume a polynomial variation of the unknown displacements and, therefore, obtain polynomial variations for strain and stress also. Among the other known properties that we may optionally represent by a polynomial variation are the material properties, $[C]$, the body forces, $\{\bar{X}\}$, and the surface tractions, $\{\bar{T}\}$. In addition, certain other quantities, such as temperature, thermal coefficient, and various geometric quantities like the thickness of a plate or shell and the width or depth of a beam, can be expressed in terms of polynomial variations.[8] Some examples of elements with variable properties will appear below.

For an element with variable properties, the order of the polynomials to be integrated in equations (5.45) and (5.46) will increase. The integration formulae, such as equations (5.17e), (5.18f), and (5.20e), are useful for these integrations. However, a widely used alternative approach for elements with complicated geometry and properties is numerical integration.

Numerical Integration to Obtain Element Stiffnesses and Loads

In the case of elements that are expressed in terms of curvilinear coordinates, such as the cylindrical system, r, θ, z, commonly used for axisymmetric elements, we are often faced with performing integrations of expressions other than polynomials, such as trigonometric functions. It may be too difficult or impractical to integrate these expressions in closed form.[8] Therefore, instead of obtaining literal expressions for the stiffness and loads, we take advantage of the computer's skill in arithmetic and perform an approximate numerical integration. There are several schemes available, but only one, *Gauss' Formula*, will be given here. For an arbitrary interval of integration from a to b, this formula is[18]

$$\int_a^b f(y)\,dy \cong \left(\frac{b-a}{2}\right)\sum_{i=1}^{n} w_i f(y_i)$$
$$y_i = \left(\frac{b-a}{2}\right)x_i + \left(\frac{b+a}{2}\right) \tag{5.52a}$$

where n is the number of terms in the summation approximating the integration, and x_i and w_i are abscissas and weight functions, which are tabulated for various values of n in Reference 18 and in other texts on numerical methods. An example to illustrate the use of equation (5.52a) is included at the end of this chapter.

For triangular natural coordinates, Gauss' Formula is[8,19]

$$\frac{1}{A}\iint_A f(L_1, L_2, L_3)\, dA \cong \sum_{i=1}^{n} W_i f(L_i) \tag{5.52b}$$

where $f(L_i)$ indicates that the function f is evaluated at the ith integration point, P_i. P_i and W_i are integration points and weight functions that are tabulated for various values of n in References 8 and 19.

Sample Problem 1 (Continued)

Recapitulating previous results for this problem, we have already derived the following results:

$$\underset{1\times 4}{[N]} = [L_1^2(3 - 2L_1), L_1^2 L_2 \ell, L_2^2(3 - 2L_2), -L_1 L_2^2 \ell] \tag{5.26}$$

$$\underset{1\times 4}{[B]} = \frac{1}{\ell^2}[(6 - 12L_1), \ell(2L_2 - 4L_1), (6 - 12L_2), \ell(4L_2 - 2L_1)] \tag{5.33a}$$

$$\underset{1\times 1}{[C]} = EI \tag{5.33b}$$

$$\underset{1\times 4}{[\phi]} = [1 \;\; \bar{x} \;\; \bar{x}^2 \;\; \bar{x}^3] \tag{5.13}$$

$$\underset{1\times 4}{[B_\alpha]} = [0 \;\; 0 \;\; 2 \;\; 6\bar{x}] \tag{5.32a}$$

$$\underset{4\times 4}{[A^{-1}]} = \begin{bmatrix} 1 & 0 & 0 & 0 \\ 0 & 1 & 0 & 0 \\ -3/\ell^2 & -2/\ell & 3/\ell^2 & -1/\ell \\ 2/\ell^3 & 1/\ell^2 & -2/\ell^3 & 1/\ell^2 \end{bmatrix} \tag{5.14b}$$

Additionally, we need the prescribed body forces and surface tractions for the beam, which we shall denote as follows:

$$\underset{1\times 1}{\{\bar{X}\}} = X \tag{5.53a}$$

$$\underset{1\times 1}{\{\bar{T}\}} = p \tag{5.53b}$$

where p is the surface load per unit area. For beam elements of constant geometry and material properties, we may therefore directly proceed to utilize equations (5.45) and (5.17e) to obtain the element stiffness and load vector.

The stiffness is

$$[k] = \int_\ell [B]^T[C][B]\,d\ell$$

or

$$[k] = \frac{EI}{\ell^4}\int_\ell$$

$$\times \begin{bmatrix} (6-12L_1)^2 & \ell(6-12L_1)(2L_2-4L_1) & (6-12L_1)(6-12L_2) & \ell(6-12L_1)(4L_2-2L_1) \\ & \ell^2(2L_2-4L_1)^2 & \ell(2L_2-4L_1)(6-12L_2) & \ell^2(2L_2-4L_1)(4L_2-2L_1) \\ & & (6-12L_2)^2 & \ell(6-12L_2)(4L_2-2L_1) \\ \text{Symmetrical} & & & \ell^2(4L_2-2L_1)^2 \end{bmatrix} d\ell$$

The integrations involved in the evaluation of $[k]$ can be accomplished by using equation (5.17e). For instance, integration for the first element of $[k]$ is

$$\int_\ell (36 - 144L_1 + 144L_1^2)\,d\ell = 36\ell - 144\frac{1!}{2!}\ell + 144\frac{2!}{3!}\ell = 12\ell$$

and so on. The resulting stiffness matrix is

$$[k] = \frac{EI}{\ell^3}\begin{bmatrix} 12 & 6\ell & -12 & 6\ell \\ & 4\ell^2 & -6\ell & 2\ell^2 \\ & \text{symm.} & 12 & -6\ell \\ & & & 4\ell^2 \end{bmatrix} \tag{5.54}$$

For uniform loading, we obtain the following load vector:

$$\{Q\} = bhX\int_\ell [N]^T\,d\ell + bp\int_\ell [N]^T\,d\ell = b(hX+p)\int_\ell \begin{Bmatrix} L_1^2(3-2L_1) \\ L_1^2L_2\ell \\ L_2^2(3-2L_2) \\ -L_1L_2^2\ell \end{Bmatrix} d\ell$$

$$= \frac{b(hX+p)\ell}{12}[6 \;\; \ell \;\; 6 \;\; -\ell]^T \tag{5.55}$$

The results in equations (5.54) and (5.55) are obtained by using the previously adopted interpolation model. Identical matrices arise from the generalized coordinate model, but the integrations are less systematic. In addition we need to compute $[A^{-1}]$ to transform the intermediate results to nodal displacements.

Comparing the above results with those of the direct formulation, equations (5.37) and (5.38), we see that the stiffness is the same for both derivations, but that the loads are significantly different from their lumped counterparts.† These differences will be discussed below in Section 5-8.

Variable Element Properties. To illustrate how variable properties may be taken into account, assume that the width of the beam, b, and the transverse surface loading, p, both vary linearly along the length of the element [Figure 5-18(a)]. We can represent their variations in terms of their nodal values by using linear interpolation

$$b = b_1 L_1 + b_2 L_2 = [L_1 \ L_2]\begin{Bmatrix} b_1 \\ b_2 \end{Bmatrix} = [N_b]\{b_n\} \tag{5.56a}$$

$$p = p_1 L_1 + p_2 L_2 = [L_1 \ L_2]\begin{Bmatrix} p_1 \\ p_2 \end{Bmatrix} = [N_p]\{p_n\} \tag{5.56b}$$

where the subscripts 1 and 2 indicate known values of b and p at these nodes. We now proceed with a formal application of equations (5.45a) and (5.45b). The stiffness is

$$[k] = \frac{Eh^3}{12\ell^4}\int_\ell (b_1 L_1 + b_2 L_2)[B]^T[B]\, d\ell$$

$$= \frac{EI_1}{\ell^3}\begin{bmatrix} 6 & 4\ell & -6 & 2\ell \\ & 3\ell^2 & -4\ell & \ell^2 \\ & \text{symm.} & 6 & -2\ell \\ & & & \ell^2 \end{bmatrix} + \frac{EI_2}{\ell^3}\begin{bmatrix} 6 & 2\ell & -6 & 4\ell \\ & \ell^2 & -2\ell & \ell^2 \\ & \text{symm.} & 6 & -4\ell \\ & & & 3\ell^2 \end{bmatrix} \tag{5.56c}$$

and the loads are

$$\{Q\} = hX\int_\ell (b_1 L_1 + b_2 L_2)[N]^T\, d\ell + \int_\ell (b_1 L_1 + b_2 L_2)(p_1 L_1 + p_2 L_2)[N]^T\, d\ell$$

$$= \frac{hXb_1\ell}{60}\begin{Bmatrix} 21 \\ 3\ell \\ 9 \\ -2\ell \end{Bmatrix} + \frac{hXb_2\ell}{60}\begin{Bmatrix} 9 \\ 2\ell \\ 21 \\ -3\ell \end{Bmatrix} + \frac{p_1 b_1\ell}{60}\begin{Bmatrix} 16 \\ 2\ell \\ 4 \\ -\ell \end{Bmatrix}$$

$$+ \frac{(p_1 b_2 + p_2 b_1)\ell}{60}\begin{Bmatrix} 5 \\ \ell \\ 5 \\ -\ell \end{Bmatrix} + \frac{p_2 b_2\ell}{60}\begin{Bmatrix} 4 \\ \ell \\ 16 \\ -2\ell \end{Bmatrix} \tag{5.56d}$$

We note that if $p_1 = p_2$ and $b_1 = b_2$, the width and surface loading are uniform, and equations (5.56c) and (5.56d) reduce to equations (5.54) and (5.55), respectively.

† Note that the loads given by equation (5.55) are the negative *fixed end forces* that are commonly used in the matrix analysis of framed structures.

Alternative Approach. Using the alternative approach given by equations (5.51), we shall now recompute the stiffness matrix for the beam element. The interpolation models for the element stresses and strains are linear.

$$\{\varepsilon\} = [L_1 \;\; L_2]\begin{Bmatrix} w_{xx1} \\ w_{xx2} \end{Bmatrix} = [N_\varepsilon]\{\varepsilon_n\}$$

$$\{\sigma\} = [L_1 \;\; L_2]\begin{Bmatrix} M_1 \\ M_2 \end{Bmatrix} = [N_\sigma]\{\sigma_n\}$$

Moreover, from equation (5.33) we can construct the matrix $[B_n]$ by substituting the nodal values of the natural coordinates

$$\{\varepsilon_n\} = \frac{1}{\ell^2}\begin{bmatrix} -6 & -4\ell & 6 & -2\ell \\ 6 & 2\ell & -6 & 4\ell \end{bmatrix}\{q\} = [B_n]\{q\}$$

Finally, the nodal stress-strain matrix is given by

$$\{\sigma_n\} = EI\begin{bmatrix} 1 & 0 \\ 0 & 1 \end{bmatrix}\{\varepsilon_n\} = EI\{\varepsilon_n\} = [C_n]\{\varepsilon_n\}$$

We can now apply the procedure given by equation (5.51).

$$[D] = \int_\ell \{N_\varepsilon\}^{\mathrm{T}}\{N_\sigma\}\,d\ell = \int_\ell \begin{bmatrix} L_1^2 & L_1 L_2 \\ L_1 L_2 & L_2^2 \end{bmatrix} d\ell = \frac{\ell}{6}\begin{bmatrix} 2 & 1 \\ 1 & 2 \end{bmatrix}$$

$$[k] = [B_n]^{\mathrm{T}}[D][C_n][B_n] = \frac{EI}{6\ell^3}\begin{bmatrix} -6 & 6 \\ -4\ell & 2\ell \\ 6 & -6 \\ -2\ell & 4\ell \end{bmatrix}\begin{bmatrix} 2 & 1 \\ 1 & 2 \end{bmatrix}\begin{bmatrix} -6 & -4\ell & 6 & -2\ell \\ 6 & 2\ell & -6 & 4\ell \end{bmatrix}$$

$$= \frac{EI}{\ell^3}\begin{bmatrix} 12 & 6\ell & -12 & 6\ell \\ & 4\ell^2 & -6\ell & 2\ell^2 \\ & & 12 & -6\ell \\ \text{Symm.} & & & 4\ell^2 \end{bmatrix} \qquad (5.57)$$

This is the same result obtained above in equation (5.54), but the integration effort has been substantially reduced.

The alternative approach is also readily adaptable to the case of variable geometry and material properties. Here, if both the width and elastic modulus vary linearly, the stresses will vary quadratically.

$$\{\varepsilon\} = [L_1 \;\; L_2]\begin{Bmatrix} w_{xx1} \\ w_{xx2} \end{Bmatrix} = [N_\varepsilon]\{\varepsilon_n\}$$

$$\{\sigma\} = [L_1(2L_1 - 1), L_2(2L_2 - 1), 4L_1L_2]\begin{Bmatrix} M_1 \\ M_2 \\ M_3 \end{Bmatrix} = [N_\sigma]\{\sigma_n\}$$

$$b = [L_1 \ L_2]\begin{Bmatrix} b_1 \\ b_2 \end{Bmatrix} = [N_b]\{b_n\}$$

$$E = [L_1 \ L_2]\begin{Bmatrix} E_1 \\ E_2 \end{Bmatrix} = \{N_E\}\{E_n\}$$

$$\{\sigma_n\} = \frac{h^3}{12}\begin{bmatrix} E_1 & 0 & 0 \\ 0 & E_2 & 0 \\ 0 & 0 & \dfrac{E_1 + E_2}{2} \end{bmatrix}\begin{bmatrix} 1 & 0 \\ 0 & 1 \\ \dfrac{1}{2} & \dfrac{1}{2} \end{bmatrix}\{\varepsilon_n\} = [C_n]\{\varepsilon_n\}$$

The second matrix in the definition of $[C_n]$ is merely $[N_\varepsilon]$ evaluated at the three nodal points at which $\{\sigma_n\}$ is defined. Then the stiffness matrix is given by

$$[D] = \int_\ell [N_\varepsilon]^{\mathrm{T}}[N_b]\{b_n\}[N_\sigma]\, d\ell$$

$$[k] = [B_n]^{\mathrm{T}}[D][C_n][B_n] \tag{5.58}$$

Sample Problem 2 (Continued)

Recapitulating the interpolation model results for this problem, we have already derived the following matrices.

$$\underset{2\times 6}{[N]} = \begin{bmatrix} L_1 & L_2 & L_3 & 0 & 0 & 0 \\ 0 & 0 & 0 & L_1 & L_2 & L_3 \end{bmatrix} \tag{5.27}$$

$$\underset{3\times 6}{[B]} = \frac{1}{2A}\begin{bmatrix} b_1 & b_2 & b_3 & 0 & 0 & 0 \\ 0 & 0 & 0 & a_1 & a_2 & a_3 \\ a_1 & a_2 & a_3 & b_1 & b_2 & b_3 \end{bmatrix} \tag{5.35a}$$

$$[C] = \frac{E}{(1+\nu)(1-2\nu)}\begin{bmatrix} 1-\nu & \nu & 0 \\ \nu & 1-\nu & 0 \\ 0 & 0 & \dfrac{1-2\nu}{2} \end{bmatrix} \tag{5.34c}$$

We shall denote the prescribed body force of the plane strain element by

$$\{\overline{X}\}^{\mathrm{T}} = [\overline{X} \ \ \overline{Y}] \tag{5.59a}$$

where $\overline{X}$ and $\overline{Y}$ are the components of the body forces. For gravity loads, $\overline{X} = 0$ and $\overline{Y} = -\rho g$. The in-plane surface tractions are

$$\{\overline{T}\}^{\mathrm{T}} = [p_x \ p_y] \tag{5.59b}$$

where p_x and p_y are the prescribed in-plane load intensities which are positive if applied in the positive coordinate directions (Figure 5-5).

The stiffness is obtained from equation (5.45a)

$$[k] = h\ [B]^{\mathrm{T}}[C][B] \iint_A dA = hA\ [B]^{\mathrm{T}}[C][B] \tag{5.60a}$$

Because the material properties and strains are constant over the element, we see that the integration is trivial and the computation of the stiffness is only a matrix multiplication.

Finally, the load vector is obtained from equation (5.45b).

$$\{Q\} = h \iint_A [N]^{\mathrm{T}}\{\overline{X}\}\, dA + h \int_{S_1} [N]^{\mathrm{T}}\{\overline{T}\}\, d\ell_1 = \frac{hA}{3}\begin{Bmatrix} \overline{X} \\ \overline{X} \\ \overline{X} \\ \overline{Y} \\ \overline{Y} \\ \overline{Y} \end{Bmatrix} + \frac{h}{2}\begin{Bmatrix} p_x^{(2)}\ell_2 + p_x^{(3)}\ell_3 \\ p_x^{(1)}\ell_1 + p_x^{(3)}\ell_3 \\ p_x^{(1)}\ell_1 + p_x^{(2)}\ell_2 \\ p_y^{(2)}\ell_2 + p_y^{(3)}\ell_3 \\ p_y^{(1)}\ell_1 + p_y^{(3)}\ell_3 \\ p_y^{(1)}\ell_1 + p_y^{(2)}\ell_2 \end{Bmatrix} \tag{5.61a}$$

where ℓ_i is the length of the ith side of the triangle, that is, the side opposite to node i; and where the superscripts on $p_x^{(\ )}$ and $p_y^{(\ )}$ indicate the side to which the uniformly distributed load is applied. The distributed loads are the load intensity per unit thickness and per unit length measured along the side of the element. The loads in equation (5.61a) are simply the static resultants of the forces.[8] However, it is instructive to note how the formula of equation (5.45b) is applied to compute equation (5.61a). As an example of the computation of a few terms, let us consider a load intensity $p_x^{(1)}$ to be uniformly distributed along side 1, Figure 5-11(c). Along this side of the element, $L_1 = 0$ and $L_3 = 1 - L_2$. Moreover, $dS_1 = \ell_1 dL_2$. Hence we may write the necessary line integral as

$$\{Q\} = h \int_{S_1} [N]^{\mathrm{T}}\{\overline{T}\}\, dS_1 = h\ell_1 \int_0^1 \begin{bmatrix} 0 & 0 \\ L_2 & 0 \\ (1 - L_2) & 0 \\ 0 & 0 \\ 0 & L_2 \\ 0 & (1 - L_2) \end{bmatrix} \begin{Bmatrix} p_x^{(1)} \\ 0 \end{Bmatrix} dL_2$$

$$= \frac{hp_x^{(1)}\ell_1}{2}\begin{Bmatrix} 0 \\ 1 \\ 1 \\ 0 \\ 0 \\ 0 \end{Bmatrix}$$

If the body forces are due to gravity only, $\overline{X} = 0$ and $\overline{Y} = -\rho g$. Hence, the first term in equation (5.61a) becomes

$$h \iint_A [N]^T\{\overline{X}\}\, dA = -\frac{\rho A h g}{3} \begin{Bmatrix} 0 \\ 0 \\ 0 \\ 1 \\ 1 \\ 1 \end{Bmatrix} \qquad (5.61b)$$

It is apparent that the weight of the element, ρAhg, is distributed equally at the three nodes.

If we consider a variable thickness element,† equation (5.60a) is readily modified to give

$$[k] = A\ [B]^T[C][B]\left(\frac{1}{A}\iint_A h\, dA\right) = \bar{h}A\,[B]^T[C][B] \qquad (5.60b)$$

where $\bar{h}$ is the average thickness over the element.[8]

Comparing the element characteristics derived above, equations (5.60a) and (5.61a), with their counterparts derived by the direct method in the previous section, equations (5.37) and (5.40), we see that the variational approach is much simpler for the computation of the element stiffness.

5-8 CONSISTENT LOADS VS. LUMPED FORMULATION

For complex elements it is clearly advantageous to utilize the variational approach to obtain the element stiffness matrix. However, lumped loads are relatively easy to assemble, so their utility is worth investigating. The principal consideration in this decision is that only for a formulation using consistent loads can it be proven that solutions provide lower bounds to the deflections, (Section 5-3). In some cases, however, the lumped characteristics may give results closer to the exact solution. The problem with the lumped properties is that we cannot generally be certain whether our solution for displacements is too great or too small.

In usual applications, the use of lumped loads provides little, if any, advantage over the consistent load case. Admittedly, the calculation of consistent loads may be more difficult, especially if numerical integration is

† For plane strain, we use elements with a constant unit thickness. However, when the CST element is adapted for plane stress, we may want to incorporate variable thickness. See Example 5-3, Section 5-10.

necessary, equation (5.45b). However, the use of consistent loads is generally recommended because it provides more rational analysis.

5-9 CONDENSATION OF INTERNAL DEGREES OF FREEDOM

Internal nodes, such as node 7 of Figure 5-7(a) and node 5 of Figure 5-7(e), do not connect with the adjoining elements in the assemblage. Therefore, the degrees of freedom of such nodes do not appear in the compatibility conditions that are used to formulate the overall equations for the structure (Chapter 6). We may eliminate such internal degrees of freedom from the equilibrium equations of each element so that these extra unknowns do not increase the number of the overall equations.[16]

We rearrange and partition the equilibrium equations, equation (5.44),

$$\left[\begin{array}{c|c}[k_{11}] & [k_{12}] \\ \hline [k_{21}] & [k_{22}]\end{array}\right]\left\{\begin{array}{c}\{q_1\} \\ \hline \{q_2\}\end{array}\right\} = \left\{\begin{array}{c}\{Q_1\} \\ \hline \{Q_2\}\end{array}\right\} \tag{5.62a}$$

where $\{q_2\}$ is the vector of displacements, and $\{Q_2\}$ is the vector of loads at the internal nodes. These equations may be rewritten as follows:

$$[k_{11}]\{q_1\} + [k_{12}]\{q_2\} = \{Q_1\} \tag{5.62b}$$

$$[k_{21}]\{q_1\} + [k_{22}]\{q_2\} = \{Q_2\} \tag{5.62c}$$

Solving equation (5.62c) for $\{q_2\}$, we obtain

$$\{q_2\} = -[k_{22}]^{-1}[k_{21}]\{q_1\} + [k_{22}]^{-1}\{Q_2\} \tag{5.63a}$$

Substituting this result into equation (5.62b) and collecting terms, we can write the condensed equilibrium equations.

$$[\bar{k}]\{q_1\} = \{\bar{Q}\} \tag{5.63b}$$

$$[\bar{k}] = [k_{11}] - [k_{12}][k_{22}]^{-1}[k_{21}] \tag{5.63c}$$

$$\{\bar{Q}\} = \{Q_1\} - [k_{12}][k_{22}]^{-1}\{Q_2\} \tag{5.63d}$$

where $[\bar{k}]$ is the *effective stiffness matrix* and $\{\bar{Q}\}$ is the *effective load vector* corresponding to external nodes of the condensed element. This process is called *static condensation.*[12]

The condensation procedure as specifically outlined by equations (5.63) is not efficient for digital computation. The same condensation can be carried out more efficiently by a symmetric backward Gaussian elimination.[8,12] Let r be the number of external degrees of freedom, and s the total number of degrees of freedom for the element. Then the condensation of the internal degrees of freedom is carried out by the following operation for $p = s, s-1, s-2, \ldots, (r+1)$.

$$(k_{ij})_{\text{new}} = (k_{ij})_{\text{old}} - k_{ip}\frac{k_{pj}}{k_{pp}} = (k_{ij})_{\text{old}} - k_{ip}c_{pj} \tag{5.64a}$$

where $i, j = 1, 2, 3, \ldots, (p-1)$. The multipliers c_{pj} are stored in the locations k_{pj} to give the submatrix $[c]$

$$\begin{bmatrix} \underset{r\times r}{[\bar{k}]} & \underset{r\times(s-r)}{[k_{12}]} \\ \underset{(s-r)\times r}{[c]} & \underset{(s-r)\times(s-r)}{[d]} \end{bmatrix} \tag{5.64b}$$

where the lower triangle of $[d]$ is $[L_{22}][D_2]$, the result of the symmetric Gauss-Doolittle decomposition of the form of equation (2.2):

$$[k_{22}] = [L_{22}][D_2][L_{22}]^{\mathrm{T}} \tag{5.64c}$$

Load vectors may be condensed during the same backward decomposition, or they may be processed later in the program. In any event, the result for the loads is accomplished by performing the following for $p = s, s-1, s-2, \ldots, (r+1)$:

$$(Q_i)_{\text{new}} = (Q_i)_{\text{old}} - g_{(p-r)i}Q_p \tag{5.64d}$$

where

$$[g] = [[c] \mid [d]]$$

and $i = 1, 2, 3, \ldots, (p-1)$. Subsequently,

$$f_{(p-r)} = Q_p/d_{(p-r)(p-r)} \tag{5.64e}$$

Equations (5.64d) and (5.64e) are equivalent to

$$\begin{Bmatrix} \underset{r\times 1}{\{\bar{Q}\}} \\ \underset{(s-r)\times 1}{\{f\}} \end{Bmatrix} = \begin{Bmatrix} \{Q_1\} - [L_{22}]^{-1}[c]^{\mathrm{T}}\{Q_2\} \\ [D_2]^{-1}[L_{22}^{\mathrm{T}}]^{-1}\{Q_2\} \end{Bmatrix} \tag{5.64f}$$

The matrices $[c]$ and $[d]$, and the vector $\{f\}$ must be saved in order to compute the internal displacements $\{q_2\}$, after the solution for $\{q_1\}$ has been found. This is done by a forward substitution giving

$$\{q_2\} = [L_{22}]^{-1}(\{f\} - [c]\{q_1\}) \tag{5.64g}$$

A FORTRAN version of the above condensation procedure can be found in the computer program at the end of this book.

5-10 SUMMARY AND EXAMPLES

The definitions of the important symbols utilized in this chapter are given in Table 5-2. The analysis of a single finite element is summarized in Table 5-3 and may be recapitulated by the following steps:

1. Select a basic element configuration (Section 5-2).

2. Select a displacement model and adopt suitable nodal displacements as the amplitudes of the model (Sections 5-3 and 5-4).

3. Using the appropriate strain-displacement and stress-strain relations, write the element strains and stresses in terms of the nodal displacements (Section 5-5).

4. On the basis of the direct method or of the variational principle governing the problem, formulate the element stiffness and loads (Sections 5-6, 5-7, and 5-8).

5. If necessary, condense to eliminate internal degrees of freedom (Section 5-9).

TABLE 5-2 SUMMARY OF DEFINITIONS FOR AN INDIVIDUAL ELEMENT

Category and Symbol	*Definition*
Displacements	
$\{u\}$	Vector of displacements at any point within the element
$\{q\}$	Vector of nodal displacements
$\{\alpha\}$	Vector of generalized coordinates
Displacement models	
$[N]$	Interpolation model
$[\phi]$	Generalized coordinate model
$[A^{-1}]$	Displacement transformation relating $\{\alpha\}$ to $\{q\}$
Strains and stresses	
$\{\varepsilon\}$	Vector of strains at any point within the element
$\{\sigma\}$	Vector of stresses at any point within the element
$[B]$	Strain-displacement matrix for interpolation models
$[B_\alpha]$	Strain-displacement matrix for generalized coordinates
$[C]$	Stress-strain matrix
Stiffnesses	
$[k]$	Element stiffness
$[k_\alpha]$	Element stiffness with respect to generalized coordinates
$[\bar{k}]$	Condensed element stiffness
Loads	
$\{\bar{X}\}$	Vector of known body force intensities
$\{\bar{T}\}$	Vector of known surface traction intensities
S_1	Portion of element surface over which known tractions occur
$\{Q\}$	Element load vector
$\{Q_\alpha\}$	Element load vector with respect to generalized coordinates
$\{\bar{Q}\}$	Condensed element load vector

TABLE 5-3 SUMMARY OF THE ELEMENT ANALYSIS PROCEDURE

Step in the Element Analysis Procedure	*Section*	*Equations*			
		Generalized Coordinate Method		*Interpolation Model Method*	
1 Element configuration	5-2	—		—	
2 Displacement model	5-3, 5-4	$\{u\} = [\phi]\{\alpha\}$ $\{q\} = [A]\{\alpha\}$ $\{\alpha\} = [A^{-1}]\{q\}$	(5.5) (5.6) (5.7)	$\{u\} = [N]\{q\}$	(5.23)
3 Element strains and stresses	5-5	$\{\varepsilon\} = [B_\alpha]\{\alpha\}$ $\{\sigma\} = [C][B_\alpha]\{\alpha\}$	(5.28a) (5.29*a*)	$\{\varepsilon\} = [B]\{q\}$ $\{\sigma\} = [C][B]\{q\}$	(5.28b) (5.29b)
4 Element stiffness and loads	5-7	$[k_\alpha] = \iiint_V [B_\alpha]^T[C][B_\alpha]\,dV$ $\{Q_\alpha\} = \iiint_V [\phi]^T\{\bar{X}\}\,dV + \iint_{S_1}[\phi]^T\{\bar{T}\}\,dS_1$ $[k] = [A^{-1}]^T[k_\alpha][A^{-1}]$ $\{Q\} = [A^{-1}]^T\{Q_\alpha\}$	(5.46a) (5.46b) (5.47a) (5.47b)	$[k] = \iiint_V [B]^T[C][B]\,dV$ $\{Q\} = \iiint_V [N]^T\{\bar{X}\}\,dV + \iint_{S_1}[N]^T\{\bar{T}\}dS_1$	(5.45a) (5.45b)
5 Condensation	5-9	$[\bar{k}] = [k_{11}] - [k_{12}][k_{22}]^{-1}[k_{21}]$ $\{\bar{Q}\} = \{Q_1\} - [k_{12}][k_{22}]^{-1}\{Q_2\}$ $\{q_2\} = -[k_{22}]^{-1}([k_{21}]\{q_1\} - \{Q_2\})$	(5.63c) (5.63d) (5.63a)		
6 Element equilibrium	5-7 5-9	$[k]\{q\} = \{Q\}$ $[\bar{k}]\{q\} = \{\bar{Q}\}$	(5.44) (5.63b)		

To demonstrate the above five steps, we shall now consider a few examples. In each of these element formulations, the solution will be keyed to the above summary by appropriate numbers.

Because all the computations involved in the element analysis are performed by the computer, these examples are presented in general matrix notation. However, in Example 5-2 we have worked out the results using numerical values in order that the inquisitive reader may see what goes on in the computer.

Example 5–1: Axial Load Member. Consider an axially loaded member, Figure 5-20, but neglect the possibility of buckling. Let the member have piecewise linear variations of cross-sectional area.

1. Using the element and natural coordinate shown in Figure 5-12(a), we can express the cross-sectional area of the element in terms of the linear model given in Figure 5-12(b) and equation (5.21):

$$A = [\tfrac{1}{2}(1 - L) \;\; \tfrac{1}{2}(1 + L)]\begin{Bmatrix} A_1 \\ A_2 \end{Bmatrix} = [N]\{A\}$$

2. Using the same model for the displacements, we obtain

$$u = [\tfrac{1}{2}(1 - L) \;\; \tfrac{1}{2}(1 + L)]\begin{Bmatrix} u_1 \\ u_2 \end{Bmatrix} = [N]\{q\}$$

where we have selected the axial displacements of the nodes 1 and 2 as the nodal degrees of freedom.

3. The strain displacement equations are

$$\varepsilon = \frac{du}{dx} = \frac{2}{\ell}\frac{du}{dL} = \frac{1}{\ell}[-1 \;\; 1]\{q\} = [B]\{q\}$$

and the stresses are

$$\sigma = E\varepsilon = E[B]\{q\}$$

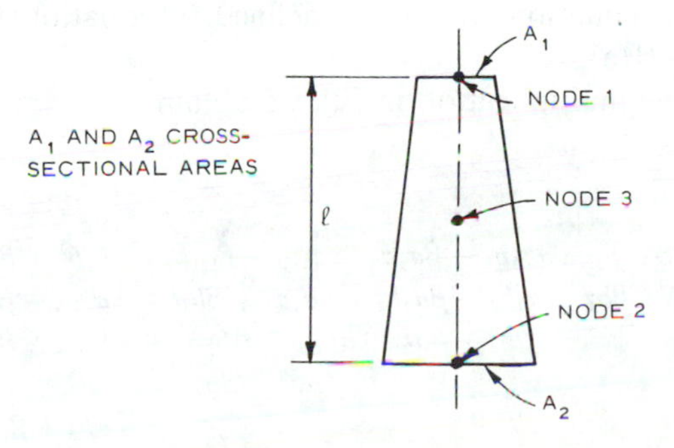

Figure 5-20 Axially loaded element.

4. Now the stiffness matrix for the element is given by direct application of the basic formula, equation (5.45a):

$$[k] = \int_{-1}^{1} [B]^{\mathrm{T}} EA [B] \frac{\ell}{2}\, dL = \frac{E\ell}{2} \int_{-1}^{1} [B]^{\mathrm{T}}([N]\{A\})[B]\, dL$$

$$= \frac{E}{4\ell} \begin{bmatrix} 1 & -1 \\ -1 & 1 \end{bmatrix} \int_{-1}^{1} \{A_1(1-L) + A_2(1+L)\}\, dL$$

$$[k] = \frac{E(A_1 + A_2)}{2\ell} \begin{bmatrix} 1 & -1 \\ -1 & 1 \end{bmatrix} = \frac{E\bar{A}}{\ell} \begin{bmatrix} 1 & -1 \\ -1 & 1 \end{bmatrix}$$

where $\bar{A}$ is the average area of the element. The load vector is given by equation (5.45b)

$$\{Q\} = \int_{-1}^{1} [N]^{\mathrm{T}}\{\bar{X}\} \frac{A\ell}{2}\, dL + \int_{-1}^{1} [N]^{\mathrm{T}}\{\bar{T}_x\} \frac{\ell}{2}\, dL$$

If $\bar{X}$ is a known constant body force and $\bar{T}_x$ is a known linear varying distributed axial load (per unit length) expressed in terms of the same interpolation function $[N]$, the loads are given by

$$\{Q\} = \frac{\bar{X}\ell}{6} \begin{Bmatrix} 2A_1 + A_2 \\ A_1 + 2A_2 \end{Bmatrix} + \frac{\ell}{6} \begin{Bmatrix} 2\bar{T}_{x1} + \bar{T}_{x2} \\ \bar{T}_{x1} + 2\bar{T}_{x2} \end{Bmatrix}$$

Example 5-2: Plane Strain Triangle—Numerical Results. In this example we shall obtain numerical results for the stiffness matrix, $[k]$, and the load vector, $\{Q\}$, of the constant strain triangle (CST) of Sample Problem 2.

Equation (5.45a) gives the matrix equation for computing $[k]$:

$$[k] = \iiint_V [B]^{\mathrm{T}}[C][B]\, dV \tag{5.45a}$$

The expressions for $[B]$ and $[C]$ are given in equations (5.35a) and (5.34c). Let us consider a simple triangular element shown in Figure 5-21(b). The notation utilized in equations (5.35a) is defined in equation (5.18c) and is shown in Figure 5-21(a).

By performing the multiplication in $[k]$, we obtain

$$[k] = \frac{Eh}{4A(1+\nu)(1-2\nu)}$$

$$\times \begin{bmatrix} \alpha b_1^2 + \beta a_1^2 & (\nu+\beta)a_1 b_1 & \alpha b_1 b_2 + \beta a_1 a_2 & \nu a_2 b_1 + \beta a_1 b_2 & \alpha b_1 b_3 + \beta a_1 a_2 & \nu a_3 b_1 + \beta a_1 b_3 \\ & \alpha a_1^2 + \beta b_1^2 & \nu a_1 b_2 + \beta a_2 b_1 & \alpha a_1 a_2 + \beta b_1 b_2 & \nu a_1 b_3 + \beta a_3 b_1 & \alpha a_1 a_3 + \beta b_1 b_3 \\ & & \alpha b_2^2 + \beta a_2^2 & (\nu+\beta)a_2 b_2 & \alpha b_2 b_3 + \beta a_2 a_3 & \nu a_3 b_2 + \beta a_2 b_3 \\ & & & \alpha a_2^2 + \beta b_2^2 & \nu a_2 b_3 + \beta a_3 b_2 & \alpha a_2 a_3 + \beta b_2 b_3 \\ & \text{Symmetrical} & & & \alpha b_3^2 + \beta a_3^2 & (\nu+\beta)a_3 b_3 \\ & & & & & \alpha a_3^2 + \beta b_3^2 \end{bmatrix}$$

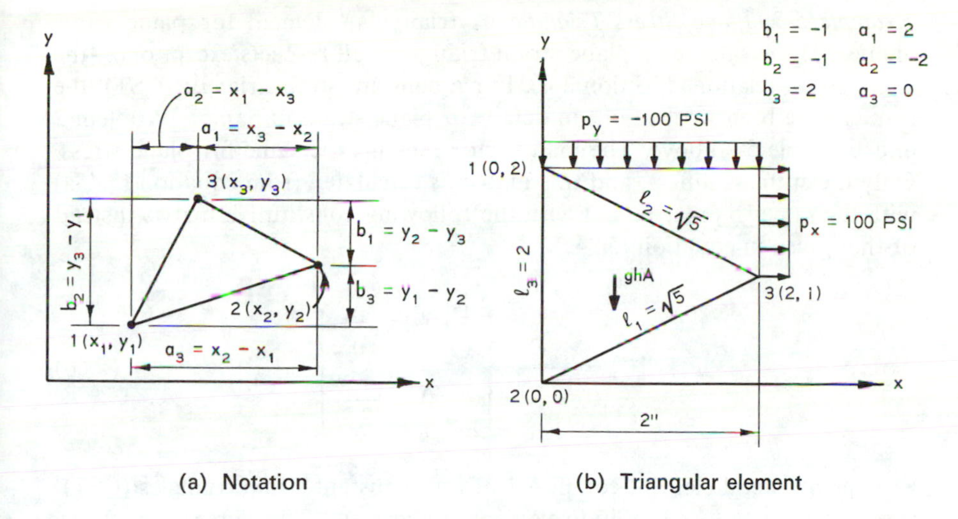

(a) Notation (b) Triangular element

Figure 5-21 Constant strain triangle.

where now a permuted version of equation (5.27) is used:

$$\{q\}^{\mathrm{T}} = [u_1 \ v_1 \ u_2 \ v_2 \ u_3 \ v_3]$$

and where $\alpha = (1 - \nu)$, $\beta = (1 - 2\nu)/2$, A is the area of the element, and h is the thickness. Note that since this is a CST, the integrand of equation (5.45) is constant; and the construction of the element stiffness consists simply of the multiplication, $Ah[B]^{\mathrm{T}}[C][B]$. Substituting the values of b_i and a_i shown in Figure 5-21(b), and assuming the Poisson's ratio and thickness to be $\nu = 0.3$ and $h = 1$, we obtain

$$[k] = \frac{E}{4.16}\begin{bmatrix} 1.5 & -1.0 & -0.1 & 0.2 & -1.4 & 0.8 \\ & 3.0 & -0.2 & -2.6 & 1.2 & -0.4 \\ & & 1.5 & 1.0 & -1.4 & -0.8 \\ & \text{symmetrical} & & 3.0 & -1.2 & -0.4 \\ & & & & 2.8 & 0.0 \\ & & & & & 0.8 \end{bmatrix} \tag{5.65}$$

To illustrate computation of the load vector, let us assume that the side 2 of the element, Figure 5-21(b), is subjected to $p_x^{(2)} = 100$ psi and $p_y^{(2)} = -100$ psi and that the mass density of the material is unity, $\overline{Y} = -g$. By using equation (5.61a) we can now write the load vector as

$$\{Q\}^{\mathrm{T}} = \frac{ghA}{3}[0 \ -1 \ 0 \ -1 \ 0 \ -1]$$

$$+ \frac{h}{2}[100\sqrt{5} \ -100\sqrt{5} \ 0 \ 0 \ 100\sqrt{5} \ -100\sqrt{5}]$$

Example 5-3: Plane Stress Triangle. A triangular element for plane stress analysis is the same as a plane strain triangle in all respects except one—the stress-strain relations (Section 3-4). For a constant strain triangle (CST), the results have been worked out in detail for plane strain in Sample Problem 2 and Example 5-2 above. The load vector remains the same for plane stress. Only the stiffness differs, and this matrix is calculated from equation (5.45a) with the same $[B]$ matrix, but with the following constitutive matrix instead of that given in equation (5.34c).

$$[C] = \frac{E}{1 - \nu^2}\begin{bmatrix} 1 & \nu & 0 \\ \nu & 1 & 0 \\ 0 & 0 & \dfrac{1-\nu}{2} \end{bmatrix} \tag{5.66}$$

A quadrilateral element composed of four constant strain triangles (CST), is sometimes employed in finite element applications. The element equations for the quadrilateral are obtained in three steps. First, the properties of the four CST's are computed. Next, they are added together. Finally, the degrees of freedom at the internal node are eliminated by condensation. Since the step of addition essentially involves use of the direct stiffness method discussed in Chapter 6, we shall delay consideration of this element until Example 6-2.

Example 5-4: Linear Strain Triangle (LST). To achieve a linear strain distribution in a triangular element which may be used for plane stress or plane strain, we must have a quadratic displacement model. Otherwise, the element analysis could be completely analogous to Sample Problem 2. However, it is advantageous to apply the alternative approach to construct the stiffness matrix, equations (5.51). The derivation and results in this example follow Felippa.[8]

1. The basic element is shown in Figure 5-13(a). The natural triangular coordinates given by Figure 5-11(c) and equation (5.18) will be used.

2. The displacement model is given in Figures 5-13(b) and 5-13(c) and by equation (5.23). The nodal degrees of freedom are the u and v displacement components at each of the six nodes.

3. The strain-displacement equations for either plane strain or plane stress are given by equation (5.35a). Hence we may write, using equation (5.23),

$$\underset{3\times 1}{\{\varepsilon\}} = \begin{bmatrix} \{B_1\}^{\mathrm{T}} & \{0\}^{\mathrm{T}} \\ \{0\}^{\mathrm{T}} & \{B_2\}^{\mathrm{T}} \\ \{B_2\}^{\mathrm{T}} & \{B_1\}^{\mathrm{T}} \end{bmatrix} \begin{Bmatrix} \{q_u\} \\ \{q_v\} \end{Bmatrix} = \underset{3\times 12}{[B]}\ \underset{12\times 1}{\{q\}}$$

where

$$\{\varepsilon\}^{\mathrm{T}} = [\varepsilon_x \;\; \varepsilon_y \;\; \gamma_{xy}]$$

$$\{B_1\}^{\mathrm{T}} = \frac{1}{2A}[(4L_1 - 1)b_1, (4L_2 - 1)b_2, (4L_3 - 1)b_3, 4(L_2 b_1 + L_1 b_2), 4(L_3 b_2 + L_2 b_3), 4(L_3 b_1 + L_1 b_3)]$$

$$\{B_2\}^{\mathrm{T}} = \frac{1}{2A}[(4L_1 - 1)a_1, (4L_2 - 1)a_2, (4L_3 - 1)a_3, 4(L_2 a_1 + L_1 a_2), 4(L_3 a_2 + L_2 a_3), 4(L_3 a_1 + L_1 a_3)]$$

$$\{q_u\}^{\mathrm{T}} = [u_1 \;\; u_2 \;\; \cdots \;\; u_6], \quad \{q_v\}^{\mathrm{T}} = [v_1 \;\; v_2 \;\; \cdots \;\; v_6]$$

The strains may be written by linear interpolation using the three corner nodal points.

$$\{\varepsilon_n\}^{\mathrm{T}} = [\varepsilon_{x1} \;\; \varepsilon_{x2} \;\; \varepsilon_{x3} \;\; \varepsilon_{y1} \;\; \varepsilon_{y2} \;\; \varepsilon_{y3} \;\; \gamma_{xy1} \;\; \gamma_{xy2} \;\; \gamma_{xy3}]$$

Now applying equation (5.49), we obtain

$$\underset{9\times 1}{\{\varepsilon_n\}} = \begin{bmatrix} [B_{n1}] & [0] \\ [0] & [B_{n2}] \\ [B_{n2}] & [B_{n1}] \end{bmatrix} \underset{12\times 1}{\{q\}} = [B_n]\{q\}$$

$$[B_{n1}] = \frac{1}{2A}\begin{bmatrix} 3b_1 & -b_2 & -b_3 & 4b_2 & 0 & 4b_3 \\ -b_1 & 3b_2 & -b_3 & 4b_1 & 4b_3 & 0 \\ -b_1 & -b_2 & 3b_3 & 0 & 4b_2 & 4b_1 \end{bmatrix}$$

$$[B_{n2}] = \frac{1}{2A}\begin{bmatrix} 3a_1 & -a_2 & -a_3 & 4a_2 & 0 & 4a_3 \\ -a_1 & 3a_2 & -a_3 & 4a_1 & 4a_3 & 0 \\ -a_1 & -a_2 & 3a_3 & 0 & 4a_2 & 4a_1 \end{bmatrix}$$

We shall assume that the material properties are constant over the element. Therefore, the stresses also vary linearly, and the nodal stress vector is defined as follows:

$$\{\sigma_n\}^{\mathrm{T}} = [\sigma_{x1} \;\; \sigma_{x2} \;\; \sigma_{x3} \;\; \sigma_{y1} \;\; \sigma_{y2} \;\; \sigma_{y3} \;\; \tau_{xy1} \;\; \tau_{xy2} \;\; \tau_{xy3}]$$

The interpolation models for the stress and strain are identical, equations (5.48).

$$\underset{3\times 9}{[N_\varepsilon]} = \underset{3\times 9}{[N_\sigma]} = \begin{bmatrix} \{N_1\}^{\mathrm{T}} & \{0\}^{\mathrm{T}} & \{0\}^{\mathrm{T}} \\ \{0\}^{\mathrm{T}} & \{N_1\}^{\mathrm{T}} & \{0\}^{\mathrm{T}} \\ \{0\}^{\mathrm{T}} & \{0\}^{\mathrm{T}} & \{N_1\}^{\mathrm{T}} \end{bmatrix}$$

$$\{N_1\}^{\mathrm{T}} = [L_1 \;\; L_2 \;\; L_3]$$

Finally, for plane strain or plane stress, the stress-strain matrix is of the following form, see equations (5.34c) and (5.66).

$$\{\sigma\} = \begin{bmatrix} C_{11} & C_{12} & C_{13} \\ & C_{22} & C_{23} \\ \text{Symm.} & & C_{33} \end{bmatrix} \{\varepsilon\} = [C]\{\varepsilon\}$$

Hence, the matrix $[C_n]$ of equation (5.50) is given by

$$[C_n] = \begin{bmatrix} C_{11}[I] & C_{12}[I] & C_{13}[I] \\ & C_{22}[I] & C_{23}[I] \\ \text{Symmetrical} & & C_{33}[I] \end{bmatrix}$$

where $[I]$ is the 3×3 identity matrix.

4. The only step remaining in the formulation of the stiffness matrix is to calculate the matrix $[D]$ of equation (5.51b):

$$[D] = h \iint_A [N_\varepsilon]^T [N_\sigma] \, dA = \begin{bmatrix} [D_1] & [0] & [0] \\ [0] & [D_1] & [0] \\ [0] & [0] & [D_1] \end{bmatrix}$$

$$[D_1] = h \iint_A \{N_1\}\{N_1\}^T \, dA = \frac{Ah}{12} \begin{bmatrix} 2 & 1 & 1 \\ 1 & 2 & 1 \\ 1 & 1 & 2 \end{bmatrix}$$

For variable thickness we would express the thickness in terms of an interpolation model and modify our calculation of $[D]$ as follows:

$$h = \{N_h\}^T \{h_n\}$$

$$[D] = \iint_A [N_\varepsilon](\{N_h\}^T \{h_n\})[N_\sigma] \, dA$$

In any case, the stiffness matrix is given by equation (5.51a)

$$\underset{12 \times 12}{[k]} = [B_n]^T [D][C_n][B_n] \tag{5.51a}$$

For the constant thickness case, the components of the element stiffness matrix are given by direct multiplication of the matrices $[B_n]$, $[D]$, and $[C]$ derived above.

The calculation of the consistent loads is routine, and only the results are given here. For a linearly distributed load acting in the x direction on side 1-4-2, Figure 5-13(a), of the element, the contribution to the load vector is

$$\begin{Bmatrix} Q_{x1} \\ Q_{x2} \\ Q_{x3} \end{Bmatrix} = \frac{\ell_3}{6} \begin{bmatrix} 1 & 0 \\ 2 & 2 \\ 0 & 1 \end{bmatrix} \begin{Bmatrix} \bar{p}_{x1} \\ \bar{p}_{x2} \end{Bmatrix}$$

$$\ell_3 = \sqrt{(x_2 - x_1)^2 + (y_2 - y_1)^2}$$

Similar contributions arise from distributed surface loads on other sides, in either the x or y direction. For uniform body forces the contribution to the load vector is

$$\{Q_u\}^{\mathrm{T}} = \frac{\bar{X}Ah}{3}[0 \quad 0 \quad 0 \quad 1 \quad 1 \quad 1]$$

$$\{Q_v\}^{\mathrm{T}} = \frac{\bar{Y}Ah}{3}[0 \quad 0 \quad 0 \quad 1 \quad 1 \quad 1]$$

Example 5-5: Linear Isoparametric Quadrilateral.[20]

1. A quadrilateral element and the associated natural coordinate system are shown in Figure 5-11(d) and are described in equation (5.19).

2. The linear displacement model for this isoparametric element is analogous to equation (5.19a). However, it is advantageous to write the model in the terminology of Table 5-1.

$$\{u\} = \begin{Bmatrix} u \\ v \end{Bmatrix} = \begin{bmatrix} N_1 & 0 & N_2 & 0 & N_3 & 0 & N_4 & 0 \\ 0 & N_1 & 0 & N_2 & 0 & N_3 & 0 & N_4 \end{bmatrix}\{q\} = [N]\{q\}$$

Here

$$N_i = \tfrac{1}{4}(1 + ss_i)(1 + tt_i)$$

and

$$\{q\}^{\mathrm{T}} = [u_1 \; v_1 \; u_2 \; v_2 \; u_3 \; v_3 \; u_4 \; v_4]$$

3. The strain-displacement equations for either plane stress or plane strain are given by equation (3.17a). Hence[20]

$$\{\varepsilon\} = \begin{Bmatrix} \dfrac{\partial u}{\partial x} \\ \dfrac{\partial v}{\partial y} \\ \dfrac{\partial u}{\partial y} + \dfrac{\partial v}{\partial x} \end{Bmatrix} = [[B_1][B_2][B_3][B_4]]\{q\}$$

where the submatrices $[B_i]$ are given by

$$[B_i] = \begin{bmatrix} \dfrac{\partial N_i}{\partial x} & 0 \\ 0 & \dfrac{\partial N_i}{\partial y} \\ \dfrac{\partial N_i}{\partial y} & \dfrac{\partial N_i}{\partial x} \end{bmatrix}$$

Because the displacement model is formulated in terms of the natural coordinates, s and t, it is necessary to relate the last equation to the derivatives with respect to these local coordinates. Using equation (5.19b) we therefore obtain

$$[B_i] = \begin{bmatrix} 1 & 0 & 0 & 0 \\ 0 & 0 & 0 & 1 \\ 0 & 1 & 1 & 0 \end{bmatrix} \begin{bmatrix} [J]^{-1} & [0] \\ [0] & [J]^{-1} \end{bmatrix} \begin{bmatrix} \{N_i'\} & \{0\} \\ \{0\} & \{N_i'\} \end{bmatrix}$$

Here the 2×2 matrix $[J]$ is obtained numerically from equation (5.19c), and

$$\{N_i'\}^{\mathrm{T}} = \begin{bmatrix} \dfrac{\partial N_i}{\partial s} & \dfrac{\partial N_i}{\partial t} \end{bmatrix} = \tfrac{1}{4}[s_i(1 + tt_i) \;\; t_i(1 + ss_i)]$$

The stress-strain equations for plane strain are given by equation (5.34c) and for plane stress by equation (5.66).

4. The element stiffness and loads are computed from the standard formulae, equations (5.45a) and (5.45b)

$$[k] = h \int_{-1}^{1} \int_{-1}^{1} [B]^{\mathrm{T}}[C][B] \det([J]) \, ds \, dt$$

$$\{Q\} = h \int_{-1}^{1} \int_{-1}^{1} [N]^{\mathrm{T}}\{\bar{X}\} \det([J]) \, ds \, dt + h \int_{S_1} [N]^{\mathrm{T}}\{\bar{T}\} \, dS_1$$

where h is the constant thickness of the element. These integrations are performed numerically. The determinant of $[J]$ can be evaluated from the expression

$$\det([J]) = \{(x_{13}y_{24} - x_{24}y_{13}) + s(x_{34}y_{12} - x_{12}y_{34}) + t(x_{23}y_{14} - x_{14}y_{23})\}/8$$

where

$$x_{ij} = x_i - x_j, \; y_{ij} = y_i - y_j.$$

Example 5-6: Axisymmetric Ring Element.

1. An axisymmetric ring element with a triangular cross section is shown in Figure 5-7(f). We can utilize the same natural coordinate system as we employed for the CST of Sample Problem 2, equation (5.18). However, instead of the cartesian coordinates x and y, we utilize the cylindrical coordinates r and z.

2. For axisymmetric deformation, the only relevant displacements are u and w in the r and z directions, respectively, because, due to the symmetry, the displacements v in the θ direction are zero. Therefore, we can adopt the same linear displacement model as that used for the CST of Sample Problem 1, equation (5.27).

3. There are four relevant strains for the axisymmetric case, equation (3.18a). Therefore the element strains are given by

$$\{\varepsilon\} = [B]\{q\} = \begin{Bmatrix} \varepsilon_r \\ \varepsilon_\theta \\ \varepsilon_z \\ \gamma_{rz} \end{Bmatrix} = \begin{Bmatrix} \dfrac{\partial u}{\partial r} \\ \dfrac{u}{r} \\ \dfrac{\partial w}{\partial z} \\ \dfrac{\partial u}{\partial z} + \dfrac{\partial w}{\partial r} \end{Bmatrix} = \frac{1}{2A} \begin{bmatrix} b_1 & b_2 & b_3 & 0 & 0 & 0 \\ \dfrac{2AL_1}{r} & \dfrac{2AL_2}{r} & \dfrac{2AL_3}{r} & 0 & 0 & 0 \\ 0 & 0 & 0 & a_1 & a_2 & a_3 \\ a_1 & a_2 & a_3 & b_1 & b_2 & b_3 \end{bmatrix} \{q\}$$

The constitutive relations are given by equation (3.18b).

$$\{\sigma\} = \begin{Bmatrix} \sigma_r \\ \sigma_\theta \\ \sigma_z \\ \tau_{rz} \end{Bmatrix} = \frac{E}{(1+v)(1-2v)} \begin{bmatrix} 1-v & v & v & 0 \\ v & 1-v & v & 0 \\ v & v & 1-v & 0 \\ 0 & 0 & 0 & \dfrac{1-2v}{2} \end{bmatrix} \{\varepsilon\} = [C]\{\varepsilon\} \tag{3.18b}$$

4. The above results may be substituted into the formulae for the stiffness and the load vector. However, we should note that the volume and surface differentials are given by

$$dV = 2\pi r\, dA = 2\pi r\, dr\, dz$$
$$dS_1 = 2\pi r\, d\ell_1$$

Hence the coordinate r enters all of our integrations and we cannot use the convenient formula, equation (5.18f). Numerical integration is appropriate in this case.

An alternative approach to this problem is to employ generalized coordinates as follows:

1. Use the same element, but employ only the global coordinate system, $r - z$.

2. The linear displacement model is completely analogous to the one given by equation (5.15).

$$\{u\} = \begin{bmatrix} 1 & r & z & 0 & 0 & 0 \\ 0 & 0 & 0 & 1 & r & z \end{bmatrix} \{\alpha\} = [\phi]\{\alpha\}$$

The transformation matrix A is therefore given by

$$\{q\} = \begin{bmatrix} 1 & r_1 & z_1 & 0 & 0 & 0 \\ 1 & r_2 & z_2 & 0 & 0 & 0 \\ 1 & r_3 & z_3 & 0 & 0 & 0 \\ 0 & 0 & 0 & 1 & r_1 & z_1 \\ 0 & 0 & 0 & 1 & r_2 & z_2 \\ 0 & 0 & 0 & 1 & r_3 & z_3 \end{bmatrix} \{\alpha\} = [A]\{\alpha\}$$

where (r_1, z_1), (r_2, z_2), and (r_3, z_3) are the coordinates of the nodes. The matrix $[A]$ can be inverted by using equation (5.18d):

$$[A^{-1}] = \begin{bmatrix} [A_1^{-1}] & [0] \\ [0] & [A_1^{-1}] \end{bmatrix}$$

$$[A_1^{-1}] = \frac{1}{2A} \begin{bmatrix} r_2 z_3 - r_3 z_2 & r_3 z_1 - r_1 z_3 & r_1 z_2 - r_2 z_1 \\ z_2 - z_1 & z_3 - z_1 & z_1 - z_2 \\ r_3 - r_2 & r_1 - r_3 & r_2 - r_1 \end{bmatrix}$$

where

$$2A = r_2(z_3 - z_1) + r_1(z_2 - z_3) + r_3(z_1 - z_2)$$

3. The strains are given by

$$\{\varepsilon\} = \begin{bmatrix} 0 & 1 & 0 & 0 & 0 & 0 \\ 1/r & 1 & z/r & 0 & 0 & 0 \\ 0 & 0 & 0 & 0 & 0 & 1 \\ 0 & 0 & 1 & 0 & 1 & 0 \end{bmatrix} \{\alpha\} = [B_\alpha]\{\alpha\}$$

and the stress-strain equations remain the same as given above by equation (3.18b).

4. Using equation (5.47a), we can now compute $[k]$. The product $[k_\alpha]$ in $[k]$ that requires integrations is given by

$$[k_\alpha] = \iint_A \begin{bmatrix} \dfrac{C_{22}}{r^2} & \dfrac{C_{12} + C_{22}}{r} & C_{22}\dfrac{z}{r^2} & 0 & 0 & \dfrac{C_{23}}{r} \\ & \begin{matrix} C_{11} + 2C_{12} \\ + C_{22} \end{matrix} & (C_{12} + C_{22})\dfrac{z}{r} & 0 & 0 & C_{13} + C_{23} \\ & & C_{22}\dfrac{z^2}{r^2} + C_{44} & 0 & C_{44} & C_{23}\dfrac{z}{r} \\ & & & 0 & 0 & 0 \\ & \text{Symmetrical} & & & C_{44} & 0 \\ & & & & & C_{33} \end{bmatrix} 2\pi r \, dA$$

We find that all our integrals in $[k_\alpha]$ (and in $\{Q\}$ below) involve polynomials in r and z. It is possible, but laborious, to obtain closed form integrations for these polynomials. Numerical integration, such as the Gaussian quadrature formulae, are fruitfully employed. Once $[k_\alpha]$ is obtained, the element stiffness follows as

$$[k] = [A^{-1}]^{\mathrm{T}}[k_\alpha][A^{-1}] \tag{5.49a}$$

If $\bar{R}$ and $\bar{Z}$ denote the components of the body forces and $\bar{p}_n$ and $\bar{p}_s$ denote the applied normal pressure and shear on the surface 1-2 of the element, the nodal load vector $\{Q\}$ is given by

$$\{Q\} = \iiint_V \begin{Bmatrix} r\bar{R} \\ r^2\bar{R} \\ rz\bar{R} \\ \bar{Z} \\ r\bar{Z} \\ z\bar{Z} \end{Bmatrix} dV$$

$$+ \iint_{S_1} \begin{Bmatrix} -\bar{p}_n(z_1 - z_2)\left(\frac{r_1}{2} + \frac{r_2 - r_1}{6}\right) + \bar{p}_s(r_2 - r_1)\left(\frac{r_1}{2} + \frac{r_2 - r_1}{6}\right) \\ \bar{p}_n(r_2 - r_1)\left(\frac{r_1}{2} + \frac{r_2 - r_1}{6}\right) + \bar{p}_s(z_1 - z_2)\left(\frac{r_1}{2} + \frac{r_2 - r_1}{6}\right) \\ -\bar{p}_n(z_1 - z_2)\left(\frac{r_1}{2} + \frac{r_2 - r_1}{3}\right) + \bar{p}_s(r_2 - r_1)\left(\frac{r_1}{2} + \frac{r_2 - r_1}{3}\right) \\ \bar{p}_n(r_2 - r_1)\left(\frac{r_1}{2} + \frac{r_2 - r_1}{3}\right) + \bar{p}_s(z_1 - z_2)\left(\frac{r_1}{2} + \frac{r_2 - r_1}{3}\right) \\ 0 \\ 0 \end{Bmatrix} dS_1$$

Example 5-7: Constant Strain Tetrahedron Element (CSTh). First we shall analyze the element shown in Figure 5-8(a) using a generalized coordinate approach.

1. The tetrahedron element is shown in Figure 5-8(a).
2. The linear displacement model follows from equation (5.2c)

$$\left.\begin{aligned} u &= \alpha_1 + \alpha_2 x + \alpha_3 y + \alpha_4 z \\ v &= \alpha_5 + \alpha_6 x + \alpha_7 y + \alpha_8 z \\ w &= \alpha_9 + \alpha_{10} x + \alpha_{11} y + \alpha_{12} z \end{aligned}\right\} \text{or} \quad \{u\} = [\phi]\{\alpha\}$$

$$\{q\} = [A]\{\alpha\}, \{\alpha\} = [A^{-1}]\{q\}$$

$$\{q\}^T = [u_1 \; u_2 \; u_3 \; u_4 \; v_1 \; v_2 \; v_3 \; v_4 \; w_1 \; w_2 \; w_3 \; w_4]$$

$$[A^{-1}] = \begin{bmatrix} [A_1^{-1}] & [0] & [0] \\ [0] & [A_1^{-1}] & [0] \\ [0] & [0] & [A_1^{-1}] \end{bmatrix}$$

and $[A_1^{-1}]$ is of the same form as in Example 5-6.

3. The strain-displacement relations from equation (3.11b) give

$$\{\varepsilon\} = \begin{bmatrix} 0 & 1 & 0 & 0 & 0 & 0 & 0 & 0 & 0 & 0 & 0 & 0 \\ 0 & 0 & 0 & 0 & 0 & 0 & 1 & 0 & 0 & 0 & 0 & 0 \\ 0 & 0 & 0 & 0 & 0 & 0 & 0 & 0 & 0 & 0 & 0 & 1 \\ 0 & 0 & 1 & 0 & 0 & 1 & 0 & 0 & 0 & 0 & 0 & 0 \\ 0 & 0 & 0 & 0 & 0 & 0 & 0 & 1 & 0 & 0 & 1 & 0 \\ 0 & 0 & 0 & 1 & 0 & 0 & 0 & 0 & 0 & 1 & 0 & 0 \end{bmatrix} \{\alpha\} = [B_\alpha]\{\alpha\}$$

or

$$\{\varepsilon\} = [B_\alpha][A^{-1}]\{q\}.$$

For an elastic, isotropic material, the stress-strain equations are given by equation (3.12f)

$$\{\sigma\} = \begin{Bmatrix} \sigma_x \\ \sigma_y \\ \sigma_z \\ \tau_{xy} \\ \tau_{yz} \\ \tau_{zx} \end{Bmatrix} = \frac{E}{(1+\nu)(1-2\nu)} \begin{bmatrix} 1-\nu & \nu & \nu & 0 & 0 & 0 \\ & 1-\nu & \nu & 0 & 0 & 0 \\ & & 1-\nu & 0 & 0 & 0 \\ & & & \frac{1-2\nu}{2} & 0 & 0 \\ & \text{Symmetrical} & & & \frac{1-2\nu}{2} & 0 \\ & & & & & \frac{1-2\nu}{2} \end{bmatrix} \{\varepsilon\}$$

$$= [C]\{\varepsilon\} \quad (3.12\text{f})$$

4. Use of equations (5.46) and (5.47) now permit the computation of $[k]$ and $\{Q\}$.

Because the computations of the element analysis with the use of the above generalized coordinates is laborious, let us consider the interpolation model approach for the same element. This is the most elementary three-dimensional isoparametric element. The derivation and results below follow Clough.[13]

1. The tetrahedral element shown in Figure 5-11(e) has a suitable nodal system for the constant strain case. Moreover, we use the natural coordinate system described by equations (5.20).

2. For a constant strain element, we require a linear variation of displacements, which is given by the following interpolation model:

$$\{u\} = \begin{Bmatrix} u \\ v \\ w \end{Bmatrix} = \begin{bmatrix} \{N_1\}^{\mathrm{T}} & \{0\}^{\mathrm{T}} & \{0\}^{\mathrm{T}} \\ \{0\}^{\mathrm{T}} & \{N_1\}^{\mathrm{T}} & \{0\}^{\mathrm{T}} \\ \{0\}^{\mathrm{T}} & \{0\}^{\mathrm{T}} & \{N_1\}^{\mathrm{T}} \end{bmatrix} \begin{Bmatrix} \{q_u\} \\ \{q_v\} \\ \{q_w\} \end{Bmatrix} = [N]\{q\}$$

$$\{N_1\}^T = [L_1 \; L_2 \; L_3 \; L_4]$$
$$\{q_u\}^T = [u_1 \; u_2 \; u_3 \; u_4]$$
$$\{q_v\}^T = [v_1 \; v_2 \; v_3 \; v_4]$$
$$\{q_w\}^T = [w_1 \; w_2 \; w_3 \; w_4]$$

3. The strain-displacement equations are taken from linear three-dimensional theory of elasticity, equation (3.11b).

$$\{\varepsilon\} = \begin{Bmatrix} \varepsilon_x \\ \varepsilon_y \\ \varepsilon_z \\ \gamma_{xy} \\ \gamma_{yz} \\ \gamma_{zx} \end{Bmatrix} = \begin{Bmatrix} \dfrac{\partial u}{\partial x} \\ \dfrac{\partial v}{\partial y} \\ \dfrac{\partial w}{\partial z} \\ \dfrac{\partial u}{\partial y} + \dfrac{\partial v}{\partial x} \\ \dfrac{\partial v}{\partial z} + \dfrac{\partial w}{\partial y} \\ \dfrac{\partial w}{\partial x} + \dfrac{\partial u}{\partial z} \end{Bmatrix} = \begin{bmatrix} \{a\}^T & \{0\}^T & \{0\}^T \\ \{0\}^T & \{b\}^T & \{0\}^T \\ \{0\}^T & \{0\}^T & \{c\}^T \\ \{b\}^T & \{a\}^T & \{0\}^T \\ \{0\}^T & \{c\}^T & \{b\}^T \\ \{c\}^T & \{0\}^T & \{a\}^T \end{bmatrix} \{q\} = [B]\{q\}$$

$$\{a\}^T = \frac{1}{6V}[a_1 \; a_2 \; a_3 \; a_4]$$

$$\{b\}^T = \frac{1}{6V}[b_1 \; b_2 \; b_3 \; b_4]$$

$$\{c\}^T = \frac{1}{6V}[c_1 \; c_2 \; c_3 \; c_4]$$

The a, b, and c quantities are defined by equation (5.20c). The stress-strain equations remain the same as given by equation (3.12c).

4. The formulae for the stiffness and the load vector, equation (5.45), may be directly applied. We see that the integrand for the stiffness matrix is constant, so the calculation of the stiffness is performed merely by a matrix multiplication

$$[k] = V[B]^T[C][B]$$

where V is the volume of the element. The consistent load vector requires simple integrations evaluated according to the formula, equation (5.20e).

Clough has also detailed the formulation for a Linear Strain Tetrahedron (LSTh).[13]

Example 5-8: *Elementary Hexahedral Element.*[10] The simplest hexahedral isoparametric element is sometimes known as the Zienkiewicz-Irons Brick with 8 nodes (ZIB8).[13] This is the three-dimensional element corresponding to the quadrilateral of Example 5-5.

1. The element is shown in Figure 5-8(c). The natural coordinates are r, s, and t with the origin of the system taken as the centroid of the element. This local system is related to the global cartesian coordinates as follows:

$$\begin{Bmatrix} x \\ y \\ z \end{Bmatrix} = \begin{bmatrix} \{N\}^T & \{0\}^T & \{0\}^T \\ \{0\}^T & \{N\}^T & \{0\}^T \\ \{0\}^T & \{0\}^T & \{N\}^T \end{bmatrix} \begin{Bmatrix} \{x_n\} \\ \{y_n\} \\ \{z_n\} \end{Bmatrix}$$

where

$$\{N\}^T = [N_1 \ N_2 \ \cdots \ N_8]$$

and

$$\{x_n\}^T = [x_1 \ x_2 \ \cdots \ x_8]$$

etc. The shape functions, N_i, are obtained from Table 5-1 and are

$$N_i = \tfrac{1}{8}(1 + rr_i)(1 + ss_i)(1 + tt_i)$$

2. The displacement model parallels the above representation of the geometry

$$\{u\} = \begin{Bmatrix} u \\ v \\ w \end{Bmatrix} = \begin{bmatrix} N_1 & 0 & 0 & N_2 & 0 & 0 & \cdots & N_8 & 0 & 0 \\ 0 & N_1 & 0 & 0 & N_2 & 0 & \cdots & 0 & N_8 & 0 \\ 0 & 0 & N_1 & 0 & 0 & N_2 & \cdots & 0 & 0 & N_8 \end{bmatrix} \{q\} = [N]\{q\}$$

where

$$\{q\}^T = [u_1 \ v_1 \ w_1 \ u_2 \ v_2 \ w_2 \ \cdots \ u_8 \ v_8 \ w_8]$$

3. The strain-displacement equations are given by equation (3.11b) and are the same as for the CSTh. Hence

$$\{\varepsilon\} = \begin{Bmatrix} \varepsilon_x \\ \varepsilon_y \\ \varepsilon_z \\ \gamma_{xy} \\ \gamma_{yz} \\ \gamma_{zx} \end{Bmatrix} = \begin{Bmatrix} \partial u/\partial x \\ \partial v/\partial y \\ \partial w/\partial z \\ \partial u/\partial y + \partial v/\partial x \\ \partial v/\partial z + \partial w/\partial y \\ \partial w/\partial x + \partial u/\partial z \end{Bmatrix} = [[B_1] \ [B_2] \ \ldots \ [B_8]]\{q\} = [B]\{q\}$$

where

$$[B_i] = \begin{bmatrix} \partial N_i/\partial x & 0 & 0 \\ 0 & \partial N_i/\partial y & 0 \\ 0 & 0 & \partial N_i/\partial z \\ \partial N_i/\partial y & \partial N_i/\partial x & 0 \\ 0 & \partial N_i/\partial z & \partial N_i/\partial y \\ \partial N_i/\partial z & 0 & \partial N_i/\partial x \end{bmatrix}$$

To obtain necessary differentiation formulae, we write

$$\begin{Bmatrix} \partial N_i/\partial x \\ \partial N_i/\partial y \\ \partial N_i/\partial z \end{Bmatrix} = [J]^{-1} \begin{Bmatrix} \partial N_i/\partial r \\ \partial N_i/\partial s \\ \partial N_i/\partial t \end{Bmatrix}$$

where $[J]$ is the 3×3 Jacobian matrix obtained numerically from

$$\underset{3 \times 3}{[J]} = \underset{3 \times 8}{\begin{bmatrix} \dfrac{\partial \{N\}^{\mathrm{T}}}{\partial r} \\ \dfrac{\partial \{N\}^{\mathrm{T}}}{\partial s} \\ \dfrac{\partial \{N\}^{\mathrm{T}}}{\partial t} \end{bmatrix}} [\{x_n\} \; \underset{8 \times 3}{\{y_n\}} \; \{z_n\}]$$

Therefore

$$[B_i] = \begin{bmatrix} 1 & 0 & 0 & 0 & 0 & 0 & 0 & 0 & 0 \\ 0 & 0 & 0 & 0 & 1 & 0 & 0 & 0 & 0 \\ 0 & 0 & 0 & 0 & 0 & 0 & 0 & 0 & 1 \\ 0 & 1 & 0 & 1 & 0 & 0 & 0 & 0 & 0 \\ 0 & 0 & 0 & 0 & 0 & 1 & 0 & 1 & 0 \\ 0 & 0 & 1 & 0 & 0 & 0 & 1 & 0 & 0 \end{bmatrix} \begin{bmatrix} [J]^{-1} & [0] & [0] \\ [0] & [J]^{-1} & [0] \\ [0] & [0] & [J]^{-1} \end{bmatrix} \begin{bmatrix} \{N_i'\} & \{0\} & \{0\} \\ \{0\} & \{N_i'\} & \{0\} \\ \{0\} & \{0\} & \{N_i'\} \end{bmatrix}$$

where

$$\{N_i'\}^{\mathrm{T}} = \left[\frac{\partial N_i}{\partial r} \quad \frac{\partial N_i}{\partial s} \quad \frac{\partial N_i}{\partial t} \right]$$
$$= \tfrac{1}{8}[r_i(1 + ss_i)(1 + tt_i) \;\; s_i(1 + rr_i)(1 + tt_i) \;\; t_i(1 + rr_i)(1 + ss_i)]$$

The stress-strain equations are given by equation (3.12f) as repeated in the previous example.

4. The element stiffness and loads are evaluated by numerical integrations of the standard formulae, equations (5.45). For example, the stiffness is obtained from

$$[k] = \int_{-1}^{1} \int_{-1}^{1} \int_{-1}^{1} [B]^{\mathrm{T}}[C][B] \det([J]) \, dr \, ds \, dt$$

Example 5-9: *Quadrilateral Element for Plate Bending*. One compatible and complete plate bending element is the rectangle originally formulated by Bogner *et al.*[21]

1. The rectangular element is shown in Figure 5-22. By using quadrilateral coordinates, equations (5.19), it is possible to generalize the element analysis for an arbitrary quadrilateral. However, for simplicity we will limit our attention to the rectangle.

2. From our experience with the beam bending element of Sample Problem 1, we know that we will need a cubic model for the transverse displacements of the plate. The nodal displacements necessary will therefore include the slopes. In addition, an important constant strain state of plates is the uniform twisting given by $\partial^2 w/\partial x\,\partial y$. Hence the appropriate degrees of freedom at each node are four

$$w,\ \theta_x = \frac{\partial w}{\partial x},\ \theta_y = \frac{\partial w}{\partial y},\ \theta_{xy} = \frac{\partial^2 w}{\partial x\,\partial y}$$

giving a total of 16 degrees of freedom for each element. We employ cubic Hermitian polynomials for our interpolation model. These are defined as

$$\left.\begin{aligned} N_{x1} &= 1 - 3s^2 + 2s^3 \\ N_{x2} &= s^2(3 - 2s) \\ N_{x3} &= as(s-1)^2 \\ N_{x4} &= as^2(s-1) \end{aligned}\right\} \quad \begin{aligned} & s = x/a \\ & 0 \le s \le 1 \\ & \frac{\partial}{\partial x} = \frac{1}{a}\frac{\partial}{\partial s} \end{aligned}$$

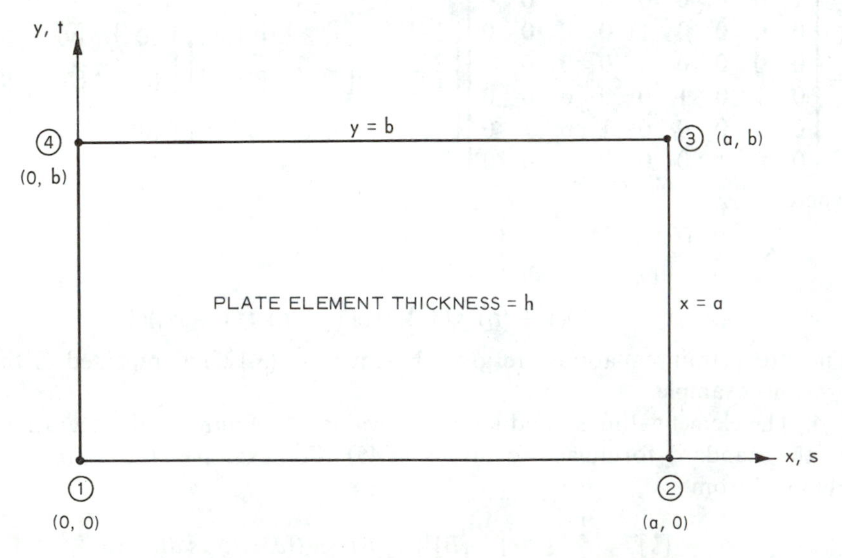

Figure 5-22 Quadrilateral element for plate bending.[21]

$$\left.\begin{aligned} N_{y1} &= 1 - 3t^2 + 2t^3 \\ N_{y2} &= t^2(3 - 2t) \\ N_{y3} &= bt(t-1)^2 \\ N_{y4} &= bt^2(t-1) \end{aligned}\right\} \quad \begin{aligned} & t = y/b \\ & 0 \le t \le 1 \\ & \frac{\partial}{\partial y} = \frac{1}{b}\frac{\partial}{\partial t} \end{aligned}$$

By using the shape functions we have just defined, the displacement model for the plate bending problem is

$$\begin{aligned} w = {} & N_{x1}N_{y1}w_1 + N_{x2}N_{y1}w_2 + N_{x2}N_{y2}w_3 + N_{x1}N_{y2}w_4 \\ & + N_{x3}N_{y1}\theta_{x1} + N_{x4}N_{y1}\theta_{x2} + N_{x4}N_{y2}\theta_{x3} + N_{x3}N_{y2}\theta_{x4} \\ & + N_{x1}N_{y3}\theta_{y1} + N_{x2}N_{y3}\theta_{y2} + N_{x2}N_{y4}\theta_{y3} + N_{x1}N_{y4}\theta_{y4} \\ & + N_{x3}N_{y3}\theta_{xy1} + N_{x4}N_{y3}\theta_{xy2} + N_{x4}N_{y4}\theta_{xy3} + N_{x3}N_{y4}\theta_{xy4} \\ = {} & [N]\{q\} \end{aligned}$$

3. Again relying on our experience gained by the beam element, Sample Problem 1, we recall that it is more convenient to work in terms of curvatures and moments than in terms of strains and stresses. For the pure bending of homogeneous plates, the curvature-displacement relations are

$$w_{xx} = \frac{\partial^2 w}{\partial x^2}, \; w_{yy} = \frac{\partial^2 w}{\partial y^2}, \; w_{xy} = w_{yx} = \frac{\partial^2 w}{\partial x \, \partial y}$$

and the moment-curvature equations are

$$\begin{aligned} M_{xx} &= D(w_{xx} + \nu w_{yy}) \\ M_{yy} &= D(w_{yy} + \nu w_{xx}) \\ M_{xy} &= M_{yx} = D(1 - \nu)w_{xy} \end{aligned}$$

where

$$D = \frac{Eh^3}{12(1 - \nu^2)}$$

and the moments are defined in equations (3.19c). The strain energy due to bending of plates is given by

$$U = \frac{D}{2} \iint_A [w_{xx}^2 + 2\nu w_{xx} w_{yy} + w_{yy}^2 + 2(1 - \nu)w_{xy}^2] \, dx \, dy$$

By using the above relations and displacement models, it is possible to write both the element curvatures and moments in terms of the nodal displacements. We will omit this lengthy process.

4. Bogner *et al.* have applied the principle of minimum potential energy to obtain the element stiffness and consistent loads in the usual fashion, equations (5.45). These results are tabulated in Reference 21 and will not be

repeated. However, it is worth noting that the alternative method of formulation for the stiffness matrix, equations (5.51), would definitely be less difficult than the standard interpolation model approach, equation (5.45a), for this higher-order (cubic) element.

Bogner *et al.* have also formulated a rectangular plate bending element with a quintic model and 36 nodal degrees of freedom.[21]

Example 5-10: Condensation of Internal Degrees of Freedom. We shall consider a simple example to illustrate the procedure of condensing the internal degrees of freedom, Section 5-9.

1. We shall adopt the axially loaded element, Figure 5-20.

2. We retain the linear model for the variation of cross-sectional areas. However, we introduce an internal node, 3, and adopt the quadratic interpolation model, equation (5.22),

$$u = [\tfrac{1}{2}L(L-1) \;\; \tfrac{1}{2}L(L+1) \;\; (1-L^2)]\begin{Bmatrix} u_1 \\ u_2 \\ u_3 \end{Bmatrix} = [N]\{q\}$$

where u_3 is the displacement at the internal node.

3. The strain-displacement equations are

$$\varepsilon = \frac{du}{dx} = \frac{1}{\ell}[(2L-1), (2L+1), -4L]\{q\} = [B]\{q\}$$

and the stresses are

$$\sigma = E\varepsilon = E[B]\{q\}$$

4. By using equation (5.45a), we can now evaluate the stiffness matrix as:

$$[k] = \int_\ell [B]^T[C][B]\, d\ell = \int_{-1}^{1} [B]^T(EA)[B]\frac{\ell}{2}\, dL$$

$$= \frac{E}{4\ell}\int_{-1}^{1}\begin{bmatrix} 4L^2-4L+1 & 4L^2-1 & -8L^2+4L \\ & 4L^2+4L+1 & -8L^2-4L \\ \text{Symmetrical} & & 16L^2 \end{bmatrix} \times [(1-L)A_1 + (1+L)A_2]\, dL$$

By performing the necessary integrations, we obtain

$$[k] = \frac{E}{6\ell}\left(A_1\begin{bmatrix} 11 & 1 & -12 \\ & 3 & -4 \\ \text{Symm.} & & 16 \end{bmatrix} + A_2\begin{bmatrix} 3 & 1 & -4 \\ & 11 & -12 \\ \text{Symm.} & & 16 \end{bmatrix}\right)$$

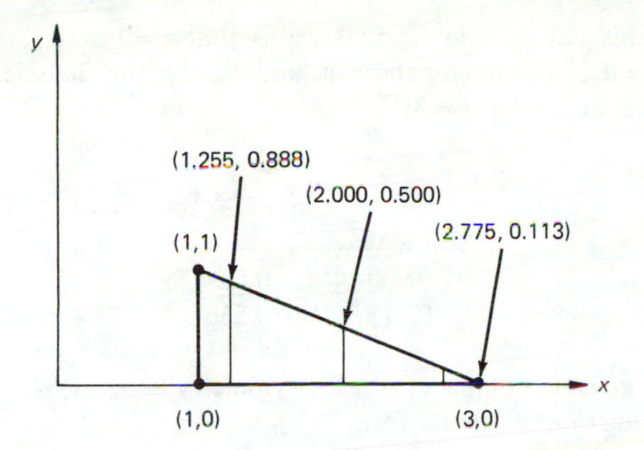

Figure 5-23 Integration for triangular area, Example 5-11.

If we assume that $A_2 = 2A_1$, we obtain

$$[k] = \frac{EA_1}{6\ell}\begin{bmatrix} 17 & 3 & -20 \\ & 25 & -28 \\ \text{Symm.} & & 48 \end{bmatrix}$$

We can now eliminate the internal degree of freedom, u_3, by using equation (5.64a). For instance, here we have $r = 2$ and $s = 3$. Hence, the first element of the condensed matrix is

$$\frac{6\ell}{EA_1}\bar{k}_{11} = (k_{11})_{\text{old}} - k_{13}\frac{k_{31}}{k_{33}} = 17 - \frac{(-20)(-20)}{48} = \frac{26}{3}$$

and so on. Finally,

$$[\bar{k}] = \frac{EA_1}{9\ell}\begin{bmatrix} 13 & -13 \\ -13 & 13 \end{bmatrix}$$

Condensation for the load vector is left to the reader as an exercise.

Example 5-11: Numerical Integration Using Gauss' Formula. To illustrate the application of Gauss' integration formula, equation (5.52a), we shall consider a simple example. Let us compute the area of the triangle shown in Figure 5-23 using three point integration, $n = 3$. The integral for this area is

$$A = \int_1^3 F(x)\,dx = \int_1^3 y(x)\,dx$$

Comparing this with equation (5.52a), we see that $a = 1$, $b = 3$, $f(y) = F(x) = y(x)$ and $dy = dx$. In a table of abscissas and weights for Gaussian integration we find the following for $n = 3$:[18]

$\pm x_i$	w_i
$n = 3$	
0.000	0.889
0.775	0.556

This table is given in compact form for symmetry about $x_i = 0$. We proceed with our calculation in tabular form

i	x_i	y_i or x	$f(y_i)$ or $y(x)$	w_i	$w_i f(y_i)$ or $w_i\, y(x)$
1	−0.775	1.225	0.888	0.556	0.494
2	0.000	2.000	0.500	0.889	0.445
3	0.775	2.775	0.113	0.556	0.063
				Total	1.002

We see that the error in our approximate integration is small, 0.2%.

REFERENCES

(1) Hrennikoff, A., "Solution of Problems in Elasticity by the Framework Method," *J. Appl. Mech.*, Vol 8 (1941). This paper is generally recognized as the earliest effort in formalization of the discrete element methods.

(2) Hudson, W. R., and Matlock, H., "Discrete-Element Analysis for Discontinuous Plates," *Proc. ASCE, J. ST Dn*, Vol 94, ST10 (1968). Presents a discrete element analysis that accounts for complex variations in stiffness, discontinuities, and variations of foundation support in pavement slabs.

(3) See Reference 10, Chapter 1.

(4) Fraeijs de Veubeke, B., "Upper and Lower Bounds in Matrix Structural Analysis," *Matrix Methods of Struct. Anal.*, AGARDograph 72, edited by B. Fraeijes de Veubeke, New York, MacMillan, 1964. Includes a proof that purely compatible methods and purely equilibrium methods produce upper and lower bounds for the stiffness influence coefficients.

(5) Irons, B. M. R., and Draper, K. J., "Inadequacy of Nodal Connections in a Stiffness Solution of Plate Bending," *AIAA J.*, Vol 3, No. 5 (May 1965). Concise statement of the completeness and compatability conditions for displacement models as applied to plate bending.

(6) Gallagher, R. H., "Analysis of Plate and Shell Structures," *Symp. FEM.* In addition to a review and convergence comparison of available finite element methods for the analysis of plates and shells, this paper has a useful summary of the principles of selecting displacement functions.
(7) Dunne, P., "Complete Polynomial Displacement Fields for the Finite Element Method," *The Aeronautical Journal*, Vol 72 (March 1968). Comments, Vol 72 (August 1968). Discussion of the completeness of polynomial models with respect to geometric isotropy.
(8) See Reference 13, Chapter 2.
(9) Pian, T. H. H., "Derivation of Element Stiffness Matrices," *AIAA J.*, Vol 2, No. 3 (March 1964). Describes how static condensation can be applied to improve stiffness matrices through the use of excess generalized coordinates.
(10) See Reference 9, Chapter 1.
(11) Sokolnikoff, I. S., and Redheffer, R. M., *Mathematics of Physics and Modern Engineering*, New York, McGraw-Hill Book Co., 1958. A good textbook of applied mathematics.
(12) See Reference 8, Chapter 1.
(13) Clough, R. W., "Comparison of Three-Dimensional Finite Elements," *Symp. FEM.* A description and comparison of various three-dimensional finite elements, including natural coordinates and interpolation functions for tetrahedra and hexahedra. Finds that the Zienkiewicz-Irons Brick (ZIB) family of isoparametric elements is superior to hexahedra formed from CSTh or LSTh elements.
(14) Jones, R., and Strome, D., "A Survey of the Analysis of Shells by the Direct Stiffness Method," *First Conf.* A comparison of curved and flat elements, and of "exact" or direct stiffness and stiffnesses from assumed displacement models.
(15) See Reference 6, Chapter 3.
(16) See Reference 4, Chapter 1.
(17) See Reference 7, Chapter 1.
(18) Abramowitz, M., and Stegun, I. A. (ed), *Handbook of Mathematical Functions with Formulas, Graphs, and Mathematical Tables*, National Bureau of Standards, Applied Math. Series 55, Washington, U. S. Govt. Printing Office, 1964. Also, New York, Dover Books, 1965. A useful and inexpensive 1000-page compendium.
(19) Hammer, P. C., and Stroud, A. H., "Numerical Evaluation of Multiple Integrals," *Math. Tables Aids Comp.*, Vol 12 (1958).
(20) Ergatoudis, B., Irons, B. M., and Zienkiewicz, O. C., "Curved, Isoparametric, 'Quadrilateral' Elements for Finite Element Analysis," *Int. J. Solids Struct.*, Vol. 4, No. 1 (1968). A basic paper on isoparametric elements which also contains a description of the necessary natural coordinate systems and interpolation functions for quadrilateral elements.
(21) Bogner, F. K., Fox, R. L., and Schmit, L. A., Jr., "The Generation of Inter-Element-Compatible Stiffness and Mass Matrices by the Use of Interpolation Formulas," *First Conf.* The addendum to this paper contains a description of complete and compatible quadrilateral plate bending elements with 16 and 36 degrees of freedom.

FURTHER READING

Argyris, J. H., Buck, K. E., Grieger, I., and Mareczek, G., "Application of the Matrix Displacement Method to the Analysis of Pressure Vessels," *Proc. ASME, J. Engr. for Ind.*, Vol. 92, No. 2 (May 1970). Includes a description of 17 of the 42 different types of finite elements included in the general purpose structural program ASKA—Automatic System for Kinematic Analysis.

Fried, I., "Some Aspects of the Natural Coordinate System in the Finite Element Method," *AIAA J.*, Vol. 7, No. 7 (July 1969). Presents derivation of basic properties of the natural system and some formulae for straight- and curved-sided elements. Derives general formulae for integration of polynomial terms.

Irons, B. M., "Engineering Applications of Numerical Integration in Stiffness Methods," *AIAA J.*, Vol. 4, No. 11 (Nov. 1966).

Khanna, J., and Hooley, R. F., "Comparison and Evaluation of Stiffness Matrices," *AIAA J.*, Vol. 4, No. 12 (Dec. 1966). Presents theory for evaluating the quality and adequacy of stiffnesses by computing the eigenvalues of the element stiffness matrix. Reference 13 of the present chapter contains examples of how this technique can be used to compare formulations for elements of the same shape and with the same number of degrees of freedom.

Luft, R. W., Roesset, J. M., and Conner, J. J., "Automatic Generation of Finite Element Matrices," *Proc. ASCE, J. ST Dn.*, Vol. 97, ST1 (Jan. 1971). Reformulates the element analysis procedure in terms of dimensionless coordinates. Introduces the concept of intrinsic matrices. Describes a code which automatically generates stiffness and mass matrices for arbitrary order displacement models.

Silvester, P. "Symmetric Quadrature Formulae for Simplexes," *Math. Comp.*, Vol. 24, No. 109 (Jan. 1970).

EXERCISES

5-1 Sketch a figure similar to Figure 5-8(d) showing a hexahedral element composed of five tetrahedra and having an internal node at its centroid.

5-2 Write seven different quintic polynomial displacement models, equation (5.4), in such a way that geometric isotropy is satisfied.

5-3 Check the following plate-bending displacement models for completeness and isotropy:

$$(i)\quad \mathrm{w} = \alpha_1 + \alpha_2 x + \alpha_3 y + \alpha_4 x^2 + \alpha_5 xy + \alpha_6 y^2 + \alpha_7 x^3 + \alpha_8 x^2 y + \alpha_9 y^3.$$

$$(ii)\quad \mathrm{w} = (\alpha_1 + \alpha_2 x + \alpha_3 x^2 + \alpha_4 x^3)(\beta_1 + \beta_2 y + \beta_3 y^2 + \beta_4 y^3).$$

$$(iii)\quad \mathrm{w} = \alpha_1 + \alpha_2 x + \alpha_3 y + \alpha_4 x^2 + \alpha_5 y^2 + \alpha_6 xy + \alpha_7 x^2 y + \alpha_8 xy^2 + \alpha_9 x^3 + \alpha_{10} y^3 + \alpha_{11} x^3 y + \alpha_{12} xy^3.$$

$$(iv)\quad \mathrm{w} = \alpha_1 + \alpha_2 x + \alpha_3 y + \alpha_4 x^2 + \alpha_5 xy + \alpha_6 y^2 + \alpha_7 x^3 + \alpha_8 x^2 y + \alpha_9 xy^2 + \alpha_{10} y^3 + \alpha_{11} x^4 + \alpha_{12} x^3 y + \alpha_{13} x^2 y^2 + \alpha_{14} xy^3 + \alpha_{15} y^4 + \alpha_{16} x^5 + \alpha_{17} x^4 y + \alpha_{18} x^3 y^2 + \alpha_{19} x^2 y^3 + \alpha_{20} xy^4 + \alpha_{21} y^5.$$

Solution: (i) complete, nonisotropic; (ii) isotropic, incomplete; (iii) complete, isotropic; (iv) complete, isotropic (see Reference 6).

5-4 (a) Construct a cubic, generalized-coordinate polynomial displacement model (quadratic strain variation) for a plane strain/stress triangular element with three primary external nodes and one centroidal internal node. (*Hints*: We need ten terms for a cubic polynomial, equation (5.4). Hence, a total of twenty displacement quantities are required. Each side requires four nodal quantities for defining cubic variation. Adopt displacements and their derivatives as unknowns at the primary nodes, and only displacements at the internal node. This is a QST, Quadratic Strain Triangle, see Reference 12.)

(b) Construct a cubic interpolation model for a plane strain/stress triangular element with three primary external nodes, six secondary external nodes and one centroidal internal node. Use Table 5-1.

5-5 Consider the u component of displacement for an isoparametric element.

$$u = \{N\}^T\{u_n\} = \Sigma N_i u_i .$$

$$\begin{Bmatrix} x \\ y \\ z \end{Bmatrix} = \begin{bmatrix} \{N\}^T & \{0\}^T & \{0\}^T \\ \{0\}^T & \{N\}^T & \{0\}^T \\ \{0\}^T & \{0\}^T & \{N\}^T \end{bmatrix} \begin{Bmatrix} \{x_n\} \\ \{y_n\} \\ \{z_n\} \end{Bmatrix}$$

(a) Show that the condition to be satisfied for rigid body displacements to be possible is $\Sigma N_i = 1$.

(b) Show that, if rigid body displacements are possible in the natural coordinate system interpolation model, constant strains are guaranteed in the global coordinate system. (Hint: If constant strains are included, one can write $u = a + bx + cy + dz$, where a,b,c, and d are arbitrary constants. See Reference 10).

5-6 Assuming a single degree of freedom at each node and using the following model,

$$u(x, y) = \alpha_1 + \alpha_2 x + \alpha_3 y + \alpha_4 xy$$

compute matrix $[A]$ in equation (5.6) and the displacement transformation matrix $[A]^{-1}$ in equation (5.7) for the square element shown.

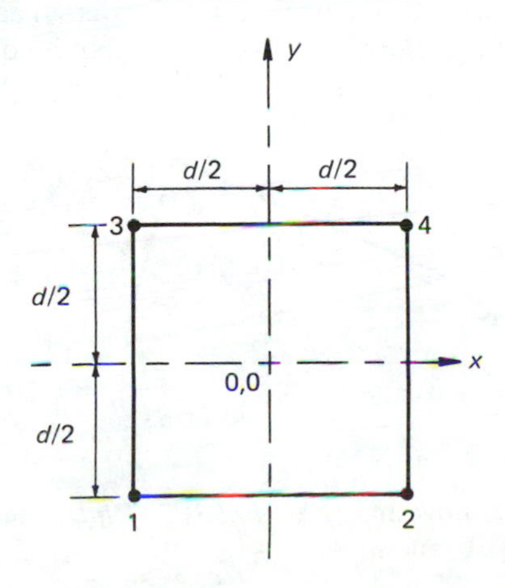

Solution:

$$[A]=\frac{d}{2}\begin{bmatrix} 2/d & -1 & -1 & d/2 \\ 2/d & 1 & -1 & -d/2 \\ 2/d & -1 & 1 & -d/2 \\ 2/d & 1 & 1 & d/2 \end{bmatrix}, \quad [A]^{-1}=\frac{1}{d}\begin{bmatrix} d/4 & d/4 & d/4 & d/4 \\ -1/2 & 1/2 & -1/2 & 1/2 \\ -1/2 & -1/2 & 1/2 & 1/2 \\ 1/d & -1/d & -1/d & 1/d \end{bmatrix}$$

5-7 (a) Carry out the complete procedure for derivation of stiffness matrix of a triangular element, Sample Problem 2, with *plane stress* conditions, Example 5-3. Use the stress-strain law given by equation (5.66). Specialize the stiffness matrix for the equilateral triangle element shown.

(b) Show that the resulting stiffness matrix is singular. (*Hint*: Prove that its determinant vanishes.)

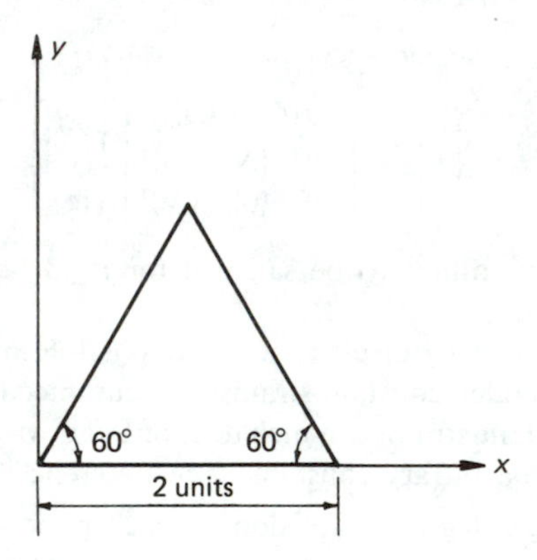

5-8 Obtain a quadratic interpolation model for the triangular element shown. It has three primary external nodes and two secondary external nodes. (*Hint*: Variation of displacements is quadratic along sides 1-2 and 2-3, but is constrained to be linear along side 1-3.)

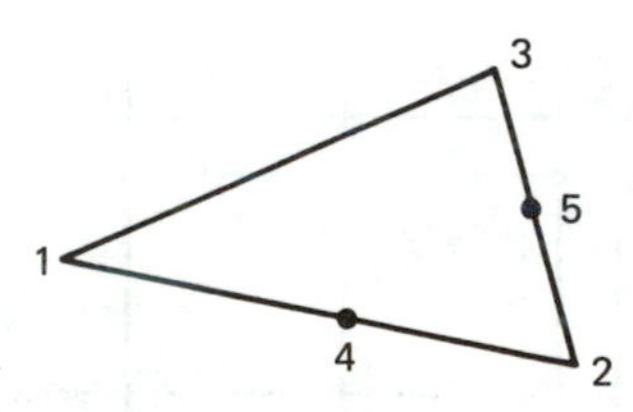

5-9 In equation (5.39b) show that $[F]=hA[B]^{T}$. (*Hint*: Use equations (5.18c) to show that $a_1+a_2+a_3=0$, and $b_1+b_2+b_3=0$.)

5-10 Compute the plane strain stiffness matrix in terms of the ratio $r = a/b$ for the rectangular element shown. The element has unit thickness. Use the following displacement model. Specialize the matrix for $\nu = 0.2$ and $r = 1$. (See Reference 12, Chapter 10.)

$$u = \alpha_1 + \alpha_2 x + \alpha_3 y + \alpha_4 xy$$
$$v = \alpha_5 + \alpha_6 x + \alpha_7 y + \alpha_8 xy$$

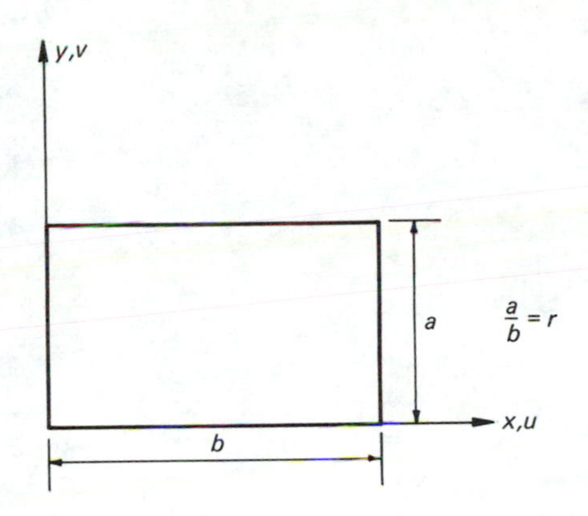

Solution:

$$[k] = \frac{E}{43.2} \begin{bmatrix} 22 & -13 & 2 & -11 & 7.5 & -1.5 & 1.5 & -7.5 \\ & 22 & -11 & 2 & 1.5 & -7.5 & 7.5 & -1.5 \\ & & 22 & -13 & -1.5 & 7.5 & -7.5 & 1.5 \\ & & & 22 & -7.5 & 1.5 & -7.5 & 1.5 \\ \text{Symmetrical} & & & & 22 & 2 & -13 & -11 \\ & & & & & 22 & -11 & -13 \\ & & & & & & 22 & 2 \\ & & & & & & & 22 \end{bmatrix}$$

$(\nu = 0.2,\ r = 1)$

5-11 Introduce four secondary external nodes (one per side) to the rectangular element in Exercise 5-10 above and obtain its stiffness equations. Adopt the following quadratic displacement model:

$$u = \alpha_1 + \alpha_2 x + \alpha_3 y + \alpha_4 x^2 + \alpha_5 xy + \alpha_6 y^2 + \alpha_7 x^2 y + \alpha_8 xy^2$$
$$v = \alpha_9 + \alpha_{10} x + \alpha_{11} y + \alpha_{12} x^2 + \alpha_{13} xy + \alpha_{14} y^2 + \alpha_{15} x^2 y + \alpha_{16} xy^2$$

6

THE OVERALL PROBLEM—AN ASSEMBLAGE OF ELEMENTS

In Chapter 5 we discussed in detail the various aspects of the analysis of a single finite element. Before we can analyze an assemblage of elements representing a complete body or structure, we must also consider the actual process by which the structure is subdivided into an adequate number of elements. After a thorough treatment of this discretization procedure in the first section of this chapter, we shall be ready to "go from part to whole," that is, to use our knowledge of the individual element to analyze the assemblage with which we are concerned.

6-1 DISCRETIZATION OF A BODY OR STRUCTURE

The subdivision process is essentially an exercise of engineering judgment. We have to decide the number, shape, size, and configuration of the elements in such a way that the original body is simulated as closely as possible. The general objective of such a discretization is to divide the body into elements sufficiently small so that the simple displacement models can adequately approximate the true solution. At the same time, we must remember that too fine a subdivision will lead to extra computational effort.

Figure 6-1(a) shows a section of a concrete gravity dam. Let us assume that it rests on a rigid foundation. The external loads on the dam consist of the weight of concrete and the horizontal forces resulting from the water in the reservoir. In addition, loadings may be caused by temperature gradients between the body of the dam and the surrounding environment.

If the dam is long, we can consider only a slice of unit thickness and assume plane strain conditions (Chapter 3). A rather coarse subdivision of the dam is shown in Figure 6-1(b) in which we have used only triangular elements. The external loading is converted into equivalent static resultants at the nodal points, Figure 6-1(b).

In our discretized assemblage, Figure 6-1(b), we can see that the curved

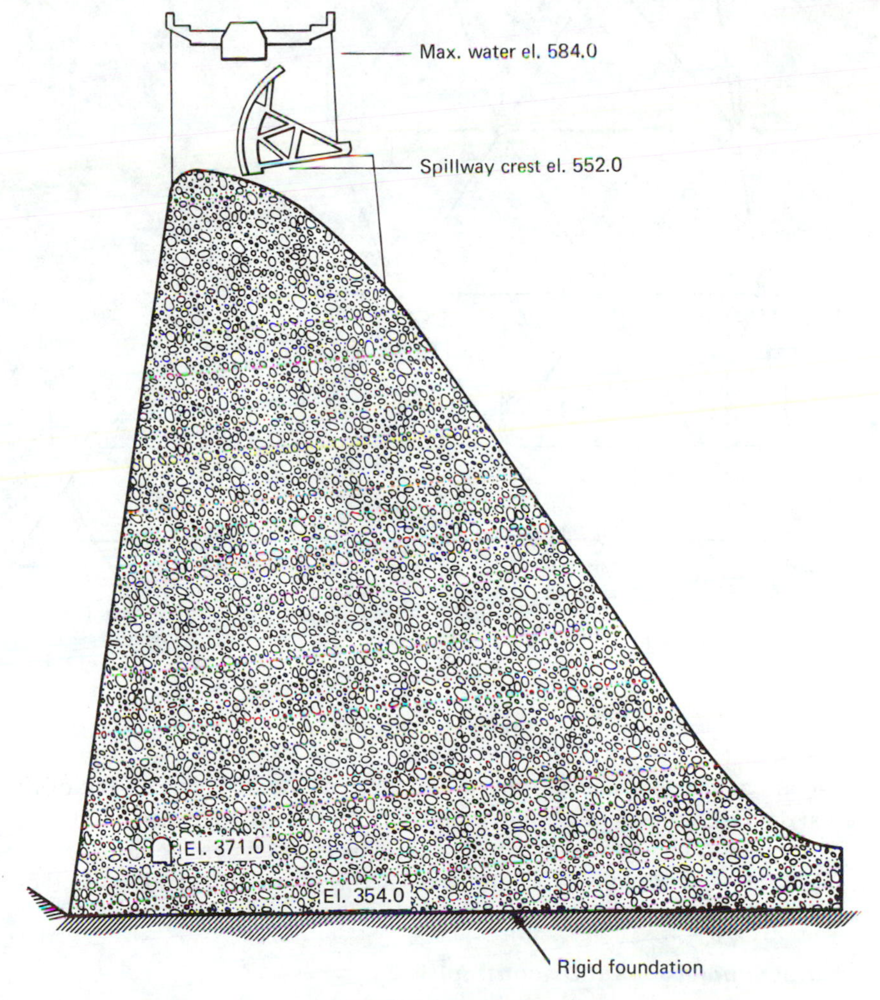

(a) A section of Norfork Dam[1]

Figure 6-1 Discretization of a dam.

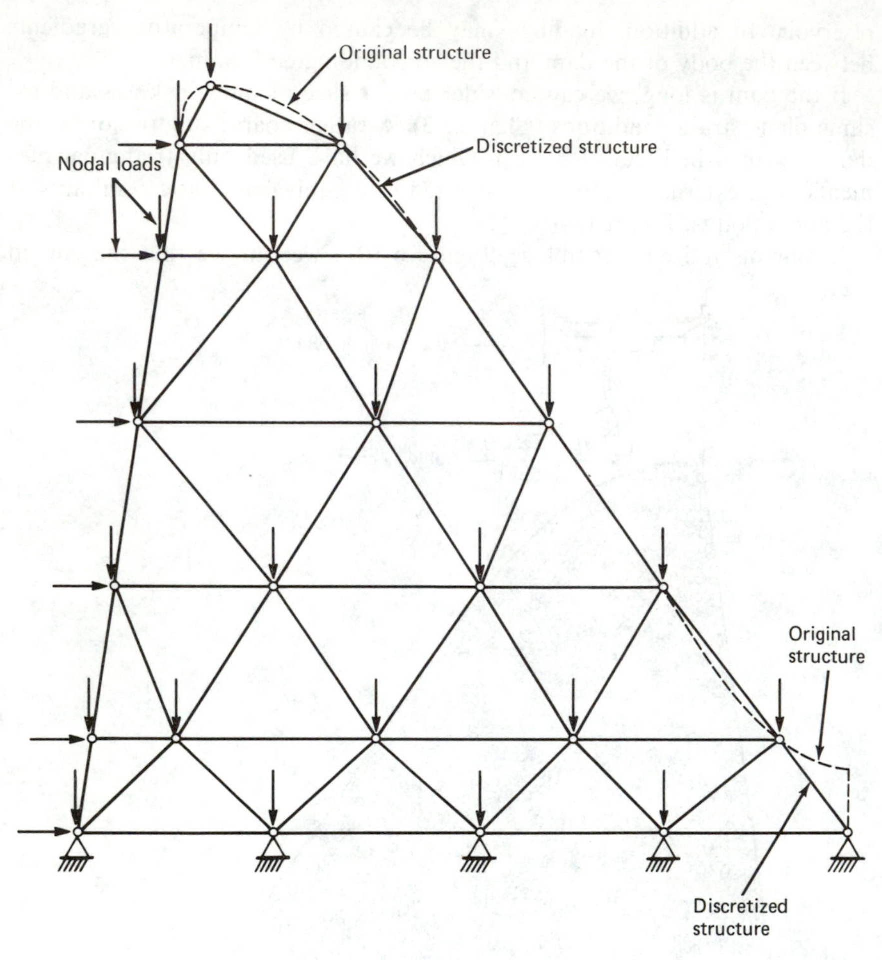

(b) Discretization of the dam section

Figure 6-1 (*continued*)

boundary at the crest and the sharp boundary at the downstream toe of the actual dam section are only approximated. Such curved boundaries and other irregularities in real structures introduce various special considerations to the idealization process. In the remainder of this section we shall attend to these considerations.

Natural Subdivisions at Discontinuities

The most obvious locations for nodes or subdivision lines and planes are places where abrupt changes in geometry, loading, and material properties occur. Examples of these natural subdivisions are shown in Figure 6-2 for various two-dimensional problems. In Figure 6-2(a), a node must occur at the point of application of a concentrated load because all loads are converted into equivalent nodal-point loads. Even for distributed loads we compute equivalent nodal loads (Sections 5-6 and 5-7). To facilitate this process, it is logical to select as nodal points any location at which there is an abrupt change of distributed loads, Figure 6-2(b). Similar considerations apply for discontinuities in geometry, such as the discontinuity in plate thickness shown in plan and section views in Figure 6-2(c). If the nodal line, that is, the line joining any two nodes, is selected as shown, each plate element will have a constant thickness. Problems involving nonhomogeneous materials also have natural locations for nodal lines, such as the interface between materials of different properties indicated in Figure 6-2(d), which represents a layered soil medium. Figure 6-2(e) shows an existing crack in a dam. In a stress analysis to determine the effect of the presence of such a crack, the mesh should be chosen in such a way that each side of the crack is considered an external boundary, and therefore nodes should be provided along both sides of the crack.[1] Figure 6-2(f) shows a different type of geometric discontinuity, a re-entrant corner. The essential requirement here is that a primary external node should occur at this corner to avoid having an element which itself has a re-entrant corner, causing possible difficulty in the calculation of the element stiffness.

Refined Mesh for Steep Gradients

The easiest mesh to construct is a completely regular one, that is, a subdivision having the same shape and size of element throughout. However, we do not often encounter problems for which such a mesh is satisfactory, because most bodies have zones in which pronounced variations in the stresses and strains occur. The most familiar examples of such regions of steep gradients of the variables are locations at which stress concentrations occur. In order to obtain a useful approximate solution for these concentrations, a finer subdivision is necessary in regions where stress concentrations are expected. One example of this is the re-entrant corner shown in Figure 6-2(f). Another example is the plate with a slot indicated in Figure 6-3. At the apex of the slot, stress concentrations will occur and a refined mesh should be provided.

Geometric Approximations

If straight-sided elements are employed, curved boundaries are approximated as piecewise linear by the sides of the elements adjacent to the boundary. Care should be taken to ascertain that the curved boundary is approximated as

(a) Concentrated load

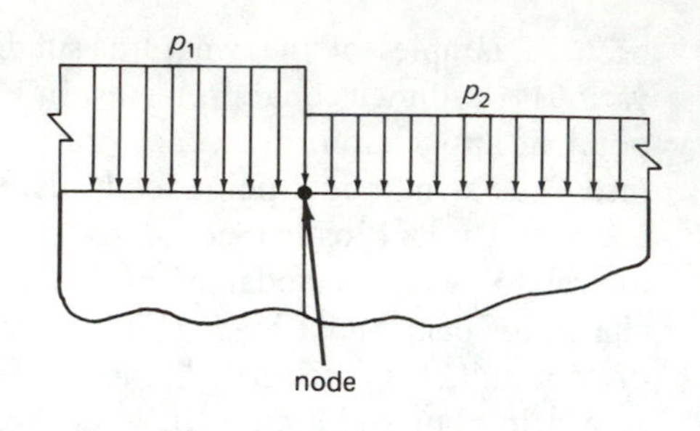

(b) Abrupt change in distributed load

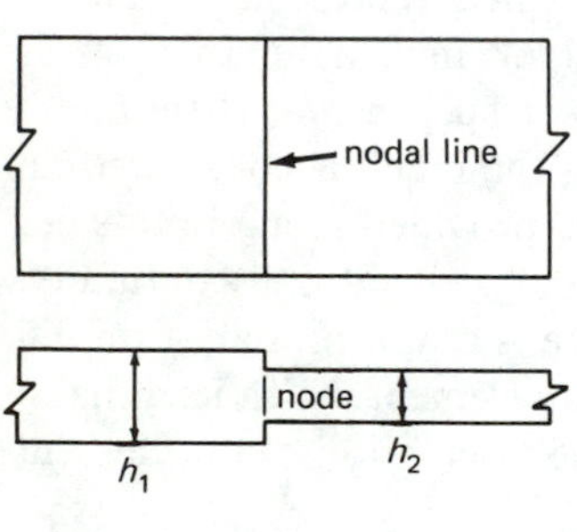

(c) Abrupt change in plate thickness

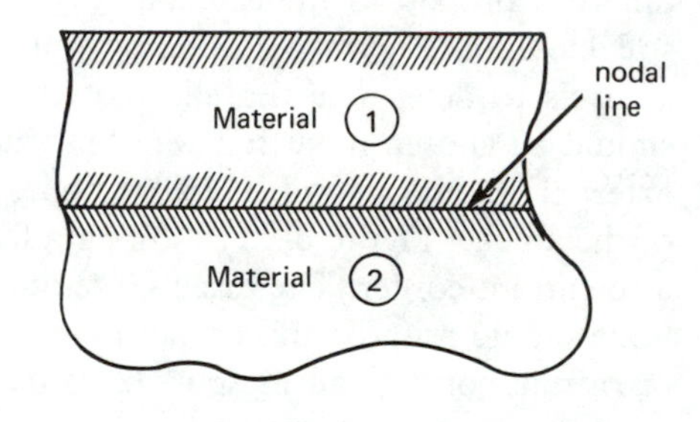

(d) Abrupt change of material properties

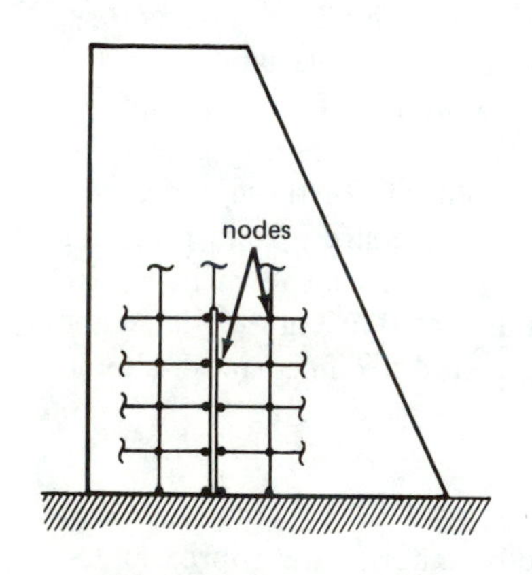

(e) Existing crack in a dam

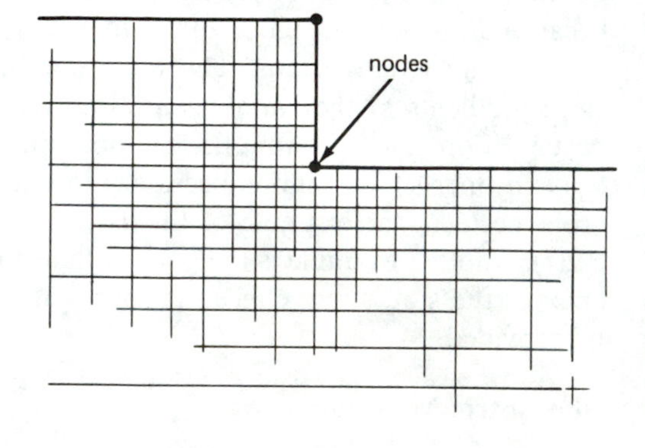

(f) Re-entrant corner

Figure 6-2 Natural subdivisions at discontinuities.

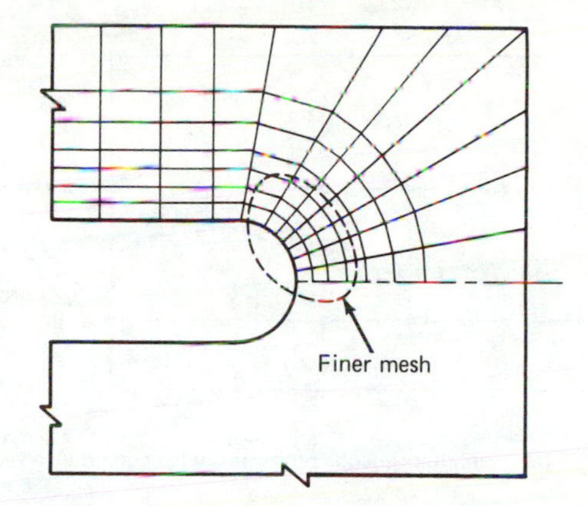

Figure 6-3 Plate with stress concentration at slot.

closely as possible, perhaps by refinement of the mesh in the vicinity of the most pronounced boundary curvature. If isoparametric elements with curved sides are used, it is possible to represent exactly many geometries that can be only approximated by conventional elements. However, even isoparametric elements may fall short of exactly representing a highly complex configuration.

Some examples of geometric approximations are shown in Figures 6-1 and 6-4. Figure 6-1 shows a two-dimensional body with a curved boundary as approximated by straight-sided elements. A similar problem is shown in Figure 6-4(a), in which the interface between two materials is curved. It is apparent from this figure that refinement of the mesh is required at certain locations to approximate adequately the interface by straight-sided elements. An additional case of geometric approximation is illustrated in Figure 6-4(b). Here an axisymmetric shell with an arbitrarily curved meridian is represented by an assemblage of conical frustra.

Many of the problems we attempt to solve by the finite element method have no closed-form solutions because of their complicated geometry. Therefore, we are often faced with the necessity of employing such geometric approximations as those just discussed. A substantial body of experience indicates that useful engineering solutions can be obtained with these approximations, provided reasonable care is exercised in discretizing the body. In interpreting our results, however, we must recognize that we may have introduced such geometric approximations.

Notation for Subdivided Continua

In order to facilitate manipulations in a computer program or code, we must have a systematic method of labeling elements and nodes. Thus an essential

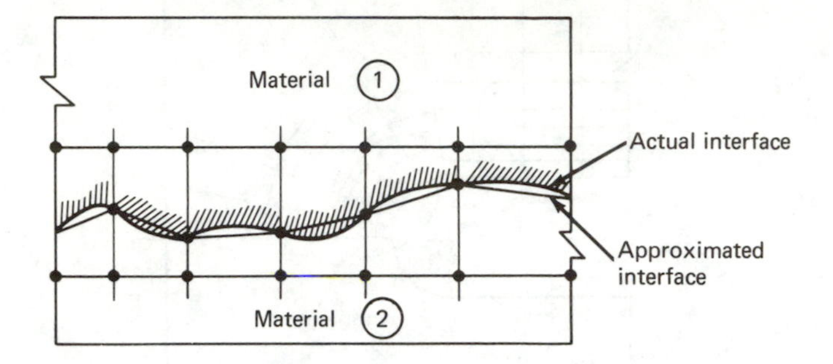

(a) Nonhomogeneous properties with curved interface

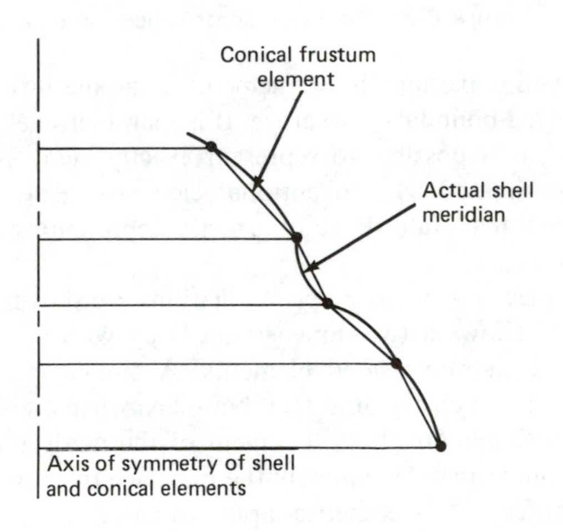

(b) Axisymmetric shell with arbitrarily curved meridian

Figure 6-4 Examples of geometric approximation.

aspect of the discretization process is the designation of a numbering system for the nodes and elements. Figure 6-5 shows two of the most common methods for numbering a two-dimensional mesh. The extension to a three-dimensional mesh is not difficult. In the first method, Figure 6-5(a), the nodes are numbered consecutively from left to right (or from bottom to top) for the right-handed coordinate system. A separate, similar numbering system is used for the elements. Additional data must be tabulated which gives the node numbers for each particular element. For example, the nodes of element 3 are 3, 4, 8, and 9. In the second method, Figure 6-5(b), each node and element is identified by an ordered pair of positive integers or indices. In the case of the

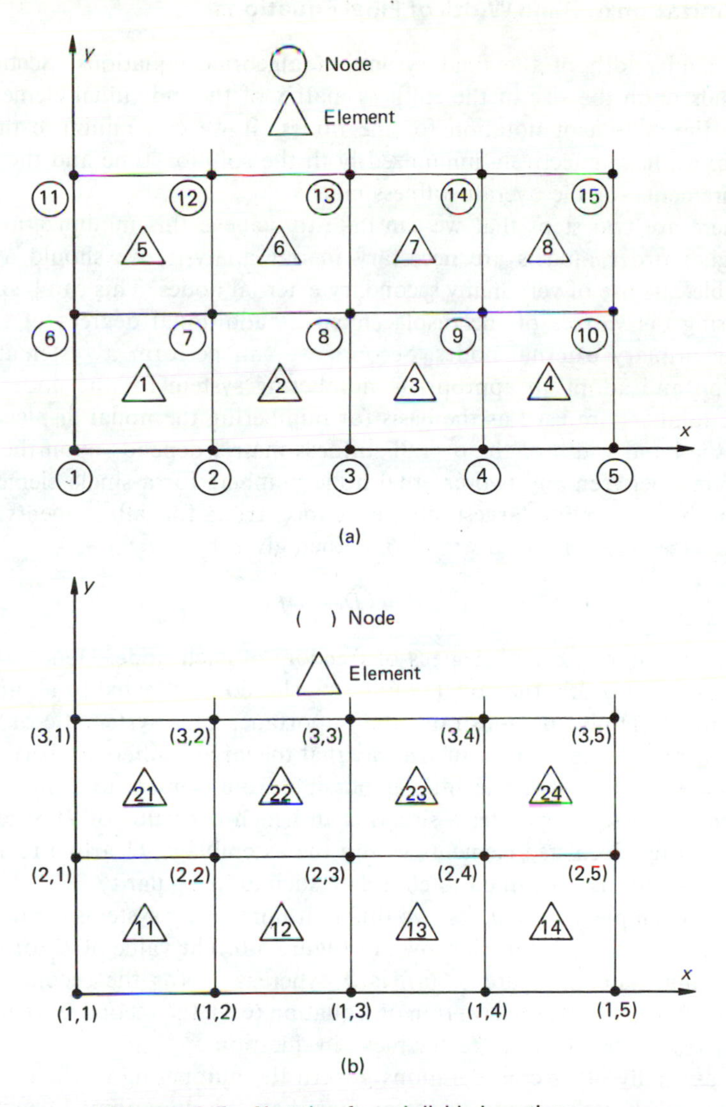

Figure 6-5 Notation for subdivided continua.

nodes, the first number designates the row number and the second the column number. The element indices are similarly assigned, and have the same value as the number of its south-west node, that is, the node closest to the origin. Both schemes have been successfully used and each has its advantages and disadvantages.[2,3] Since the first scheme is simple and is the most widely employed, we shall henceforth utilize this approach.

Minimization of Band Width of Final Equations

The band width of the final system of algebraic equations, Section 2-2, depends upon the size of the stiffness matrix of the individual elements and upon the system of notation for the nodes. If we can minimize the band width, we have effectively minimized both the solution time and the storage requirements for the overall stiffness matrix.

There are two steps that we can take to achieve this minimization. First if higher order models are necessary in our analysis, we should avoid, if possible, the use of very many secondary external nodes. This can be done by choosing derivatives of the displacement as additional degrees of freedom at the primary external nodes. Second, we can perform a systematic subdivision and adopt an appropriate numbering system for the nodes. If the node numbers are used as the basis for numbering the nodal displacements, then the band width of the overall stiffness matrix depends upon the largest difference between any two external node numbers for a single element. Let D be the maximum largest difference occurring for all elements of the assemblage. The semiband width, B, is then given by

$$B = (D + 1)f \tag{6.1}$$

where f is the number of degrees of freedom at each node. Hence, to minimize the band width, the nodal numbering should be selected to minimize D. Equation (6.1) also demonstrates the importance of a systematic or regular subdivision into elements in such a way that the largest difference between any two node numbers varies as little as possible from element to element. If the numbering system generates a situation in which the value of D is governed by only a few localized elements within the assemblage, clearly a revision of the subdivision is warranted to eliminate such an irregularity.

As an example, consider the two different numbering systems for the nodes of a simple rectangular mesh shown in Figure 6-6. The value of D for the first numbering system, Figure 6-6(a), is 9, whereas D for the second system, Figure 6-6(b), is 6. By the criterion of Equation (6.1), the second system clearly minimizes the band width for the mesh in question.

Occasionally other considerations govern the numbering of the nodes and we must work with a band width greater than the minimum. For example, there are some cases of nonhomogeneous problems for which the minimum band width would give an ill-conditioned set of equations. Generally we can expect little difficulty in solving the set of the symmetric algebraic equations if the assemblage stiffness matrix is *diagonally predominant*, that is, if the coefficients of largest magnitude occur on or near the principal diagonal. The axisymmetric analysis of a concrete pile in soft clay is one example of a case which may require a labeling system in which nodes are numbered in the

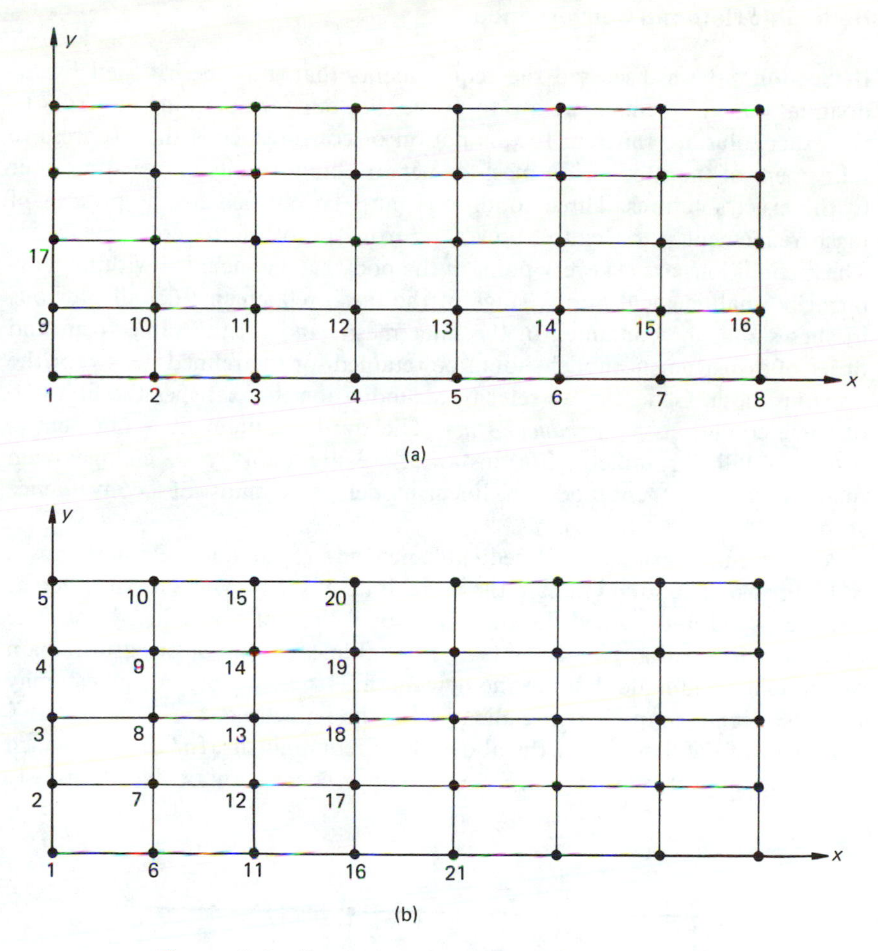

Figure 6-6 Numbering of nodes to reduce band width.

vertical direction to obtain a diagonally predominant matrix. Unfortunately, such a numbering system also gives a large band width.

If we utilize a wavefront method of processing our equations for solution (Section 2-2), a criterion different from minimization of the band width may govern the numbering of the nodes. In fact, one of the advantages of the wavefront technique is that it can overcome the storage inefficiencies of a mesh poorly labeled with respect to band width. However, a numbering system that does minimize the band width usually does not severely affect the efficiency of the wavefront method and may also produce a favorable, although not a minimum, wavefront size.[4]

Reducible Nets and Convergence

In Section 5-3 we discussed the requirements that must be satisfied by individual finite elements if we are to prove that our formulation converges to the exact solution. Inherent to this notion of convergence is the progressive refinement of the subdivision mesh or net to obtain results closer and closer to the exact solutions. Three conditions must be satisfied by the process of mesh refinement in order to allow a rigorous proof of the convergence.[5,6] These conditions are (1) every point in the body can be included within an arbitrarily small element at any stage of the mesh refinement; (2) all previous meshes should be contained in the finer mesh; and (3) the same form and order of displacement model should be retained for the refined net as for the previous mesh. Generally we refer to a subdivision that satisfies the first two of these conditions as a *reducible net*. The third requirement is one that is usually fulfilled routinely. For instance, we are unlikely to change from quadratic displacement models to linear models in the midst of a convergence study.

An elementary example of a reducible net for a rectangular region is shown in Figure 6-7. The solid lines are the nodal lines of the previous mesh, whereas both the solid and dashed lines are the nodal lines of the reduced net. The key point here is that nodal points and nodal lines in the former grid remain nodal points and nodal lines in the new mesh. Hence, the previous mesh may be considered a simplified specialization of the refined net.

In view of the first two of the above three requirements, for cases in which geometric approximation is required, convergence cannot be rigorously

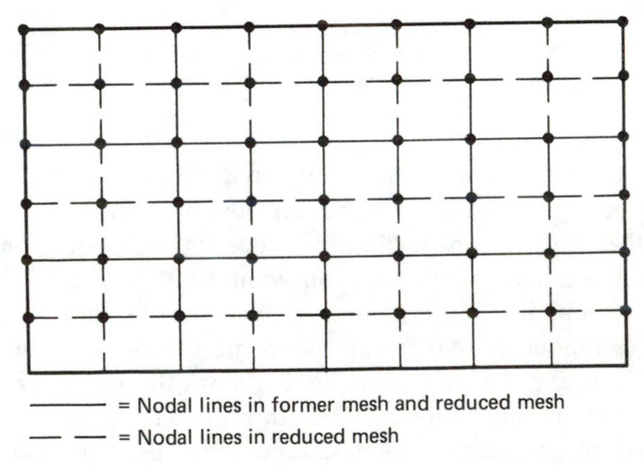

Figure 6-7 Reducible net for rectangular region.

proved. However, experience has shown that discretizations involving geometric approximation do yield useful engineering solutions.

If we perform a finite element analysis of a difficult equilibrium or eigenvalue problem, a single solution will not be sufficient to prove convergence. We must repeat the solution process using a reduced net and compare the two solutions. If the difference is appreciable, we may not be approaching the exact solution for the problem, and further subdivision may be required. A study of the two initial solutions may also give us some indications about the necessity for local refinement of the mesh in regions of steep gradients of the unknowns. For a propagation problem the same type of convergence investigation is required, but we must consider in addition the effect of changing the time increment.

Example of a Reducible Net

Let us consider a simple example of the pure bending of a beam, Figure 6-8(a). The beam is subjected to a bending stress distribution with a maximum intensity of 1000 psi as shown in Figure 6-8(a). Because of symmetry and antisymmetry about the y and x axes, respectively, only the shaded quadrant of the beam is discretized. Sides $0y$ and $0x$ are both restrained against horizontal movement. Figure 6-8(b) shows three different meshes, I, II, and III. The meshes are refined progressively from I through III, and the conditions for reducible net are observed.

A finite element program for plane stress (Appendix I) was used to obtain solutions for the three cases. This program utilizes quadrilateral finite elements composed of four CST's (Example 6-2), that is, a linear displacement model is used for each triangular element.

Table 6-1 shows the finite element solutions in comparison with the exact solution. The solution exhibits obvious improvement with progressive

TABLE 6-1 RESULTS FOR REDUCIBLE NET

Case	*No. of Nodes*	*No. of Elements*	*Displacements in inches* ($\times 10^{-4}$)			
			Point A (4.5, 3.0)		*Point B* (2.25, 1.5)	
			u	v	u	v
I	9	4	1.361444	−1.161280	0.3482609	−0.2936806
II	25	16	1.455171	−1.239908	0.3679722	−0.3123025
III	81	64	1.486224	−1.264569	0.3732400	−0.3171139
Exact, equation (6.2)			1.500000	−1.275000	0.3750000	−0.3187500

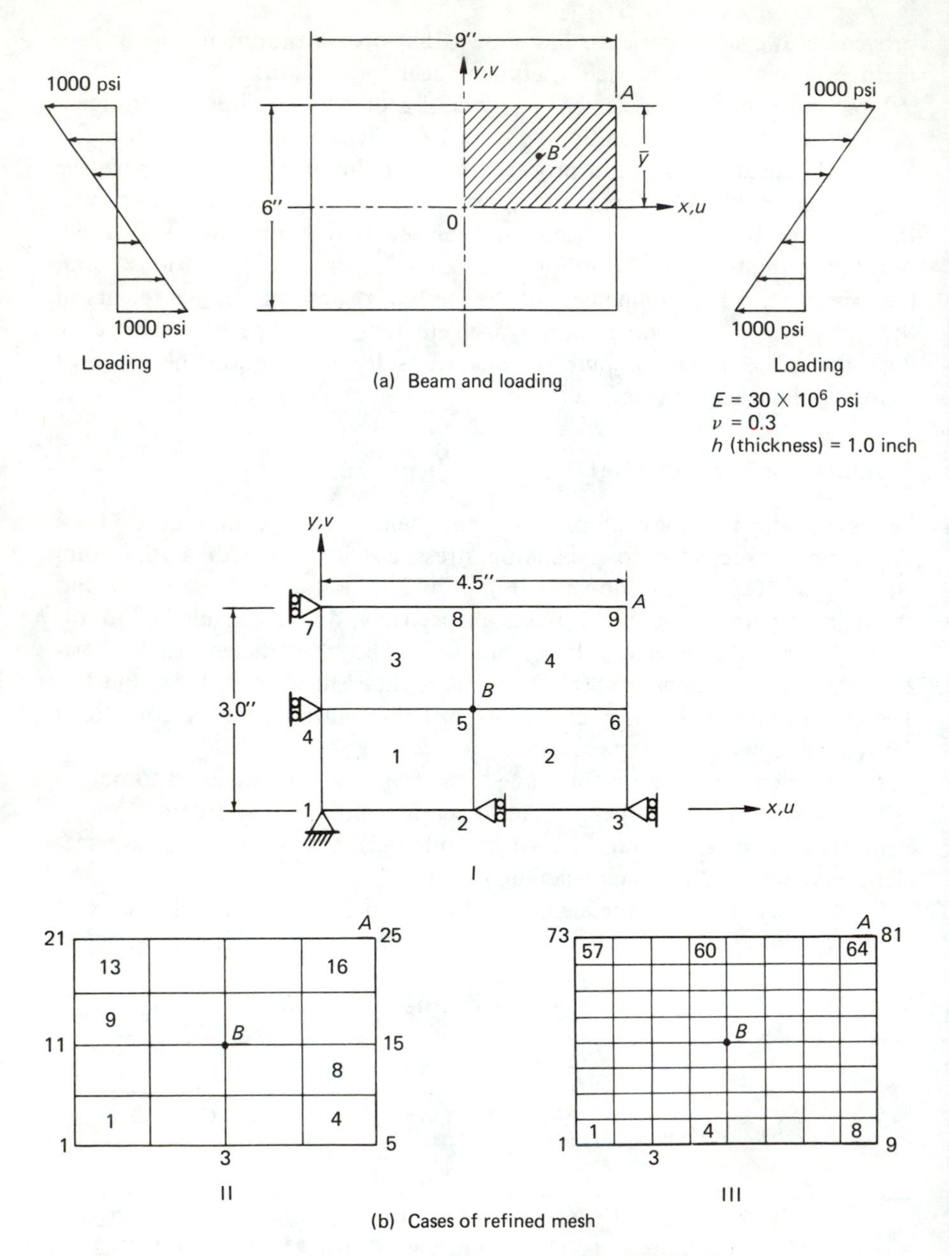

(a) Beam and loading

(b) Cases of refined mesh

Figure 6-8 Beam bending problem: reducible net.

refinement of the mesh. The exact solution was obtained from the following formulae:[7]

$$u = \frac{p}{E\bar{y}} xy$$
$$v = -\frac{p}{2E\bar{y}} (x^2 + \nu y^2) \qquad (6.2)$$

where $\bar{y}$ is the distance from the neutral axis to the top fiber.

Effects of Element Aspect Ratio

Another characteristic of the discretization that affects a finite element solution is the *aspect ratio* of the elements. The aspect ratio describes the shape of the element in the assemblage. For two-dimensional elements, this parameter is conveniently defined as the ratio of the largest dimension of the element to the smallest dimension.

The optimum aspect ratio at any location within the grid depends largely upon the difference in rate of change of displacements in different directions. If the displacements vary at about the same rate in each direction, the closer the aspect ratio to unity, the better the quality of the solution. In other words, generally we wish to avoid long narrow elements.

For an illustration, we consider the beam bending example in Figure 6-8. The problem has been solved for the five different meshes with rectangular elements shown in Figure 6-9(a). In selecting these meshes we retain about the same number of nodes and elements in order to isolate the effects of the changing aspect ratio. Table 6-2 compares the five finite element solutions

TABLE 6-2 COMPARISON OF RESULTS: ASPECT RATIO

Case	*Aspect Ratio*	*No. of Nodes*	*No. of Elements*	*Displacement in inches* ($\times 10^{-4}$)			
				Point A (4.5, 3.0)		*Point C* (4.5, 0.0)	
				u	v	u	v
I	8.00	26	12	1.296567	−1.102082	0.0	−0.9724254
II	2.00	21	12	1.442210	−1.227177	0.0	−1.091019
III	1.125	20	12	1.452951	−1.237485	0.0	−1.098896
IV	2.00	20	12	1.427559	−1.218161	0.0	−1.083542
V	4.5	21	12	1.354237	−1.159768	0.0	−1.033323
Exact, equation (6.2)				1.500000	−1.275000	0.0	−1.125000

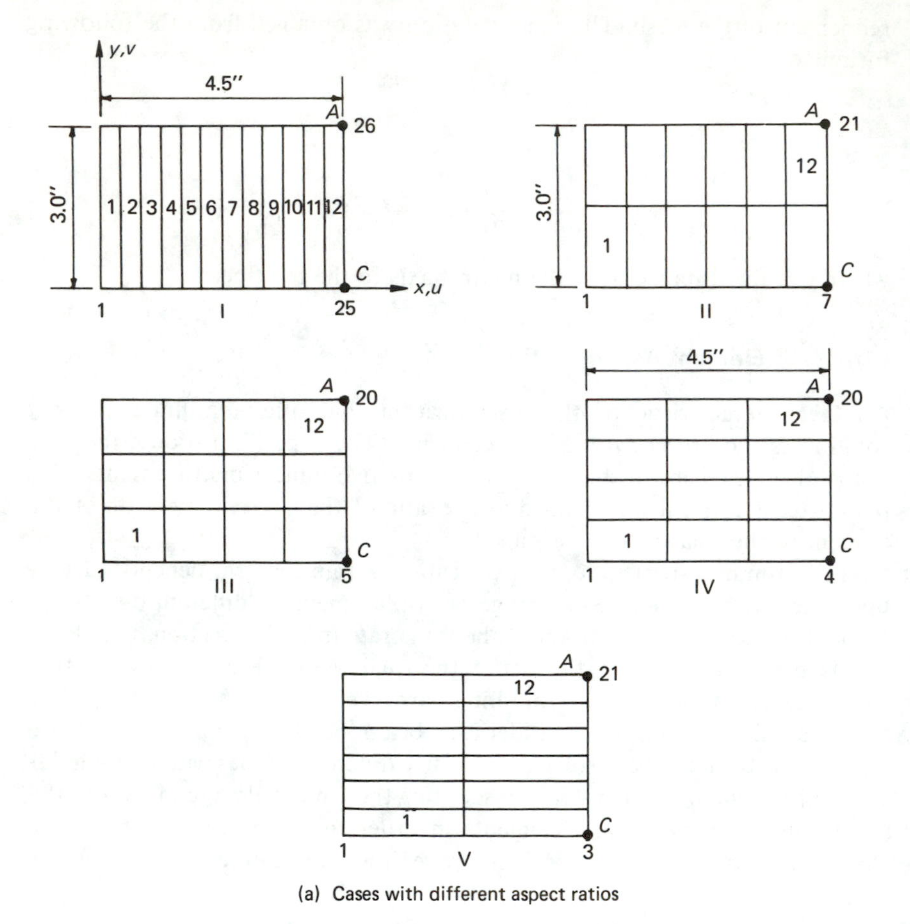

(a) Cases with different aspect ratios

Figure 6-9 Effects of aspect ratio.

with the exact solution determined from equation (6.2). Figure 6-9(b) shows a plot of the inaccuracy of the solution for the horizontal displacement of the point $A(4.5'', 3.0'')$ versus the aspect ratio. It can be seen that the aspect ratios nearer to unity yield better solutions.

Discretization of Very Large Bodies

In some problems the number of finite elements required may be so large that available computer facilities cannot conveniently accomodate the problem. In such cases the engineer can resort to one of two procedures to overcome this difficulty: the coarse-to-fine subdivision method, and the substructure method.

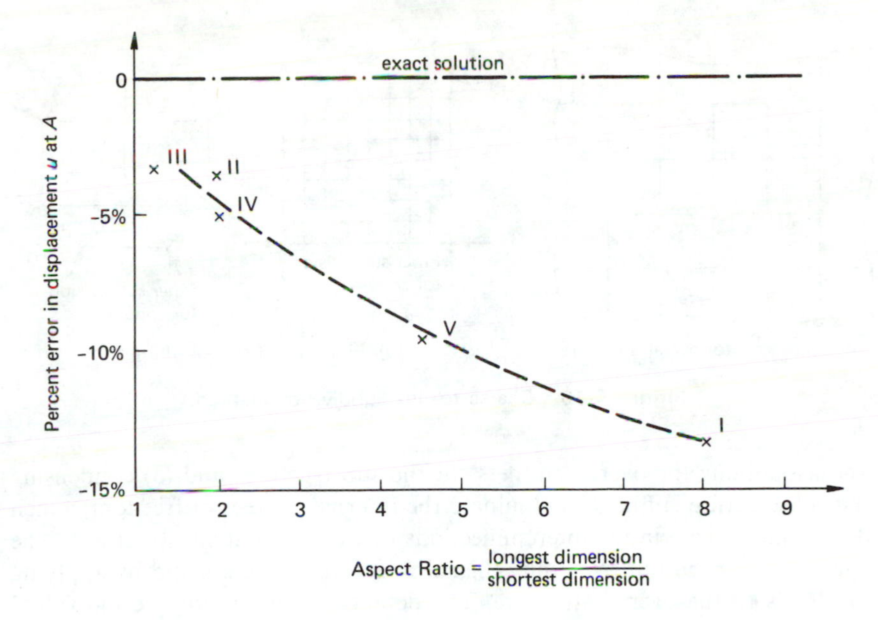

(b) Inaccuracy of solution as a function of aspect ratio

Figure 6-9 (*continued*)

Two steps are involved in the *coarse-to-fine subdivision method.* First, a large continuum is subdivided into a coarse mesh, Figure 6-10(a), consisting of a number of elements that can be accommodated conveniently. A finite element analysis is performed to obtain the displacements and stresses. Second, zones of particular interest, for example zones A and B in Figure 6-10(b), are isolated and subdivided into finer meshes. Loads and/or displacements resulting from the first analysis are now used as input to the second finite element analysis, which provides the detailed results required. An example of a zone of special interest is a region of stress concentration such as a notch or hole.

The *substructure method* is usually applied to such large structures as aircraft frames, multistory building frames and ships. Each of these can be considered to consist of a number of major components or substructures. For example, an aircraft frame might be considered to consist of wing, tail, and fuselage components. The various components or substructures may be treated as if they were complicated "finite elements," interconnected at external nodes to form the overall structure. The stiffness matrix for each substructure is determined by (1) subdividing the substructure into a number of smaller, simple finite elements; (2) computing the simple element stiffnesses

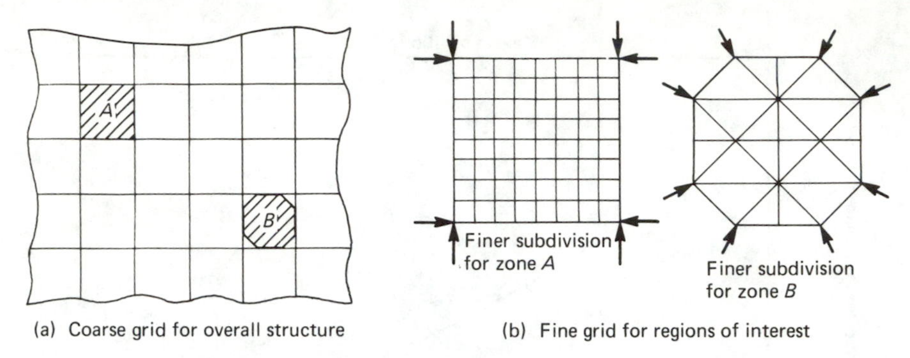

(a) Coarse grid for overall structure

(b) Fine grid for regions of interest

Figure 6-10 Coarse-to-fine subdivision method.

and assembling the overall stiffness for the substructure; and (3) condensing the substructure stiffness to eliminate the internal degrees of freedom, which do not participate in the interconnections of the different substructures. The analysis is carried out by adding the substructure stiffnesses and by applying the loads to the overall structure. The detailed solutions for the individual substructures are then obtained by performing finite element analyses for each of them with input loads and/or displacements derived from the analysis of the parent structure. An example of a substructure used in analysis of a ship is given in Chapter 9.

Finite Representation of Infinite Bodies

In many problems, such as the analysis of plates or shells, the boundaries of the structure or body are clearly defined. Hence the entire structure is included in the finite element assemblage. However, some classes of problems involve continua for which all the boundaries are not defined. The most common of these are certain plane strain problems, and soil and foundation problems. In particular, the latter type of problem is typically concerned with a semi-infinite soil mass. Some examples of finite element representations of semi-infinite soil masses are shown in Figures 5-3 and 6-11.

Because we cannot include an infinite body in our assemblage, we must limit it or make it finite. Fortunately this is feasible, since the effect of the loading (or disturbance) decreases with increasing distance from the point of its application. It is possible to determine the significant extent of the medium to be modeled by a trial and error procedure which varies the extent and computes the resulting effect upon the displacement solutions. Alternatively, we may assimilate the experience gained by other investigators considering similar problems.

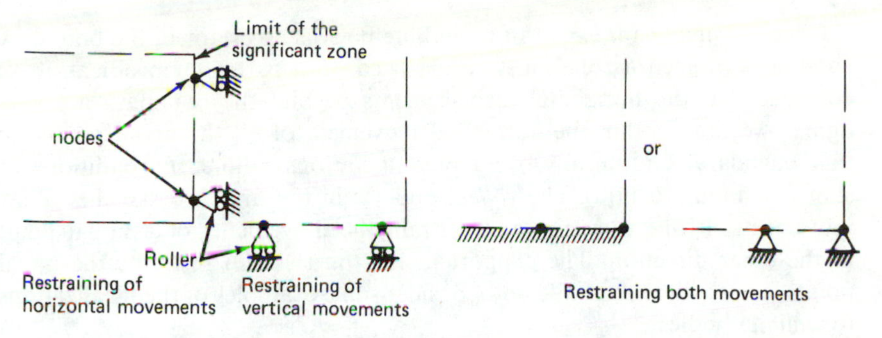

(a) Idealization of boundaries of infinite bodies

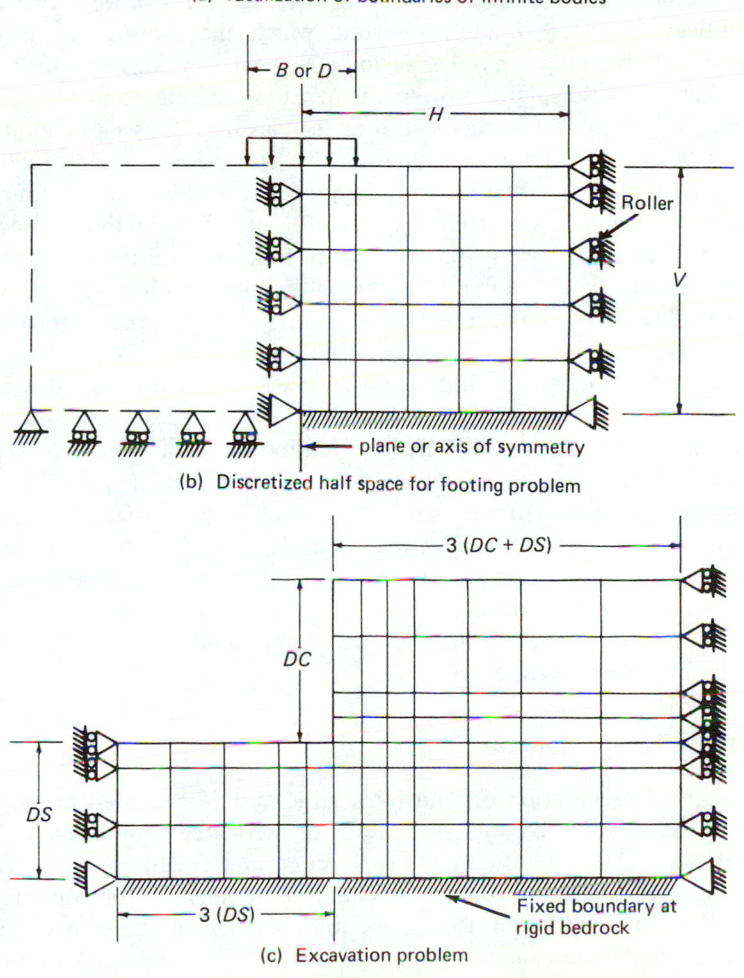

(b) Discretized half space for footing problem

(c) Excavation problem

Figure 6-11 Finite representation of infinite bodies.[9,10]

Once the significant extent of an infinite body is ascertained, the boundary conditions of such a zone must be idealized. In a particular problem, if we conclude that the horizontal displacements beyond the boundary are negligible, we can restrain the horizontal movement of all the nodal points on that boundary. Graphically, we represent various boundary conditions as shown in Figure 6-11(a). The *roller* supports in the first two sketches allow movements in one direction and restrain the movements of a nodal point in the other direction. The support in the third sketch restrains the nodal points in both directions. We now consider some examples of the idealizations for infinite bodies.

For the circular or rectangular footing in Figure 6-11(b), we may ascertain the critical distances H and V beyond which the footing has negligible influence by obtaining a number of finite element solutions in which H and V are varied. These solutions are compared to each other and to the results of simplified analytical methods such as the classical Boussinesq solution. A number of investigators have done this.[8,9,10] For a homogeneous soil mass, the influence of the footing becomes insignificant if H is taken as four to six times the width B or diameter D of the footing. The minimum necessary value of V is ten to twelve times B or D. Side boundaries are usually restrained against horizontal movement or are placed on rollers as shown. If the problem is symmetric or axisymmetric, it is necessary to discretize the continuum only on one side of the center line. In this case, the center line is also restrained against horizontal displacements. Finally, the bottom boundary can be either completely fixed, or constrained only against vertical movement, as shown in the right and left portions of Figure 6-11(b), respectively. Total restraint is often used if the lower boundary is taken at the known location of a bedrock surface.

Figure 6-11(c) depicts an assemblage of elements for an excavation in a soil or rock mass. Here it has been found that the left boundary should be at a distance of about three times the thickness DS of the layer below the excavation. The distance to the right boundary should be about three times the total depth to bedrock, $DC + DS$.[10]

Automated Mesh Generation

With manual preparation of data for a large and complicated problem, the processes of subdivision and generation of error-free input may be much more costly and time-consuming than the computer execution of the problem itself. Many computer programs include some facility for automatically subdividing the continuum, for numbering the nodes and elements, and for computing nodal coordinates. However, these are not functions that can be assumed completely by the computer; hence, some type of man-machine interaction is necessary to permit the exercise of engineering judgment. A detailed discussion of automated mesh generation is beyond the scope of this work. Generally, the criteria above govern the process, whether it is performed by hand or by machine. Existing mesh generation techniques are rather individualized and, with a few exceptions,[11] not many have been formally published.

6-2 MESH REFINEMENT VS. HIGHER ORDER ELEMENTS

The accuracy of a finite element solution can be improved by one of two methods: either by refining the mesh, or by selecting higher order displacement models. The first method relies on an increased number of piecewise displacement models of simple form to represent a complex exact solution. On the other hand, the latter technique utilizes an improved polynomial approximation, illustrated schematically in Figure 5-9.

A simple practical example that indicates the success of either approach is shown in Figure 6-12. In this figure, the fraction of transverse shear taken by the core of a cantilever sandwich beam is shown as a function of the dimensionless length. This is essentially a second order effect in the behavior of

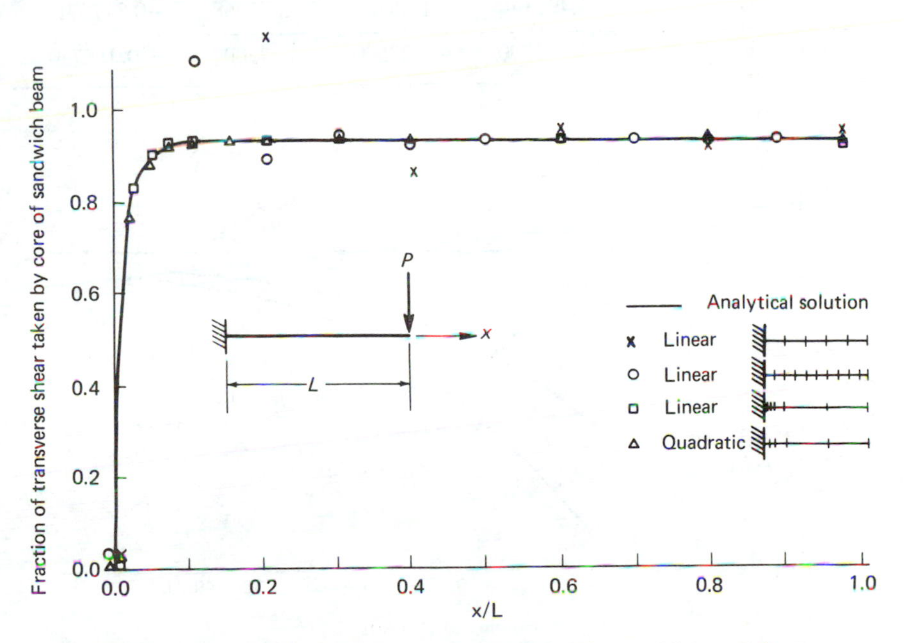

Figure 6-12 Mesh refinement vs. higher order elements for a cantilever sandwich beam (after Reference 12).

the beam; hence, the fraction varies very rapidly near the fixed support. The ×, ○ and □ symbols show the improvements achieved by successive and local mesh refinement using a linear model for the shear rotations. The △ symbols show the improvement attained by utilizing a quadratic model for this displacement parameter.

Another simple example is the beam bending problem, Figure 6-8. Previously, we obtained solutions for progressive mesh refinement for the quadrilateral element composed of four constant strain triangles (CST's). Solutions were also obtained for this problem with the same meshes, but with quadrilaterals having a quadratic internal displacement model and constrained to have linear variations along the edges. These results are tabulated in Table 6-3 and are plotted in Figure 6-13 in comparison with the

TABLE 6-3 BEAM BENDING WITH LINEAR STRAIN QUADRILATERAL (LSQ)

Case	*No. of Nodes*	*No. of Elements*	*Displacements in inches* ($\times 10^{-4}$)			
			Point A (4.5, 3.0)		*Point B* (2.25, 1.5)	
			u	v	u	v
I	9	4	1.417668	−1.207257	0.358015	−0.301722
II	25	16	1.474063	−1.254745	0.370798	−0.314339
III	81	64	1.492148	−1.269067	0.373968	−0.317571
Exact, equation (6.2)			1.500000	−1.275000	0.375000	−0.318750

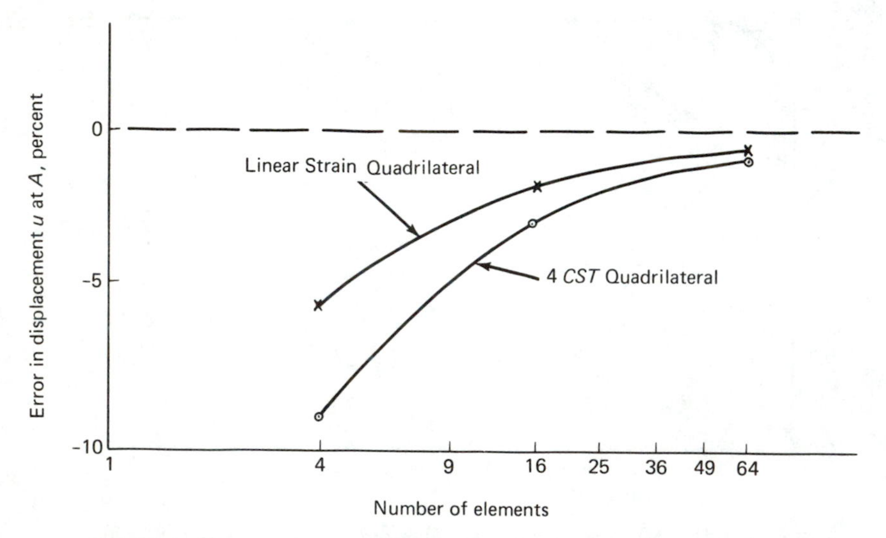

Figure 6-13 Mesh refinement vs. higher order elements: beam bending.

results from Table 6-1. We see that, although the difference between the two solutions for the refined mesh with 64 elements is quite small, the higher order model is the more accurate for each mesh.

Tradeoffs Between the Methods

The question that concerns engineers is which of the two methods for improving accuracy is the most economical. Unfortunately, this question has not been answered definitely in the general sense, although a few restricted and specific studies have been performed. There are tradeoffs in time and cost as well as accuracy that must be evaluated if the question is to be answered properly. In the final analysis, what should govern the selection of an approach is the degree of accuracy obtained per unit cost.[13] This cost should include the expense of preparing the input, the charges for digital computation time, and the amortization of software development.[14]

For the sake of discussion, consider a problem that has been solved with two different finite element discretizations, one of which employs displacement models of higher order than the other. Further assume that the discretizations are such that the same total number, N, of algebraic equations results for the assemblage from both discretizations. Then the higher order model will in general give the more accurate solution. However, the higher order model will also give a greater degree of connectivity; that is, f and/or D in equation (6.1) are increased, thus increasing the half band width, B, of the equation set. Because the number of operations in the decomposition process of the solution (Section 2-2) is proportional to NB^2, the equations for the higher order model will take longer to solve. In addition, although the higher order assemblage will involve a smaller number of elements, the computation time consumed in formulating the stiffness matrix and load vector for each element will be greater for the higher order models; therefore, there is a tradeoff affecting the total formulation and assembly time. When consideration of input preparation and software amortization costs are added to this picture, it is apparent that the issue is clouded indeed.

In one limited study, Clough compared the accuracy achieved per unit solution time for two hexahedral elements, utilizing linear and quadratic displacement models.[13] He found that the quadratic-model element was more efficient for the three-dimensional analysis of a slender cantilever beam (100″ span, 10″ depth), but that the linear model was preferable for short, deep cantilevers (10″ span, 10″ depth). Thus, in addition to the factors discussed in the preceding paragraph, we see that the type of problem considered affects the tradeoff between mesh refinement and higher order models.

Finally, the basic philosophy of the finite element method is that *simple* but relatively complete models are used to give approximate solutions to complicated problems. Nevertheless, higher order elements have been developed to improve the accuracy of the method, although the tradeoffs involved have not been definitively studied. One notable example of elements using higher order models is the TUBA family of plate bending elements, which employ fifth, sixth, and seventh order polynomials as displacement models.[15]

6-3 INTERCONNECTIONS AT NODES

The process of constructing the algebraic equations for the assemblage from the equations for the individual elements is routine. Nodal compatibility is used as the basis for this process. This simple requirement states that all elements adjacent to a particular node must have the same displacement at that node. Here our notion of displacement is a generalized one which includes translations and rotations. For higher order models, it may also include strains, curvatures, and other derivatives of the translations with respect to the spatial coordinates.

From the physical point of view, the imposition of nodal compatibility represents the construction of the assemblage by rigidly joining the pieces or elements together at certain preselected joining points. Under the mathematical interpretation of the finite element method (Section 5-1) the compatibility criterion is the recognition of the nodes as the intersections among the various regions or zones in the assemblage. These intersections govern the interdependence of the patchwork of displacement models.

Assembly Rules

The rules that govern the assembly process correspond directly to the procedures used in the matrix analysis of framed structures and discrete networks. These rules can be summarized by the statement that, because the displacements are matched at the nodes, the loads and stiffnesses are added at these locations.

One way that we may derive a mathematical statement of the assembly rules is to use the variational concept of the finite element method. In Section 5-7 we noted that instead of applying the variational principal to the entire assemblage, we apply it to the individual elements. We will now apply it to the assemblage to obtain the assembly rules. Assume that the total number of elements in the assemblage is E and that N is the total number of nodal degrees of freedom, that is, the total number of equations to be solved for the assemblage. Further, assume that we know the stiffness matrix and load vector for each element and that the element load vectors include all the loading on the body. The nodal displacements, that is, the unknowns for the entire assemblage, may be written as an $N \times 1$ vector $\{r\}$. If we let a subscript e denote the element number, and if we use a node labeling system such as the one selected in Section 6-1, we can write the expanded element stiffnesses and loads as $N \times N$ square matrices $[K_e]$, and $N \times 1$ vectors $\{R_e\}$, respectively. These are constructed by inserting the known stiffness coefficients and loads in their proper locations and filling the remaining locations with zeroes. For example, the $N \times 1$ load vector, $\{R_e\}$, for the eth element, corresponding to the element load vector $\{Q_e\}$ of Chapter 5, is expressed as

$$\{R_e\}^{\text{T}} = [\{0\}^{\text{T}}\ \{0\}^{\text{T}}\ \ldots\ \{0\}^{\text{T}}\ \{Q_e\}^{\text{T}}\ \{0\}^{\text{T}}\ \ldots\ \{0\}^{\text{T}}]$$

The loads $\{Q_e\}$ are placed in positions in the total load vector $\{R_e\}$ corresponding to the position in $\{r\}$ of the nodal displacements for element e, $\{q_e\}$. A similar construction is performed for the stiffness. Therefore, the total potential for the element can be written

$$\Pi_e = \tfrac{1}{2}\underset{1\times n}{\{q_e\}^{\text{T}}}\underset{n\times n}{[k_e]}\underset{n\times 1}{\{q_e\}} - \underset{1\times n}{\{q_e\}^{\text{T}}}\underset{n\times 1}{\{Q_e\}} = \tfrac{1}{2}\underset{1\times N}{\{r\}^{\text{T}}}\underset{N\times N}{[K_e]}\underset{N\times 1}{\{r\}} - \underset{1\times N}{\{r\}^{\text{T}}}\underset{N\times 1}{\{R_e\}}$$

where n and N are the number of nodal displacements for a single element and for the assemblage, respectively.

Now the total potential of the assemblage may be written as

$$\Pi = \sum_{e=1}^{E} \Pi_e = \frac{1}{2}\sum_{e=1}^{E}\{r\}^{\text{T}}[K_e]\{r\} - \sum_{e=1}^{E}\{r\}^{\text{T}}\{R_e\}$$

$$= \frac{1}{2}\{r\}^{\text{T}}\left\{\left(\sum_{e=1}^{E}[K_e]\right)\{r\} - \sum_{e=1}^{E}\{R_e\}\right\}$$

We now apply the principle of minimum potential energy to the assemblage as follows

$$\delta\Pi = \{\delta r\}^{\text{T}}\left\{\left(\sum_{e=1}^{E}[K_e]\right)\{r\} - \sum_{e=1}^{E}\{R_e\}\right\} = 0$$

This extremization condition is satisfied by causing the expression in braces to vanish, giving

$$\left(\sum_{e=1}^{E}[K_e]\right)\{r\} = \sum_{e=1}^{E}\{R_e\}$$

These are the *equilibrium equations for the assemblage*, which may be abbreviated in the following form

$$[K]\{r\} = \{R\} \tag{1.7}$$

From the above derivation, we can state the *assembly rules* for the assemblage stiffness matrix and load vector as follows:

$$[K] = \sum_{e=1}^{E} [K_e] \tag{6.3}$$

$$\{R\} = \sum_{e=1}^{E} \{R_e\} \tag{6.4}$$

6-4 EFFECT OF DISPLACEMENT MODELS ON INTERELEMENT COMPATIBILITY

For an assemblage of one-dimensional elements, the enforcement of nodal compatibility provides sufficient continuity among the components of the assemblage. However, when we consider a two- or three-dimensional body, compatibility at a discrete number of nodal points does not necessarily guarantee continuity of the body. In the two-dimensional case, we have interfaces between adjacent elements which are called *nodal lines*. Similarly in the three-dimensional case, the interfaces include *nodal planes*. Interelement compatibility or continuity at such interfaces is affected both by nodal compatibility and by the nature of the assumed displacement models.

Compatibility at Element Interfaces

If three conditions are met, displacement compatibility will occur at the interface between two typical elements in the assemblage. These three conditions are (1) the same isotropic displacement model must be used in both elements; (2) for each element the displacements on the interface must depend only upon the nodal displacements occurring on that interface; and (3) interelement nodal compatibility must be enforced. The last condition is normally satisfied by the standard assembly procedure. The first two conditions are not difficult to satisfy for most classes of problems, and they are usually included in the selection of a displacement model (Section 5-3) and in ensuring the possibility of a reducible net (Section 6-1).

Some simple examples for the two-dimensional case will illustrate the occurrence of interelement compatibility. The extension to the three-dimensional case is rudimentary. Figure (6-14a) shows an isometric view of a two-element assemblage. Linear and quadratic displacement models which are compatible at the interface are shown in Figures 6-14(b) and 6-14(c), respectively. In each case, the u displacements along the interface 2-5-3 are governed by the order of the model and by the magnitudes of the nodal displacements on the interface, u_2, u_5, and u_3. Specifically, in Figure 6-14(b)

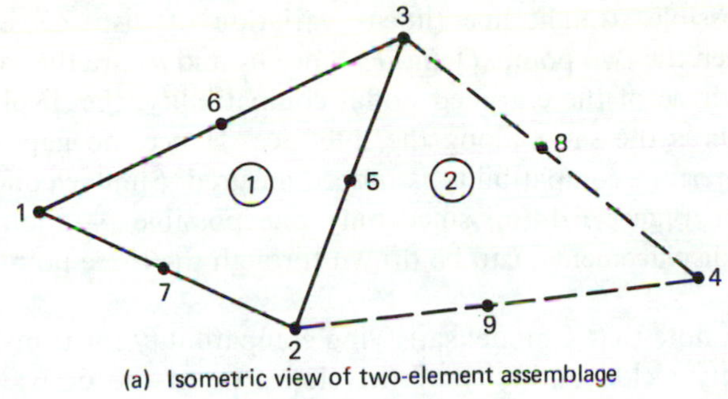

(a) Isometric view of two-element assemblage

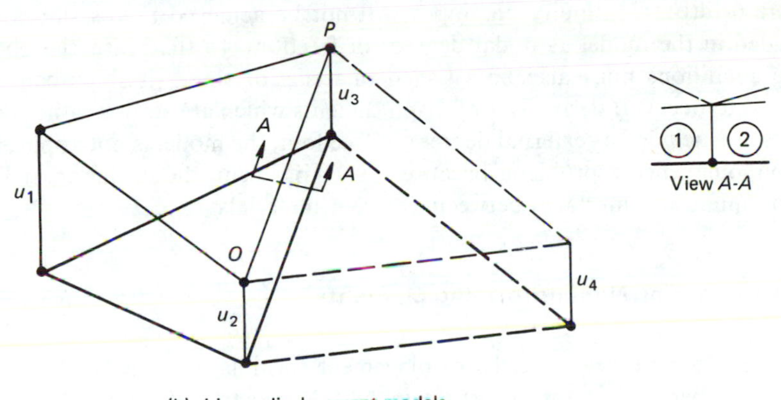

(b) Linear displacement models

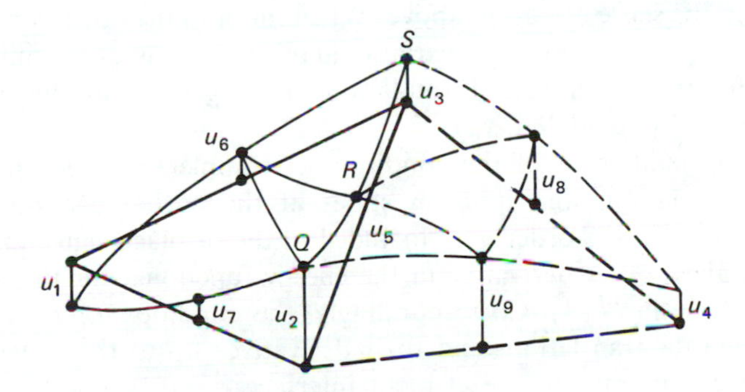

(c) Quadratic displacement models

Figure 6-14 Compatibility at element interfaces.

only one possible straight line (linear variation of displacements) can be drawn between the two points O and P. Since u_2 and u_3 are the same for each element by virtue of the enforced nodal compatibility, the displacement for both elements is the same along the interface. Hence, no gaps or overlaps occur and interface compatibility is indeed ensured. Similar conclusions can be drawn for Figure 6-14(c), since only one possible parabola (quadratic variation of displacements) can be drawn through the three points Q, R, and S.

We should note that a model satisfying compatibility for translations does not necessarily yield continuity of the slopes or of the derivatives of displacements across the element interface. This is shown by view *A-A* in Figure 6-14(b). Continuity on slopes will not be achieved unless slopes are included in the model as nodal degrees of freedom. In that case the above three conditions must also be satisfied in terms of slopes if slope compatibility is to occur. If derivatives of displacements which are strains rather than slopes are included as external degrees of freedom, the model is not applicable to nonhomogeneous problems because in such problems the strains are in fact discontinuous at interfaces between different materials.

Conforming and Nonconforming Elements

In Section 5-3 we saw that the displacement formulation can be shown to give an upper bound for the stiffness of the structure if conforming and complete elements are used. Among the requirements for a conforming element is that the displacements must be compatible between adjacent elements in the sense discussed above. When this and the other conditions are fulfilled, it is possible to use the variational interpretation of the finite element method to prove that the formulation converges monotonically to the minimum total potential energy.[5,6]

A rigorous statement of the condition of completeness requires that the uniform translations and uniform states of the partial derivatives of the translations up to the order n be included in the displacement model, where n is the highest order derivative in the energy functional for the variational principle, equation (4.1). Correspondingly, the condition of conformability requires that the translations and their derivatives up to the order $n - 1$ be finite and continuous on the element interfaces.[5] For example, in a plane strain formulation such as that detailed in Sample Problem 2 of Chapter 5, the highest order of derivatives in the potential energy functional, equation (4.12), is one. Hence, only the translational displacements need satisfy interelement compatibility. On the other hand, for the plate bending problem of Example 5-9, the energy formulation is written in terms of curvatures, so

$n = 2$ and both transverse displacement and slopes must be compatible at the nodes and interfaces.

We also saw in Section 5-3 that successful solutions can be obtained by use of nonconforming elements.[16] Of course such elements satisfy nodal compatibility by virtue of the application of the assembly rules. However, interface compatibility is not satisfied in general by such elements. Consequently, nonconforming elements are not consistent with the mathematical interpretation of the finite element method which enables a rigorous proof of an upper bound to the stiffness. Nevertheless, nonconforming elements which are more flexible than conforming elements are being used increasingly, because the former usually provide more accurate results for the same discretization. We shall see a detailed example of this in Section 9-7.

6-5 CONSTRUCTION OF STIFFNESS MATRIX AND LOADS FOR THE ASSEMBLAGE

The assembly rules given in Section 6-3 are the necessary and sufficient basis for the construction of the stiffness matrix and load vector for the assemblage. However, the details of the construction play a significant role in the efficiency of any finite element algorithm. In this section, we shall concentrate on the most widely used and efficient technique, the *direct stiffness method.*[17]

Transformation to Global Coordinates

Recall that a local coordinate system is defined for a particular element, whereas a global coordinate system refers to the entire assemblage. It is usually possible to adopt local displacement directions that coincide with the global coordinates. All the sample problems and examples presented in Chapter 5 follow this procedure. However, for some types of problems we may find it expedient to perform the analysis of the individual elements with a local displacement system which has directions different from those in the global system. In such cases, before we can construct our algebraic equations for the assemblage, we must transform our element stiffnesses and loads to a common frame of reference, the global coordinate system.†

If we adopt the notation that the subscripts ℓ and g signify local and global reference systems, respectively, we can write the relation between the local and global element displacements as follows:

$$\{u_\ell\} = [t]\{u_g\} \tag{6.5}$$

† One advantage of the isoparametric element approach is that we always use global displacements in the element formulation and hence avoid the necessity of transforming to global coordinates at a later time.

This is a simple *displacement transformation* involving the direction cosines which relate the local and global systems. Using equation (6.5), we may construct the transformation for the nodal displacements.

$$\{q_\ell\} = [T]\{q_g\} \tag{6.6}$$

Here the matrix $[T]$ is usually of the form

$$[T] = \begin{bmatrix} [t] & [0] & \dots & [0] \\ [0] & [t] & \dots & [0] \\ \cdot & \cdot & & \cdot \\ \cdot & \cdot & & \cdot \\ \cdot & \cdot & & \cdot \\ [0] & [0] & \dots & [t] \end{bmatrix}$$

where the number of matrices $[t]$ in $[T]$ equals the number of element nodes. If we now substitute equation (6.6) into equation (5.42) and apply the principle of minimum potential energy, we obtain the equilibrium equation in global coordinates

$$[k_g]\{q_g\} = \{Q_g\} \tag{6.7}$$

where

$$[k_g] = [T]^{\mathrm{T}}[k_\ell]\ [T] \tag{6.8}$$

$$\{Q_g\} = [T]^{\mathrm{T}}\{Q_\ell\} \tag{6.9}$$

Equations (6.8) and (6.9) can also be derived directly from the equilibrium equations, equation (5.44), by substituting equation (6.6) and premultiplying by $[T]^{-1}$. We note, however, that the transformation matrix $[T]$ is always an *orthogonal matrix*; that is, its inverse is equal to its transpose.

For a specific example of equations (6.5) and (6.6), consider Fig. 6-15. This figure shows a conical frustrum element, a type of element for which local coordinates s and t are commonly used. An assemblage of such elements is used to represent an axisymmetric thin shell, Figure 6-4(b). One useful set of global coordinates is the radial and axial directions shown, r and z. Then equation (6.5) may be written for element i as follows

$$\{u_\ell\} = \begin{Bmatrix} u_s \\ u_t \end{Bmatrix} = \begin{bmatrix} \cos\alpha_i & \sin\alpha_i \\ -\sin\alpha_i & \cos\alpha_i \end{bmatrix} \begin{Bmatrix} u_r \\ u_z \end{Bmatrix} = [t_i]\{u_g\}$$

Note that $[t_i]^{-1} = [t_i]^{\mathrm{T}}$, i.e., $[t]$ is orthogonal. Equation (6.6) for element i is

$$\begin{Bmatrix} u_{s,i} \\ u_{t,i} \\ u_{s,i+1} \\ u_{t,i+1} \end{Bmatrix} = \begin{bmatrix} [t] & [0] \\ [0] & [t] \end{bmatrix} \begin{Bmatrix} u_{r,i} \\ u_{z,i} \\ u_{r,i+1} \\ u_{z,i+1} \end{Bmatrix}$$

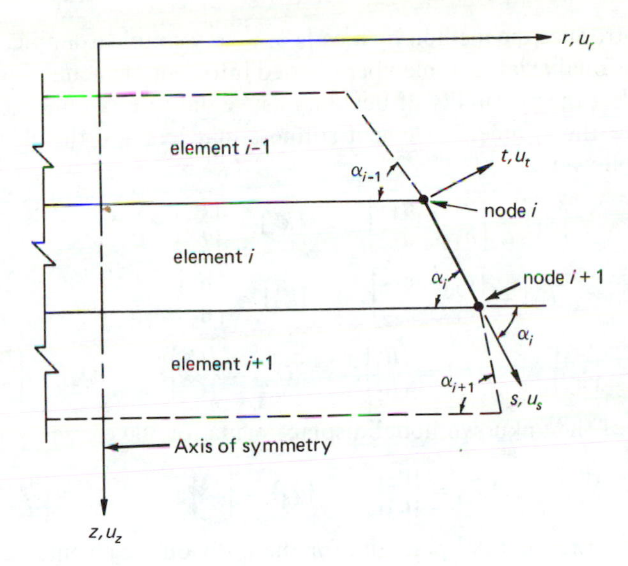

Figure 6-15 Global and local coordinates for axisymmetric shell.

So far, we have omitted the rotational degrees of freedom, that is the "slopes" θ_i and θ_{i+1}, which are essential displacement parameters for the shell bending problem; however, these rotations are unaffected by the transformation from local to global coordinates. Hence the complete form of equation (6.6) for this example is

$$\{q_\ell\} = \begin{Bmatrix} u_{s,i} \\ u_{t,i} \\ \theta_i \\ u_{s,i+1} \\ u_{t,i+1} \\ \theta_{i+1} \end{Bmatrix} = \begin{bmatrix} \cos\alpha_i & \sin\alpha_i & 0 & 0 & 0 & 0 \\ -\sin\alpha_i & \cos\alpha_i & 0 & 0 & 0 & 0 \\ 0 & 0 & 1 & 0 & 0 & 0 \\ 0 & 0 & 0 & \cos\alpha_i & \sin\alpha_i & 0 \\ 0 & 0 & 0 & -\sin\alpha_i & \cos\alpha_i & 0 \\ 0 & 0 & 0 & 0 & 0 & 1 \end{bmatrix} \begin{Bmatrix} u_{r,i} \\ u_{z,i} \\ \theta_i \\ u_{r,i+1} \\ u_{z,i+1} \\ \theta_{i+1} \end{Bmatrix}$$

$$= [T_i]\{q_g\}$$

We shall henceforth omit the subscripts g or ℓ from $[k]$, $\{q\}$, and $\{Q\}$ and assume that such nonsubscripted symbols refer to element characteristics which are already expressed in global coordinates.

Theory of the Direct Stiffness Method

The direct stiffness method is employed almost universally for assembling the algebraic equations in finite element applications. Its popularity arises from the storage economy of the method and the ease of coding the technique for the computer.

We can introduce the method by means of a very simple example. Consider a one-dimensional axial-load member divided into four elements, Figure 6-16. We will neglect the possibility of buckling as we did in Example 5-1. Assume that we know the symmetric element stiffness matrices and the element load vectors as follows:

$$[k_1] = \begin{bmatrix} a_{11} & a_{12} \\ a_{12} & a_{22} \end{bmatrix}, \qquad [k_2] = \begin{bmatrix} b_{11} & b_{12} \\ b_{12} & b_{22} \end{bmatrix}$$

$$[k_3] = \begin{bmatrix} c_{11} & c_{12} \\ c_{12} & c_{22} \end{bmatrix}, \qquad [k_4] = \begin{bmatrix} d_{11} & d_{12} \\ d_{12} & d_{22} \end{bmatrix} \tag{6.10}$$

$$\{Q_1\} = \begin{Bmatrix} A_1 \\ A_2 \end{Bmatrix}, \qquad \{Q_2\} = \begin{Bmatrix} B_1 \\ B_2 \end{Bmatrix}, \qquad \{Q_3\} = \begin{Bmatrix} C_1 \\ C_2 \end{Bmatrix}, \qquad \{Q_4\} = \begin{Bmatrix} D_1 \\ D_2 \end{Bmatrix}$$

The vectors of the unknown nodal displacements for the elements are

$$\{q_1\} = \begin{Bmatrix} u_1 \\ u_2 \end{Bmatrix}, \qquad \{q_2\} = \begin{Bmatrix} u_2 \\ u_3 \end{Bmatrix}, \qquad \{q_3\} = \begin{Bmatrix} u_3 \\ u_4 \end{Bmatrix}, \qquad \{q_4\} = \begin{Bmatrix} u_4 \\ u_5 \end{Bmatrix} \tag{6.11}$$

Then we can write the total potential for the individual elements as follows

$$\begin{aligned} \Pi_1 &= \tfrac{1}{2}\{q_1\}^{\mathrm{T}}[k_1]\{q_1\} - \{q_1\}^{\mathrm{T}}\{Q_1\} \\ &= \tfrac{1}{2}(a_{11}u_1^2 + 2a_{12}u_1u_2 + a_{22}u_2^2) - A_1u_1 - A_2u_2 \end{aligned}$$

and so forth. The total potential for the assemblage is therefore given by the sum of the individual element potentials.

$$\begin{aligned} \Pi = \tfrac{1}{2}[a_{11}u_1^2 + a_{12}u_1u_2] - A_1u_1 \\ + \tfrac{1}{2}[a_{12}u_1u_2 + (a_{22} + b_{11})u_2^2 + b_{12}u_2u_3] - (A_2 + B_1)u_2 \\ + \tfrac{1}{2}[b_{12}u_2u_3 + (b_{22} + c_{11})u_3^2 + c_{12}u_3u_4] - (B_2 + C_1)u_3 \\ + \tfrac{1}{2}[c_{12}u_3u_4 + (c_{22} + d_{11})u_4^2 + d_{12}u_4u_5] - (C_2 + D_1)u_4 \\ + \tfrac{1}{2}[d_{12}u_4u_5 + d_{22}u_5^2] - D_2u_5 \end{aligned}$$

By applying the principle of minimum potential energy, we obtain

$$\frac{\partial \Pi}{\partial u_1} = a_{11}u_1 + a_{12}u_2 - A_1 = 0$$

$$\frac{\partial \Pi}{\partial u_2} = a_{12}u_1 + (a_{22} + b_{11})u_2 + b_{12}u_3 - (A_2 + B_1) = 0$$

$$\frac{\partial \Pi}{\partial u_3} = b_{12}u_2 + (b_{22} + c_{11})u_3 + c_{12}u_4 - (B_2 + C_1) = 0$$

$$\frac{\partial \Pi}{\partial u_4} = c_{12}u_3 + (c_{22} + d_{11})u_4 + d_{12}u_5 - (C_2 + D_1) = 0$$

$$\frac{\partial \Pi}{\partial u_5} = d_{12}u_4 + d_{22}u_5 - D_2 = 0$$

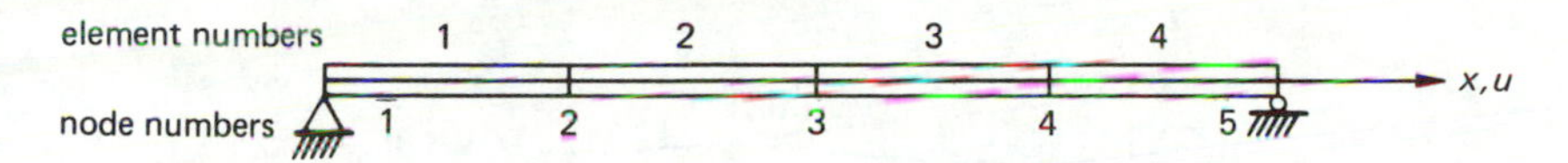

Figure 6-16 One-dimensional axial load member.

The above equilibrium equations for the assemblage can be written in matrix form as

$$\begin{bmatrix} a_{11} & a_{12} & 0 & 0 & 0 \\ a_{12} & (a_{22} + b_{11}) & b_{12} & 0 & 0 \\ 0 & b_{12} & (b_{22} + c_{11}) & c_{12} & 0 \\ 0 & 0 & c_{12} & (c_{22} + d_{11}) & d_{12} \\ 0 & 0 & 0 & d_{12} & d_{22} \end{bmatrix} \begin{Bmatrix} u_1 \\ u_2 \\ u_3 \\ u_4 \\ u_5 \end{Bmatrix} = \begin{Bmatrix} A_1 \\ (A_2 + B_1) \\ (C_1 + B_2) \\ (C_2 + D_1) \\ D_2 \end{Bmatrix} \tag{6.12}$$

or

$$[K]\{r\} = \{R\} \tag{1.7}$$

The direct stiffness method of assembling the algebraic equations becomes apparent by comparing the equations for the assemblage, equation (6.12), with those for the individual elements, equations (6.10). We see that it is possible to add directly the individual stiffnesses and loads to locations in the overall matrices $[K]$ and $\{R\}$, in conformity with the requirement of one-to-one correspondence between the nodes of the elements and those of the assemblage. The result in equation (6.12) is also consistent with the assembly rules, equations (6.3) and (6.4). By recognizing the possibility of this direct addition, we avoid the necessity for either the reapplication of the variational principle, as in the example above, or the individual expansion of nth order matrices to N^{th} order, as implied in Section 6-3.

The procedure sketched above is straightforward and can easily be extended to two- and three-dimensional assemblages. We will now consider a more general example to clarify the method further. Here we examine the two-dimensional assemblage shown in Figure 6-17. Although this assemblage is much smaller than might be used in a practical problem, the numbering shown in Figure 6-17 is like that which would be used in an actual problem. In this representation the directions of the local and global displacements are the same, so no coordinate transformations are required. The stiffness matrix and load vector for two selected elements is shown schematically in Figure

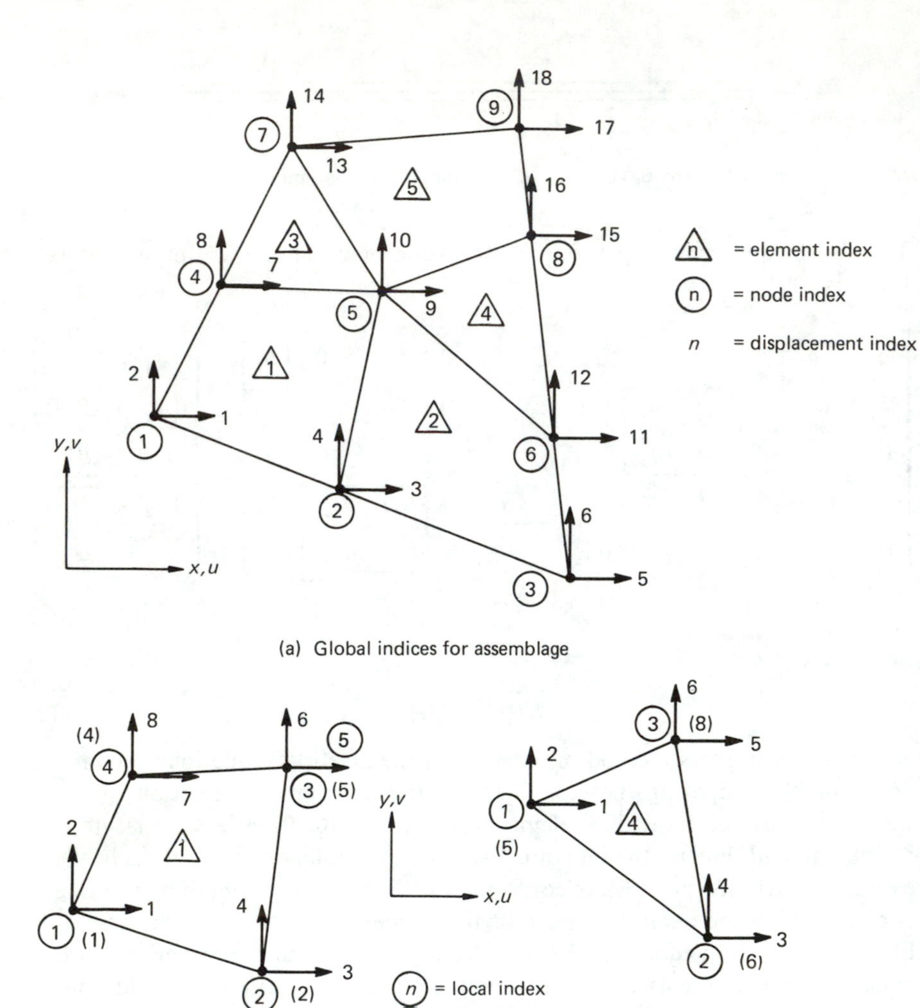

Figure 6-17 Two-dimensional assemblage to illustrate direct stiffness method (Figure 6-18).

6-18(a), whereas Figure 6-18(b) shows these quantities for the assemblage. We note that because there are two displacement degrees of freedom at each node, we can imagine the stiffness matrices to be partitioned into 2×2 submatrices and the load vector into 2×1 subvectors. The different types of notation for the components of the element submatrices and load subvectors in Figure 6-18(a) indicate the corresponding locations in the assemblage

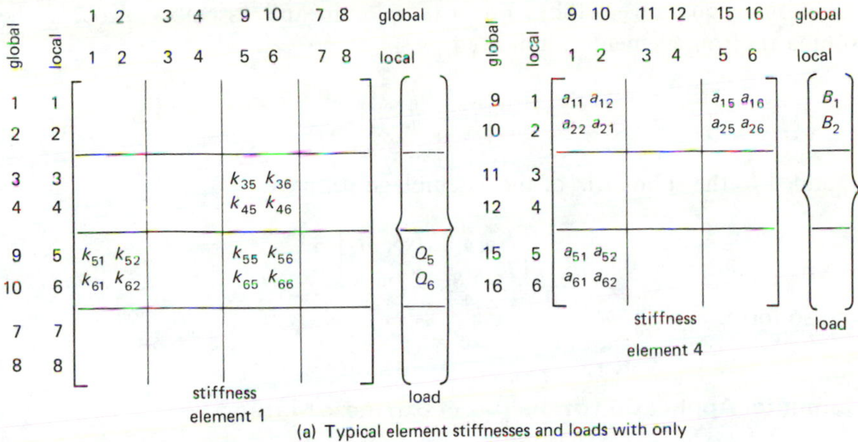

(a) Typical element stiffnesses and loads with only selected nonzero entries shown

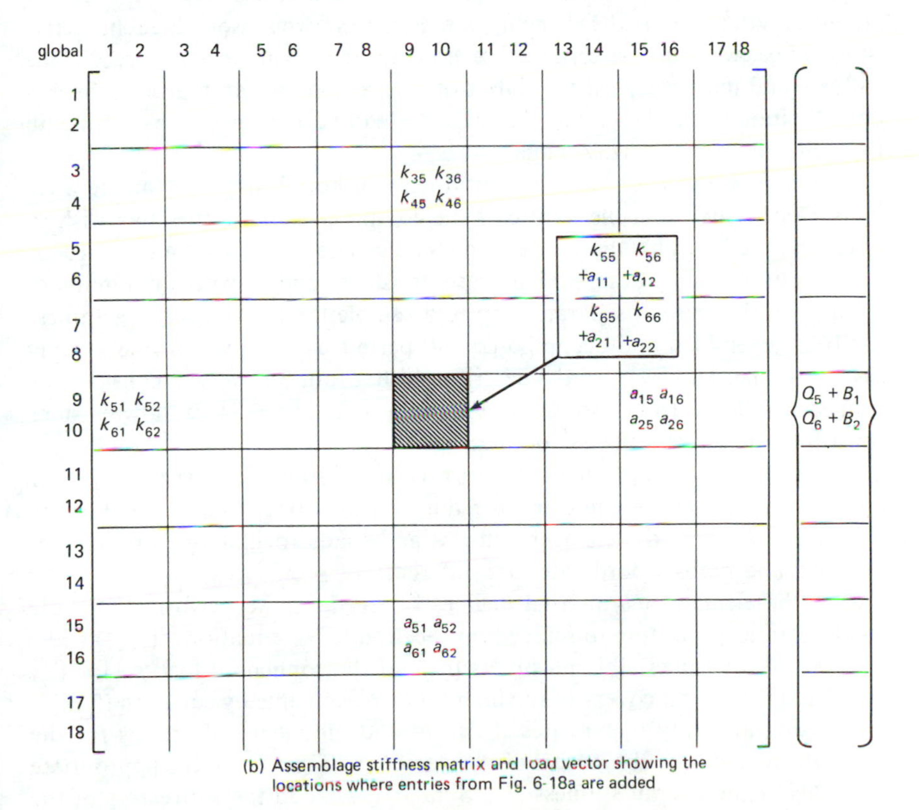

(b) Assemblage stiffness matrix and load vector showing the locations where entries from Fig. 6-18a are added

Figure 6-18 Schematic representation of addition in direct stiffness method.

equations, equation (6.18b). For example, in the assembly process the submatrix from element 1, indicated by

$$\begin{bmatrix} k_{55} & k_{56} \\ k_{65} & k_{66} \end{bmatrix}$$

is added to the submatrix of the assemblage stiffness:

$$\begin{bmatrix} K_{9,9} & K_{9,10} \\ K_{10,9} & K_{10,10} \end{bmatrix}$$

and so forth.

Computer Application of the Direct Stiffness Method

We will now outline some of the practical aspects of the application of the direct stiffness method. The first feature of the method that requires our attention is that we *add* the element characteristics to the assemblage characteristics. Thus, we must be certain that no extraneous values are stored in the stiffness and load arrays at the outset of a particular solution process. This is done by initializing the global stiffness and load vector, that is, by setting all the elements of these arrays equal to zero.

The main portion of the direct stiffness method consists of an iterative loop. Depending upon the nature of the computer program we may wish to have separate loops for the stiffness and load generation. This is advantageous if, for example, many different load cases must be handled with a minimum of storage. In this case, a separate loop to assemble the loads after the stiffness has been assembled and decomposed will permit us to solve for one loading case at a time. However, in the description that follows, we will consider the stiffness and loads to be assembled within the same loop. The following steps constitute the direct stiffness routine:

1. We compute the element stiffness matrix $[k]$ and load vector $\{Q\}$ for one element. Generally, these are stored in an $n \times n$ array and an $n \times 1$ array.

2. We transform the element stiffness and loads to global coordinates if the local and global coordinates are different.

3. If the element has internal degrees of freedom, we perform the static condensation procedure discussed in Section 5-9, equations (5.64). The matrices of multipliers $[c]$ and pivots $[d]$ and the condensed load vector $\{f\}$ are stored for later recovery of the internal displacements by equation (5.64g).

4. Using an array that relates the local and global nodal indices for the element, we add the submatrices of the element stiffness $[k]$ to the appropriate locations of the overall stiffness $[K]$. Similarly, we add the subvectors of the element loads $\{Q\}$ to the proper positions of the overall load vector $\{R\}$.

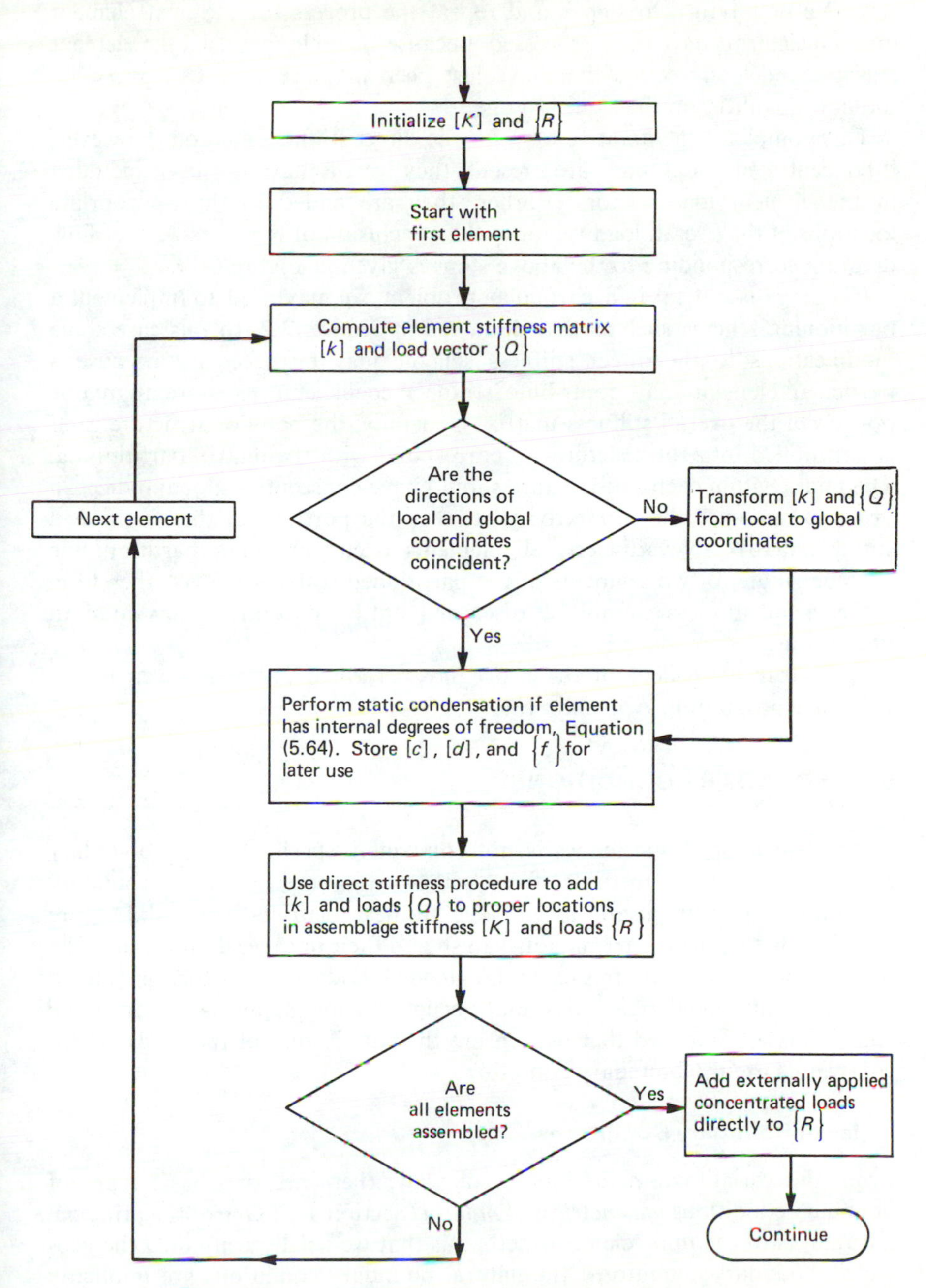

Figure 6-19 Iterative loop for direct stiffness procedure.

5. We now return to step 1 and repeat the process for the next element until all elements have been processed. Because we no longer need the element stiffness and load vectors that have just been processed, we can overwrite the new quantities in the same storage.

This completes the iterative loop for the direct stiffness method. However, if concentrated nodal loads are present, they usually have not been included in the element load vectors. Rather, they are added to the appropriate locations of the overall load vector at the conclusion of the loop. A brief flow diagram corresponding to the above steps is given in Figure 6-19.

If storage is critical in a particular problem, we may need to implement a partitioning scheme such as that discussed in Section 2-2. In this case some modifications to the direct stiffness scheme may be necessary because a particular element may contribute stiffness coefficients to more than one portion of the overall stiffness matrix. Generally, the body or structure itself is partitioned into substructures to correspond with the matrix partitioning. The nodes within each substructure should have consecutive global indices in order to minimize the connectivity between the portions of the partitioned stiffness matrix. Nevertheless, all elements occurring on a partition line will contribute to two segments of the partitioned stiffness matrix; therefore, the element stiffness cannot be discarded until the contributions to both portions are recorded.

A Fortran equivalent of the procedure described above is given in the computer program in Appendix I.

6-6 BOUNDARY CONDITIONS

A problem in solid mechanics is not completely specified unless boundary conditions are prescribed. In fact, without the imposition of boundary conditions, the element and assemblage stiffness matrices, $[k]$ and $[K]$, are *singular*; that is, their determinants vanish and their inverses do not exist. The physical significance of this is that a loaded body or structure is free to experience unlimited rigid body motion unless some supports or kinematic constraints are imposed that will ensure the equilibrium of the loads. These constraints are the boundary conditions.

Types of Boundary Conditions

From the variational-method point of view, there are two basic types of boundary conditions, *geometric* and *natural* (Section 4-1). One of the principal advantages of the finite element method is that we need specify only the geometric boundary conditions; the natural boundary conditions are implicitly satisfied in the solution procedure as long as we employ a suitable, valid variational principle. So far, we have limited our discussion to the displacement method of finite element analysis, in which we assume tentative displacement solutions. In this case, our geometric boundary conditions consist only of kinematic constraints or displacement boundary conditions. Traction boundary conditions are incorporated into the load vector, $\{Q\}$, equation (5.45b), and $\{R\}$, equation (6.4).

As a simple example, consider the cantilever beam shown in Figure 6-20. By applying the principle of minimum potential energy and the Euler-Lagrange formulation of Section 4-1, we obtain the following Euler equation:

$$(EIw'')'' = p$$

This is the familiar governing differential equation for beam bending where the primes indicate differentiation with respect to x. The boundary conditions are given by

$$[EIw''\ \delta w']_0^L = 0$$
$$[(EIw'')'\ \delta w]_0^L = 0$$

Thus, in a variational solution, the geometric boundary conditions we must satisfy are $\delta w = \delta w' = 0$ at $x = 0$, which correspond to specified values of w and w' at $x = 0$. The natural boundary conditions which will be satisfied implicitly by the approximate variational solution are

$$EIw'' = (EIw'')' = 0 \qquad \text{at} \qquad x = L.$$

Hence, in a finite element analysis of the beam, the only boundary conditions we need enforce are that the transverse deflection and slope be zero at the fixed end.

In the displacement method of finite element analysis, we can further categorize geometric boundary conditions as being *homogeneous* or *nonhomogeneous*, and as *normal* or *skewed*. Homogeneous conditions are most common and occur at locations that are completely constrained against movement; that is, their displacements are zero. Conversely, finite nonzero values may be specified at some points; these are nonhomogeneous boundary

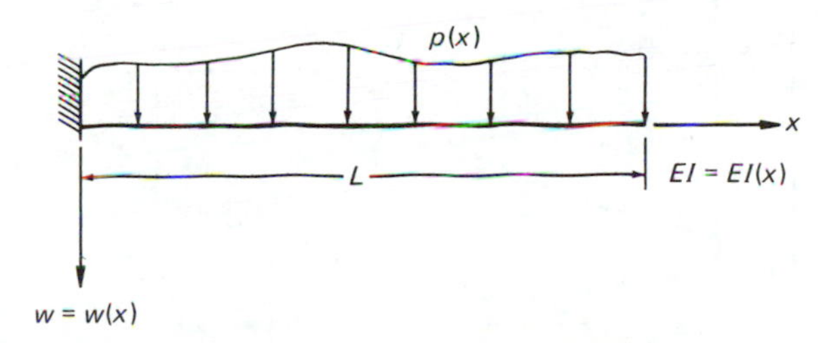

Figure 6-20 Cantilever beam.

conditions. For example, these arise in the analysis of a structure to determine the stresses caused by the settlement of a support or the differential settlement of foundations. The distinction between normal and skewed conditions arises at locations on the boundary at which only some components of the displacement are restrained. If the restrained components are parallel to the global coordinates, the conditions are normal; otherwise they are skewed. An example of a skewed boundary condition is shown in Figure 6-21. It is apparent that, if the angle α is 0 degrees or any multiple of 90 degrees, the constraint is no longer skewed but becomes normal.

Finally, it is noteworthy that *elastic supports* can be readily accommodated by the finite element method. However, they do not introduce a different type of boundary condition into the analysis. Rather, the deformable portion of such supports is included as finite elements in the structure or body that is discretized. The conventional geometric boundary conditions are then applied at the point where elastic supports are grounded. In practice, we do not add a new equation for this grounding point; rather, the appropriate matrix element on the principal diagonal of the stiffness matrix is merely modified by adding the support stiffness to it.

Modification of the Equilibrium Equations

The first step in the modification of the equilibrium equations, equations (1.7), to account for the boundary conditions is to transform for skewed supports, if necessary. In order that the skewed conditions may be treated as normal constraints, we transform the coordinates of those nodes where skewed conditions are specified. This transformation procedure is analogous to that described in Section 6-5 for the transformation from local to global coordinates.

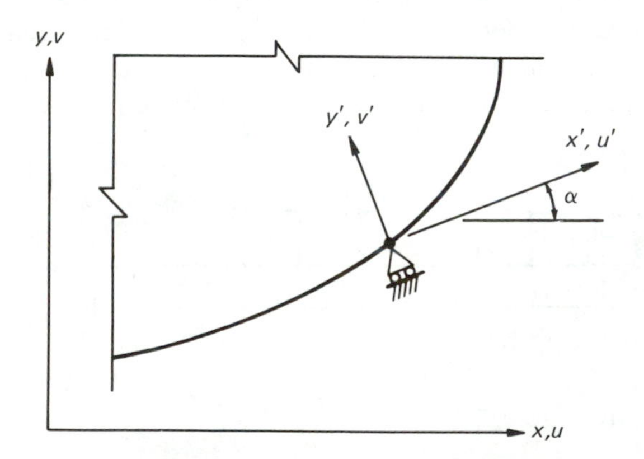

Figure 6-21 Skewed kinematic constraint.

If we adopt the notation that a prime indicates the skewed coordinate system (Fig. 6-21), we can write the transformation for displacements at the ith node as

$$\{r_i\} = [s_i]\{r_i'\} \tag{6.13}$$

$[s_i]$ is a simple point transformation involving the direction cosines which relate the global and skewed systems. $[s_i]$ is of the same form as $[t]$ in equation (6.5). By using equation (6.13) we may write the transformation for the entire nodal displacement vector as

$$\{r\} = [S]\{r'\} \tag{6.14}$$

Here the matrix $[S]$ is of the form

$$[S] = \begin{bmatrix} [I] & & & & 0 \\ & [I] & & & \\ & & \ddots & & \\ & & & [s_i] & \\ & & & & \ddots & \\ 0 & & & & & [I] \end{bmatrix}$$

where $[I]$ is an identity matrix of the same order as $[s_i]$, and the number of submatrices on the diagonal of $[S]$ is equal to the number of nodes in the assemblage. It follows that the order of $[s_i]$ and $[I]$ is equal to the number of displacements degrees of freedom at each node. Note that there is one $[s_i]$ matrix in $[S]$ for each node with skewed constraints.

Following the same procedure used in Section 6-5, the resulting transformations for the stiffness, loads, and consistent masses are

$$[K'] = [S]^{\mathrm{T}}[K][S] \tag{6.15}$$

$$\{R'\} = [S]^{\mathrm{T}}\{R\} \tag{6.16}$$

Note that $[K']$ like $[K]$ is symmetric.

The procedure to transform skewed boundaries into normal boundaries, as outlined by equations (6.14) through (6.16), is often performed before the individual element stiffness and loads are assembled. In this case, the above equations apply for $\{q_g\}$, $[k_g]$, and $\{Q_g\}$ rather than for $\{r\}$, $[K]$, and $\{R\}$, and the order of $[S]$ is n rather than N.

We will now drop the primed notation and assume either that all boundary conditions are normal, or that they have been transformed to a skewed system in which they may be treated as normal. However, if skewed boundaries are present, we observe that we must either store the submatrices $[s]$ or recompute them if we are to recover our displacements in global coordinates at the conclusion of the solution process.

The prescribing of geometric boundary conditions can be explained mathematically by partitioning the global equilibrium equation as follows.[5]

$$\left[\begin{array}{c|c} [K_{11}] & [K_{12}] \\ \hline [K_{12}]^{\mathrm{T}} & [K_{22}] \end{array}\right] \left\{\begin{array}{c} \{r_1\} \\ \hline \{r_2\} \end{array}\right\} = \left\{\begin{array}{c} \{R_1\} \\ \hline \{R_2\} \end{array}\right\} \tag{6.17a}$$

where $\{r_1\}$ is the vector of unconstrained or free displacements, and $\{r_2\}$ is the vector of specified displacements. The problem then reduces to the solution of the first set of equations in equation (6.17a):

$$[K_{11}]\{r_1\} = \{R_1\} - [K_{12}]\{r_2\} \tag{6.17b}$$

Here $[K_{11}]$ is no longer singular, so the system can be solved. The reactions at the constrained displacements can be computed from the second set of equations

$$\{R_2\} = [K_{12}]^{\mathrm{T}}\{r_1\} + [K_{22}]\{r_2\} \tag{6.17c}$$

In the case of homogeneous boundary conditions, $\{r_2\}$ is null and the procedure is considerably simplified.

A more practical way of forming the modified equilibrium equations is to write equations (6.17) in the following partitioned form[5]

$$\left[\begin{array}{c|c} [K_{11}] & [0] \\ \hline [0] & [I] \end{array}\right] \left\{\begin{array}{c} \{r_1\} \\ \hline \{r_2\} \end{array}\right\} = \left\{\begin{array}{c} \{R_1\} - [K_{12}]\{r_2\} \\ \hline \{r_2\} \end{array}\right\} \tag{6.18a}$$

In order to preserve the banded nature of the equations, the process in equation (6.18a) is performed without the reordering of equations implied by the partitioning. Rather, the contributions to the subvector $\{R_1\} - [K_{12}]\{r_2\}$ are first constructed for each nonhomogeneous condition. Then the row and column of $[K]$ corresponding to that condition are made null with the exception of the diagonal element, which is made unity. Finally, the prescribed value of the displacement is inserted in the load vector. For a specified displacement r_j occurring at the jth degree of freedom, the above process is summarized as

$$\begin{aligned} R_i &= R_i - K_{ij}r_j \quad \text{for} \quad i = 1, 2, 3, \ldots, N \quad \text{if} \quad r_j \neq 0 \\ K_{jm} &= K_{mj} = 0 \quad \text{for} \quad m = 1, 2, 3, \ldots, N \\ K_{jj} &= 1 \\ R_j &= r_j \end{aligned} \tag{6.18b}$$

Note that these operations ensure that the equilibrium equations remain symmetrical.

6-7 SOLUTION OF THE OVERALL PROBLEM

Now that we have modified the equilibrium equations to account for the boundary conditions, we are ready to solve the equations for the unknown displacements. Once we obtain the displacements, we can proceed to evaluate whichever element stresses and/or strains we shall need to complete our analysis. In Chapter 2 we outlined the procedures for solving the three basic types of problems: equilibrium, eigenvalue, and propagation. We shall not go into any more detail on these procedures, because a more complete treatment is beyond the scope of this work. It is sufficient to note that many standardized computer codes are available for the solution of the three types of problems. However, we must exercise some judgment in utilizing such standard programs if we are to obtain efficient and accurate finite element production programs.

Solution for Element Stresses or Strains

The solution of the overall equations yields displacements at all the inter-element nodes in the assemblage. Thus, the stress and strain components can be computed at any location within an element by using equations (5.29b) and (5.28b). Or, if the alternative stiffness formulation has been used, we can employ equations (5.50) and (5.49). Before going into detail on the practical aspects of computing the stresses or strains, it is important to note that the element stresses do not satisfy the equilibrium conditions for the individual elements.[18] In applying the principle of minimum potential energy, we approximate the overall equilibrium of the body, but do not provide for inter-element equilibrium. Nevertheless, as our approximation to the total potential and to the displacement solution is improved either by using refined elements or by reducing the mesh, we also obtain improved results for the element stress components.

We can illustrate this observation by considering the analysis of a beam with elements derived in Sample Problem 1 of Chapter 5. Recall that the stress resultants of interest, that is, the bending moments, are proportional to the curvature, d^2w/dx^2. However, the highest displacement derivative for which interelement compatibility is maintained is the slope, dw/dx. Hence, we see that it does not necessarily follow that the bending moments are continuous across the element interfaces. It is clear that this possible discontinuity of bending moment is a violation of local equilibrium. Even if our consistent load formulation gives rise to concentrated moments applied at the nodes, in the most general case we find that the internal stresses and applied loads do not balance at any particular nodal point.

Because of the approximation involved, it is logical to use some average value of the stress (or strain) as representative for the element. The most common method of averaging is to use the stress at the centroid of the element. This quantity can be computed directly by using equation (5.29b), or it can be calculated as some average value of the nodal stresses. For many applications we use linear displacement models; and as a result, the stresses are constant over the element. In this case it is usual to assign the stress to the centroid of the element. For example, Figure 6-22 shows the axial stresses σ_x computed from the finite element analysis for the case II of the beam bending problem shown in Figure 6-8. The computed stresses are constant over the elements. It can be seen that the exact solution[7] is quite close to the line joining the constant stresses at the centroids.

For higher order elements, an alternative procedure is admissible. Instead of averaging the stress over the element to obtain a centroidal value, the stress at a particular node may be taken as the average value at that node among all the adjacent elements.

In a computer procedure, the stresses (or strains) are usually computed element-by-element. The first step is to recover the displacements of any

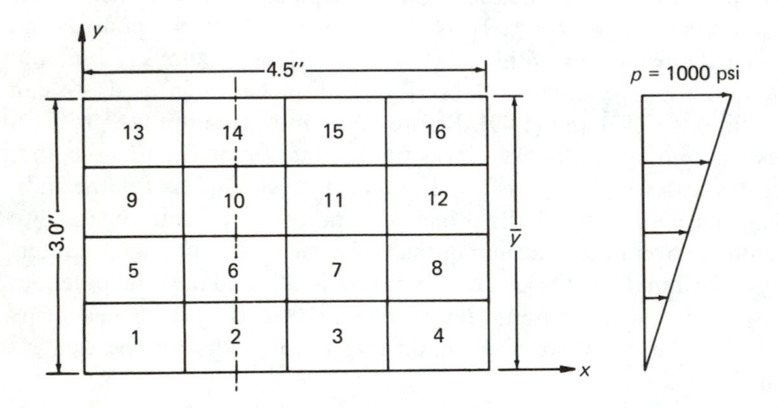

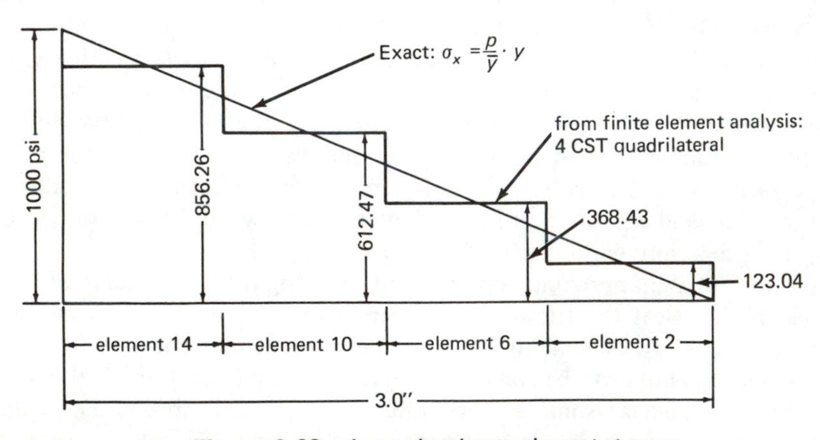

Figure 6-22 Approximation to element stresses.

internal nodes for the element by using equation (5.64g). If desired, the element displacements may be printed at this stage. Next, the stress components at the centroid of the element are computed, either by direct evaluation of equation (5.29b) at that point or by averaging the nodal stresses in a manner consistent with the displacement model. In two- or three-dimensional problems, the principal stresses at the centroid are usually computed once the stress components are known. The element stresses may be printed at this time, or they may be stored for unified display at the completion of computation.

Display of Results

The nodal displacements and the desired stress and strain results can be obtained from the computer as a printed output in tabular form. However, many problems have so many elements in the assemblage that the interpretation and reduction of the resulting extensive tabular values are laborious and time-consuming. This is particularly true when there are many load cases for an equilibrium problem or many eigenvectors for an eigenvalue problem. However, the most critical case is certainly the propagation problem, in which a great quantity of data may be generated at each time step.

For engineering analysis and design, it is usually convenient to reduce the data, such as displacements or principal stresses, to a graphical form. It is possible to automate this process so that the computer can directly produce such graphs using either on-line or off-line equipment. The plotting can be accomplished by standard plot routines developed by hardware manufacturers or software companies. Alternatively, special plotting routines may be written that are particularly suited to finite element applications.[19]

The various types of equipment that can be employed to produce graphical display of results include the line printer, the incremental plotter, and the cathode ray tube plotter. The line printer is usually considered an expedient to be used when other equipment is unavailable, since the quality of the resulting graphs is relatively crude, and since the necessary software is difficult to devise. The incremental plotter produces high-quality ink drawings that may be suitable for inclusion in engineering reports. However, plotting by this technique is performed off-line; and although it is certainly faster than manual graphing, it is relatively time-consuming compared to the entire computation process. Finally, the quickest graphical display device is the cathode ray tube. Plotting of results is practically instantaneous. Hard copy can be obtained by microfilming the image on the tube, and standard prints may be taken from the microfilm negatives. An additional advantage of the cathode ray tube is that, for propagation problems, the results at many of the time steps may be plotted and a motion picture taken of the time-varying

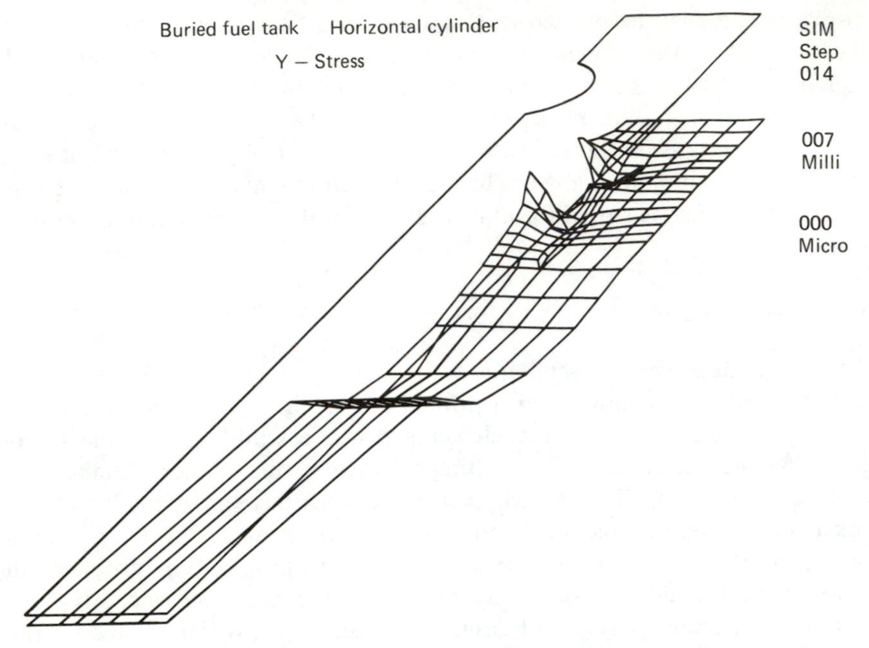

Figure 6-23 Example of a cathode ray tube display.[20]

display. This results in an effective visualization of the dynamic process and permits detailed study of key frames.

An example of one frame of a cathode ray tube plot for a dynamic stress analysis problem is shown in Figure 6-23.[20] This is an isometric projection of a two-dimensional domain with the instantaneous vertical stress plotted in the vertical direction. The registers on the right margin indicate that the frame represents the solution after the fourteenth time step, which corresponds to a time of 0.0070 seconds. This process of isometric projection has been performed automatically from a two-dimensional plot of the finite element mesh.

A more conventional display of displacements or principal stresses is a two-dimensional contour plot. Other common types of display are graphs of critical parameters as functions of space or time.

6-8 EXAMPLES

Because the assembly and solution processes usually entail a large amount of data manipulation, it is impractical to present detailed numerical examples of actual applications. However, to clarify some of the basic principles presented in this chapter, we shall present a few simple examples. The extension to real problems is not difficult, because the assembly and solution processes are readily programmed for the computer. A further example accompanies the computer program in Appendix I.

Example 6-1: Simply Supported Beam. A simply supported beam made of aluminum ($E = 10^7$ psi) is shown in Figure 6-24. It has a span of 10″ and a cross section of 1″ × 1″, and it is loaded with a linearly varying distributed load as shown. The beam has been subdivided into two finite elements. The characteristics of these elements were derived in Sample Problem 1 of Chapter 5. Now we will solve for the transverse deflections and rotations.

The element stiffness is the same for both elements and is obtained from equation (5.54)

$$[k_1] = [k_2] = \frac{EI}{125}\begin{bmatrix} 12 & 30 & -12 & 30 \\ & 100 & -30 & 50 \\ \text{symm.} & & 12 & -30 \\ & & & 100 \end{bmatrix}$$

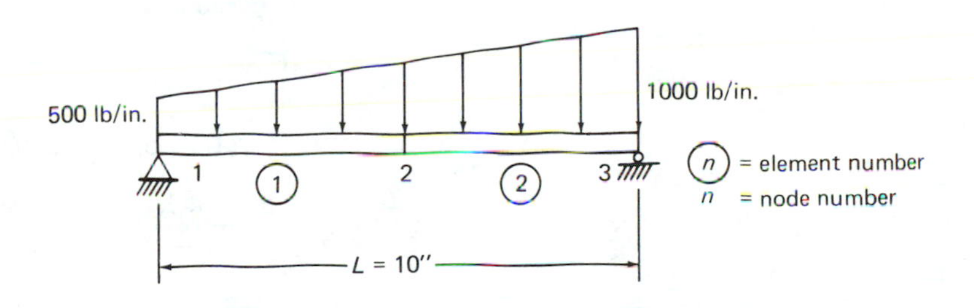

(a) Simply supported beam

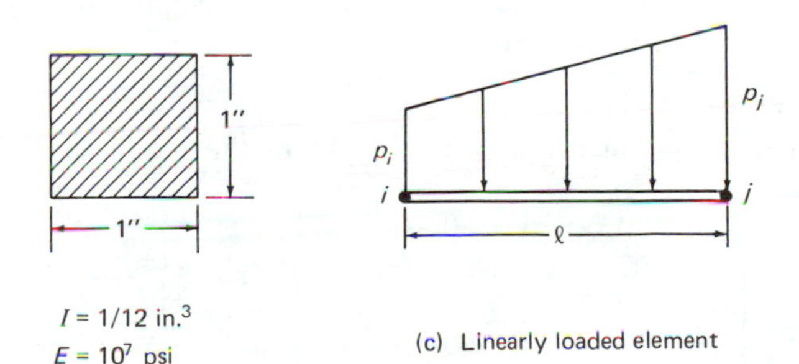

(b) Cross section

(c) Linearly loaded element

Figure 6-24 Example 6-1.

The load vector for a typical element is obtained from equation (5.56d)

$$\{Q\}^{\mathrm{T}} = \frac{\ell}{60}\,[(21p_i + 9p_j) \quad \ell(3p_i + 2p_j) \quad (9p_i + 21p_j) \quad -\ell(2p_i + 3p_j)]$$

and for the two elements these vectors are

$$\{Q_1\}^{\mathrm{T}} = [1437.5 \quad 1250.0 \quad 1687.5 \quad -1354.1]$$
$$\{Q_2\}^{\mathrm{T}} = [2062.5 \quad 1770.1 \quad 2312.5 \quad -1875.0]$$

By using the direct stiffness method, the equilibrium equations for the assemblage are obtained as

$$\frac{EI}{125}\begin{bmatrix} 12 & 30 & -12 & 30 & 0 & 0 \\ 30 & 100 & -30 & 50 & 0 & 0 \\ -12 & -30 & (12+12) & (-30+30) & -12 & 30 \\ 30 & 50 & (-30+30) & (100+100) & -30 & 50 \\ 0 & 0 & -12 & -30 & 12 & -30 \\ 0 & 0 & 30 & 50 & -30 & 100 \end{bmatrix}\begin{Bmatrix} w_1 \\ \theta_1 \\ w_2 \\ \theta_2 \\ w_3 \\ \theta_3 \end{Bmatrix} = \begin{Bmatrix} 1437.5 \\ 1250.0 \\ (1687.5 + 2062.5) \\ (-1354.1 + 1770.1) \\ 2312.5 \\ -1875.0 \end{Bmatrix}$$

After adding as indicated by the dashed lines, we can introduce the boundary conditions $w_1 = w_3 = 0$. For instance, we can follow the procedure described in equation (6.18b) for $w_1 = 0$, and set

$$k_{11} = 1.0,\ k_{12} = k_{13} = \cdots = k_{16} = 0,\ k_{21} = k_{31} = \cdots = k_{61} = 0,\ R_1 = w_1 = 0$$

After a similar operation for $w_3 = 0$, we obtain the modified assemblage equations as

$$\frac{10^7}{(12)(125)}\begin{bmatrix} 1 & 0 & 0 & 0 & 0 & 0 \\ 0 & 100 & -30 & 50 & 0 & 0 \\ 0 & -30 & 24 & 0 & 0 & 30 \\ 0 & 50 & 0 & 200 & 0 & 50 \\ 0 & 0 & 0 & 0 & 1 & 0 \\ 0 & 0 & 30 & 50 & 0 & 100 \end{bmatrix}\begin{Bmatrix} w_1 \\ \theta_1 \\ w_2 \\ \theta_2 \\ w_3 \\ \theta_3 \end{Bmatrix} = \begin{Bmatrix} 0 \\ 1250.0 \\ 3750.0 \\ 416.0 \\ 0 \\ -1875.0 \end{Bmatrix}$$

The solution of the above equations and the corresponding analytical solution from strength of materials are

$$\{r\}_{\text{FEM}} = \begin{Bmatrix} 0 \\ 0.0366665 \text{ rad.} \\ 0.1171185 \text{ inch} \\ 0.0007295 \text{ rad.} \\ 0 \\ -0.0383335 \text{ rad.} \end{Bmatrix}, \quad \{r\}_{\text{SM}} = \begin{Bmatrix} 0 \\ 0.0366666 \text{ rad.} \\ 0.1171875 \text{ inch} \\ 0.0007292 \text{ rad.} \\ 0 \\ -0.0383333 \text{ rad.} \end{Bmatrix}$$

We see that with the coarsest of subdivisions we obtain acceptable results for this simple problem.

Example 6-2: Quadrilateral Composed of 4-*CST's*. When CST elements are used, it is common practice to group four elements to form a quadrilateral element. Such a quadrilateral involves a fewer number of unknowns than the four triangles. In addition it ensures that the solution is independent of the *skew* of a subdivision mesh. Figure 6-25(a) shows two different oversimplified meshes of CST's which do not necessarily give the same solution because they have different skews. A 4-CST quadrilateral, Figure 6-25(b), avoids the skewness by producing the mesh shown in Figure 6-25(c), which has the same number of external nodes as the meshes in Figure 6-25(a).

We will now show schematically how the stiffness for the element is constructed from the stiffness of the individual triangles. This procedure essentially involves two steps: (1) addition of the element equations of the CST's by using the direct stiffness method; and (2) condensation of the internal degrees of freedom by using the procedure of Section 5-9.

We derived a general expression for the CST stiffness in Sample Problem 2 and Example 5-2 of Chapter 5. The stiffness matrix for a typical triangle, say element 1 with nodes 1, 2, and 5, is given as follows:

$$\underset{6\times 6}{[k^{(1)}]} = \left[\begin{array}{c|c|c} [k_{11}^{(1)}] & [k_{12}^{(1)}] & [k_{15}^{(1)}] \\ \hline [k_{21}^{(1)}] & [k_{22}^{(1)}] & [k_{25}^{(1)}] \\ \hline [k_{51}^{(1)}] & [k_{52}^{(1)}] & [k_{55}^{(1)}] \end{array}\right]$$

where the superscript indicates the element number, and the subscripts of the partitioned matrices indicate the nodes. Each of the submatrices is a 2×2 matrix with influence coefficients for the degrees of freedom at the two nodes. For example, two of these submatrices are

$$[k_{11}^{(1)}] = \begin{bmatrix} k_{11}^{(1)} & k_{12}^{(1)} \\ k_{21}^{(1)} & k_{22}^{(1)} \end{bmatrix}, \qquad [k_{12}^{(1)}] = \begin{bmatrix} k_{13}^{(1)} & k_{14}^{(1)} \\ k_{23}^{(1)} & k_{24}^{(1)} \end{bmatrix}$$

where the subscripts in the 2×2 submatrices correspond to the indices of degrees of freedom. The stiffness matrix for the quadrilateral will be 10×10 and is given in Figure 6-26. The double partition lines indicate the arrangement of the matrix to facilitate condensation of the degrees of freedom occurring at the internal node 5. The condensation is performed as described in Section 5-9.

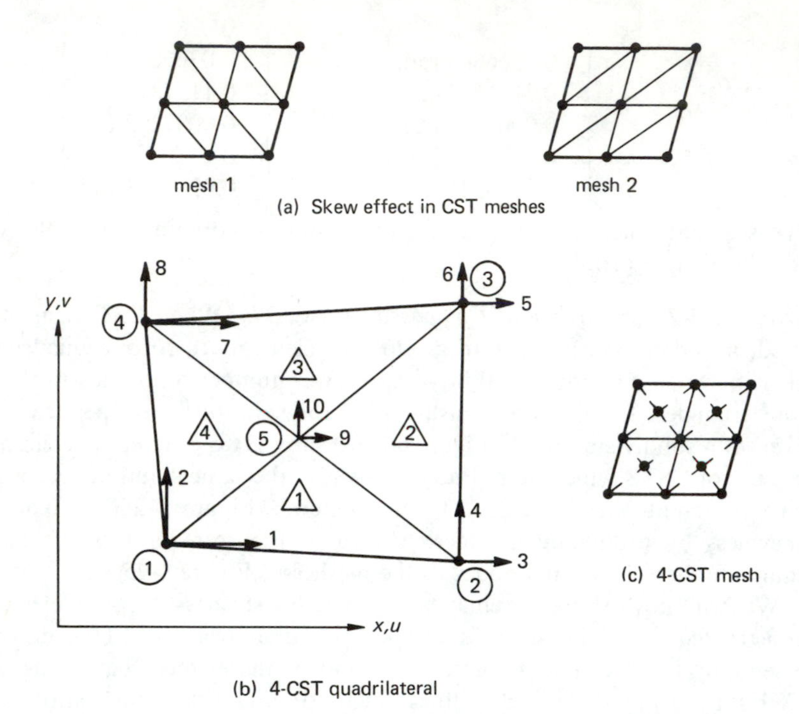

Figure 6-25 Quadrilateral constructed from four CST's.

We will now compute the numerical results for a specific version of a quadrilateral composed of four CST elements—a rectangle for plane strain analysis. This rectangle is shown in Figure 6-27. We have derived the stiffness matrix for the individual triangles in Example 5-2, where we have also computed the stiffness matrix for elements 4 and 2 as

$$[k^{(2)}] = [k^{(4)}] = \frac{E}{4.16}\begin{bmatrix} 1.5 & -1.0 & -0.1 & 0.2 & -1.4 & 0.8 \\ & 3.0 & -0.2 & -2.6 & 1.2 & -0.4 \\ & & 1.5 & 1.0 & -1.4 & -0.8 \\ & & & 3.0 & -1.2 & -0.4 \\ \text{Symmetrical} & & & & 2.8 & 0.0 \\ & & & & & 0.8 \end{bmatrix}$$

$$\{q^{(2)}\} = \begin{Bmatrix} u_2 \\ v_2 \\ u_3 \\ v_3 \\ u_5 \\ v_5 \end{Bmatrix}, \quad \{q^{(4)}\} = \begin{Bmatrix} u_4 \\ v_4 \\ u_1 \\ v_1 \\ u_5 \\ v_5 \end{Bmatrix}$$

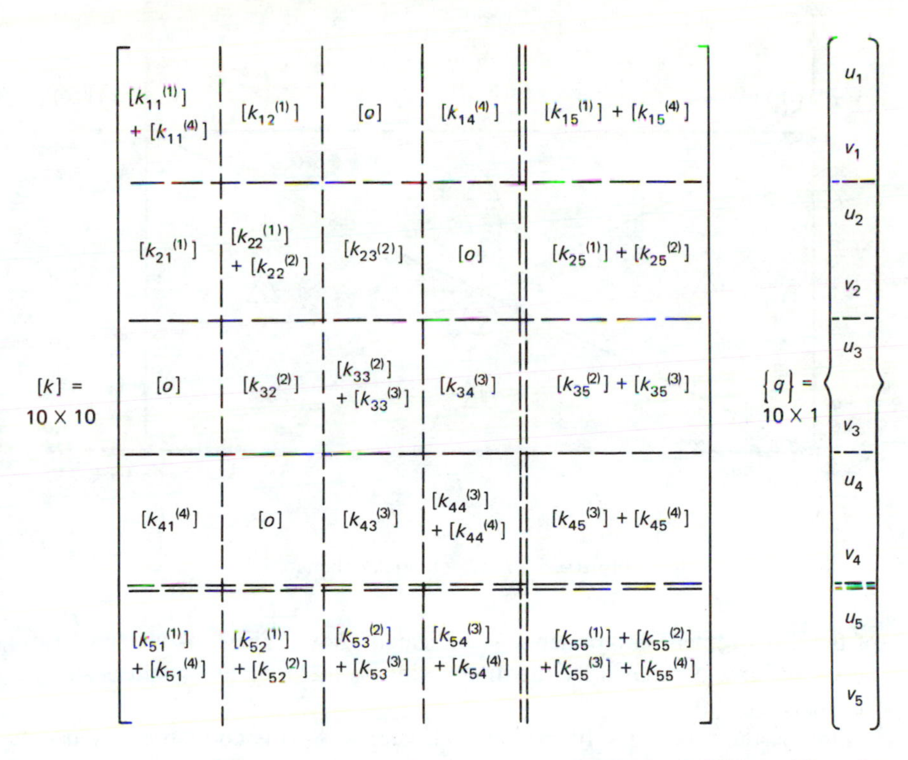

Figure 6-26 Stiffness matrix and nodal displacement vector for 4-CST quadrilateral.

Similarly, we compute $[k^{(1)}]$ and $[k^{(3)}]$ as

$$[k^{(1)}] = [k^{(3)}] = \frac{E}{4.16}\begin{bmatrix} 1.5 & 1.0 & 0.1 & 0.2 & -1.6 & -1.2 \\ & 3.0 & -0.2 & 2.6 & -0.8 & -5.6 \\ & & 1.5 & -1.0 & -1.6 & 1.2 \\ & & & 3.0 & 0.8 & -5.6 \\ \text{Symmetrical} & & & & 3.2 & 0.0 \\ & & & & & 11.2 \end{bmatrix}$$

$$\{q^{(1)}\} = \begin{Bmatrix} u_1 \\ v_1 \\ u_2 \\ v_2 \\ u_5 \\ v_5 \end{Bmatrix}, \quad \{q^{(3)}\} = \begin{Bmatrix} u_3 \\ v_3 \\ u_4 \\ v_4 \\ u_5 \\ v_5 \end{Bmatrix}$$

Note that we have rearranged the stiffness to correspond with the ordering

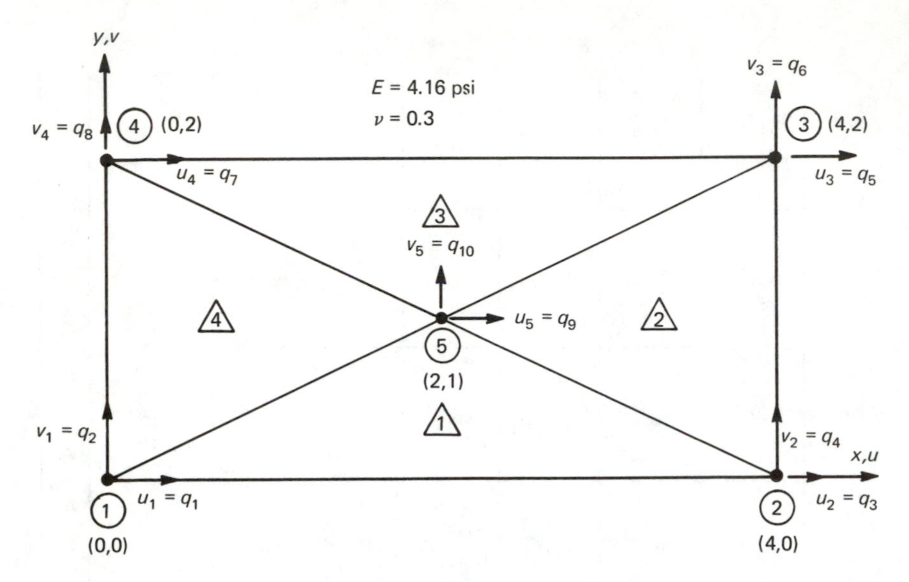

Figure 6-27 Rectangular element.

of the displacement components as indicated above. The four element stiffnesses are now combined by the direct stiffness method. The result is shown in Figure 6-28(a).

Finally, the total 10 × 10 matrix of Figure 6-28(a) is condensed by using the procedure in Section 5-9; hence, the internal degrees of freedom (u_5, v_5) at node 5 are eliminated. The condensed 8 × 8 matrix is shown in Figure 6-28(b).

Example 6-3: Assemblage of Three Elements. An assemblage of three rectangular plane strain elements is shown in Figure 6-29. The dimensions of all three of the elements are the same, that is, width = 4″ and height = 2″. Using the element stiffness matrix evaluated in Example 6-2, we shall assemble the global stiffness matrix. We shall then introduce the following boundary conditions to the unloaded assemblage

$$u_1 = v_1 = v_3 = 0$$
$$u_3 = 0.1 \text{ in.}$$

The element stiffness matrix is given in Figure 6-28. We can generate a table from Figure 6-29 which contains the node information for each element. This is shown as Table 6-4, and is typical of the array which is a necessary part of the input data for a finite element code. Note that the element node indices (1, 2, 3, and 4) are taken counterclockwise, as shown in Figure 6-27.

$$
\underset{10 \times 10}{[k]} = \left[\begin{array}{cccccccc|cc}
3.0 & 2.0 & 0.1 & 0.2 & 0.0 & 0.0 & -0.1 & -0.2 & -3.0 & -2.0 \\
 & 6.0 & -0.2 & 2.6 & 0.0 & 0.0 & 0.2 & -2.6 & -2.0 & -6.0 \\
 & & 3.0 & -2.0 & -0.1 & 0.2 & 0.0 & 0.0 & -3.0 & 2.0 \\
 & & & 6.0 & -0.2 & -2.6 & 0.0 & 0.0 & 2.0 & -6.0 \\
 & & & & 3.0 & 2.0 & 0.1 & 0.2 & -3.0 & -2.0 \\
 & & & & & 6.0 & -0.2 & 2.6 & -2.0 & -6.0 \\
 & \text{Symmetrical} & & & & & 3.0 & -2.0 & -3.0 & 2.0 \\
 & & & & & & & 6.0 & 2.0 & -6.0 \\
\hline
 & & & & & & & & 12.0 & 0.0 \\
 & & & & & & & & & 24.0
\end{array}\right], \quad \{q\} = \left\{\begin{array}{c} u_1 \\ v_1 \\ u_2 \\ v_2 \\ u_3 \\ v_3 \\ u_4 \\ v_4 \\ \hline u_5 \\ v_5 \end{array}\right\}
$$

(a) Assembled quadrilateral stiffness

$$
\underset{8 \times 8}{[k]} = \left[\begin{array}{cccccccc}
2.08 & 1.00 & -0.48 & 0.20 & -0.92 & -1.00 & -0.68 & -0.20 \\
 & 4.17 & -0.20 & 1.43 & -1.00 & -1.83 & 0.20 & -3.77 \\
 & & 2.08 & -1.00 & -0.68 & 0.20 & -0.92 & 1.00 \\
 & & & 4.17 & -0.20 & -3.77 & 1.00 & -1.83 \\
 & & & & 2.08 & 1.00 & -0.48 & 0.20 \\
 & \text{Symmetrical} & & & & 4.17 & -0.20 & 1.43 \\
 & & & & & & 2.08 & -1.00 \\
 & & & & & & & 4.17
\end{array}\right], \quad \{q\} = \left\{\begin{array}{c} u_1 \\ v_1 \\ u_2 \\ v_2 \\ u_3 \\ v_3 \\ u_4 \\ v_4 \end{array}\right\}
$$

(b) Condensed quadrilateral stiffness

Figure 6-28 Four-CST quadrilateral matrices.

Once the element stiffness and load vectors are known, the array in Table 6-4 contains necessary information for the assembly process. We note that the indices of the displacements at any node can be related directly to the node index. Specifically, the indices of the displacements at any node i are given by $2i - 1$ and $2i$. This is true for both global and local systems.

We can compute the semiband width from Table 6-1. The maximum difference between global indices in any row of the table is $D = 4$. For this problem $f = 2$, so by using equation (6.1) we obtain $B = 2(4 + 1) = 10$.

Using the direct stiffness method illustrated in Figure 6-18 and the information in Table 6-4, we can now assemble the global stiffness matrix. The necessary portion of the 16×16 stiffness matrix for the assemblage is given in Figure 6-30. Here we have assumed a Poisson's ratio of $\nu = 0.3$ and a unit thickness.

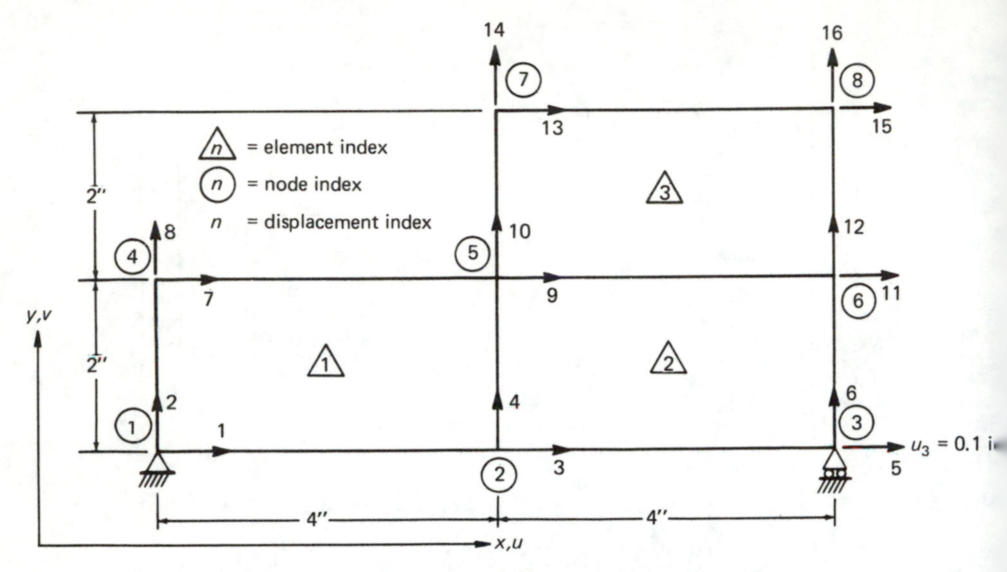

Figure 6-29 Example 6-3.

TABLE 6-4 ARRAY CONTAINING ELEMENT INFORMATION

Element Index	*Global Index of Element Nodes*			
	Node 1	*Node 2*	*Node 3*	*Node 4*
1	1	2	5	4
2	2	3	6	5
3	5	6	8	7

We can now modify the stiffness matrix and load vector to account for the boundary conditions according to equations (6.18b). Since we have one nonhomogeneous boundary condition, our previously null load vector will now have some nonzero elements. The modified stiffness and load vector are shown in Figure 6-31.

REFERENCES

(1) Clough, R. W., "The Stress Distribution of Norfork Dam," Structures and Materials Research, Department of Civil Engineering, Series 100, Issue 19, University of California, Berkeley (March 1962). One of the earliest practical applications of the finite element method. Plane stress analysis of a gravity dam including the effects of existing cracks, hydrostatic pressure and uplift, temperature gradients, and foundation deformation. Also includes a comparison with the discrete element method.

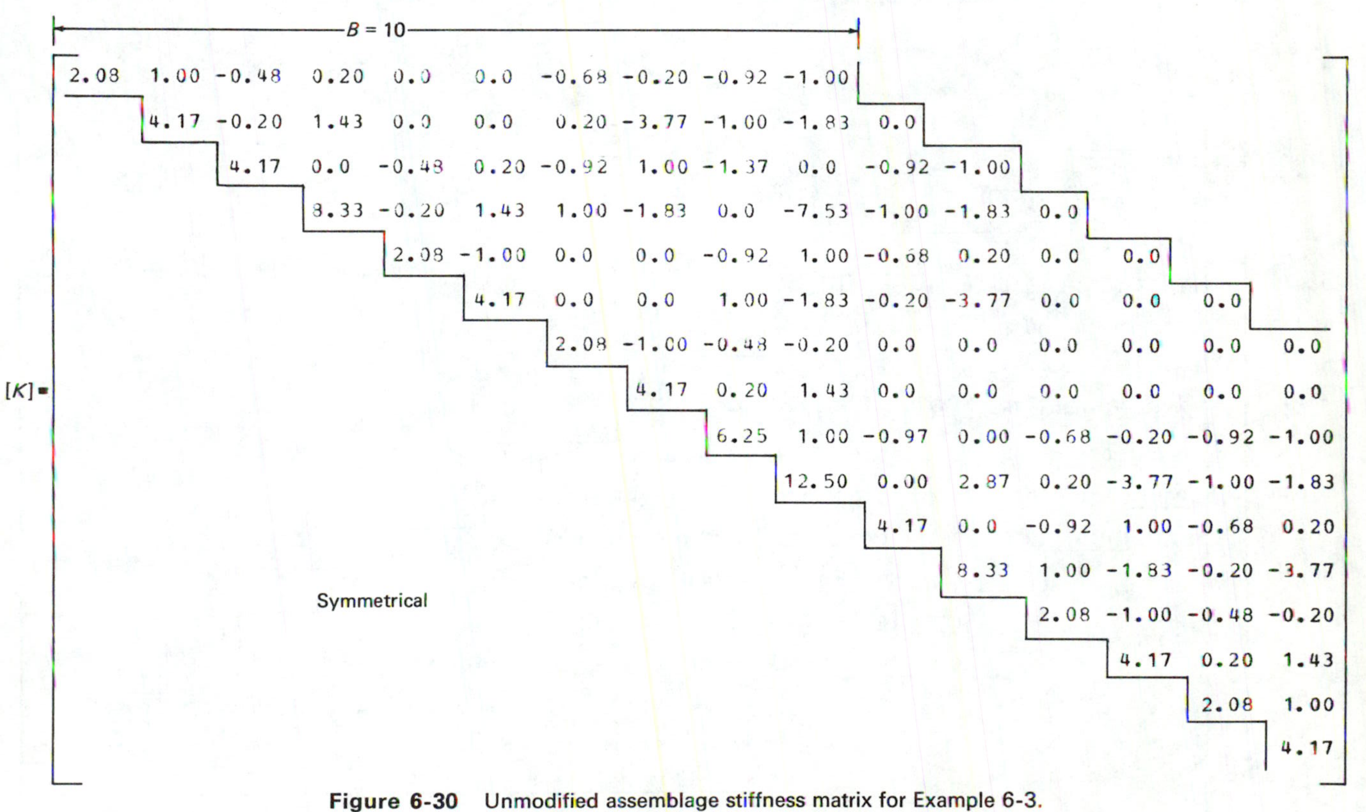

$[K] =$ (Symmetrical; $B = 10$)

2.08	1.00	-0.48	0.20	0.0	0.0	-0.68	-0.20	-0.92	-1.00						
	4.17	-0.20	1.43	0.0	0.0	0.20	-3.77	-1.00	-1.83	0.0					
		4.17	0.0	-0.48	0.20	-0.92	1.00	-1.37	0.0	-0.92	-1.00				
			8.33	-0.20	1.43	1.00	-1.83	0.0	-7.53	-1.00	-1.83	0.0			
				2.08	-1.00	0.0	0.0	-0.92	1.00	-0.68	0.20	0.0	0.0		
					4.17	0.0	0.0	1.00	-1.83	-0.20	-3.77	0.0	0.0	0.0	
						2.08	-1.00	-0.48	-0.20	0.0	0.0	0.0	0.0	0.0	0.0
							4.17	0.20	1.43	0.0	0.0	0.0	0.0	0.0	0.0
								6.25	1.00	-0.97	0.00	-0.68	-0.20	-0.92	-1.00
									12.50	0.00	2.87	0.20	-3.77	-1.00	-1.83
										4.17	0.0	-0.92	1.00	-0.68	0.20
											8.33	1.00	-1.83	-0.20	-3.77
												2.08	-1.00	-0.48	-0.20
													4.17	0.20	1.43
														2.08	1.00
															4.17

Figure 6-30 Unmodified assemblage stiffness matrix for Example 6-3.

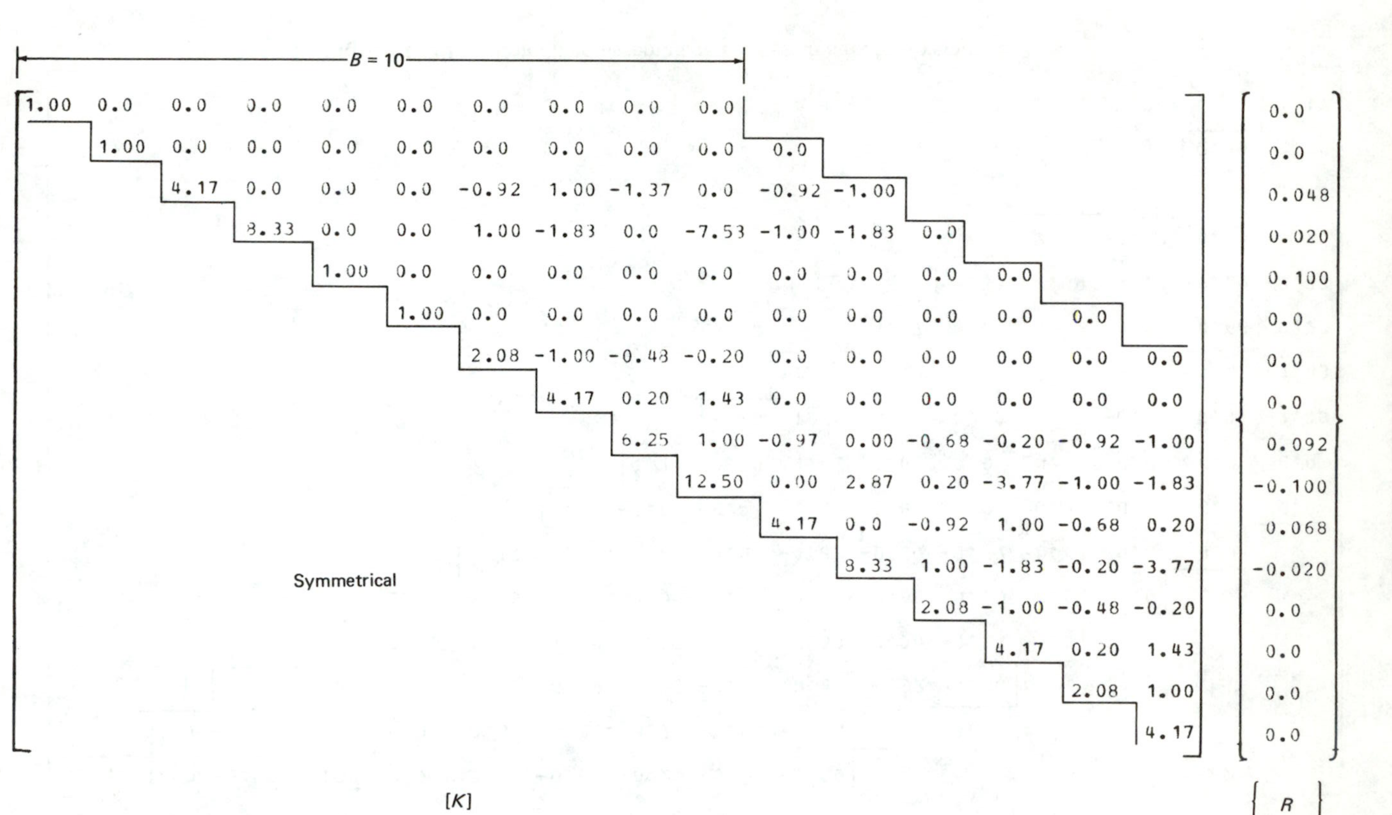

$B = 10$

Row	1	2	3	4	5	6	7	8	9	10	11	12	13	14	15	16	{R}
1	1.00	0.0	0.0	0.0	0.0	0.0	0.0	0.0	0.0	0.0							0.0
2		1.00	0.0	0.0	0.0	0.0	0.0	0.0	0.0	0.0	0.0						0.0
3			4.17	0.0	0.0	0.0	-0.92	1.00	-1.37	0.0	-0.92	-1.00					0.048
4				8.33	0.0	0.0	1.00	-1.83	0.0	-7.53	-1.00	-1.83	0.0				0.020
5					1.00	0.0	0.0	0.0	0.0	0.0	0.0	0.0	0.0	0.0			0.100
6						1.00	0.0	0.0	0.0	0.0	0.0	0.0	0.0	0.0	0.0		0.0
7							2.08	-1.00	-0.48	-0.20	0.0	0.0	0.0	0.0	0.0	0.0	0.0
8								4.17	0.20	1.43	0.0	0.0	0.0	0.0	0.0	0.0	0.0
9									6.25	1.00	-0.97	0.00	-0.68	-0.20	-0.92	-1.00	0.092
10										12.50	0.00	2.87	0.20	-3.77	-1.00	-1.83	-0.100
11											4.17	0.0	-0.92	1.00	-0.68	0.20	0.068
12												8.33	1.00	-1.83	-0.20	-3.77	-0.020
13													2.08	-1.00	-0.48	-0.20	0.0
14														4.17	0.20	1.43	0.0
15															2.08	1.00	0.0
16																4.17	0.0

Symmetrical

$[K]$ $\{R\}$

Figure 6-31 Modified assemblage stiffness and load vector for Example 6-3.

(2) See Reference 13, Chapter 2.
(3) Becker, E. B., and Brisbane, J. J., "Application of the Finite Element Method to the Stress Analysis of Solid Propellant Rocket Grains," Rohm & Haas Co., Report S-76, Vol. I, Nov. 1965, NTIS, Springfield, Va., AD 474031. Linear thermoelastic analysis of incompressible propellants and their casings. Includes a presentation of basic finite element techniques.
(4) See Reference 11, Chapter 2.
(5) See Reference 8, Chapter 1.
(6) de Arentes e Oliveira, E. R., "Completeness and Convergence in the Finite Element Method," *Second Conf.* A study of convergence generated by elements with decreasing size.
(7) Connor, J., and Will, G., "Computer-Aided Teaching of the Finite Element Displacement Method," Research Report 69-23, Department of Civil Engineering, MIT, Cambridge (Feb. 1969). Solves a number of problems from structural mechanics.
(8) Duncan, J. M., Monismith, C. L., and Wilson, E. L., "Finite Element Analyses of Pavements," *Highway Research Record* No. 228, HRB (1968). Multilayered pavement system is analyzed by iterative nonlinear analysis. Concept of "resilient modulus" derived from repeated triaxial tests is used. Effects of seasonal temperature variation is also considered. Good comparisons are obtained with field data on Gonzales Bypass in California.
(9) Dunlop, P., Duncan, J. M., and Seed, H. B., "Finite Element Analyses of Slopes in Soils," Contract Report S-68-6, USAEWES (May 1968). Also published in (1) *Proc. ASCE, J. SM&F Dn*, Vol. 96, SM2 (Mar. 1970); (2) *Proc. 7th Int. Conf. on SM&FE*, Mexico City (Aug. 1969). Details in Chapter 10.
(10) Duncan, J. M., and Goodman, R. E., "Finite Element Analysis of Slopes in Jointed Rock," Contract Report No. S-68-3, USAEWES (Feb. 1968). Details in Chapter 10.
(11) Frederick, O. C., Wong, Y. C., and Edge, F. W., "Two-Dimensional Automatic Mesh Generation for Structural Analysis," *IJNME*, Vol. 2, No. 1 (1970). Presents partially automated method of discretizing irregular and nonhomogeneous two-dimensional continua into triangular elements.
(12) Abel, J. F., and Popov, E. P., "Static and Dynamic Finite Element Analysis of Sandwich Structures," *Second Conf.* Extension of the finite element method to the analysis of multilayer beams and shells wherein transverse shearing deformation is significant. Includes example showing effect of mesh refinement vs. higher order models.
(13) See Reference 13, Chapter 5.
(14) See Reference 6, Chapter 5.
(15) Argyris, J. H., Buck, K. E., Fried, I., Hilber, H. M., Mareczek, G., and Scharpf, D. W., "Some New Elements for the Matrix Displacement Method," *Second Conf.* Presents families of higher order elements for three-dimensional problems, plates, and shells. Also discusses advances in elastoplastic and large displacement analysis.
(16) Bazeley, G. P., Cheung, Y. K., Irons, B. M., and Zienkiewicz, O. C., "Triangular Elements in Plate Bending—Conforming and Nonconforming Solutions,"

First Conf. Discusses the convergence characteristics of complete but nonconforming elements.
(17) See Reference 5, Chapter 1.
(18) See Reference 7, Chapter 1.
(19) NASTRAN (Level 12) developed by National Aeronautics and Space Administration. A number of computer plotting options are available with this finite element system. NASTRAN can be used to solve various static, dynamic, eigenvalue, and nonlinear problems.
(20) Official U.S. Navy photograph, courtesy of U.S. Naval Civil Engineering Laboratory, Port Hueneme, Calif.

FURTHER READING

Akyuz, F. A., "FEDGE—A General-Purpose Computer Program for Finite Element Data Generation," Tech. Memo. 33-431, Vol. I and II, JPL, Calif. Inst. of Tech., Pasadena, Calif. (Sept. 1969). Develops an algorithm for automatic discretization of one-, two-, and three-dimensional bodies. It is based on the concept of natural coordinate systems and suitable classification of the topological properties of complex geometries.

Akyuz, F. A., "Natural Coordinate Systems—An Automatic Input Data Generation Scheme for a Finite-Element Method," *Nuclear Engg. and Design*, Amsterdam, North-Holland Publishing Co., Vol. 11 (1970).

Kavlie, D., and Powell, G. H., "Efficient Reanalysis of Modified Structures," *Proc. ASCE, J. ST Dn.*, Vol. 97, ST1 (Jan. 1971). Discusses and compares resolution techniques for the displacement method when the member dimensions of statically indeterminate structures are progressively modified in a design process.

Taig, I. C., "Automated Stress Analysis Using Substructures," *First Conf.*

See Reference 4, Chapter 8.

See Reference 16, Chapter 8.

See Reference 46, Chapter 8.

See Reference 46, Chapter 9.

EXERCISES

6-1 Using equation (6.1) or similar concepts, compute the semi-band widths for the two meshes shown for the following cases:

(a) One degree of freedom (DOF) at primary nodes, and no DOF at secondary nodes.

(b) Two DOF at primary, no DOF at secondary.

(c) Three DOF at primary, no DOF at secondary.

(d) Three DOF at primary, one DOF at secondary. (*Hint:* Equation (6.1) does not apply.)

(e) One DOF at primary, three DOF at secondary. (*Hint:* Equation (6.1) does not apply.)

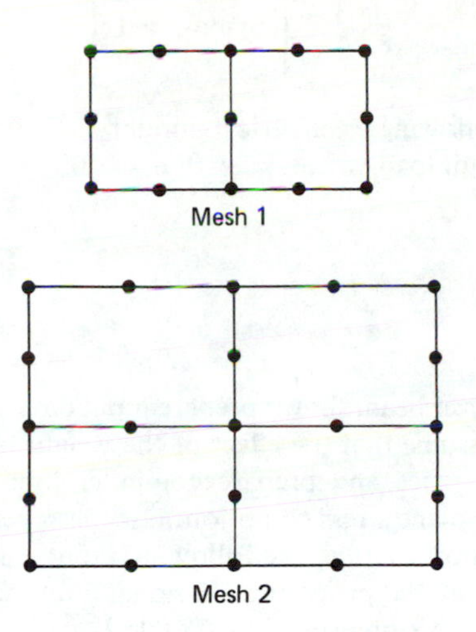

Solution:

Case	*Mesh 1*	*Mesh 2*
a	4	5
b	8	10
c	12	15
d	16	21
e	16	23

6-2 (a) Use the direct stiffness method to assemble the equilibrium equations for the assemblage shown. Assume a single degree of freedom at each node and adopt the following symbolic stiffness matrix for the ith typical element:

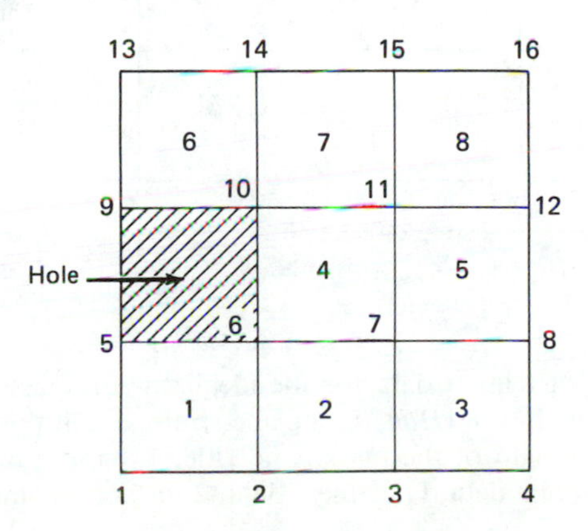

$$[k_i] = \begin{bmatrix} 4 & 1 & 1 & 1 \\ & 4 & 1 & 1 \\ \text{Symm.} & & 4 & 1 \\ & & & 4 \end{bmatrix}$$

(b) Introduce the following geometric boundary conditions in the assemblage equations. Assume a null load vector. $u_4 = 0$, $u_{16} = 0.1$

6-3 (a) For the cantilever beam shown prepare input data suitable for the computer code in Appendix I. Assume that the effect of the weight of the beam is negligible.

(b) Refine the mesh twice and prepare computer input data for both refined meshes. (*Hint:* Assume plane stress condition. Use mesh generation capabilities. By using the mesh generation routine, the following input cards are required: 1 title, 1 basic parameters, 1 material properties, 12 nodal point data, 5 element data, and 1 blank card at the end to stop execution. Total: 21 cards for part (a).)

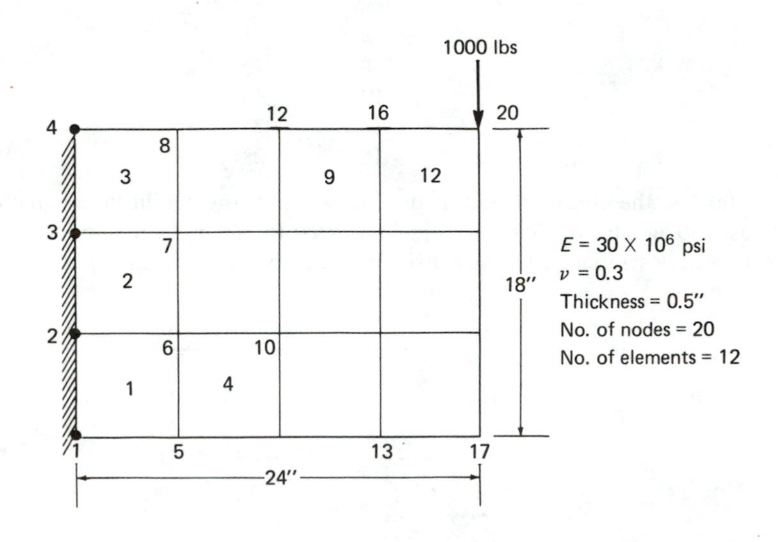

6-4 Prepare computer input data for the idealized soil mass subjected to a long strip footing load as shown. (*Hint:* Use plane strain, a unit thickness of the assemblage, and neglect weight of the mass. One title, 1 basic parameters, 1 material property, 39 nodal point data, 11 element data, 2 surface loading, and 1 blank cards are required.)

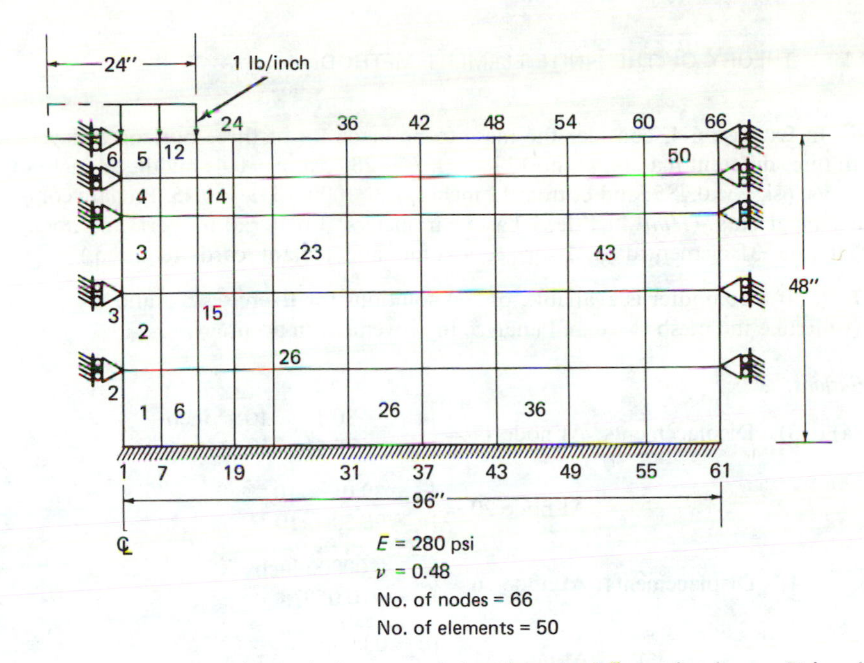

6-5 Number the nodes and elements of the discretized flat strip shown. It is subjected to the axial in-plane load as indicated. The properties of the strip are $E = 2 \times 10^6$ psi, $\nu = 0.35$, and thickness $= 0.25$ inch. Prepare input data for the computer program in Appendix I.

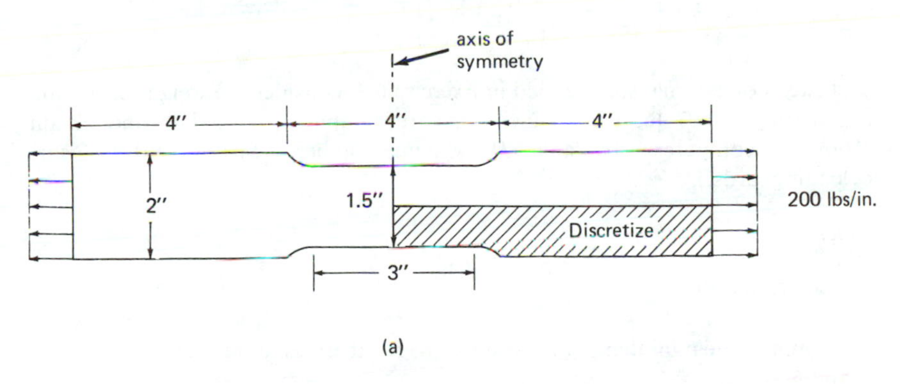

(a)

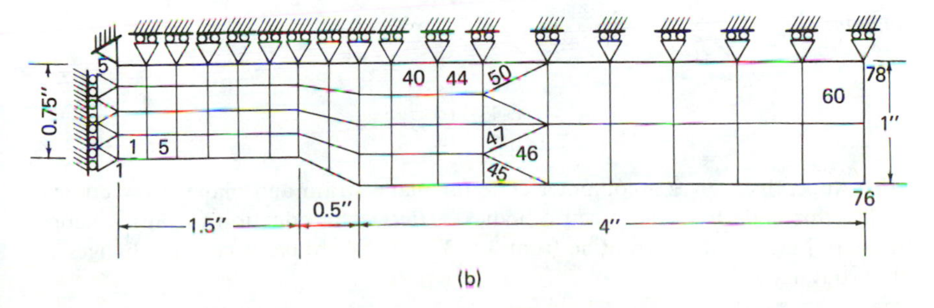

(b)

6-6 In Exercise 6-4, consider the mass to be made up of three horizontal layers with three different materials: top 12 inches, $E = 280$ psi, $\nu = 0.4$; middle 24 inches, $E = 560$ psi, $\nu = 0.485$; and bottom 12 inches, $E = 1000$ psi, $\nu = 0.45$. Prepare computer input data. (*Hints:* 1 title, 1 basic parameters, 3 material property, 39 nodal point data, 31 element data, 2 surface loading, and 1 blank cards required.)

6-7 (a) If a computer is available, obtain solutions for Exercises 6-3 and 6-4.
(b) Refine the mesh twice and analyze improvement in accuracy.

Solutions:

(a) (6-3) Displacements: At node 17 $= \begin{cases} u = -30.15 \times 10^{-5} \text{ inch} \\ v = -75.51 \times 10^{-5} \end{cases}$

At node 20 $= \begin{cases} u = 39.05 \times 10^{-5} \\ v = 96.50 \times 10^{-5} \end{cases}$

(6-4) Displacements: At node 6 $= \begin{cases} u = 0.00000 \text{ inch} \\ v = -0.05978 \end{cases}$

At node 54 $= \begin{cases} u = 0.00606 \\ v = 0.00854 \end{cases}$

Stresses: Element 5 $\begin{cases} \sigma_x = 0.554 \text{ psi} \\ \sigma_y = -0.999 \end{cases}$

Element 14 $\begin{cases} \sigma_x = -0.219 \\ \sigma_y = -0.341 \end{cases}$

6-8 Instead of external surface load in Exercise 6-4, consider the weight of the soil mass, density $= 0.07$ lbs per cu inch. Prepare computer input data and obtain solutions if a computer is available. (*Hints:* Same number of cards except the 2 surface loading cards.)

Solution:

Displacements at node 6: $\begin{cases} u = 0.00000 \text{ inch.} \\ v = 0.03291 \end{cases}$

6-9 Compute the minimum semi-band widths for the meshes in Exercises 6-2, 6-3, 6-4, and 6-5.

Solution:

6-2. 6
6-3. 12
6-4. 16
6-5. 14

6-10 Appendix I gives a computer code for plane strain and plane stress bodies. Modify this code for axisymmetric bodies. (*Hints:* Use equation (3.18b). Change matrix [C] in QUAD subroutine from 3×3 to 4×4. Make necessary changes in other routines.)

7

TECHNIQUES FOR NONLINEAR ANALYSIS

All phenomena in solid mechanics are nonlinear. In many applications, however, it is practical and convenient to use linear formulations of problems to obtain engineering solutions. On the other hand, some problems definitely require nonlinear analysis if realistic results are to be obtained. Some examples of situations in the latter category include the postyielding and large deflection behavior of structures, the postbuckling deformations of beams, plates, and shells, and nearly all problems in soil and rock mechanics.

One of the most valuable applications of the finite element method is to nonlinear problems. In this chapter we shall discuss the basic techniques of nonlinear finite element analysis. We shall preface the discussion with a treatment of the accommodation of initial strains. Not only is this subject utilized in several nonlinear analysis methods, but it is also an important aspect of the linear displacement method described in Chapters 5 and 6.

7-1 INITIAL STRAINS

Initial strains may occur in a body because of various influences such as temperature changes, creep, shrinkage, crystal growth, and *in situ* conditions.†

† Stresses (strains) arising from *in situ* conditions occur mainly in the case of soils and rocks and will be discussed in Chapter 10.

The effects of all these influences can be considered together by using the notation $\{\varepsilon_o\}$ for the total initial strain field.

A finite element formulation for small initial strains in linear elasticity can be obtained by noting that the elastic stresses correspond to the effective elastic strain $\{\varepsilon^e\}$, which is the difference between the total strain $\{\varepsilon\}$ and the initial strain $\{\varepsilon_o\}$. Hence the elastic stress-strain relations, equations (3.12b), can be expressed as

$$\{\sigma\} = [C]\{\varepsilon^e\} = [C](\{\varepsilon\} - \{\varepsilon_o\}) \tag{7.1}$$

Since the stresses and strains contribute only to the strain energy portion of the total potential functional, equation (5.42), we need consider only the strain energy term of the functional. This term can now be rewritten as

$$U = \frac{1}{2}\iiint_V \{\varepsilon\}^{\mathrm{T}}[C]\{\varepsilon\}\,dV - \iiint_V \{\varepsilon\}^{\mathrm{T}}[C]\{\varepsilon_o\}\,dV + \frac{1}{2}\iiint_V \{\varepsilon_o\}^{\mathrm{T}}[C]\{\varepsilon_o\}\,dV \tag{7.2}$$

The net effect on the resulting equilibrium equations for the finite element is the addition of a load term $\{Q_o\}$ which represents the contribution of the initial strains

$$[k]\{q\} = \{Q\} + \{Q_o\} \tag{7.3a}$$

Here the definition of $\{Q_o\}$ follows from equation (7.2) and equation (5.28b) and is

$$\{Q_o\} = \iiint_V [B]^{\mathrm{T}}[C]\{\varepsilon_o\}\,dV \tag{7.3b}$$

The concept of the *additional* or *correction load* $\{Q_o\}$ permits the effects of all initial strains to be included in the formulation. This initial strain approach is also used for nonlinear analysis and will be discussed later in this chapter.

Example 7-1: Thermal Expansion Strains. A common type of initial strain encountered in engineering problems is the strain associated with a linear thermal expansion caused by uniform temperature change. In that case, only volumetric strains result:

$$\varepsilon_{xo} = \varepsilon_{yo} = \varepsilon_{zo} = \int_{T_o}^{T} \alpha(T)\,dT \tag{7.4a}$$

T is the temperature, T_o is some base or initial temperature, and $\alpha(T)$ is the coefficient of thermal expansion. The coefficient may be temperature dependent. We may rewrite equation (7.4a) as

$$\{\varepsilon_o\} = \left(\int_{T_0}^{T} \alpha(T)\,dT\right)\{J\} \tag{7.4b}$$

where the 6 × 1 vector $\{J\}$ is defined as

$$\{J\}^T = [1\ 1\ 1\ 0\ 0\ 0] \tag{7.5}$$

Example 7-2: Stress Analysis of Porous Media. A similar approach can be utilized to include an initial stress effect in the stress analysis of porous media if both the porosity and the pore pressure are known at every point in the medium.[1] Using Biot's[2] theory of deformation of porous elastic solids, we may write the volumetric stress associated with the pore pressure as

$$\sigma_f = -fp \tag{7.6a}$$

where f is the porosity of the material and p is the pore pressure. Under the assumption that the pore pressure is independent of the deformation of the solids, the stress-strain relations are given by

$$\{\sigma\} = [C]\{\varepsilon\} + \sigma_f\{J\} \tag{7.6b}$$

where $[C]$ is the constitutive matrix obtained from tests on the porous material. In these tests the gross area of the porous section is considered. The strain energy is

$$U = \iiint_V \tfrac{1}{2}(\{\varepsilon\}^T[C]\{\varepsilon\} + 2\sigma_f\{\varepsilon\}^T\{J\})\, dV \tag{7.6c}$$

Hence, using equation (5.28b) and applying the variational principles, we obtain the correction load as

$$\{Q_o\} = -\iiint_V \sigma_f[B]^T\{J\}\, dV \tag{7.6d}$$

Because of some limitations of the definition of fluid stress in equation (7.6a), recent finite element formulations have employed an alternative approach based on the theory of mixtures (Chapters 8 and 13).

7-2 NONLINEAR PROBLEMS

In the displacement method of finite element analysis, nonlinearities occur in two different forms. The first is *material* or *physical nonlinearity*, which results from nonlinear constitutive laws (Section 3-5). The second is *geometric nonlinearity*, which derives from finite changes in the geometry of the deforming body (Section 3-2).

Categories of Nonlinear Problems

Depending on the sources of nonlinearities, we can divide nonlinear problems into three categories. In brief, these categories are problems involving material nonlinearity alone; problems involving geometric nonlinearity alone; and problems involving both material and geometric nonlinearities.

The first category, material nonlinearity alone, is the easiest to visualize. It encompasses problems in which the stresses are not linearly proportional to the strains, but in which only small displacements and small strains are considered. When we speak of displacements, we refer to the changes in the overall geometry of the body, whereas strains are related to internal deformations. The word "small" usually implies infinitesimal changes in the geometry of the body. Hence, local distortions of a differential element can be ignored. For example, the areas of the original, undeformed element can be used in computing stresses. The linear strain displacement relations, equations (3.11b), are used. Many significant engineering problems fall under the first category, an example being elastic-plastic analysis of various structures.

Although linear stress-strain equations are assumed to hold in the second category, problems involving geometric nonlinearity arise both from nonlinear strain-displacement relations, such as equations (3.11a), and from finite changes in geometry.[3] In other words, this category encompasses large strains and large displacements. An important subclass of geometrically nonlinear problems is the case of small or infinitesimal strains and large or finite displacements. An example of this subclass is the elastic postbuckling behavior of structures.

Finally, the third and most general category of nonlinear problems is the combination of the first two categories. It involves nonlinear constitutive behavior as well as large strains and finite displacements. The deformation of a rubberlike material is an example of the third category.

Method of Approach

Because the usual methods of attacking all three categories of nonlinear problems are fundamentally the same, in the succeeding section of this chapter we shall concentrate on the first category, the case of material nonlinearity with small displacements and small strains. The solution techniques for this class of nonlinear problems are relatively easy to understand. We shall eventually show how these solution methods can be extended to the other two categories.[4]

Finite element analysis for material nonlinearity is still under intensive research, and our emphasis on this class of problems does not imply that the solutions are of uniformly acceptable quality. One of the principal limitations in this area is the difficulty of adequately representing material properties. In other words, better techniques of computing material parameters and of utilizing experimental data must be devised. Relatively little information for two- and three-dimensional nonlinear material behavior has been obtained. Moreover, few analytical and experimental results are available for comparison with finite element solutions for two- and three-dimensional problems involving material nonlinearity.

Our presentations are in terms of a typical element of the assemblage. This is possible because the deformation and material properties of each element are studied independently (Chapter 5). The process of assembling the elements (Chapter 6) is unaffected by the linearity or nonlinearity of the material behavior or by the specific character of the problem.

The variational principles of mechanics, in their most general form, are applicable to nonlinear material behavior and large displacements. Therefore, the basic variational techniques for obtaining element stiffnesses and loads are the same as those presented in Chapter 5. We shall not discuss the generalization of the variational methods for nonlinear problems. Details of one variational formulation for the incremental method of nonlinear analysis can be found in the works of Biot[5] and Felippa.[6]

7-3 BASIC NONLINEAR SOLUTION TECHNIQUES

The solution of nonlinear problems by the finite element method is usually attempted by one of three basic techniques: *incremental* or *stepwise procedures*, *iterative* or *Newton methods*, and *step-iterative* or *mixed procedures*. In this section, we shall present the fundamentals of each of these techniques. For simplification, we shall consider only the nonlinear equilibrium equation for a single element

$$[k]\{q\} = \{Q\} \tag{7.7a}$$

where the nonlinearity occurs in the stiffness matrix $[k]$, which is a function of nonlinear material properties $[C(\sigma)]$. We can indicate that the material parameters in $[k]$ are no longer constants by writing

$$[k] = [k(\{q\},\{Q\})] \tag{7.7b}$$

The symbolic nonlinear relationship between $\{Q\}$ and $\{q\}$ is shown in Figure 7-1(a).† Figure 7-1(b) shows the nonlinear stress-strain curve corresponding to the load, $\{Q\}$, and displacement, $\{q\}$, in Figure 7-1(a). It is on the basis of this stress-strain or constitutive law that we determine the variable matrix $[C(\bar{\sigma})]$ for the nonlinear analysis.

† In the figures of this chapter the matrix notation is dropped for clarity.

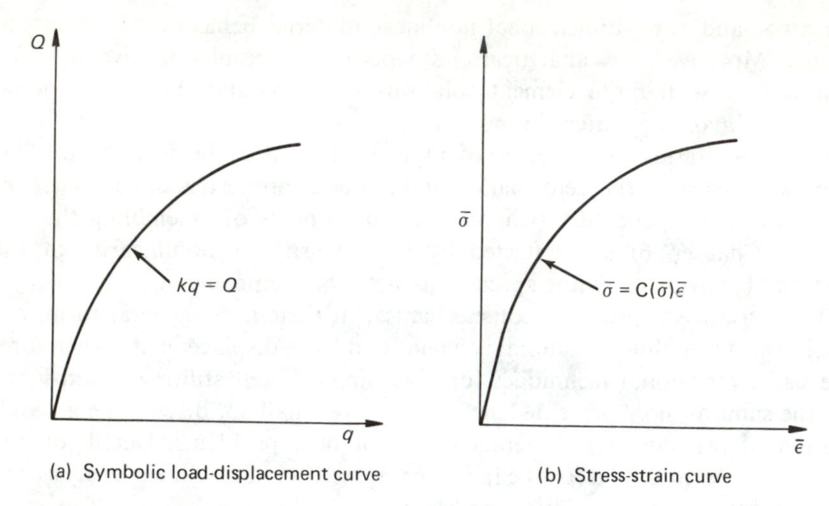

(a) Symbolic load-displacement curve (b) Stress-strain curve

Figure 7-1 Nonlinear curves.

Incremental Procedures

The basis of the incremental or stepwise procedure is the subdivision of the load into many small partial loads or increments. Usually these load increments are of equal magnitude, but in general they need not be equal. The load is applied one increment at a time, and during the application of each increment the equations are assumed to be linear. In other words, a fixed value of $[k]$ is assumed throughout each increment, but $[k]$ may take different values during different load increments. The solution for each step of loading is obtained as an increment of the displacements $\{q\}$. These displacement increments are accumulated to give the total displacement at any stage of loading, and the incremental process is repeated until the total load has been reached.

Essentially, the incremental procedure approximates the nonlinear problem as a series of linear problems, that is, the nonlinearity is treated as *piecewise linear* (Chapter 3).

In writing equations for the incremental method, let the initial or reference state of the body be given by the initial loads and displacements, $\{Q_o\}$ and $\{q_o\}$. Usually, $\{Q_o\}$ and $\{q_o\}$ are null vectors because we start from the undeformed state of the body. We can, however, specify any initial equilibrium state of $\{Q_o\}$ and $\{q_o\}$. We divide the total load into M increments, so the total effective load is

$$\{Q\} = \{Q_o\} + \sum_{j=1}^{M} \{\Delta Q_j\} \tag{7.8a}$$

where the Δ notation is used to indicate a finite increment. Hence, after the application of the ith increment, the load is given by

$$\{Q_i\} = \{Q_o\} + \sum_{j=1}^{i} \{\Delta Q_j\} \tag{7.8b}$$

where $\{Q_M\} = \{Q\}$. We adopt a similar notation for the displacements, so that after the ith step the displacements are

$$\{q_i\} = \{q_o\} + \sum_{j=1}^{i} \{\Delta q_j\} \tag{7.9}$$

To compute the increment of displacements, we use a fixed value of the stiffness, which is evaluated at the end of the previous increment

$$[k_{i-1}]\{\Delta q_i\} = \{\Delta Q_i\} \text{ for } i = 1,2,3,\ldots, M \tag{7.10a}$$

where

$$[k_{i-1}] = [k_{i-1}(\{q_{i-1}\}, \{Q_{i-1}\})] \tag{7.10b}$$

and where $[k_o]$ is the initial value of the stiffness. $[k_o]$ is computed from material constants derived from the given stress-strain curves at the start of the loading. Equations (7.10) give the basic incremental method, and equations (7.8) and (7.9) are essential auxiliary relations. The incremental procedure is schematically indicated in Figure 7-2. Usually, in the incre-

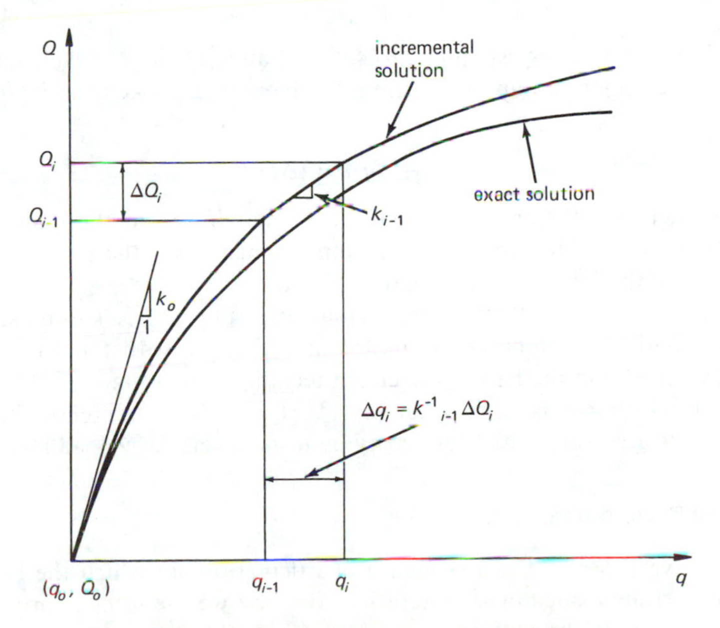

Figure 7-2 Basic incremental procedure.

mental procedure the tangent moduli (Chapter 3) are used to formulate $[C(\sigma)]$ and to compute the stiffness matrix $[k]$ in equation (7.10). This matrix is often referred to as the *tangent stiffness matrix.*

The incremental method is analogous to the numerical methods used for the integration of systems of linear or nonlinear differential equations, such as the Euler method and Runge-Kutta techniques.

The accuracy of the incremental procedure can be improved by taking smaller increments of the load, say by adopting half of the load increment. However, since a new incremental stiffness matrix $[k_{i-1}]$ must be computed for each step, we see that the increased accuracy is purchased at the cost of additional computational effort.

The *midpoint Runge-Kutta scheme* is one common modification of the incremental method which utilizes the additional computational effort to better advantage than the device of simply halving the load increment. Here two cycles of analysis are performed for each load increment. In the first cycle half of the increment $\{\Delta Q_i\}$ is applied, and a temporary displacement increment $\{\Delta q^*_{i-1/2}\}$ is computed from

$$[k_{i-1}]\{\Delta q^*_{i-1/2}\} = \{\Delta Q_i\}/2 \tag{7.11a}$$

where the superscript * indicates a temporary status. The displacements at the midpoint of the increment are now computed

$$\{q^*_{i-1/2}\} = \{q_{i-1}\} + \{\Delta q^*_{i-1/2}\} \tag{7.11b}$$

The stiffness matrix corresponding to $\{q^*_{i-1/2}\}$ and $\{Q_{i-1/2}\}$ is evaluated and utilized to compute an approximation to the full displacement increment, as follows:

$$[k_{i-1/2}]\{\Delta q_i\} = \{\Delta Q_i\} \tag{7.11c}$$

Effectively, $[k_{i-1/2}]$ is an approximation of the stiffness at the midpoint of the ith increment. Therefore, it is an improvement over the stiffness at the beginning of the increment in equation (7.10a).

The midpoint Runge-Kutta scheme is shown in Figure 7-3. This illustration also schematically compares the midpoint technique with the basic incremental method and the halved increment technique by using the artifice that each method has the same result at $\{q_{i-1}\}$, $\{Q_{i-1}\}$. The differences between the methods are exaggerated for clarity by using a very large load increment.

Iterative Procedures

The iterative procedure is a sequence of calculations in which the body or structure is fully loaded in each iteration. Because we use some approximate, constant value of the stiffness in each step, equilibrium is not necessarily

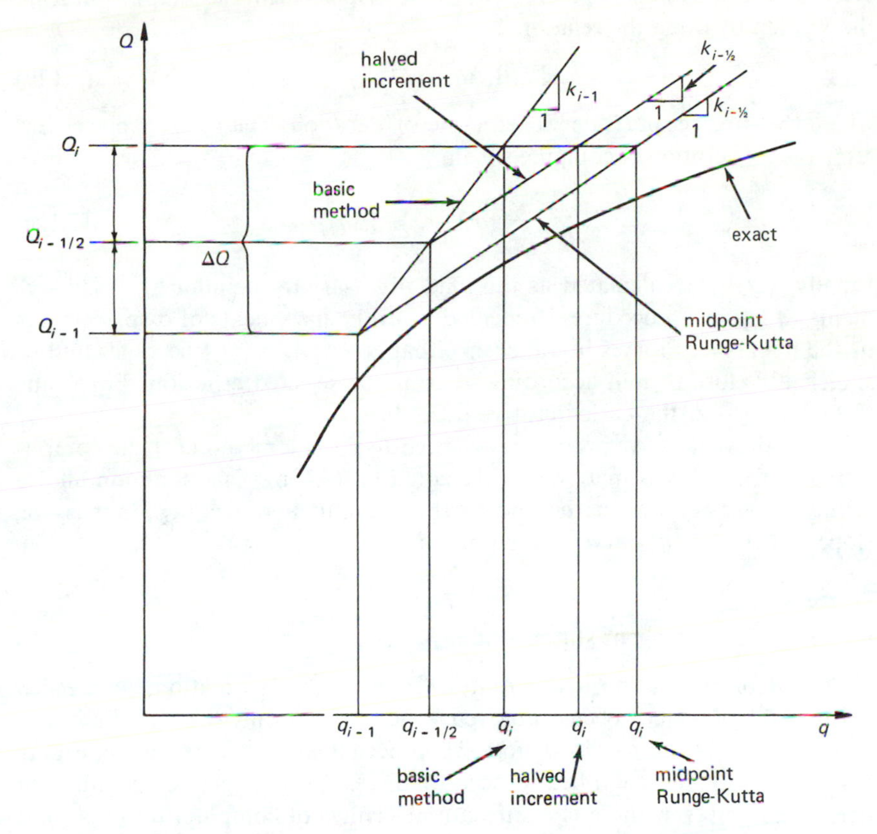

Figure 7-3 Midpoint Runge-Kutta incremental scheme.

satisfied. After each iteration, the portion of the total loading that is not balanced is calculated and used in the next step to compute an additional increment of the displacements. This process is repeated until equilibrium is approximated to some acceptable degree. Essentially, the iterative procedure consists of successive corrections to a solution until equilibrium under the total load $\{Q\}$ is satisfied.

Let $\{Q_o\}$ and $\{q_o\}$ be the initial loads and displacements in our nonlinear problem. $\{q_o\}$ and $\{Q_o\}$ are not necessarily null in the general case. For the ith cycle of the iteration process, the necessary load is determined by

$$\{Q_i\} = \{Q\} - \{Q_{e,\, i-1}\} \tag{7.12a}$$

where $\{Q\}$ is the total load to be applied and $\{Q_{e,\, i-1}\}$ is the load equilibrated after the previous step. An increment to the displacements is computed during the ith step by using the relation

$$[k^{(i)}]\{\Delta q_i\} = \{Q_i\} \tag{7.12b}$$

where the superscript (i) denotes a cycle of iteration. The total displacement after the ith iteration is computed from

$$\{q_i\} = \{q_o\} + \sum_{j=1}^{i} \{\Delta q_j\} \tag{7.12c}$$

Finally, $\{Q_{e,i}\}$ is calculated as the load necessary to maintain the displacements $\{q_i\}$.† The procedure is repeated until the increments of displacements or the unbalanced forces become zero, that is, $\{\Delta q_i\}$ or $\{Q_i\}$ becomes null or sufficiently close to null according to some preselected criterion. Equations (7.12) summarize the basic iterative procedure.

In the above iterative procedure we need to select a method for the computation of the stiffness matrix $[k^{(i)}]$ in equation (7.12b). One common choice is the *tangent stiffness* at the end of the previous iterative step, that is, the slope of the $\{Q\}$-$\{q\}$ curve at the point $\{q_{i-1}\}$, $\{Q_{i-1}\}$

$$[k^{(i)}] = [k_{i-1}] \tag{7.13}$$

where $[k_o]$ is the tangent stiffness at $\{q_o\}$, $\{Q_o\}$.

Instead of computing a different stiffness for each iteration, a *modified iterative technique* has been employed which utilizes only the initial stiffness $[k_o]$. The basic and modified iterative procedures are illustrated in Figure 7-4. Obviously the modified procedure necessitates a greater number of iterations; however, there is a substantial savings of computation because it is not necessary to invert a new stiffness at each cycle.[4]

The iterative method described here is analogous to various iterative schemes used for the numerical solutions of nonlinear equations, such as the Newton and Newton-Raphson methods.

Mixed Procedures

The step-iteration or mixed procedures utilize a combination of the incremental and iterative schemes. One such method is shown in Figure 7.5. Here the load is applied incrementally, but after each increment successive iterations are performed. Apparently, the method yields higher accuracy at the price of more computational effort.

† $\{Q_{e,i}\}$ is computed by first obtaining the strains, $\{\varepsilon_i\}$, from equation (5.28) and then substituting these strains into equation (7.3b).

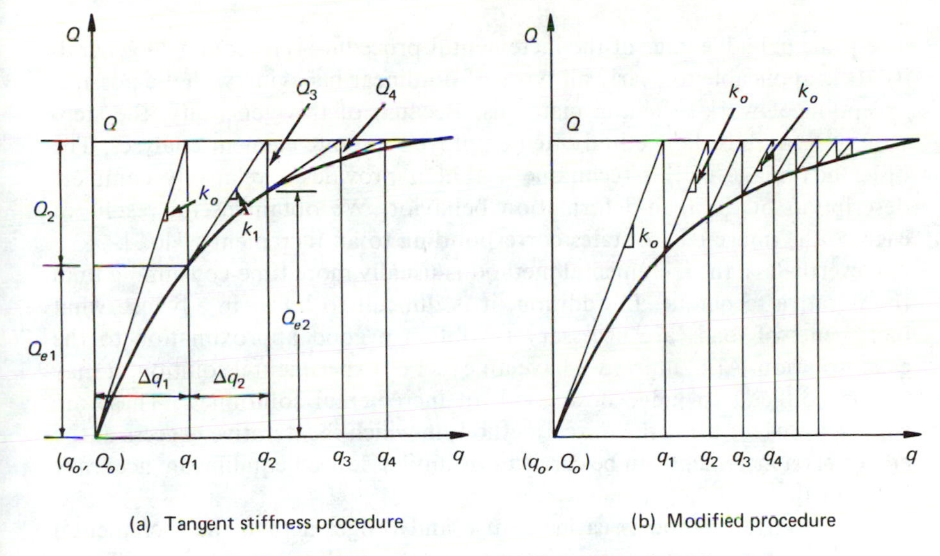

Figure 7-4 Iterative or Newton procedures.[4]

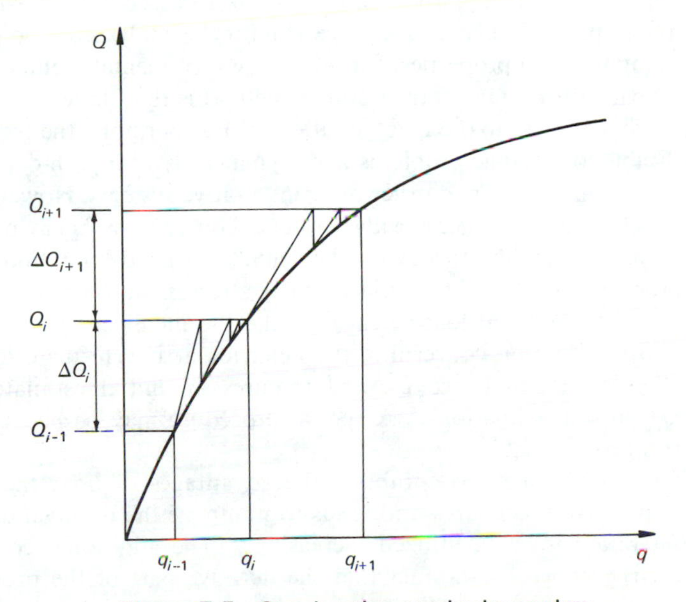

Figure 7-5 Step-iteration or mixed procedure.

Comparison of the Basic Procedures

The principal advantage of the incremental procedure is its complete generality. It is applicable to nearly all types of nonlinear behavior, with the possible exception of work-softening materials. Because of this generality, the stepwise procedure is the method often employed in finite element analyses. The other advantage of this technique is that it provides a relatively complete description of the load-deformation behavior. We obtain useful results at each of the intermediate states corresponding to an increment of load.

Nevertheless, the incremental method is usually more time-consuming than the iterative technique. In addition, it is difficult to know in advance what increments of loads are necessary to obtain a good approximation to the exact solution. Also, unless we have an exact or experimental solution, it may become difficult to judge how good an incremental solution is. These are some reasons for using the mixed method, in which the iterative process at the end of each increment can be carried out until a desired equilibrium accuracy is attained.

The iterative method is easier to use and program than the incremental method; and it is faster, provided we need to analyze only a few different loadings. It has been found useful in the case in which the materials have different elastic properties in tension and compression, the so-called bimodular materials.[7] Finally, the iterative procedure, combined with the secant stiffness approach, may prove successful for the analysis of bodies with work-softening material properties, for which the incremental method fails.

The principal disadvantage of the iterative method is that there is no assurance that it will converge to the exact solution.[6,8] Furthermore, the technique is not applicable to dynamic problems and to materials with path-dependent behavior, such as hysteritic and other nonconservative systems. However, the approach has been applied successfully to pseudoconservative behavior, as in the case of the deformation theory of plasticity.[6,7] A third limitation of the iterative procedure is that the displacements, stresses, and strains are determined for only the total load; hence, unlike the incremental and mixed solutions, no information concerning the behavior at intermediate loads is obtained. Finally, the iterative method requires an initial estimate of a nonzero $\{q_o\}$ in some situations.[4] A special procedure may be necessary to establish this first estimate.[9]

Because the mixed method combines the advantages of both the incremental and iterative procedures and tends to minimize the disadvantages of each, step-iteration is being utilized increasingly. The additional computational effort is justified by the fact that the iterative part of the procedure permits one to assess the quality of the approximate equilibrium at each stage.

7-4 COMPUTER ASPECTS OF NONLINEAR TECHNIQUES

In physically nonlinear problems the material properties vary. An important aspect of our analysis, therefore, is a knowledge of the stress-strain curves or the constitutive laws. In Chapter 3 we described the two basic forms for expressing constitutive laws for computational purposes: the tabular or digital form, and the functional form. Further aspects of these representations for computer input are given in Chapter 10. However, regardless of the mode of representation, we can express the constitutive law symbolically as

$$\{\sigma\} = f(\{\sigma\}, \{\varepsilon\}) = [C(\{\sigma\})]\{\varepsilon\} \tag{7.14}$$

Evaluation of Stiffness

At every stage of an iterative or incremental procedure, we usually must compute new element stiffness matrices. A prerequisite to this computation is the current value of the material parameters $[C]$. For example, in the case of isotropic elastic behavior we need two material parameters, such as E and ν (or λ and μ, or K and G), which are functions of the state of stress within the finite element. The tangent moduli and the secant moduli are the two common approaches used to compute the constitutive behavior of physically nonlinear materials (Section 3-5).

Whether we use the tabular or the functional representation of the constitutive law, we should be careful to ensure that the real behavior is consistently simulated. Inconsistent results may be obtained from an unrealistic simulation of the material properties, because the tangent or secant modulus is essentially computed on the basis of the first derivatives of the simulated curves.

Once the increments of displacements are obtained, the increments of the strains and stresses may be evaluated by using the proper strain-displacement equations and the current stress-strain law. For the small displacement, small strain case of nonlinear elasticity, for instance, the incremental forms of the strain-displacement equation (5.28), and the stress-strain, equation (3.12b), are

$$\{\Delta\varepsilon_i\} = [B]\{\Delta q_i\} \tag{7.15}$$

$$\{\Delta\sigma_i\} = [C(\{\sigma_{i-1}\})]\{\Delta\varepsilon_i\} \tag{7.16}$$

For small displacements, the stresses and strains may be cumulatively added, so at the end of the ith stage

$$\{\varepsilon_i\} = \{\varepsilon_o\} + \sum_{j=1}^{i} \{\Delta\varepsilon_j\} \tag{7.17}$$

$$\{\sigma_i\} = \{\sigma_o\} + \sum_{j=1}^{i} \{\Delta\sigma_j\} \tag{7.18}$$

where $\{\varepsilon_o\}$ and $\{\sigma_o\}$ are the initial strains and stresses at $\{q_o\}$, $\{Q_o\}$.

Once the current state, $(\{\sigma_i\}, \{\varepsilon_i\})$, is computed, we can enter the given constitutive law, such as a uniaxial test curve, Figure 3-5(a), a triaxial test curve, Figure 3-5(b), or an equivalent stress-strain curve, equation (3.35), to compute the appropriate moduli.

Elementary Flow Charts for the Basic Procedures

Flow charts for the basic incremental and iterative procedures are given in Figures 7-6 and 7-7, respectively. These charts are based on the equations

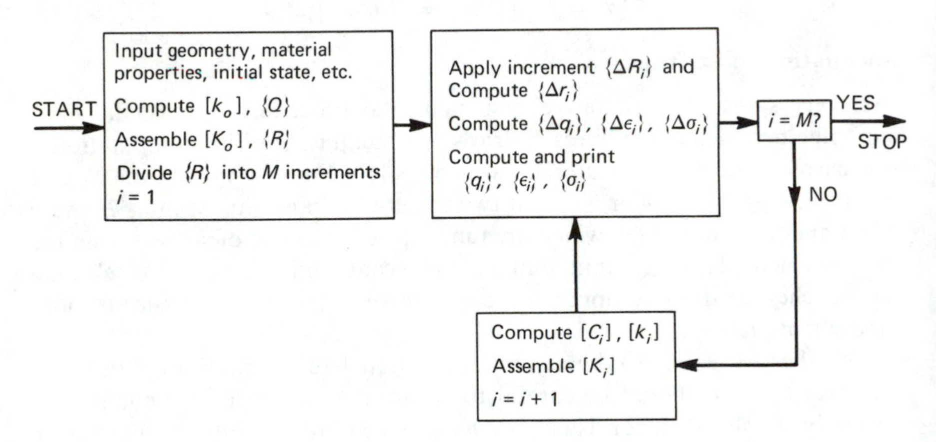

Figure 7-6 Flow chart for basic incremental procedure.

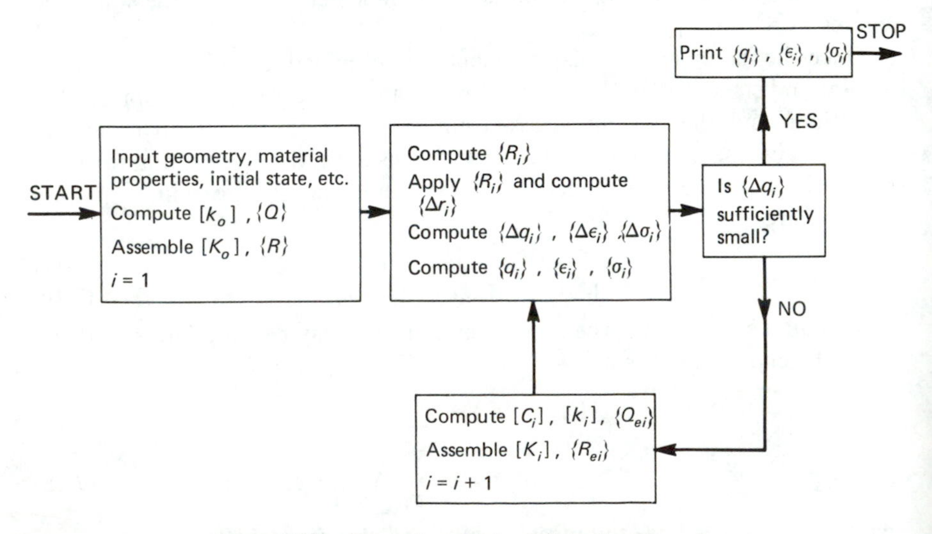

Figure 7-7 Flow chart for basic iterative procedure.

given in Section 7-3 and are generalized for an assemblage of elements. The assembly process implied in these figures is the direct stiffness method described in Chapter 6.

7-5 PROBLEMS INVOLVING MATERIAL NONLINEARITY

We will now discuss some specific versions of the nonlinear analysis procedures that are used for the first category of nonlinear problems, material nonlinearity. We will consider only nonlinear elastic and elastic-plastic behavior; general viscoelastic problems, which include the effects of time (the rate of loading) and of the previous history of material deformation, are more complicated.

Both incremental and iterative methods are used for nonlinear elasticity and for elastoplasticity. In addition, two modified methods are also commonly employed for elastic-plastic analysis: the *initial strain method*, and the *initial stress method*. After a brief treatment of nonlinear elasticity, we shall discuss these four procedures for elastic-plastic behavior.

Nonlinear Elastic Behavior

A common solution procedure for elastic problems is the incremental method utilizing the tangent stiffness concept. The iterative technique is occasionally employed in conjunction with the secant stiffness approach. In either case, the solution procedure is straightforward if no unloading or hysteritic elasticity occurs. If they do occur, the incremental method must be used and special modifications to the procedure may become necessary.

The stiffness matrix for the nonlinear analysis is computed from the usual relationship, equation (5.45a)

$$[k] = \iiint_V [B]^{\mathrm{T}}[C(\sigma)][B]\, dV \tag{5.45a}$$

The matrix $[C(\sigma)]$ is now variable and may need updating at each step of the procedure.

Incremental Method for Elastic-Plastic Behavior

This method is essentially the procedure given in equations (7.8) through (7.10). It is the most general technique available and is commonly used, particularly with the flow theory of plasticity.[10–16]

In applying the incremental procedure, the elastic-plastic constitutive law, equation (3.32b), can be expressed in terms of finite increments as

$$\{\Delta\sigma\} = [C^{ep}]\{\Delta\varepsilon\} \tag{7.19}$$

The matrix $[C^{ep}]$ is updated for each increment of load by modifying its components, as explained in Section 3-5. In other words, the tangent stiffness is computed at the end of each increment and used for the succeeding increment, equation (7.10b). The increments of plastic strain $\{\Delta\varepsilon^P\}$ required for the modification of $[C^{ep}]$ are computed from equation (3.31a). Once $[C^{ep}]$ is obtained, the element stiffness can be computed by writing equation (5.45a) as

$$[k] = \iiint_V [B]^T[C^{ep}][B]\, dV \tag{7.20}$$

The principal drawback of the incremental method is the necessity of computing new constitutive matrices and element stiffness and of assembling and inverting an overall stiffness for each step of the analysis.

Iterative Method for Elastic-Plastic Behavior

The iterative method is used for elastic-plastic behavior when the deformation theory of plasticity is employed. Usually, a bilinear approximation to the stress-strain relation is adopted, with a reduced, but nonzero (strain-hardening) modulus in the plastic zone.[7,8] A schematic representation of the iterative procedure is shown in Figure 7-8. Here the stiffness is modified after each step on the basis of a secant modulus. The total load is applied in each iteration

$$[k_i]\{q_i\} = \{Q\} \tag{7.21}$$

and the process is repeated until the displacements converge.

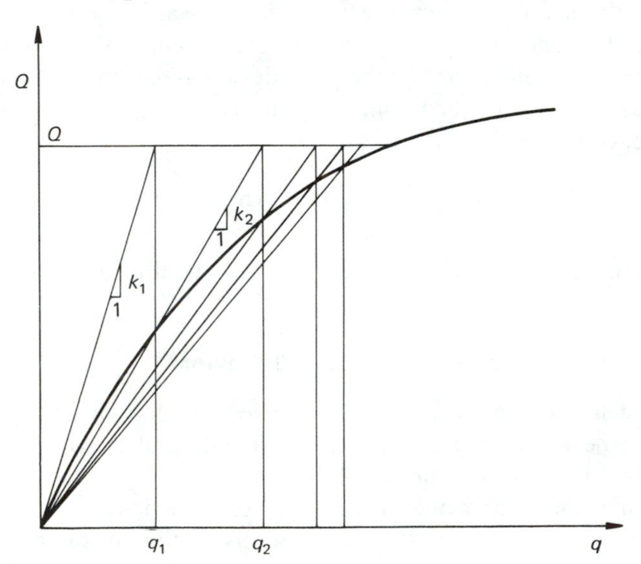

Figure 7-8 Iterative procedure for elastic-plastic behavior.

Although the nonlinear elastic-plastic constitutive relationship is used to compute the secant moduli in each iteration, it is important to note that each stiffness $[k_i]$ is utilized as if it described a linear elastic body. In a sense, therefore, each step is equivalent to an elastic analysis for the body under full loading. The computed secant moduli for each iteration are treated as if they were elastic constants. No separate consideration of elastic and plastic stresses and strains is necessary.

One disadvantage of the iterative method given by equation (7.21) is that the first few applications of the full load may induce a stress-strain state distant from the actual stress-strain curve. To avoid this possibility, a mixed procedure may be necessary.

Initial Strain Method for Elastic-Plastic Behavior

Both the iterative and incremental procedures for elastic-plastic behavior described above require the recomputation of the element stiffness matrices and solution of assemblage equations at every stage of the calculations. A modified method, the *initial strain method*, which utilizes the same stiffness throughout, has been developed to shorten the computational time.[16–19] The basic idea of this procedure parallels the concept of the additional or correction loads described in Section 7-1. Figure 7-9 shows the basic principle of the method. Under a load $\{Q\}$ with elastic stiffness $[k]$, we solve for a displacement corresponding to a point A. However, the correct equilibrium state for the load $\{Q\}$ is the point B on the nonlinear curve. The difference between the correct displacement $\{q_B\}$ and the computed displacement

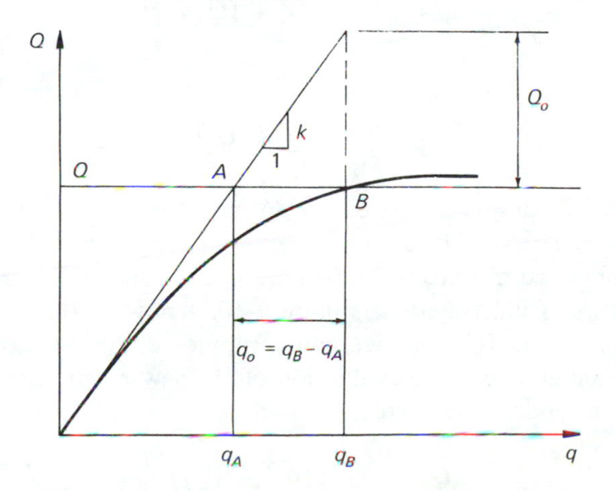

Figure 7-9 Basis of the initial strain method.

$\{q_A\}$ is $\{q_o\}$. The "initial strains" $\{\varepsilon_o\}$ corresponding to $\{q_o\}$ can be computed by the usual means

$$\{\varepsilon_o\} = [B]\{q_o\}$$

and a correction load $\{Q_o\}$ is found by using equation (7.3b)

$$\{Q_o\} = \iiint_V [B]^{\mathrm{T}}[C]\{\varepsilon_o\}\, dV \tag{7.3b}$$

This load is approximately the load necessary to offset the initial strains and to correct the calculated displacements from A to B.[17,18]

When the initial strain method is applied to the analysis of plastic behavior, the plastic strains $\{\varepsilon^P\}$ are analogous to the initial strains $\{\varepsilon_o\}$. In a manner similar to that described in Section 7-1, the elastic strain $\{\varepsilon^e\}$ can be expressed as the difference between the total strain $\{\varepsilon\}$ and the plastic strain $\{\varepsilon^P\}$. Hence, in analogy to equation (7.1),

$$\{\sigma\} = [C]\{\varepsilon^e\} = [C](\{\varepsilon\} - \{\varepsilon^P\}) \tag{7.22}$$

The final equilibrium equations are of the same form as equation (7.3a). The only difference is that the correction load vector $\{Q_o\}$ arises from the plastic strains.

A number of different schemes are employed to implement the initial strain method. Three of these schemes follow: the incremental-type procedure, the direct incremental procedure, and the iterative procedure.

The *incremental-type procedure*[16,17] is a non-iterative step-by-step scheme obtained by rewriting equation (7.3a) as

$$[k]\{q_i\} = \{Q_i\} + \{Q_{o,\,i-1}\} \tag{7.23a}$$

where

$$\{Q_i\} = \sum_{j=1}^{i} \{\Delta Q_j\} \tag{7.23b}$$

Although the loading is increased by an increment at each stage, equation (7.23b), this is not strictly an incremental scheme, since the applied load $\{Q_i\}$ and computed displacement $\{q_i\}$ are not incremental. The incremental-type procedure is illustrated in Figure 7-10. Knowing the loads $\{Q_i\}$ and the correction loads $\{Q_{o,\,i-1}\}$ we can compute $\{q_i\}$ using equation (7.23a). The next essential step is the evaluation of the new plastic strain $\{\varepsilon_i^P\}$ which can be used to update the correction load:[17]

$$\{Q_{o,\,i}\} = \iiint_V [B]^{\mathrm{T}}[C^e]\{\varepsilon_i^P\}\, dV \tag{7.23c}$$

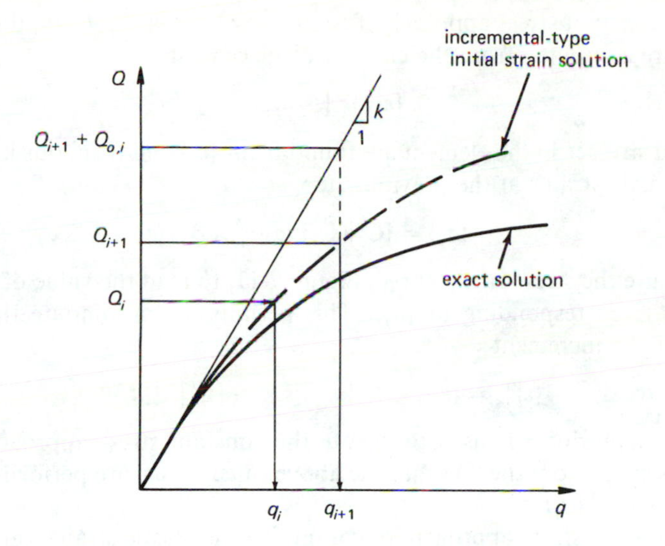

Figure 7-10 Incremental-type initial strain procedure.

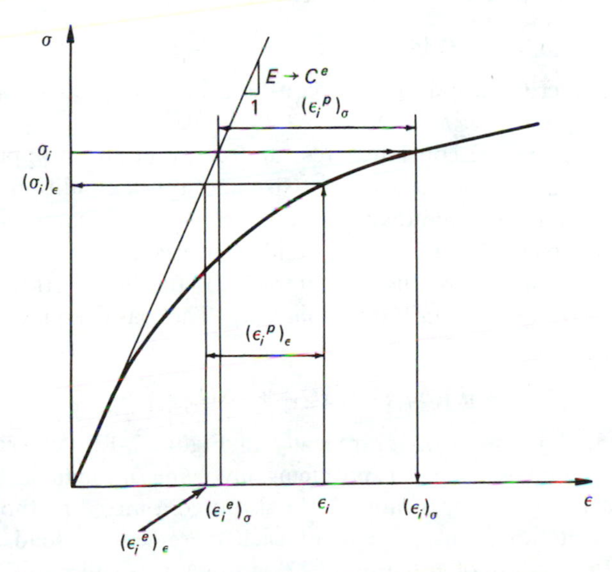

Figure 7-11 Constant stress and constant strain approaches for computing plastic strains.

Two approaches are used to compute the revised plastic strain: the *constant stress* and *constant strain* approaches.[16,17]

In the constant stress approach of computing the plastic strain, the element strains are computed from the element displacements

$$\{\varepsilon_i\} = [B]\{q_i\} \tag{7.24}$$

The total stresses in the element are found from $\{\varepsilon_i\}$ and from our knowledge of the plastic strains at the previous step

$$\{\sigma_i\} = [C^e](\{\varepsilon_i\} - \{\varepsilon_{i-1}^P\}) \tag{7.25a}$$

We next use the stress-strain curve, Figure 7-11, to read the value of the total strain $\{\varepsilon_i\}_\sigma$ corresponding to $\{\sigma_i\}$. This permits us to compute the plastic strain for the increment

$$\{\varepsilon_i^P\}_\sigma = \{\varepsilon_i\}_\sigma - \{\varepsilon_i^e\}_\sigma = \{\varepsilon_i\}_\sigma - [C^e]^{-1}\{\sigma_i\} \tag{7.25b}$$

Here the subscript σ is used to denote the constant stress approach, which draws its name from the fact that the above calculations are performed at the constant stress $\{\sigma_i\}$.

The constant strain approach for computing the plastic strain is also shown in Figure (7.11). As in the constant stress approach, the element strains $\{\varepsilon_i\}$ are computed from the element displacements, equation (7.24). We use the stress-strain curve to read the stresses $\{\sigma_i\}_\varepsilon$ corresponding to $\{\varepsilon_i\}$. This permits us to compute the revised plastic strain

$$\{\varepsilon_i^P\}_\varepsilon = \{\varepsilon_i\} - \{\varepsilon_i^e\}_\varepsilon = \{\varepsilon_i\} - [C^e]^{-1}\{\sigma_i\}_\varepsilon \tag{7.26}$$

where the subscripts ε indicate the constant strain approach based on the strain $\{\varepsilon_i\}$.

A comparison of the constant stress and constant strain approaches has shown that for a given load increment size and a well conditioned problem, the former is more accurate than the latter. However, in some problems the constant stress method may develop sudden divergence.[16]

The second basic scheme used to implement the initial strain method is the *direct incremental initial strain method*. The basic equations for this approach are

$$[k]\{\Delta q_i\} = \{\Delta Q_i\} + \{\Delta Q_{o,\,i-1}\} \tag{7.27}$$

and the procedure is shown graphically in Figure 7-12. Argyris used this direct incremental scheme for problems involving mechanical loads with plastic, thermal, and creep strains.[18] He also used a mixed method involving a number of iterations at the end of each increment of load. The direct incremental procedure of equation (7.27) is sometimes called the *incremental initial strain method*.[20]

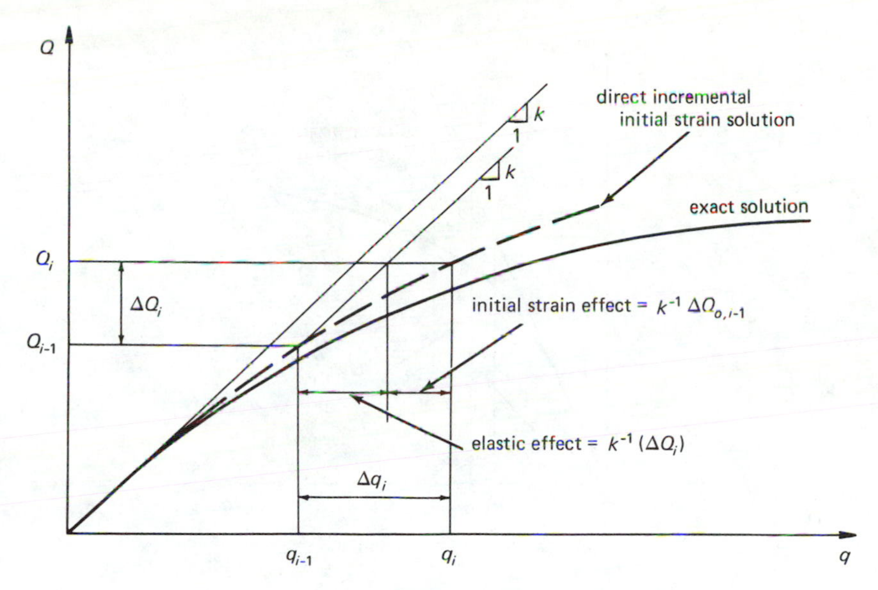

Figure 7-12 Direct incremental initial strain method.

A third initial strain method is the *iterative initial strain method*. The statement of this technique is given by

$$[k]\{q_i\} = \{Q\} + \{Q_{o,\,i-1}\} \tag{7.28}$$

which is illustrated in Figure 7-13(a). The total load $\{Q\}$ is applied for each iteration, and $\{Q_{o,o}\}$ is taken as null. To update the correction loads $\{Q_o\}$ after each iteration, an improved set of plastic strains is computed from the stress-strain law. For instance, we can use the constant stress or constant strain approach, Figures 7-13(b) and (c), respectively, for computing the plastic strains for this iterative procedure.[25]

Equation (7.28) can be combined with the incremental approach, equation (7.27), to formulate a mixed initial strain method.

If the von Mises yield criterion with the Prandtl-Reuss equations are employed, the effective stress-strain curve, equation (3.35), can be used as the basis of computations of $[C^{ep}]$ to characterize elastic-plastic behavior for the initial strain method.[18,19]

Initial Stress Method for Elastic-Plastic Behavior

The initial strain method fails to converge for perfect plasticity or for a very small degree of work hardening because of the large plastic strains that occur

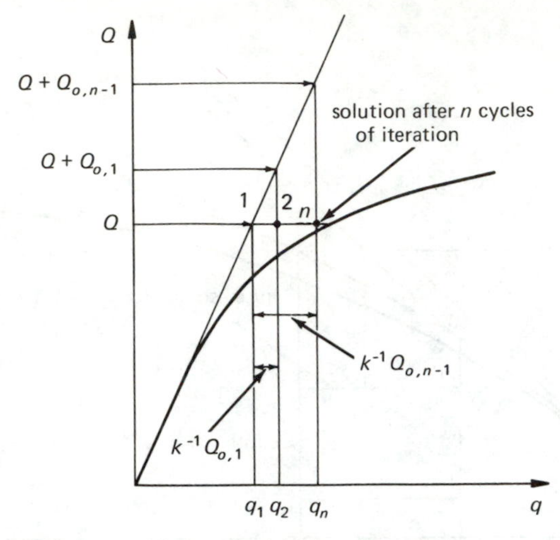

(a) Iterative initial strain method

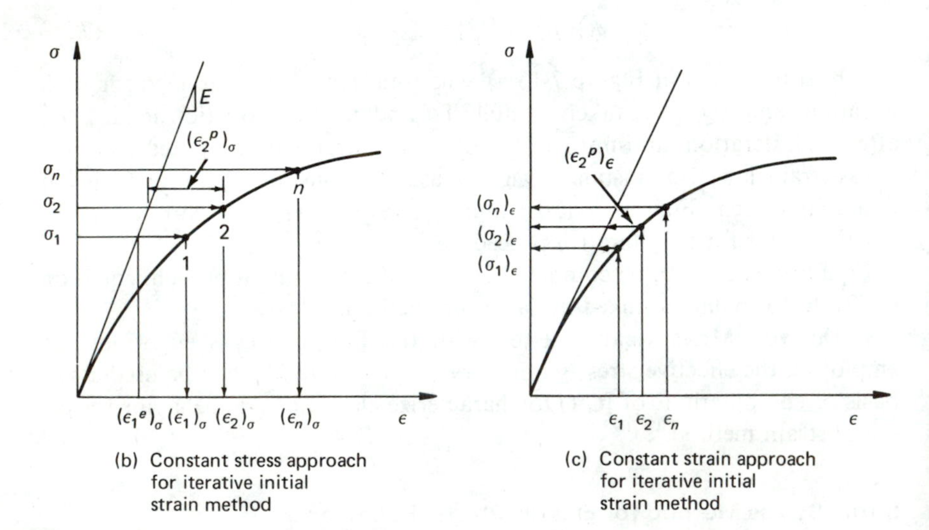

(b) Constant stress approach for iterative initial strain method

(c) Constant strain approach for iterative initial strain method

Figure 7-13 Iterative initial strain method (with constant stress and strain approaches).

in these cases. An alternative approach formulated to overcome such difficulty is the *initial stress method.*[21] This method appears to be suitable for general plastic behavior because it relies on the fact that a unique stress exists for an increment of strain. Figure 7-14 shows that the converse is not necessarily true.

The initial stress method is basically a mixed method in which the stiffness matrix is modified for each increment but is held the same for the iterations within the increment. The iterative procedure for the ith increment is given by

$$[k]\{\Delta q_i^{(j)}\} = \{\Delta Q_i\} + \{\Delta Q_{o,i}^{(j)}\} \quad \text{for} \quad j = 0, 1, 2, \ldots \tag{7.29a}$$

where

$$\{\Delta Q_{o,i}^{(o)}\} = \{0\} \tag{7.29b}$$

where i denotes a load increment and j denotes a cycle of iteration within the load increment. For $j = 0$, we apply $\{\Delta Q_i\}$ and compute increments of the displacements $\{\Delta q_i^{(j)}\}$, strain $\{\Delta \varepsilon_i^{(j)}\}$, and the elastic stress $\{\Delta \sigma_i^{(j)}\}$. Because of the nonlinearity, the stress increment $\{\Delta \sigma_i^{(j)}\}$ will not generally be the correct stress necessary to equilibrate the loads $\{\Delta Q_i\}$. If the correct stress increment is $\{\Delta \sigma_{ci}\}$, the difference between the computed and correct stress is treated as the "initial stress" and a revised correction load vector is calculated from

$$\{\Delta Q_{o,i}^{(j+1)}\} = \iiint\limits_V [B]^{\mathrm{T}}(\{\Delta \sigma_i^{(j)}\} - \{\Delta \sigma_{ci}\})\, dV \tag{7.29c}$$

This process is repeated until convergence is obtained, usually in three or four cycles.[21] Zienkiewicz *et al.*[21] proposed and used the initial stress approach

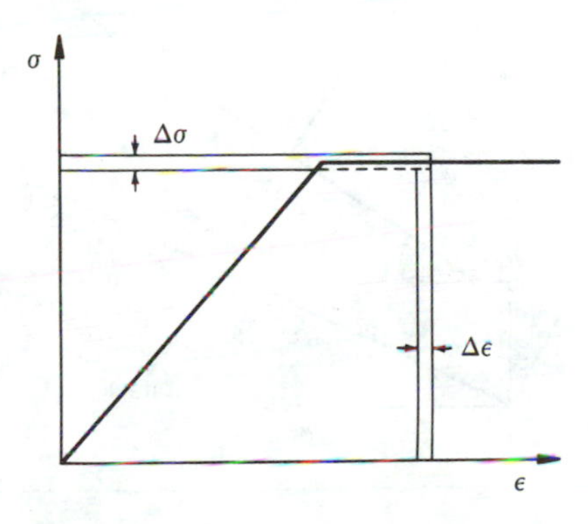

Figure 7-14 Uniqueness of stress for an increment of strain.

in which the stress increment $\{\Delta\sigma_{ci}\}$ was computed from the general elastic-plastic stress-strain relation

$$\{\Delta\sigma_{ci}\} = [C^{ep}]\{\Delta\varepsilon_i^{(j)}\} \tag{7.29d}$$

where $[C^{ep}]$ can either be derived on the basis of the Mohr-Coulomb rule, equation (3.36), or the von Mises criterion, equation (3.33).

7-6 PROBLEMS INVOLVING GEOMETRIC NONLINEARITY

The discussion in the previous sections was restricted to small strains and small displacements. In this section we shall limit ourselves to the important subclass of geometric nonlinearity in which the displacements are large but the strains are small (Section 7-2).

As usual, we solve the nonlinear problem through approximating it by a sequence of linearized subproblems. The principal effect of large displacements is that we may no longer neglect the changes of geometry brought about by the displacements. This is illustrated by the simple case of a truss member shown in Figure 7-15. Here the non-negligible change of geometry is $\Delta\theta$, which is a function of the nodal displacements u_1, u_2, v_1 and v_2. $\Delta\theta$ is a change of direction of the local coordinate system. In Section 6-5 we saw

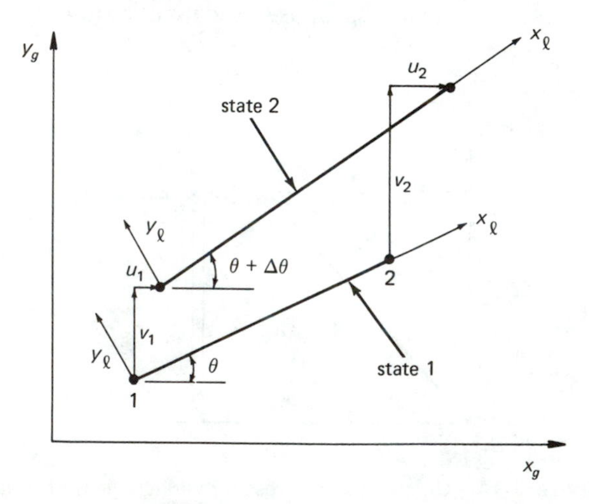

Figure 7-15 Finite displacement of truss member.

that the displacement vector, stiffness matrix, and load vector in the local system are related to those in the global system by the transformations

$$\{q_g\} = [T]^{\mathrm{T}}\{q_\ell\} \tag{6.6}$$

$$[k_g] = [T]^{\mathrm{T}}[k_\ell][T] \tag{6.8}$$

$$\{Q_g\} = [T]^{\mathrm{T}}\{Q_\ell\} \tag{6.9}$$

For the large displacement case the direction cosines in the transformation matrix $[T]$ become functions of the displacement state as well as of the initial geometry, so nonlinearities are introduced into our equilibrium equations. Symbolically, this can be denoted by

$$[T] = [T(\{q\})] \tag{7.30}$$

For the case of small strains the stiffness matrix in local coordinates $[k_\ell]$ is linear; that is, it is not a function of the displacements. Hence it remains the same for all states of deformation. Because of the nonlinearity of the transformation equation (6.8), however, the element stiffness in global coordinates $[k_g]$ varies with the displacement state.

Incremental, iterative and mixed methods can be used to deal with geometric nonlinearities.

Incremental Procedure for Geometric Nonlinearity

For an increment of load $\{\Delta Q\}$, equation (6.9) can be used to express the change in the load vector as follows:

$$\{Q_g\} + \{\Delta Q_g\} = ([T]^{\mathrm{T}} + [\Delta T]^{\mathrm{T}})(\{Q_\ell\} + \{\Delta Q_\ell\}) \tag{7.31a}$$

Here $\{Q_g\}$, $\{Q_\ell\}$ and $[T]$ are the known loads and the transformation at the beginning of the increment. Performing the matrix multiplication, we obtain

$$\{Q_g\} + \{\Delta Q_g\} = [T]^{\mathrm{T}}\{Q_\ell\} + [T]^{\mathrm{T}}\{\Delta Q_\ell\} + [\Delta T]^{\mathrm{T}}\{Q_\ell\} + [\Delta T]^{\mathrm{T}}\{\Delta Q_\ell\} \tag{7.31b}$$

The last term on the right hand side is of higher order and can be neglected for a reasonably small increment. Subtracting equation (6.9) from equation (7.31b) we obtain

$$\{\Delta Q_g\} = [T]^{\mathrm{T}}\{\Delta Q_\ell\} + [\Delta T]^{\mathrm{T}}\{Q_\ell\} \tag{7.32a}$$

The first term on the right hand side is the usual expression for the global load vector, so we may substitute the standard incremental equilibrium equation to obtain

$$\{\Delta Q_g\} = [k_g]\{\Delta q_g\} + [\Delta T]^{\mathrm{T}}\{Q_\ell\} \tag{7.32b}$$

The second term on the right represents the effect of the change of geometry on the global equilibrium equation of the element. We can rewrite this term as

$$[\Delta T]^T\{Q_\ell\} = \sum_{i=1}^{n} Q_{\ell i}\{\Delta T\}_i$$

where $\{\Delta T\}_i$ is the ith column of $[\Delta T]^T$. In analogy to the differential form

$$\{dT\}_i = \frac{\partial\{T\}_i}{\partial\{q_g\}}\{dq_g\}$$

we can write the increment of the ith column as

$$\begin{aligned}\{\Delta T\}_i &= \frac{\partial\{T\}_i}{\partial\{q_g\}}\{\Delta q_g\} \\ &= \left[\frac{\partial\{T\}_i}{\partial q_{g1}}\ \frac{\partial\{T\}_i}{\partial q_{g2}} \cdots \frac{\partial\{T\}_i}{\partial q_{gn}}\right]\{\Delta q_g\} \\ &= [G_i]\{\Delta q_g\}\end{aligned}$$

Here each matrix $[G_i]$ is symmetrical. Hence, equation (7.32b) becomes

$$[k_g]\{\Delta q_g\} + \left(\sum_{i=1}^{n} Q_{\ell i}[G_i]\right)\{\Delta q_g\} = \{\Delta Q_g\} \tag{7.33}$$

We define the quantity in parentheses as the *geometric stiffness* $[k_G]$, so the last equation can be rewritten

$$([k] + [k_G])\{\Delta q\} = \{\Delta Q\} \tag{7.34a}$$

$$[k_G] = \sum_{i=1}^{n} Q_{\ell i}[G_i] \tag{7.34b}$$

where the subscript g has been dropped. The above formulation follows Argyris.[22]

The incremental form of equation (7.34a) is

$$([k] + [k_G])_{i-1}\{\Delta q_i\} = \{\Delta Q_i\} \tag{7.35}$$

where the subscript $i-1$ indicates that stiffness matrices are evaluated for the state of displacement at the beginning of the increment. Clearly, the incremental method for large displacements requires more effort than that for material nonlinearity, because two stiffnesses must be computed at each increment. However, the effort to recompute $[k]$ is minimized by noting that the matrix $[k_\ell]$ is invariant and only the transformation in equation (6.8) changes. Equation (7.35) can be readily modified to include initial strain effects[22]

$$([k] + [k_G])_{i-1}\{\Delta q_i\} = \{\Delta Q_i\} + \{\Delta Q_{o,\,i-1}\} \tag{7.36}$$

From equation (7.34b) we see that the geometric stiffness matrix contains a contribution of the loads through the $\{Q_\ell\}$ term, which represents the loads at the beginning of the increment. Therefore, the geometric stiffness is also known as the *initial stress matrix*. In addition, by utilizing the load terms in $[k_G]$, equation (7.34a) can also be used to formulate the classical stability problem. If we can factor a load magnitude term out of $[k_G]$, for example $[k_G] = \lambda[\bar{k}_G]$, the eigenvalue problem for buckling can be obtained by considering an increment without external loading

$$([k] + \lambda[\bar{k}_G])\{\Delta q\} = 0 \tag{7.37a}$$

which has nontrivial solutions only if

$$|\,[k] + \lambda[\bar{k}_G]\,| = 0 \tag{7.37b}$$

This form can be transformed to an eigenvalue problem in standard form by the techniques presented in Chapter 2. Because of the above application, $[k_G]$ is also known as the *stability matrix*.

Iterative Procedure for Geometric Nonlinearity

The iterative method for large displacements is relatively simple. It consists of applying the total load and using the resulting displacements to revise the locations or coordinates of the nodal points at each cycle of iteration. The new geometry is used to recompute the stiffness and loads. Hence, at every stage the element characteristics are revised and a linear analysis is performed. If the strains are small, the equations for this process may be written

$$[k_{i-1}]\{q_i\} = \{Q_{i-1}\} \tag{7.38a}$$

$$\{Q_{i-1}\} = [T_{i-1}]^{\mathrm{T}}\{Q_\ell\} \tag{7.38b}$$

$$[k_{i-1}] = [T_{i-1}]^{\mathrm{T}}[k_\ell][T_{i-1}] \tag{7.38c}$$

The process is repeated until the displacements no longer change significantly. Although the convergence is usually rapid, the usual disadvantages of iterative methods apply. Specifically, the result is obtained for only the total load; and only one load vector may be applied at a time. The process is cumbersome compared to the incremental procedure.[7,20]

7-7 PROBLEMS INVOLVING BOTH MATERIAL AND GEOMETRIC NONLINEARITY

Work in the most general category of nonlinear problems is still in the development stage. Recent significant contributions to finite deformation theory for the finite element method have been made by Oden and his associates,[4] Becker,[23] Yaghmai,[24] and others.[4] A complete discussion of these problems is beyond the scope of this work; however, we shall state the general form of the incremental equations:[4]

$$([k] + [k_G] + [k_h] + [k_r])\{\Delta q\} = \{\Delta Q\} \tag{7.39}$$

Here $[k]$ is the usual *elastic stiffness matrix*, $[k_G]$ is the *geometric stiffness matrix*, $[k_h]$ is the *initial displacement matrix*, and $[k_r]$ is the *load correction matrix*. A generalized mathematical basis for the incremental and iterative procedures is given by Oden,[4] along with a comprehensive list of references.

REFERENCES

(1) Crose, J. G., "Finite Element Analysis of Porous Media," Rept. No. TR-0200 (S4816-76)-1, The Aerospace Corp., May 1969, NTIS, Springfield, Va. Presents a finite element analysis for deformations of porous elastic solids. Uses Biot's theory, Reference 2.

(2) Biot, M. A., "Theory of Elasticity and Consolidation for a Porous Anisotropic Solid," *J. Appl. Phys.*, Vol. 26, No. 2 (Feb. 1955). Develops general coupled equations of elasticity and consolidation for porous media.

(3) Powell, G. H., "Theory of Nonlinear Elastic Structures," *Proc. ASCE, J. STDn*, Vol. 95, ST12 (Dec. 1969). Separates different sources of geometric nonlinearity and presents expressions for secant and tangent stiffnesses of members.

(4) Oden, J. T., "Finite Element Applications in Nonlinear Structural Analysis," *Symp. FEM*. Presents mathematical basis of various methods of nonlinear analysis and gives a comprehensive review and list of references.

(5) Biot, M. A., *Mechanics of Incremental Deformation*, New York, John Wiley & Sons, 1965.

(6) See Reference 13, Chapter 2.

(7) Wilson, E. L., "Finite Element Analysis of Two-Dimensional Structures," Ph. D. Dissertation, University of California, Berkeley, 1963. One of the earlier applications of iterative nonlinear finite element analysis.

(8) See Reference 10, Chapter 1.

(9) Walker, A. C., "A Method of Solution for Nonlinear Simultaneous Algebraic Equations," *IJNME*, Vol. 1, No. 2 (1969). Uses Taylor expansion for moderately nonlinear structural systems and indicates its use for an initial estimate in the Newton-Raphson method.

(10) Pope, G. G., "The Application of the Matrix Displacement Method in Plane Elasto-Plastic Problems," *First Conf.* Incremental procedure with flow theory for plane stress and plane strain problems.

(11) Swedlow, T. L., Williams, M. L., and Yang, W. M., "Elasto-Plastic Stresses in Cracked Plates," Calcit Report SM 65-19, Calif. Inst. of Tech., Pasadena, 1965. Incremental procedure with flow theory of plasticity.
(12) Marcal, P. V., and King, I. P., "Elastic-Plastic Analysis of Two-Dimensional Systems by the Finite Element Method," *Int. J. Mech. Sci.*, Vol. 9 (1967). Incremental procedure for flow theory with Prandtl-Reuss equation and von Mises yield criterion. Introduces the concept of "partial stiffness coefficients," and presents a procedure for handling the transition zone from elastic to plastic states.
(13) Akyuz, F. A., and Mervin, J. E., "Solution of Nonlinear Problems of Elasto-plasticity by the Finite Element Method," *AIAA J.*, Vol. 6, No. 10 (Oct. 1968). Incremental procedure for the solution of the contact problem.
(14) Popov, E. P., Khojasteh-Bakht, M., and Yaghmai, S., "Bending of Circular Plates of Hardening Materials," *Int. J. Solids & Structures*, Vol. 3 (1967). Uses incremental procedure with von Mises yield rule and isotropic work hardening.
(15) Popov, E. P., Khojasteh-Bakht, M., and Yaghmai, S., "Analysis of Elastic-Plastic Circular Plates," *Proc. ASCE, J. EM Dn.*, Vol. 93, EM6 (Dec. 1967).
(16) Lansing, W., Jensen, W. R., and Falby, W., "Matrix Analysis Methods for Inelastic Structures," *First Conf.* Review of the initial strain method and application to the flow theory of plasticity.
(17) Gallagher, R. H., Padlog, J., and Bijlaard, P. P., "Stress Analysis of Heated Complex Shapes," *J. Am. Rocket Soc.*, Vol. 32, No. 5 (May 1962). Presents the initial strain method for elastic-plastic analysis of heated complex structures.
(18) Argyris, J. H., "Elasto-plastic Matrix Displacement Analysis of Three-Dimensional Continua," *J. Roy. Aero. Soc.*, Tech. Note ,Vol. 69 (Sept. 1965). Generalization of the matrix method for elastic-plastic behavior. Uses initial strain approach with incremental and iterative procedures.
(19) See Reference 21, Chapter 3.
(20) Zienkiewicz, O. C., and Cheung, Y. K., "Application of the Finite Element Method to Problems in Rock Mechanics," *Proc. First Int. Cong. Rock Mechs.*, Lisbon (1966). Presents methods of nonlinear analysis for elastic and elastic-plastic solutions.
(21) See Reference 23, Chapter 3.
(22) Argyris, J. H., "Continua and Discontinua," *First Conf.* Gives a broad survey of the developments of matrix methods, including linear and nonlinear analysis, and small and large deflection theories.
(23) Becker, E. B., "A Numerical Solution of a Class of Problems of Finite Elastic Deformation," Ph.D. Dissertation, University of California, Berkeley, 1966. Solves the problem of finite stretching of a square sheet by using the finite element method.
(24) Yaghmai, S., "Incremental Analyses of Large Deformations in Mechanics of Solids with Applications to Axisymmetric Shells of Revolution," Ph.D. Dissertation, University of California, Berkeley, 1968. A general incremental variational method for the analysis of geometrically and physically nonlinear problems in continuum mechanics is developed.
(25) Murray, D. W., Private communication, University of Alberta, Edmonton.

FURTHER READING

Desai, C. S., "Nonlinear Analysis Using Spline Functions," *Proc. ASCE, J. SM&F Dn.*, Vol. 97, SM 10 (Oct. 1971).

Marcal, P. V., "Finite Element Analysis with Material Nonlinearities: Theory and Practice," *Recent Advances*. A survey of finite element analysis for elastic-plastic behavior.

Yamada, Y., Yoshimura, N., and Sakurai, T., "Plastic Stress-Strain Matrix and its Application for the Solution of Elastic-Plastic Problems by the Finite Element Method," *Int. J. Mech. Sci.*, Vol. 10 (1968). Uses small and variable load increments to negotiate the transition between the elastic and plastic zones.

See Reference 19, Chapter 9.

See Reference 63, Chapter 9.

8
GENERALIZATION OF THE THEORY

To enable the beginner to obtain a gradual understanding and a physical appreciation of the basic principles of the finite element method, we restricted our attention in Chapters 1 through 7 to the displacement method of analysis which is common in solid mechanics applications. In this chapter we shall discuss some aspects of the extension of the method to such other fields of study as those in Table 1-1. We shall also summarize briefly the mathematical foundations of the method because they have contributed significantly to the generalization of potential applications of the finite element method.[1]

8-1 GENERALIZED TERMINOLOGY AND CONCEPTS

We shall adopt a terminology which will permit a general description of the method and which will facilitate the description of the diverse applications given in this chapter and in Part C. Several terms, like *elements* and *nodal points*, introduced in Chapter 1, can be retained. However, many notions associated specifically with the displacement method and with solid mechanics must be generalized or modified. For instance, our concept of a *continuum* previously was restricted to a body, structure, or a specific quantity of solid material. We must broaden this concept to include fluids or other media of interest. In addition, for some problems the relevant continuum may consist simply of a region of space in which a phenomenon occurs.

In subdividing a continuum into an *assemblage* of finite elements, we previously envisioned either a physical subdivision of the material of the body, or the inscription of dividing lines or planes onto the material (Section 5-1). In either case the nodal points were associated with a particular particle or point of matter. In other words, the nodal points in the assemblage actually move and the assemblage mesh is deformed; that is, such movements cause displacements *of* the nodes. This concept of a *material subdivision* is not always relevant for applications to fields other than solid mechanics. For instance, in fluid mechanics the assemblage represents a *spatial subdivision* rather than a material subdivision.[2] Here the finite element does not represent the fluid, but the space through which it flows. The velocity potential is a typical primary unknown in fluid mechanics, and we wish to solve for the potential *at* nodes of an assemblage that is fixed in space.† In certain coupled problems, such as fluid flow through a porous deformable medium, we deal simultaneously with displacements and potentials, and hence may adopt a combined point of view of the assemblage. This concept is satisfactory for small displacements of the nodes.

Definitions

The principal unknowns of a problem are called the *field variables*. We have heretofore dealt with displacements, but other examples of field variables are temperature, potentials, and stresses. A *field variable model* or *pattern* is the assumed function which approximates the distribution of the field variable over a finite element, for example, a displacement model. Similarly, the *amplitude of a field variable* is the value or magnitude of the variable at a node or at a particular location. We shall use the term *nodal field variable vector* to denote the vector of unknown amplitudes at the nodal points. Finally, we occasionally refer to the derivatives or rates of change of variables as the *gradients of a field variable*, corresponding to such quantities as rotations, slopes, and strains in the displacement method.

In the displacement method we obtain a set of element equilibrium equations relating the nodal displacements to the nodal forces through a stiffness matrix. For the general case, we shall use the term *finite element equations* to refer to the algebraic equations that result from the element analysis. The vector of forcing parameters analogous to loads is called the *nodal action*

† In mechanics literature, the fixing of attention on a particular point in space is known as an *Eulerian* point of view, whereas the consideration of moving particles or points of matter is called a *Lagrangian* view.

vector or the *nodal force parameter vector.* Some examples of actions are applied temperatures, prescribed displacements, and fluid flux. The other essential component of the finite element equations is the matrix of coefficients, which is the stiffness in the displacement case. This matrix is called the *property matrix* or *characteristic matrix*, because it generally characterizes the properties of the matter or material considered in the problem. Examples are conductivity, flexibility, and permeability matrices. The process of assembling the element equations to obtain assemblage equations is identical to the *direct stiffness method*, even though the property matrix involved may not be a stiffness matrix.

Governing Principles

For convenience in formulating the finite element equations, we usually employ a *variational principle* or *energy theorem*, such as the principle of minimum potential energy used in the displacement method. Other variational principles for solid mechanics, such as Hamilton's principle, were presented in Chapter 4. The functional used in such a formulation is referred to as the *associated variational functional.* Examples are the total potential for the displacement method, and the complementary energy for the equilibrium method. Occasionally it is possible to obtain an associated functional without knowledge of a specific energy theorem that governs the problem. For example if the differential equation governing the problem is the LaPlace equation

$$\frac{\partial^2 u}{\partial x^2} + \frac{\partial^2 u}{\partial y^2} + \frac{\partial^2 u}{\partial z^2} = 0 \tag{8.1}$$

it is known from variational calculus that the associated functional is

$$A = \iiint_V \frac{1}{2} \left[\left(\frac{\partial u}{\partial x}\right)^2 + \left(\frac{\partial u}{\partial y}\right)^2 + \left(\frac{\partial u}{\partial z}\right)^2 \right] dV \tag{8.2}$$

where u is the field variable.[3,4] This functional may not have an immediately clear physical meaning for the particular problem. More often the functional can be related to some work function associated with the system.[5]

Existence of an associated functional is, however, not necessary for the finite element formulations. In addition to the direct method (Section 5-6), general energy balance concepts[6,7] and residual methods[8] can be successfully used to generate the finite element equations. The residual approach will be discussed subsequently.

8-2 THE SIX-STEP FINITE ELEMENT PROCEDURE IN GENERAL TERMS

The six steps enumerated in Section 1-3 for the displacement method of finite element analysis are now briefly summarized for general applications. To illustrate the steps, the example of seepage through a rigid, porous medium is employed throughout this section.

1. *Discretization of the continuum.* In this process the continuum is divided into an equivalent system of finite elements according to the principles outlined in Sections 5-2 and 6-1 for the displacement method. However, as explained in Section 8-1, our concept of the continuum and assemblage is broadened. In the seepage problem, for example, our elements are fixed in space and do not change in size or shape while the fluid seeps through them.

2. *Selection of the field variable models.* Assumed patterns of the field variables within each element are selected, usually in polynomial form as described in Section 5-3. The unknowns of the system thereby become the amplitudes of the field variables at the nodes. For the seepage, the field variable is usually the hydraulic head.[9] This is a scalar quantity, while displacements are vectors. Hence only one unknown amplitude will occur at each node(if gradients are not used as basic unknowns).

3. *Derivation of the finite element equations.* The derivation of the finite element equations may be achieved by direct methods (Section 5-6), variational methods (Section 5-7), or the residual methods to be described in Section 8-4. The associated functional for the seepage problem is

$$A = \iiint_V \frac{1}{2}\left[k_x\left(\frac{\partial H}{\partial x}\right)^2 + k_y\left(\frac{\partial H}{\partial y}\right)^2 + k_z\left(\frac{\partial H}{\partial z}\right)^2\right] dV \tag{8.3}$$

which is a more general form of equation (8.2). Here H is the hydraulic head and the k's are the permeabilities of the medium.[9] The resulting property matrix is the permeability matrix.

4. *Assembly of the algebraic equations for the overall discretized continuum.* The assembly process is exactly analogous to that for the displacement method, that is, the direct stiffness method is used to obtain an overall permeability matrix.

5. *Solution for the nodal field variable vector.* The solution of the overall equations is achieved by the matrix methods described in Chapter 2. If the problem is nonlinear, the methods of Chapter 7 can also be applied provided the nature of the material or geometric nonlinearity is understood.

6. *Computation of the element resultants from the nodal field variable amplitudes.* Computation of the element resultants, or secondary field variables, will be governed by the type of problem being considered. However,

this process is generally analogous to the calculation of stresses and/or strains in the displacement method (Section 6-7). In the seepage problem typical element resultants desired are the fluid velocities and/or flow rates.

By comparing the above six-step procedure with the one given in Section 1-3, we can see that the basic procedure is very much the same for all types of applications. The most significant conclusion we may draw from this is that a computer program developed for one class of problem can often be applied to another class with little or no modifications required. This is a primary feature in the systematic generality of the finite element method.

8-3 FIELDS OF APPLICATION

This section consists of a brief nonmathematical introduction to the fields of application of the generalized finite element method, with continuing emphasis upon the variational approach. Specific treatment of the diverse fields, including the variational principles and examples of solutions, are included in Part C. Table 8-1 summarizes the fields of application of the method. The table lists the field variables, the actions, the essential properties that enter the finite element equations, and the associated variational functional for each type of problem. In the case of the functionals, they are indicated only by an equation number or reference number.

Structural, Soil, and Rock Mechanics

In addition to the displacement method, other finite element approaches to problems in solid mechanics are the *equilibrium method*, the *mixed method*, and the *hybrid methods*. Pian and Tong[10] have classified and contrasted these methods, and a slightly modified form of their summary is reproduced as Table 8-2. The classification is based upon the variational principle used, the field variable models, the type of interelement relations, and the unknowns in the final equations. Basically, the distinction between the displacement, equilibrium, and mixed methods is that the field variables are compatible displacements, equilibrium stresses, and both displacements and stresses, respectively. The hybrid methods utilize field variable models of one kind within each element, and functions of the opposite kind along interelement boundaries.[10] For example, the hybrid equilibrium method involves both assumed stresses within the element and assumed displacements along the element boundaries.

The equilibrium and hybrid equilibrium methods generate governing algebraic relations that are flexibility equations having the following general form:

$$[f]\{\tau\} = [G]\{q\}$$

TABLE 8-1 GENERALIZATION OF THE FINITE ELEMENT METHOD.

Field of Application	*Field Variable(s)*	*Action(s)*	*Property(ies)*	*Location of Variational Functional*
Structural, soil, and rock mechanics				
Displacement method	Displacement	Mechanical loads, prescribed displacements	Elastic or elastic-plastic stiffness	Potential energy, equation (4.11)
Equilibrium method	Stress	Same	Same	Complimentary energy, equation (4.16)
Mixed method	Displacement and stress	Same	Same	Hellinger-Reissner principle, equation (4.22)
Structural and soil dynamics	Displacement	Same	Same plus mass and viscous damping	Hamilton's principle, equation (4.19)
Torsion of prismatic bars	Warping or stress function	Torque	Elastic shear modulus	Equation (12.9g), References 2 and 4 of Chapter 12
Thermoelasticity				
Uncoupled	Displacement and/or stress	Mechanical loads, applied displacements, temperatures	Elastic or elastic-plastic stiffness, coefficient of thermal expansion	Same as structural, soil, and rock mechanics
Coupled	Temperature plus displacement and/or stress	Same	Same plus thermal conductivity	References 23 to 27, 30, and References of Chapter 13

Field of Application	*Field Variable(s)*	*Action(s)*	*Property(ies)*	*Location of Variational Functional*
Viscoelasticity	Displacement and/or stress	Mechanical loads, applied displacements, applied temperatures	Relaxation stiffness (time-dependent)	References 31 to 34 and References of Chapter 14
Seepage in rigid media	Hydraulic head or fluid potential	Sources, flux	Permeability, porosity	Equation (12.2), References 9, 19, 56, and References of Chapter 12
Seepage in deformable media				
Uncoupled	Displacements and/or stress	Mechanical loads, applied displacements, pore pressure	Elastic or elastic-plastic stiffness plus porosity	Same as structural, soil, and rock mechanics
Coupled	Displacements and/or stress plus pore pressure	Mechanical loads, applied displacements, fluid flow	Elastic or elastic-plastic stiffness plus permeability and porosity	References 18, 22 to 25, 30, and References of Chapter 13
Heat conduction	Temperature	Heat flux, boundary temperature	Conductivity, specific heat	References 9, 19, 20, 27 and References 5–12 of Chapter 12
Fluid mechanics	Potential, velocity, or displacement	Velocity, stress, pressure, gravity	Viscosity	References 2, 4, 6, 7, and References of Chapter 14

TABLE 8-1 (Continued)

Field of Application	*Field Variable(s)*	*Action(s)*	*Property(ies)*	*Location of Variational Functional*
Hydroelasticity	Potential, velocity, displacement	Mechanical loads, displacements	Mass, stiffness, viscosity	References 28, 29, Equation 13.2, and References of Chapter 13
Diffusion-convection	Concentration function	Concentration	Dispersion coefficient, porosity	References 8, 23 of Chapter 12
Other problems in mathematical physics, such as: Electrostatics Piezoelectric vibration Gas dynamics				References of Chapter 14

TABLE 8-2 CLASSIFICATION OF FINITE ELEMENT METHODS IN SOLID MECHANICS.†

Methods	*Variational Principle*	*Assumed Inside Each Element*	*Along Interelement Boundary*	*Unknowns in Final Equations*
Displacement	Minimum potential energy	Smooth displacement distribution	Continuous displacement	Nodal displacements
Equilibrium	Minimum complementary energy	Smooth and equilibrating stress distribution	Equilibrium of boundary tractions	(a) Generalized nodal displacements (b) Stress function parameters
Hybrid equilibrium	Modified complementary energy	Smooth and equilibrating stress distribution	Assumed compatible displacements	Nodal displacements
Hybrid displacement	Modified potential energy	Smooth displacement distribution	Assumed equilibrating boundary tractions	Boundary redundant forces

† Reproduced with permission from *Numerical Methods in Engineering*, Vol. 1, No. 1, 1969, Wiley-Interscience, London.

TABLE 8-2 (Continued)

Method	*Variational Principle*	*Assumed inside each element*	*Along Interelement Boundary*	*Unknowns in Final Equations*
Mixed (for plate bending problems)	Reissner's variational principle	Smooth displacement and stress distributions	(1) Continuous w, $\frac{\partial w}{\partial n}, \frac{\partial w}{\partial s}$	Nodal displacements
			(2) Continuous w and M_n	Nodal values of w and M_n
			(3) Continuous M_n, M_{ns}, $\frac{\partial M_n}{\partial n}$	Nodal values of M_n, M_{ns} and $\frac{\partial M_n}{\partial n}$
Mixed	Modified Reissner's principle	Smooth displacement and stress distribution	Assumed distribution of interelement tractions	Displacement and stress parameters and boundary tractions

Note: w and M denote the transverse displacement and bending moment of a plate, respectively. n and s indicate normal and tangential directions at the plate element boundary.

where $\{\tau\}$ is a vector of unknown stress amplitudes. However, in most solid mechanics applications, these equations are transformed to stiffness equations for final assembly and solution.

$$[k]\{q\} = \{Q\}$$

The main reason for this widespread practice is the lack of a generalized and systematic flexibility method that can compete with the efficiency of the direct stiffness method.[40]

The displacement method has been the most widely used approach in structural, soil, and rock mechanics. The equilibrium and hybrid methods have been employed by various investigators.[10,11,12] The mixed method has had limited application.[13,14,15]

A detailed presentation of the various approaches introduced above is given in Chapter 9. Chapters 9 and 10 contain several examples drawn from structural, soil, and rock mechanics.

Structural and Soil Dynamics[16,17]

Generally, the displacement method is employed for problems in structural and soil dynamics. Hamilton's principle, equation (4.19), provides a variational basis for the finite element formulation. In addition to the stiffness matrix, other property matrices that enter the formulation are the *mass matrix* and perhaps the *viscous damping matrix*. Finally, time is one of the independent variables for dynamic problems.

For free vibration problems the finite element equations for the assemblage will constitute an eigenvalue problem. In all other cases propagation problems result. The two basic approaches to solving transient or propagation problems are the direct integration of the equations with respect to time (Section 2-4), and the *normal mode* or *mode superposition method*. The latter will be considered in the chapter on applications in dynamics, Chapter 11.

Torsion, Heat Conduction, Seepage

Torsion, heat conduction, and seepage through porous media are three important phenomena governed by differential equations of similar form, specifically the LaPlace and Poisson equations. Some other examples of this type of problem, often called *field problems*, are gravitation, irrotational motion of a perfect fluid, electrostatics, magnetostatics, steady electrical currents, diffusion-convection, and surface waves on a fluid.

The following associated functional is often employed for finite element formulations of field problems[3]

$$A = \iiint_V \frac{1}{2}\left[k_x\left(\frac{\partial\psi}{\partial x}\right)^2 + k_y\left(\frac{\partial\psi}{\partial y}\right)^2 + k_z\left(\frac{\partial\psi}{\partial z}\right)^2 - C\psi\right] dV \qquad (8.4)$$

where C, k_x, k_y, and k_z are known functions of x, y, and z. If $C = 0$, the Euler equation of equation (8.4) is LaPlace's equation

$$\frac{\partial}{\partial x}\left(k_x \frac{\partial \psi}{\partial x}\right) + \frac{\partial}{\partial y}\left(k_y \frac{\partial \psi}{\partial y}\right) + \frac{\partial}{\partial z}\left(k_z \frac{\partial \psi}{\partial z}\right) = 0 \tag{8.5}$$

Otherwise, it is Poisson's equation

$$\frac{\partial}{\partial x}\left(k_x \frac{\partial \psi}{\partial x}\right) + \frac{\partial}{\partial y}\left(k_y \frac{\partial \psi}{\partial y}\right) + \frac{\partial}{\partial z}\left(k_z \frac{\partial \psi}{\partial x}\right) = C \tag{8.6}$$

A variational principle proposed by Gurtin[18] is also employed for many applications. An advantage of Gurtin's principle is that the initial conditions are introduced explicitly into the formulation. Finally, Galerkin's residual approach has been introduced for transient field problems by Zienkiewicz and Parekh.[19]

Descriptions of some specific associated functionals and applications to torsion, heat conduction, and seepage are presented in Chapter 12.

Thermoelasticity, Consolidation, Hydroelasticity

A significant category of engineering problems is one in which interaction occurs between two or more different phenomena. Some examples are thermoelasticity, consolidation, hydroelasticity, and aeroelasticity. In the most general form of each of these problems, we cannot simply superimpose the separate effects, since each is affected by the other. Therefore, these problems are called *coupled problems*.

For example, in linear uncoupled thermoelasticity the temperature changes are known and are assumed not to affect the material properties. Such thermal effects can be included directly in our displacement formulation as equivalent initial strains (Section 7-1). However, in the coupled thermoelastic problem the changing temperatures affect the mechanical properties, and the stress-deformation behavior influences the thermal properties. Similarly, in consolidation the interacting phenomena are deformation of the porous soil, and flow of the pore fluid. In hydroelasticity or aeroelasticity, they are the deformation of solids and the motion of fluids.

Biot[20,21] was the first to present theories to account for coupled problems in thermoelasticity and consolidation. In these theories, however, the definition of the relation between fluid stress and fluid strain is inadequate.[22] A more refined theory can be found in works by Crochet and Naghdi[23] and Green and Naghdi.[24] Sandu[22] has applied the finite element method to the consolidation problem.

The finite element method has been applied to transient coupled thermoelastic problems by Nickell and Sackman,[25] and by Oden and Kross.[26] The former used a variational approach based on a stationary principle formulated by Gurtin.[18,27] The latter employed a general energy balance procedure, rather than a variational approach.[6,7]

The primary application of the finite element method to hydroelasticity has been the investigation of the sloshing of a liquid in an elastic container.[28,29]

Recently Sandhu and Pister[30] have presented a general variational principle for linear coupled field problems.

Examples of applications of the finite element method to coupled problems are given in Chapter 13.

Viscoelasticity

Time enters viscoelastic problems because the stress depends upon the rate of strain as well as the strain itself. Hence, if the strain is known as a function of time, one way of writing the linear stress-strain relationship is

$$\{\sigma(t)\} = \int_{-\infty}^{t} [C^{ve}(t - t')] \frac{\partial}{\partial t'} \{\varepsilon(t')\} \, dt' \tag{8.7}$$

where $[C^{ve}]$ is the *relaxation modulus* matrix, t is time and t' is a dummy time variable for integration. Equation (8.7) is known as a *relaxation integral law.* Gurtin[31] has formulated variational principles for the linear viscoelasticity described by equation (8.7).

Taylor and Chang[32] have employed Gurtin's variational principle to devise a displacement finite element approach to the stress analysis of viscoelastic materials under steady state temperature gradients. The equilibrium equations they obtain are a set of linear integral equations that must be integrated step-by-step in time, with the stiffness changing after each time step. Among the other investigators who have applied the finite element method to problems in viscoelasticity are Zienkiewicz, *et al.*,[33] who utilized a Kelvin material model, and Watt,[34] who considered the problem of saturated incompressible soils.

Detailed examples of applications to viscoelasticity are given in Chapter 14.

Fluid Mechanics

With the exception of the seepage and hydroelastic problems mentioned above, the field of fluid mechanics have received relatively little attention in the finite element literature. However, by using the associated functional for field problems, equation (8.4), the potential flow of fluids[4] and long-period waves[35] have been studied. These and some other applications to fluid mechanics[2] are discussed in Chapter 14.

Topics of Common Interest

Although Part C is divided into chapters that describe applications of the finite element method to various specific fields of interest, there are a number of fundamental topics in the individual chapters which are sometimes useful in fields other than the particular chapter subject. The concept of separation of variables described in Chapter 9 is a valuable tool in any type of application. The most common situation where this device is employed is that in which the geometry of the problem is axisymmetric, but the properties or actions are asymmetric. Chapter 9 also includes a description of the formulation of the method for incompressible solids. This formulation is used to study the behavior of solid rocket grains, polymers, rubber-like materials, saturated cohesive soils, and other solids with a Poisson's ratio approaching 0.5. Approaches to sequential construction are outlined in Chapter 10. The formulation for field problems, Chapter 12, is directly applicable to several of the problems in Chapter 14, such as the potential flow of inviscid fluids, the seiche of lakes, and the distribution of electrical and magnetic potentials.

8-4 MATHEMATICAL BASES OF THE FINITE ELEMENT METHOD

A rigorous study of the mathematical bases of the finite element method is beyond the scope and purpose of this book. For most engineering applications such detailed knowledge is unnecessary. Nevertheless, a sound mathematical foundation for the finite element method has been essential, both in extending the analysis technique to various fields of engineering and mathematical physics, and in studying the accuracy and convergence of the method. To initiate an appreciation of the mathematical viewpoint, we shall briefly survey the background material and pertinent references. In addition, we shall discuss two specific topics: finite elements in the space-time domain, and residual methods.

Historical Survey of Finite Element Mathematics

Interestingly, the earliest publication on what is now known as the finite element was a mathematical treatment by Courant[36] in 1943. However, as mentioned in Section 1-1, the method did not come into wide use until almost 20 years later, nor were the original applications to engineering problems based on the earlier mathematical work; rather the initial developments were based on the direct methods given in Section 5-6. With the increasing use of the technique, an effort to place the method on a sound mathematical foundation has gained momentum.

Subsequent to the use of direct methods, the connection between the finite element approach and such variational procedures as the Ritz[3,37] method was recognized.[38,39,40] The basic idea of a variational method utilizing a patchwork of assumed functions may be traced to the works of Courant[36] and Synge.[41] Felippa and Clough[40] have contrasted the physical and variational interpretations of the method, and a number of relevant references are given in their paper. For a variational description of the method, various investigators have relied upon the exposition of Mikhlin.[42,43]

The development of the theory of splines[44,45] has provided the mathematical language and tools for constructing approximate functions, or field variable models, that can be used in a rigorous treatment of the variational concept of the finite element method. A *spline function* is defined as a piecewise polynomial of the nth degree which is smoothly connected to adjoining spline functions and which has $n - 1$ continuous derivatives.[22,45]

Several authors have treated the subjects of the theoretical accuracy and the convergence characteristics of the finite element method as applied to solid mechanics.[39,46,47,48,49] In the variational approach, convergence is studied by using the concept of the *energy norm*, defined as the square root of the internal strain energy of the system.[50,51] In a different approach, the order of error and convergence of the method has been analyzed by means of Taylor's series expansions.[52]

A broadened theory of finite elements for a general class of problems in mathematical physics was recently published by Oden.[6,7] The following extract from this work summarizes the generalized properties of finite elements:

1. A finite element model can be generated for any type of continuous abstract function, its values may, for example, be scalars or vectors or tensors of any order.
2. It is shown that finite element models can be developed without introducing specific coordinate systems; thus, local approximations of various fields can be described in terms of, for example, general curvilinear coordinates, and the usual rectilinear boundaries can be regarded as special cases of a more general formulation.
3. The nature and number of independent variables is arbitrary. This means that finite element models of functions of any type of independent variable, not necessarily coordinates of points in Euclidean space, can be constructed. For example, it may be possible to construct finite element models of a thermodynamical process, in which a state function is specified in terms of the temperature, motion (implicitly), time, etc. at a point in 'state space'; or one can develop finite element models in four-dimensional space-time.
4. Finite elements can be developed in non-Euclidean spaces. Thus elements can be generated directly in a curved space without approximating the space itself by piecewise 'flat' elements.
5. The familiar ideas of Lagrange and Hermite interpolation functions can be generalized.

6. The notion of 'generalized forces,' 'generalized displacements,' etc. evolve from a more general idea of conjugate fields. Here we refer to two companion fields defined over a certain space, one being the field approximated by a finite element model and the other its 'conjugate' with respect to a functional defined on the space.

7. The formulation of a finite element model of a given function is a purely topological construction and has nothing to do with variational principles. Indeed, though always convenient, it is not always necessary to formulate finite element models by using variational principles which involve taking variations of a well-defined functional.†

Space-Time Finite Elements

An immediate application of the general theory presented by Oden is the construction of finite elements in the space-time domain. Conventionally, for the time-dependent or propagation problems (Chapter 2), we first use the finite element method to formulate the solution in the physical space. In the next step we use a different scheme, such as finite differences, to obtain solutions over a range of time. Essentially, in this procedure we first adopt a field variable model of the form

$$\{u(x,y,z,t)\} = [N_1(x,y,z)]\{q(t)\} \tag{8.8}$$

where u is the field variable, $[N_1]$ is the matrix of interpolation functions in space, and $\{q\}$ is the time-dependent nodal field variable vector. In conjunction with the finite element formulation from equation (8.8), and by using the initial conditions, we utilize a finite difference scheme such as

$$\{q(t)\} = [N_2(t, \Delta t)]\{q(t - \Delta t)\} \tag{8.9}$$

where $[N_2]$ denotes an interpolation function in the time domain.

In contrast to this conventional scheme, Oden's generalization entails a field variable model of the form

$$\{u(x,y,z,t)\} = [N(x,y,z,t)]\{q\} \tag{8.10}$$

in which the interpolation functions constituting $[N]$ contain time as one of the dimensions, as well as space. A thorough investigation of the applications of this approach has not yet been reported (1971), but the concept has attracted much attention in the search for adequate general solution methods for propagation problems.

† Reproduced with permission from *Numerical Methods in Engineering*, Vol. 1, No. 2, 1969, Wiley-Interscience, London.

Residual Methods for the Formulation of Finite Element Equations

In addition to the direct and variational approaches, the finite element equations can be formulated by employing the *residual methods*, such as the *collocation*, *least squares*, and *Galerkin* methods.[8] The differential equation for a problem can be written as

$$Lu = f \tag{8.11}$$

where u is the field variable, L is some differential operator, and f is a known action function. Then the residual R for an approximate trial solution is defined as

$$R = L\,u^* - f \tag{8.12}$$

where u^* is a trial function or a field variable model. The various residual methods are based upon different techniques for minimizing the residual. The residual methods are particularly useful for problems in which a variational functional may not be available, although they may be applied to any boundary value problem with established differential equations. The residual approach to the finite element method has been used by Szabo and Lee,[53,54] Zienkiewicz and Parekh,[19] Tada and Lee,[55] and Desai.[56]

REFERENCES

(1) Zienkiewicz, O. C., "The Finite Element Method, From Intuition to Generality," *Appl. Mech. Rev.*, Vol. 23, No. 3 (March 1970). Presents an historical development of the method with a comprehensive list of references.

(2) Oden, J. T., and Somogyi, D., "Finite Element Applications in Fluid Dynamics," *Proc. ASCE, J. EM Dn*, Vol. 95, EM3 (June 1969). Derives finite element equations with fluid velocity as the field variable. Presents the concept of spatial and material subdivision.

(3) Hildebrand, F. B., *Methods of Applied Mathematics*, Englewood Cliffs, Prentice-Hall, 1965.

(4) Martin, H. C., "Finite Element Analysis of Fluid Flows," *Second Conf.* Application to the steady irrotational flow of an incompressible fluid. Discussion of boundary conditions of both Dirichlet and Neumann type.

(5) See Reference 1, Chapter 4.

(6) See Reference 11, Chapter 1.

(7) Oden, J. T., "A General Theory of Finite Elements: II. Applications," *IJNME*, Vol. 1, No. 3 (1969). Presents applications of the theory in Reference 6. A space-time finite element is developed.

(8) See Reference 2, Chapter 1.

(9) Zienkiewicz, O. C., Mayer, P., and Cheung, Y. K., "Solution of Anisotropic Seepage by Finite Elements," *Proc. ASCE, J. EM Dn*, Vol. 92, EM1 (Feb. 1966).

(10) See Reference 8, Chapter 4.

(11) Fraeijs de Veubeke, B., "Displacement and Equilibrium Models in the Finite Element Method," In *Stress Analysis*, Chapter 9, edited by O. C. Zienkiewicz and G. S. Holister, London, John Wiley & Sons, 1965. Introduces variational principles, different types of displacement and stress models, convergence, and solution bounds for the finite element method.
(12) Pian, T. H. H., "Derivation of Element Stiffness Matrices by Assumed Stress Distribution," *AIAA J.*, Vol. 2 (July 1964).
(13) Herrmann, L. R., "A Bending Analysis for Plates," *First Conf.* Develops a mixed procedure in which transverse deflections and moments are adopted as the unknown field variables. Applicable to both thin and moderately thick plates, as shear deformations are allowed.
(14) Herrmann, L. R., "Finite Element Bending Analysis for Plates," *Proc. ASCE, J. EM Dn*, Vol. 93, EM5 (Oct. 1967). Generalizes the formulation of Reference 13.
(15) Dunham, R. S., and Pister, K. S., "A Finite Element Application of the Hellinger-Reissner Variational Theorem," *Second Conf.* Uses a linear mixed model for the derivation of the stiffness of a triangle and a quadrilateral with 4 CST for the plane stress case.
(16) Przemieniecki, J. S., *Theory of Matrix Structural Analysis*, New York, McGraw-Hill Book Co., 1968. Includes solution of dynamic problems and finite element applications.
(17) Various authors, "Earthquake Engineering Research at Berkeley," *Earthquake Eng. Res. Center Report* No. 69–1, University of California, Berkeley, Jan. 1969. NTIS, Springfield, Va. Collection of papers presented at Fourth World Conference on Earthquake Engineering, Santiago, Chile, January 1969. Includes papers on dynamic finite element applications to earth dams, soil deposits, foundation-structure interaction, reservoir-dam systems, and shell roofs.
(18) Gurtin, M. E., "Variational Principles for Linear Initial Value Problems," *Quart. Appl. Math*, Vol. 23 (1964). These principles are used in many finite element applications to transient field problems. A special aspect of these formulations is that they include explicitly the initial conditions in the variational principle.
(19) Zienkiewicz, O. C., and Parekh, C. J., "Transient Field Problems—Two- and Three-Dimensional Analysis by Isoparametric Finite Elements," *IJNME*, Vol. 2 (1970). Uses Galerkin's method and isoparametric elements. Solves example problems in transient heat flow.
(20) Biot, M. A., "Thermoelasticity and Irreversible Thermodynamics," *J. Appl. Phy.*, Vol. 27, No. 3 (March 1956). Presents basic principles, equilibrium equations, and variational principles for solutions of thermoelastic and dynamic problems including porous anisotropic media.
(21) See Reference 2, Chapter 7.
(22) Sandhu, R. S., "Fluid Flow in Saturated Porous Elastic Media," Ph. D. Dissertation, University of California, Berkeley, 1968. Develops a finite element formulation for coupled problems and uses it for flow consolidation.
(23) Crochet, M. J., and Naghdi, P. M., "On Constitutive Equations for Flow of Fluid Through an Elastic Solid," *Int. J. Engr. Sci.*, Vol. 4 (1966). Develops nonlinear constitutive laws including thermodynamic effects.

(24) Green, A. E., and Naghdi, P. M., "A Theory of Mixtures," *Arch. for Rat. Mech. and Anal.*, Vol. 24 (1967). By using the principle of material objectivity, derives equations of mass, momentum, and moment of momentum for each material constituting a mixture.
(25) Nickell, R. E., and Sackman, J. L., "Approximate Solutions in Linear, Coupled Thermoelasticity," *J. Appl. Mech.*, Vol. 35, No. 2 (June 1968).
(26) Oden, J. T., and Kross, D. A., "Analysis of General Coupled Thermoelasticity Problems by the Finite Element Method," *Second Conf.* Details in Chapter 13.
(27) Gurtin, M. E., "Variational Principles of Linear Elastodynamics," *Arch. for Rat. Mech. and Anal.*, Vol. 16, No. 1(1964). Presents a solution and associated variational principles for the mixed problem of elastodynamics for linearly elastic, inhomogeneous, and anisotropic solids. The formulation permits inclusion of the initial velocity distribution in the principle itself.
(28) Guyan, R. J., Ujihara, B. H., and Welch, P. W., "Hydroelastic Analysis of Axisymmetric Systems by a Finite Element Method," *Second Conf.* Considers the axisymmetric dynamic behavior of an elastic container-fluid system (Fig. 13-5) where the fluid is inviscid and incompressible. In the approach used, the fluid region is not discretized, and only the boundary of the structural discretization is considered. The free surface is idealized as concentirc annular ring elements.
(29) Luk, C. H., "Finite Element Analysis for Liquid Sloshing Problems," ASRL TR 144-3, Cambridge, Mass. Aeroelastic and Struct. Res. Lab., MIT, May 1969. Details in Chapter 13.
(30) Sandhu, R. S., and Pister, K. S., "A Variational Principle for Linear, Coupled Field Problems in Continuum Mechanics," *Int. J. of Eng. Sci.*, Vol. 8, No. 12 (Dec. 1970).
(31) Gurtin, M. E., "Variational Principles in the Linear Theory of Viscoelasticity," *Arch. for Rat. Mech. and Anal.*, Vol. 13 (1963).
(32) Taylor, R. L., and Chang, T. Y., "An Approximate Method for Thermoviscoelastic Stress Analysis," *Nuc. Eng. & Des.*, Vol. 4 (1966).
(33) Zienkiewicz, O. C., Watson, M., and King, I. P., "A Numerical Method of Visco-elastic Stress Analysis," *Int. J. Mech. Sci.*, Vol. 10 (1968).
(34) Watt, B. J., "Analysis of Viscous Behavior in Undrained Soils," Research Report R69-70, Cambridge, Department of Civil Engineering, MIT, Nov. 1969.
(35) Taylor, C., Patil, B. S., and Zienkiewicz, O. C., "Harbour Oscillation: A Numerical Treatment for Undamped Natural Modes," *Proc. Inst. Civil Engrs.*, Vol. 43 (June 1969). Details in Chapter 14.
(36) Courant, R., "Variational Methods for the Solution of Problems of Equilibrium and Vibrations," *Bull. Am. Math. Soc.*, Vol. 49 (1943).
(37) Kantorovitch, L. V., and Krylov, V. I., *Approximate Methods of Higher Analysis*, Groningen, P. Noordhoff, Ltd., 1958.
(38) See Reference 7, Chapter 1.
(39) See Reference 4, Chapter 5.
(40) See Reference 8, Chapter 1.
(41) Synge, J. L., *The Hypercircle in Mathematical Physics*, New York, Cambridge University Press, 1957.

(42) Mikhlin, S. G., *Variational Methods in Mathematical Physics*, New York Pergamon Press, MacMillan Co., 1964.
(43) Mikhlin, S. G., *The Problem of the Minimum of a Quadratic Functional*, San Francisco, Holden-Day, Inc., 1965.
(44) Schoenberg, I. J., "Contributions to the Problem of Approximation of Equidistant Data by Analytic Functions," *Quart. Appl. Math.*, Vol. 4 (1946).
(45) Ahlberg, J. H., Nilson, E. N., and Walsh, J. L., *The Theory of Splines and Their Applications*, New York, Academic Press, 1967.
(46) Key, S. W., "A Convergence Study of the Direct Stiffness Method," Ph. D. Dissertation, University of Washington, 1966.
(47) See Reference 6, Chapter 6.
(48) Tong, P., and Pian, T. H. H., "The Convergence of the Finite Element Method in Solving Linear Elastic Problems," *Int. J. Solids Structures*, Vol. 3 (1967).
(49) Johnson, M. W., and McLay, R. W., "Convergence of the Finite Element Method in the Theory of Elasticity," *J. Applied Mech.*, Vol. 35, No. 2 (June 1968).
(50) Prager, W., and Synge, J. L., "Approximations in Elasticity Based on the Concept of Function Space," *Quart. Appl. Math.*, Vol. 5, No. 3 (Oct. 1947).
(51) de Arantes e Oliveira, E. R., "Theoretical Foundations of the Finite Element Method," *Int. J. Solids Struct.*, Vol. 4 (1968).
(52) Walz, J. E., Fulton, R. E., and Cyrus, N. J., "Accuracy and Convergence of Finite Element Approximations," *Second Conf.*
(53) Szabo, B. A., and Lee, G. C., "Stiffness Matrix for Plates by Galerkin's Method," *Proc. ASCE, J. EM Dn*, Vol. 95, EM3 (June 1969). Plate element stiffness is obtained by minimizing the residual generated by using a polynominal displacement model.
(54) Szabo, B. A., and Lee, G. C., "Derivation of Stiffness Matrices for Problems on Plane Elasticity by Galerkin Method," *IJNME*, Vol. 1, No. 3 (July 1969).
(55) Tada, Y., and Lee, G. C., "Finite Element Solution to an Elastica Problem of Beams," *IJNME*, Vol. 2, No. 2 (1970). Uses Galerkin method to obtain finite element equations for the problem of large deflections of an inextensible beam-column.
(56) Desai, C. S., "Approximate Solution for Unconfined Seepage," *Proc. ASCE, J. I&D Dn,* Vol. 99, No. IRI, March 1973. Uses Galerkin's method to derive finite element equations. Using an iterative scheme obtains solutions for locating the free (phreatic) surface for both rising and falling external fluid heads.

FURTHER READING

Frederick, D., and Chang, T. S., *Continuum Mechanics*, Boston, Allyn and Bacon, Inc., 1965. Introductory treatment including discussion of the Eulerian and Lagrangran frames of reference mentioned in Section 8.1.
Leonard, J. W., and Bramlette, T. T., "Finite Element Solutions to Differential Equations," *Proc. ASCE, J. EM Dn*, Vol. 96, EM6 (Dec. 1970). Uses Galerkin's method to derive formulation of a general class of field equations for combined initial-value and boundary-value problems in applied mathematics. Example given is Prandtl-Meyer steady compressible flow of an isentropic gas.
See Reference 46, Chapter 9.

PART C
APPLICATIONS

9

STRUCTURAL MECHANICS

Part C of this book is a compilation of solved problems that illustrate applications of the finite element method in various disciplines of engineering. In addition, each chapter includes concepts and techniques which are related to the particular application and which were not presented earlier in the text.

From its inception the finite element method demonstrated great potential for the analysis of complex structures.[1] Consequently, the method has received greater attention in structural mechanics than in any other field of engineering. Moreover, the scope of structural mechanics is broad; and the types of structural problems solved by the method are diverse.

9-1 SUMMARY OF APPLICATIONS IN STRUCTURAL MECHANICS

Because the finite element method has been applied to a large number of diverse problems, a complete survey of applications in structural mechanics is impossible here. Some other works concentrate heavily on this subject and may be consulted as references for further applications.[2,3,4,5,6,7]

Table 9-1 summarizes six major topics in structural mechanics to which the finite element method has been applied: two- and three-dimensional stress analysis, plates, shells, stability, and composite structures. The last category encompasses (1) structures idealized as assemblages of two or more different types of finite elements, and (2) structures represented by complex assemblages of one type of element. An example of the latter is a box girder, which may be

TABLE 9-1 TOPICS IN STRUCTURAL MECHANICS.

Topic†	Typical Applications	References
I Two-dimensional stress analysis		3, 43
A Plane strain	Gravity dams; buried pipes	8, 11
B Plane stress	Stretching of plates with holes, notches, etc.; shear walls; bending of deep beams	6, 7, 8, 11, 30, 44
C Axisymmetric	Thick axisymmetric shells; nuclear reactor containment vessels; rocket motors and heat shields; machine parts	9, 10, 12, 44
II Three-dimensional stress analysis	Nuclear reactor containment vessels; arch dams and their foundations; machine parts; thick shells	3, 6, 7, 13, 39, 44, 55
III Bending of plates	Floor slabs; bridge deck slabs; ship decks, hulls, bulkheads, etc.; aircraft and spacecraft panels	3, 5, 6, 7, 14, 15, 22, 23, 25, 27, 28, 29, 45–54, 61, 62
IV Bending/stretching of shells		3, 6, 7, 15, 22, 56
A Axisymmetric	Roof domes; water and fuel tanks; pressure vessels; aerospace structures	2, 16, 17, 36, 57, 59, 60
B General	Thin arch dams; shell roofs; ship hulls; aircraft and spacecraft panels; tubular joints	5, 15, 18, 56
V Stability (buckling)	Beams and columns; frames; plates; shells; stiffened panels	3, 6, 19, 20, 45
VI Composite structures	Plates with edge beams; stiffened panels and shells; building floor systems; frame-wall-floor systems; bridge deck systems; folded plate roof systems; cylindrical shell roof systems; plate and box girders; structural joints; ships; aerospace structures	2, 4, 5, 6, 7, 10, 21, 35

† Includes elastic, yielding, elastic-plastic, and thermal stress analysis.

represented by a three-dimensional assemblage of flat-plate/plane-stress elements.

Each of the topics in Table 9-1 may include problems with a variety of structural phenomena, such as material and/or geometric nonlinearity, thermal effects, material anisotropy, residual stresses, and prestressing forces. Examples in this chapter not only include problems from each topic, but also indicate how various special phenomena are included in the analysis. However, two important subjects, dynamics and viscoelasticity, are treated in Chapters 11 and 14, respectively. In addition, the subject of torsion of prismatic members is treated in Chapter 12.

9-2 FINITE ELEMENT METHODS FOR STRUCTURAL MECHANICS

The material summarized in Chapters 3 and 4, and the theory of the displacement method described in Chapters 5 through 7, were presented in terms directly relevant to structural mechanics. The displacement method is the most widely employed finite element approach for structural problems. Nevertheless, the equilibrium, hybrid equilibrium, and mixed methods (Table 8-2 and Section 8-3) are used to some extent, particularly for plate and shell problems. Therefore, we shall now summarize the formulation process for each of these methods.

Formulation of the Equilibrium Method

The presentation here follows Pian and Tong,[23] and Fraeijs de Veubeke.[24,25] The field variables for the equilibrium method are the stresses, and the field variable model is expressed in the following form:

$$\{\sigma\} = [\phi]\{\beta\} + [\phi_x]\{\beta_x\} \tag{9.1}$$

Here $\{\beta\}$ is the vector of unknown generalized stress coordinates which satisfy the stress equation of equilibrium in the absence of body forces, and $\{\beta_x\}$ is the vector of known generalized stress coordinates which equilibrate both the prescribed body forces and the prescribed surface tractions of the element. $[\phi]$ and $[\phi_x]$ are functions of the spatial coordinates. The strain-stress equations are written in the usual form:

$$\{\varepsilon\} = [D]\{\sigma\} \tag{3.12d}$$

If the generalized loads $\{Q\}$ are selected as values of the surface tractions at a number of external nodes, they can be expressed in terms of the generalized stresses by equations of the form

$$\{Q\} = [G]^{\mathrm{T}}\{\beta\} + [G_x]^{\mathrm{T}}\{\beta_x\} \tag{9.2}$$

The surface tractions $\{T\}$ at any point on the surface can be related to the generalized forces as follows:

$$\{T\} = [N_s]\{Q\} \tag{9.3}$$

where $[N_s]$ is an appropriate interpolation model derived from Equation (9.1).

The generalized displacements $\{q\}$ are chosen as weighted averages of the displacements along the boundaries to have one-to-one correspondence with the generalized forces. Hence, they may be written as†

$$\{q\} = \iint\limits_S [N_s]^{\mathrm{T}}\{u\}\, dS \tag{9.4}$$

Because each generalized displacement is defined as an average value along one side or face of the element, each element of the vector $\{q\}$ is common to two finite elements at most. Furthermore, compatibility of displacements is maintained only in an average sense.

By substituting equations (9.1) through (9.3) into the complementary energy functional, equation (4.18), we obtain

$$\Pi_c = \iiint\limits_V \tfrac{1}{2}\{\sigma\}^{\mathrm{T}}[D]\{\sigma\}\, dV - \{Q\}^{\mathrm{T}}\{q\}$$

$$= \tfrac{1}{2}\{\beta\}^{\mathrm{T}}[F]\{\beta\} + \{\beta\}^{\mathrm{T}}[F_x]\{\beta_x\} + \tfrac{1}{2}\{\beta_x\}^{\mathrm{T}}[F_{xx}]\{\beta_x\} \tag{9.5a}$$

$$- \{\beta\}^{\mathrm{T}}[G]\{q\} - \{\beta_x\}^{\mathrm{T}}[G_x]\{q\}$$

where

$$[F] = \iiint\limits_V [\phi]^{\mathrm{T}}[D][\phi]\, dV$$

$$[F_x] = \iiint\limits_V [\phi]^{\mathrm{T}}[D][\phi_x]\, dV \tag{9.5b}$$

$$[F_{xx}] = \iiint\limits_V [\phi_x]^{\mathrm{T}}[D][\phi_x]\, dV$$

The principle of minimum complementary energy gives

$$\delta\Pi_c = 0 = [F]\{\beta\} + [F_x]\{\beta_x\} - [G]\{q\} \tag{9.6a}$$

† This equation and equation (9.9) are obtained from the requirement that

$$\{Q\}^{\mathrm{T}}\{q\} = \iint\limits_S \{T\}^{\mathrm{T}}\{u\}\, dS.$$

which can be solved for $\{\beta\}$ as follows:

$$\{\beta\} = [F]^{-1}[G]\{q\} - [F]^{-1}[F_x]\{\beta_x\} \tag{9.6b}$$

Substituting equation (9.6b) into the functional, equation (9.5), and taking the variation of Π_c with respect to $\{q\}$, we obtain stiffness equations for the element.

$$[k]\{q\} = \{\bar{Q}\} \tag{9.7a}$$

$$[k] = [G]^{\mathrm{T}}[F]^{-1}[G] \tag{9.7b}$$

$$\{\bar{Q}\} = [G]^{\mathrm{T}}[F]^{-1}[F_x]\{\beta_x\} - [G_x]\{\beta_x\} \tag{9.7c}$$

Equations (9.7) can be utilized in the direct stiffness method. Once we have solved for the generalized displacements $\{q\}$, we can compute the desired stresses from equations (9.6b) and (9.1).

There are two requirements we must satisfy in selecting a stress model for the equilibrium method, equation (9.1). First, to avoid boundary displacement modes that cause no stresses in the elements, the total number of generalized stress coordinates $\{\beta\}$ for all elements should exceed the total number of unknown generalized displacements $\{q\}$. Second, to ensure that the finite element solution is independent of the particular solution which equilibrates the prescribed loads, the polynomials in $[\phi]\{\beta\}$ should be complete and of at least the same degree as those in $[\phi_x]\{\beta_x\}$.[23]

Formulation of the Hybrid Equilibrium Method

The presentation here follows Pian and Tong,[23] Pian,[26] Severn and Taylor,[27] and Allwood and Cornes.[28] In the hybrid equilibrium method the field variable model for the stresses given by equation (9.1) is still employed. However, we select our generalized displacements differently. The generalized displacements $\{q\}$ are chosen as the displacements of a number of external nodes. Hence, the displacements at any point on the element boundary can be approximated by an interpolation displacement model in terms of the generalized displacements

$$\{u\}|_s = [N_s]\{q\} \tag{9.8}$$

The generalized displacements are chosen independently from the generalized stress coordinates; hence, in the hybrid method it is easier to satisfy the requirement that the total number of generalized stress coordinates exceed the total number of generalized displacements. Moreover, it is not so difficult as in the equilibrium formulation to choose interpolation functions that will satisfy equilibrium because now $[\phi_x]\{\beta_x\}$ must equilibrate only the prescribed body forces and not the prescribed surface tractions.

The generalized nodal forces $\{Q\}$ that correspond to the generalized nodal displacements $\{q\}$ are given by a weighted average of the boundary tractions.

$$\{Q\} = \iint_S [N_s]^T\{T\}\, dS \tag{9.9}$$

Here the surface tractions $\{T\}$ are related to the generalized stress coordinates as follows:

$$\{T\} = [R]\{\beta\} + [R_x]\{\beta_x\} \tag{9.10}$$

Substituting equations (9.1), (9.9) and (9.10) into the modified complementary energy functional[23] and finding the extremum, we obtain the equation analogous to equation (9.6b)

$$\{\beta\} = [F]^{-1}[G]\{q\} - [F]^{-1}[F_x]\{\beta_x\} \tag{9.11a}$$

where now

$$[G] = \iint_S [R]^T[N_s]\, dS \tag{9.11b}$$

Finally, the stiffness equations for the element in the hybrid method correspond to equations (9.7) and are

$$[k]\{q\} = \{\bar{Q}\} \tag{9.12a}$$

$$[k] = [G]^T[F]^{-1}[G] \tag{9.12b}$$

$$\{\bar{Q}\} = [G]^T[F]^{-1}[F_x]\{\beta_x\} - [G_x]\{\beta_x\} + \iint_{S_2} [N_s]^T\{\bar{T}\}\, dS_2 \tag{9.12c}$$

where

$$[G_x] = \iint_S [R_x]^T[N_s]\, dS \tag{9.12d}$$

and $[G]$ is given by equation (9.11b).

Formulation of the Mixed Method

One advantage of the mixed method is the versatility of the associated Hellinger-Reissner functional, equation (4.22). By using various forms of this functional, we can obtain problem statements requiring interelement continuity of different combinations of stress and displacement parameters.[23] For example, mixed formulations of the plate bending problem are possible which require interelement continuity for any one of the following sets of parameters (Table 8-2):

1. w, $\partial w/\partial n$, and $\partial w/\partial s$
2. w and M_n
3. M_n, M_{ns}, and $\partial M_n/\partial n$

n and s are the normal and tangential directions along the element boundary. Herrmann[29] has utilized the second set of compatibility requirements.

Another feature of the Hellinger-Reissner principle is that the assumed stresses need not satisfy the equilibrium equations or the stress-strain-displacement equations within the element. In addition, the assumed stresses or displacements need not satisfy the specified boundary conditions. This latitude is possible because the Euler equations of the functional are the stress equations of equilibrium, the strain-stress law, and the displacement and stress boundary conditions.[30]

The following presentation follows Dunham and Pister.[30] If we choose to satisfy the displacement boundary conditions with our field variable models, the Hellinger-Reissner functional, equation (4.23), becomes

$$\begin{aligned}\Pi_R = &\iiint_V (\{\sigma\}^T\{\varepsilon\} - \tfrac{1}{2}\{\sigma\}^T[D]\{\sigma\} - \{u\}^T\{\bar{X}\})\, dV \\ &- \iint_{S_1} \{u\}^T\{\bar{T}\}\, dS_1\end{aligned} \tag{9.13}$$

Our mixed field variable model is taken in the form

$$\{u\} = [N_u]\{q\} \tag{9.14a}$$

$$\{\sigma\} = [N_\sigma]\{\tau\} \tag{9.14b}$$

where $\{\tau\}$ is the vector of nodal stresses. The elements of $\{q\}$ and $\{\tau\}$ may be selected at different nodes, and the order of the interpolation models for each unknown may be different.[29] The strains are obtained from the strain-displacement equations in the usual manner (Section 5-5)

$$\{\varepsilon\} = [B]\{q\} \tag{5.28b}$$

By substituting equations (9.14) and (5.28b) into equation (9.13), we obtain

$$\Pi_R = \{\tau\}^T[k_{\tau\tau}]\{\tau\} + \{\tau\}^T[k_{\tau u}]\{q\} - \{q\}^T\{Q\} \tag{9.15a}$$

where

$$[k_{\tau\tau}] = -\iiint_V [N_\sigma]^T[D][N_\sigma]\, dV \tag{9.15b}$$

$$[k_{\tau u}] = \iiint_V [N_\sigma]^T[B]\, dV \tag{9.15c}$$

$$\{Q\} = \iiint_V [N_u]^T\{\bar{X}\}\, dV + \iint_{S_1} [N_u]^T\{\bar{T}\}\, dS_1 \tag{9.15d}$$

The stationary condition of the functional is satisfied by the equations

$$\begin{bmatrix} [k_{\tau\tau}] & [k_{\tau u}] \\ [k_{\tau u}]^T & [0] \end{bmatrix} \begin{Bmatrix} \{\tau\} \\ \{q\} \end{Bmatrix} = \begin{Bmatrix} \{0\} \\ \{Q\} \end{Bmatrix} \tag{9.16}$$

The first of these equations is a stress-displacement relation; the second is an equilibrium equation.

Equation (9.16) may be used in the direct stiffness method where interelement compatibility is maintained for both displacements and stresses. The resulting "stiffness" matrix is not positive definite; however, Gaussian elimination may be used successfully if the equations corresponding to stress degrees of freedom are eliminated before those corresponding to displacement degrees of freedom.[30]

Comparison of the Methods

For the majority of problems in structural mechanics, one of the principal goals of the analysis is to obtain the stresses. In Section 6-7 we saw that the displacement method gives stresses that are not smooth across interelement boundaries. The fact that the equilibrium, hybrid, and mixed methods give a more accurate representation of the stress is a primary factor in favor of these approaches.

Nevertheless, the displacement method remains the most widely used approach. Compared to the displacement method, the number and band width of the stiffness equations for the other methods are usually greater for the same number of elements.[24,31] Furthermore, one difficulty encountered in constructing the equilibrium and hybrid equilibrium models is the satisfaction of equilibrium. This process is often not so easy as the formulation of compatible displacement models. Finally, the assemblage equations in the mixed method are not so well conditioned for solution as are the displacement method equations.[31]

Significantly, the equilibrium method can be proved to provide a lower bound to the element stiffness, whereas the compatible displacement method gives an upper bound[32] (Section 5-3). Recall that an upper bound to element stiffness is a numerical result that gives magnitudes of stiffness higher than the true stiffness, whereas a lower bound to stiffness yields magnitudes lower than the true stiffness. In other words, an upper bound stiffness gives a lower bound to the true solution for displacements, that is, the simulated structure is less flexible than the actual structure; but a lower bound stiffness gives an upper bound to the true solution, that is, the simulated structure is more flexible. Hence, if we solve the same problem by both methods, we can determine the proximity to the exact solution by comparing the two results, which bracket the true solution. Generally, hybrid method solutions will fall somewhere between the displacement and equilibrium solutions; therefore, the hybrid solutions are usually more accurate than the known upper or lower bounds.[23]

9-3 SPECIAL TECHNIQUES

One special technique widely used in solid mechanics and other applications of the finite element method is separation of variables. This procedure reduces the number of dimensions of the finite element discretization. A second special topic we shall consider is the analysis of incompressible solids.

Linear Analysis by Separation of Variables

For certain classes of linear problems, the required computational effort can be reduced by employing separation of variables. In a three-dimensional problem, for example, if we can use a Fourier series to represent the solution in one direction, we need only a two-dimensional discretization. Some specific types of problems for which this approach has been used are cylindrical shell roofs,[33] folded plate roofs,[34,35] and axisymmetric solids[9] and rotational shells[36] subject to asymmetric loading.

The cylindrical shell roof shown in Figure 9-1(a) is a two-dimensional problem; however, usually the displacements, w, of the shell are assumed to be of the form

$$w(s, y) = \sum_{n=1}^{N_f} W_n(s) \sin \frac{n\pi y}{L}$$

where L is the span length in the y direction. Therefore, a total of N_f one-dimensional finite element solutions are needed to obtain w. This is considerably more economical than performing a single analysis with a two-dimensional discretization of the entire shell structure, particularly since an adequate solution may often be obtained with only a relatively small value of N_f.

Similarly, for the axisymmetric solid in Figure 9-1(b), the nonsymmetric loading shown may be written as the following one-term Fourier series

$$\begin{aligned} P_r(r, \theta, z) &= F(r, z) \cos \theta \\ P_\theta(r, \theta, z) &= -F(r, z) \sin \theta \end{aligned}$$

Hence the θ dependence of all the displacements, stresses, and strains can be

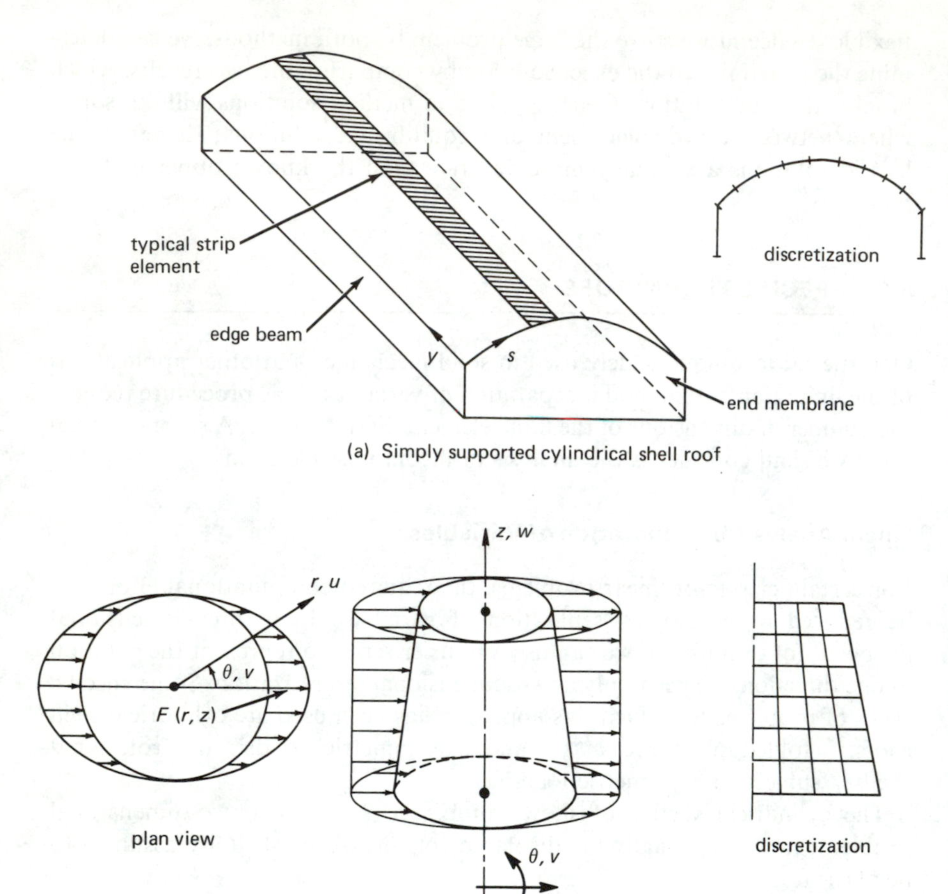

(a) Simply supported cylindrical shell roof

(b) Axisymmetric solid under asymmetric load

Figure 9-1 Problems solved by separation of variables.

expressed in terms of cos θ and sin θ. We can use the two-dimensional mesh shown in Figure 9-1(b) to solve this three-dimensional problem. The elements in this case are axisymmetric rings with quadrilateral cross sections. The formulation differs, however, from that in Example 5-6 because we now must account for the possibility of a displacement v in the θ direction in addition to u and w displacements in the r and z directions, respectively. Specific examples of problems of the axisymmetric solid type are a laterally loaded pile in a soil medium, and a rocket subject to transverse loads.

The *orthogonality* of the trigonometric functions makes the Fourier series particularly valuable in separation of variable methods:

$$\int_0^{2\pi} \cos m\theta \cos n\theta \, d\theta = \begin{cases} 2\pi & \text{for} \quad m = n = 0 \\ \pi & \text{for} \quad m = n \neq 0 \\ 0 & \text{for} \quad m \neq n \end{cases}$$

$$\int_0^{2\pi} \sin m\theta \sin n\theta \, d\theta = \begin{cases} \pi & \text{for} \quad m = n \neq 0 \\ 0 & \text{for} \quad m \neq n \text{ and } m = n = 0 \end{cases} \tag{9.17}$$

$$\int_0^{2\pi} \cos m\theta \sin n\theta \, d\theta = 0 \qquad \text{for all } m \text{ and } n$$

This property is important because it reduces the coupling of the simultaneous algebraic equations to be solved. We can show this for one general case by assuming displacements for an axisymmetric problem in cylindrical coordinates in the following form:

$$u(r, z, \theta) = \sum_{n=0}^{N_f} U_n(r, z) \cos n\theta$$

$$v(r, z, \theta) = \sum_{n=1}^{N_f} V_n(r, z) \sin n\theta \tag{9.18}$$

$$w(r, z, \theta) = \sum_{n=0}^{N_f} W_n(r, z) \cos n\theta$$

These assumptions are valid for nonaxisymmetric loads which are symmetric about a plane containing the axis of revolution.[9] For the most general case of nonaxisymmetric loading, we would need a complete Fourier expansion (sine and cosine) for each displacement. From equations (9.18), the potential energy functional will be of the form

$$\Pi = \sum_{m=0}^{N_f} \sum_{n=0}^{N_f} \left(\iiint_V P_1(n, m, r, z)\Theta_1(n, m, \theta) \, dV - \iint_{S_1} P_2(n, m, r, z)\Theta_2(n, m, \theta) \, dS_1 \right)$$

where Θ_1 and Θ_2 involve products of trigonometric functions. When performing the integrations, according to equation (9.17) we obtain nonzero results only for the terms of Θ_1 and Θ_2 which are $\cos^2 n\theta$ and $\sin^2 n\theta$. Hence, the double summation is reduced to a single summation and $N_f + 1$ independent sets of algebraic equilibrium equations result:

$$[K_n]\{r_n\} = \{R_n\} \text{ for } n = 0,1,2, \ldots, N_f \tag{9.19}$$

Each of these sets contains N equations. If the Fourier series trigonometric functions were not orthogonal, we would have a single coupled set of $N(N_f + 1)$ equations.

The application of the method of separation of variables is illustrated in Sections 9-11 and 11-6.

Example: Formulation of Axisymmetric Ring Element for Asymmetrical Loads. We shall consider only the case of nonaxisymmetric loads which are symmetric about a plane containing the axis of revolution. We must express the body forces and surface tractions by Fourier expansions in the θ direction. These equations are of the same form as equations (9.18), and N_f is the number of terms in the expansion necessary to represent adequately the loading.

In order to compare our formulation with Example 5-6, we shall consider an axisymmetric ring element with a triangular cross-section, Figure 5-7(f), and employ linear displacement models to represent the displacements in equations (9.18).

$$\{u_n\} = \begin{Bmatrix} U_n \\ V_n \\ W_n \end{Bmatrix} = \begin{bmatrix} \{N_1\}^T & \{0\}^T & \{0\}^T \\ \{0\}^T & \{N_1\}^T & \{0\}^T \\ \{0\}^T & \{0\}^T & \{N_1\}^T \end{bmatrix} \begin{Bmatrix} \{q_{un}\} \\ \{q_{vn}\} \\ \{q_{wn}\} \end{Bmatrix} = [N]\{q_n\}$$

$$\{N_1\}^T = [L_1 \ L_2 \ L_3], \ \{q_{un}\}^T = [u_{1n} \ u_{2n} \ u_{3n}]$$

$$\{q_{vn}\}^T = [v_{1n} \ v_{2n} \ v_{3n}], \ \{q_{wn}\}^T = [w_{1n} \ w_{2n} \ w_{3n}]$$

The strain-displacement equations for the three-dimensional problem in cylindrical coordinates, obtained from equations (3.11b), are

$$\varepsilon_r = \frac{\partial u}{\partial r}, \qquad \varepsilon_\theta = \frac{1}{r}\left(\frac{\partial v}{\partial \theta} + u\right), \qquad \varepsilon_z = \frac{\partial w}{\partial z}$$

$$\gamma_{r\theta} = \frac{1}{r}\left(\frac{\partial u}{\partial \theta} - v\right) + \frac{\partial v}{\partial r}, \qquad \gamma_{\theta z} = \frac{\partial v}{\partial z} + \frac{1}{r}\frac{\partial w}{\partial \theta}$$

$$\gamma_{zr} = \frac{\partial w}{\partial r} + \frac{\partial u}{\partial z}$$

By using these equations, we can construct the element strains for each harmonic.

$$\{\varepsilon_n\} = \begin{Bmatrix} \varepsilon_{rn} \\ \varepsilon_{\theta n} \\ \varepsilon_{zn} \\ \gamma_{r\theta n} \\ \gamma_{\theta zn} \\ \gamma_{zrn} \end{Bmatrix} = \frac{1}{2A}\begin{bmatrix} b_1 & b_2 & b_3 & 0 & 0 & 0 & 0 & 0 & 0 \\ c_1 & c_2 & c_3 & nc_1 & nc_2 & nc_3 & 0 & 0 & 0 \\ 0 & 0 & 0 & 0 & 0 & 0 & a_1 & a_2 & a_3 \\ nc_1 & nc_2 & nc_3 & d_1 & d_2 & d_3 & 0 & 0 & 0 \\ 0 & 0 & 0 & a_1 & a_2 & a_3 & nc_1 & nc_2 & nc_3 \\ a_1 & a_2 & a_3 & 0 & 0 & 0 & b_1 & b_2 & b_3 \end{bmatrix}\{q_n\}$$

$$= [B_n]\{q_n\}$$

where $c_i = 2AL_i/r$, $d_i = b_i - c_i$ and $i = 1,2,3$. Here the notation is the same as Sample Problem 2 of Chapter 5 and Example 5-6. The stress-strain equations are given by equation (3.12f), which are used to relate $\{\sigma_n\}$ to $\{\varepsilon_n\}$.

Now we may write the total potential for the element as a single summation by using the orthogonality conditions

$$\Pi = \pi \sum_{n=0}^{N_f} \left[\iint_A \left(\tfrac{1}{2}\{q_n\}^T[B_n]^T[C][B_n]\{q_n\} - \{q_n\}^T[N]^T\{\bar{X}_n\} \right) r\, dr\, dz - \int_{S_1} \{q_n\}^T[N]\{\bar{T}_n\}\, dS_1 \right]$$

The element equilibrium equations which result from the application of the principle of minimum potential energy are

$$[k_n]\{q_n\} = \{Q_n\} \qquad \text{for} \quad n = 0,1,2,\ldots, N_f$$

where

$$[k_n] = \pi \iint_A [B_n]^T[C][B_n] r\, dr\, dz$$

$$\text{for} \quad n = 1,2,\ldots,N_f$$

$$\{Q_n\} = \pi \iint_A [N]^T\{\bar{X}_n\} r\, dr\, dz + \pi \int_{\ell_1} [N]^T\{\bar{T}_n\}\, d\ell_1$$

The formulae for $n = 0$ have a multiplying factor of 2π and have nonzero terms only for the cosine-series quantities. Because terms of the integrands contain powers of r as well as the coordinates, L_i, we cannot use equation (5.18f) to perform the integrations. Numerical integration is necessary to evaluate the element stiffness matrix and load vector.

Using the direct stiffness method, we combine the equations for the individual elements to form assemblage equations, equations (9.19).

Mixed Method for Analysis of Incompressible Solids

Such materials as solid rocket propellants, saturated cohesive soils, and certain plastics and elastomers are incompressible or nearly incompressible. This means that the Poisson's ratio for these materials is 0.5 or very near to this value. By considering equations (3.12f), we see that the standard elastic stress-strain equations become undefined ($\nu = 0.5$) or ill-conditioned ($\nu \to 0.5$) because of the occurrance of the term $(1 - 2\nu)$ in the denominator of all nonzero elements of the constitutive matrix $[C]$.

Herrmann[37] has developed a variational principle for incompressible materials which are linearly elastic and isotropic. It is a special case of the

Hellinger-Reissner principle (Section 4-2). The functional used by Herrmann, with the thermal effects omitted, is

$$\Pi_H = \iiint\limits_V \Big(\mu[\varepsilon_x^2 + \varepsilon_y^2 + \varepsilon_z^2 + 2(\gamma_{xy}^2 + \gamma_{yz}^2 + \gamma_{zx}^2) + 2\nu H(\varepsilon_x + \varepsilon_y + \varepsilon_z) - \nu(1 - 2\nu)H^2] - \{u\}^{\mathrm{T}}\{\bar{X}\}\Big)\, dV - \iint\limits_{S_1} \{u\}^{\mathrm{T}}\{\bar{T}\}\, dS_1 \qquad (9.20a)$$

in which H is a stress parameter proportional to the hydrostatic or mean pressure, equation (3.4), and is given by

$$H = \frac{J_1}{2\mu(1 + \nu)} \qquad (9.20b)$$

The general finite element formulation for incompressible isotropic solids is as follows.[38] We first assume a mixed model for the element in interpolation form

$$\{u\} = [N]\{q\} \qquad (5.8)$$

$$H = \{N_h\}^{\mathrm{T}}\{h\} \qquad (9.21)$$

where $\{h\}$ is the vector of nodal values of H. It is convenient to select the order of the $\{N_h\}^{\mathrm{T}}$ model to be the same as that of the $[N]$ model.[39] The element strains can be computed in the usual manner

$$\{\varepsilon\} = [B]\{q\} \qquad (5.28b)$$

Now the strain terms in the associated functional, equation (9.20a), can be expressed in the notation of the strain vector $\{\varepsilon\}$ as follows:

$$\begin{aligned} \varepsilon_x^2 + \varepsilon_y^2 + \varepsilon_z^2 + 2(\gamma_{xy}^2 + \gamma_{yz}^2 + \gamma_{zx}^2) &= \{\varepsilon\}^{\mathrm{T}}[I']\{\varepsilon\} \\ \varepsilon_x + \varepsilon_y + \varepsilon_z &= \{\varepsilon\}^{\mathrm{T}}\{J\} \end{aligned} \qquad (9.22a)$$

The diagonal matrix $[I']$ and the vector $\{J\}$ are defined as

$$[I'] = \begin{bmatrix} 1 & & & & & \\ & 1 & & & 0 & \\ & & 1 & & & \\ & & & 2 & & \\ & 0 & & & 2 & \\ & & & & & 2 \end{bmatrix}, \quad \{J\} = \begin{Bmatrix} 1 \\ 1 \\ 1 \\ 0 \\ 0 \\ 0 \end{Bmatrix} \qquad (9.22b)$$

Substituting into Herrmann's functional and obtaining its stationary value from

$$\delta\Pi_H = \frac{\partial\Pi_H}{\partial\{q\}}\{\delta q\} + \frac{\partial\Pi_H}{\partial\{h\}}\{\delta h\} = 0$$

we obtain the following governing finite element equations:

$$\left[\begin{array}{c|c}[k_{11}] & [k_{12}] \\ \hline [k_{12}]^{\mathrm{T}} & [k_{22}]\end{array}\right]\left\{\begin{array}{c}\{q\} \\ \hline \{h\}\end{array}\right\} = \left\{\begin{array}{c}\{Q\} \\ \hline \{0\}\end{array}\right\} \tag{9.23a}$$

Here the various submatrices in equation (9.23a) are defined by

$$[k_{11}] = 2\mu \iiint_V [B]^{\mathrm{T}}[I'][B]\, dV \tag{9.23b}$$

$$[k_{12}] = 2\nu\mu \iiint_V [B]^{\mathrm{T}}\{J\}\{N_h\}^{\mathrm{T}}\, dV \tag{9.23c}$$

$$[k_{22}] = -\, 2\mu\nu(1-2\nu) \iiint_V \{N_h\}\{N_h\}^{\mathrm{T}}\, dV \tag{9.23d}$$

$$\{Q\} = \iiint_V [N]^{\mathrm{T}}\{\overline{X}\}\, dV + \iint_{S_1} [N]^{\mathrm{T}}\{\overline{T}\}\, dS_1 \tag{5.45b}$$

Note that a matrix form of the stress-strain relations does not enter the functional. Therefore, to evaluate the element stresses from the computed values of $\{q\}$ and $\{h\}$, we need the following modified stress-strain equations:

$$\begin{aligned}\{\sigma\} &= 2\mu[I']^{-1}\{\varepsilon\} + 2\mu\nu H\{J\} \\ &= 2\mu[I']^{-1}[B]\{q\} + 2\mu\nu\{N_h\}^{\mathrm{T}}\{h\}\{J\}\end{aligned} \tag{9.24}$$

The above formulation allows the use of all values of Poisson's ratio in the range $0 \leq \nu \leq 0.5$. To obtain results significantly more accurate than the displacement method solutions, interelement compatiblity must be maintained for both $\{q\}$ and $\{h\}$ quantities.[39,30]

Using a variational functional similar to equation (9.20), Christian[40] obtained a finite element formulation for the undrained stress analysis of saturated cohesive soils. Taylor *et al.*[41] have extended Herrmann's principle to the case of orthotropic incompressible materials. Finally, Tong[42] has formulated an analogous principle suitable for the hybrid equilibrium method of finite element analysis. In all of these formulations it is possible to include the linear, uncoupled thermal effects.[37,38,41]

9-4 EXAMPLE: PLANE STRESS ANALYSIS OF A GRAVITY DAM

Vertical cracks exist in some concrete gravity dams. In order to study the effect of such cracks on the stress distribution and on the factor of safety of a structure, Clough[11] applied the finite element method to the analysis of

Norfork Dam. This is a U.S. Army Corps of Engineers dam on the North Fork River in Arkansas. A number of vertical cracks were discovered in this dam after construction. Monolith 16, shown in Figure 6-1, had a single vertical crack at about mid-width which extended over most of the height. In the analysis various combinations of crack height, hydrostatic loading, thermal loadings, and hydrostatic uplift were considered to obtain the critical cases.

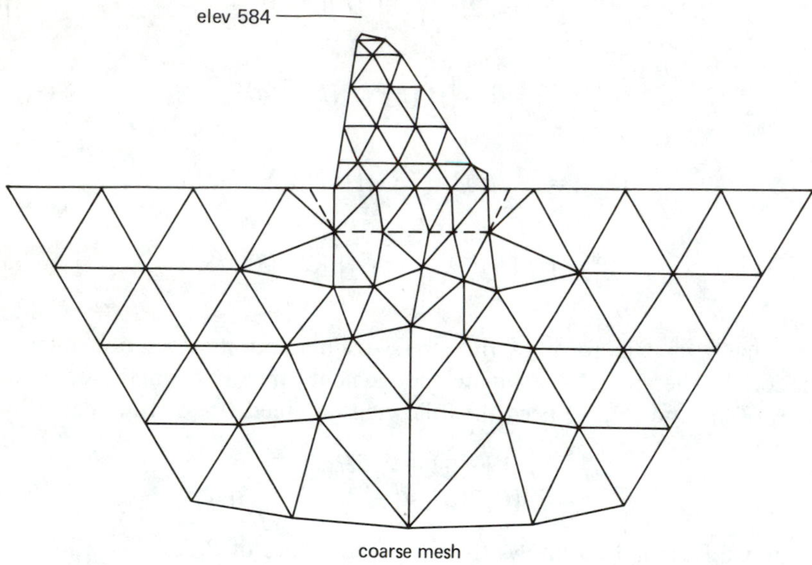

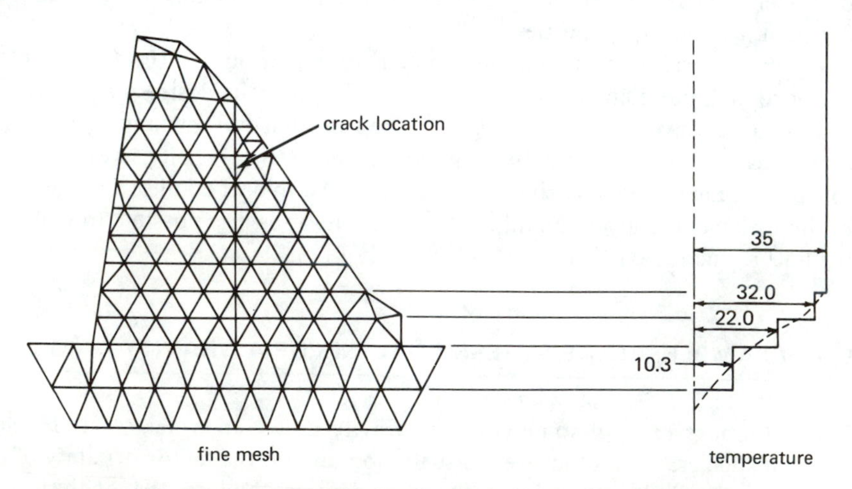

Figure 9-2 Finite element meshes for Norfork Dam (from Reference 11).

Figure 9-2 shows the finite element meshes used to analyze the structure. CST elements were employed in both meshes. For the value of Poisson's ratio typical of concrete in the dam (0.18–0.22), the difference between plane stress and plane strain is slight; therefore, plane stress conditions were arbitrarily assumed. The coarse-to-fine subdivision method (Section 6-1) was used to take into account the flexibility of the foundation. The results obtained for the coarse mesh along the dashed line were introduced as boundary conditions for the fine mesh.

The existing vertical crack was simulated by a separation of six or seven nodes up from the base of the dam along the vertical nodal line shown in the fine mesh of Figure 9-2. The crack was treated by the method illustrated in Figure 6-2(e), and loadings were applied in a sequence that prevented the crack facings from overlapping. Hydrostatic uplift force was introduced through a horizontal crack at the heel of the dam. This crack extended inward three to three and one-half nodes and was treated in the same manner as the vertical crack.

Typical results of the analysis are shown in Figure 9-3. This case includes dead loading plus hydrostatic pressure, a vertical crack length of six nodes, and orthotropic properties for the foundation. The results for this and other cases showed that, although significant stress concentrations arise at the heel of the dam and at the top and bottom of the vertical crack, the magnitudes of these stresses were moderate, not exceeding 500 psi.

This early study demonstrated the effectiveness and versatility of the finite element method for the solution of complex practical problems.

9-5 EXAMPLE: AXISYMMETRIC STRESS ANALYSIS OF A BOLT-NUT SYSTEM

Iversen[43] performed a stress analysis of a bolt-nut system using the finite element idealization shown in Figure 9-4(a). Axisymmetric elements with CST cross-sections were used. The loading was simulated by the applied displacements shown in Figure 9-4(b). One special technique was necessary to define the boundary conditions existing between the thread surfaces. As shown in Figure 9-4(b), displacements along the line *A-A* were assumed to be unrestricted. However, displacements normal to line *A-A* were required to be the same for the bolt and nut.

Results of the analysis are shown in Figure 9-4(c). Here the force distribution in the threads, computed by the finite element analysis, is compared to the theoretical distribution; and good agreement is found.

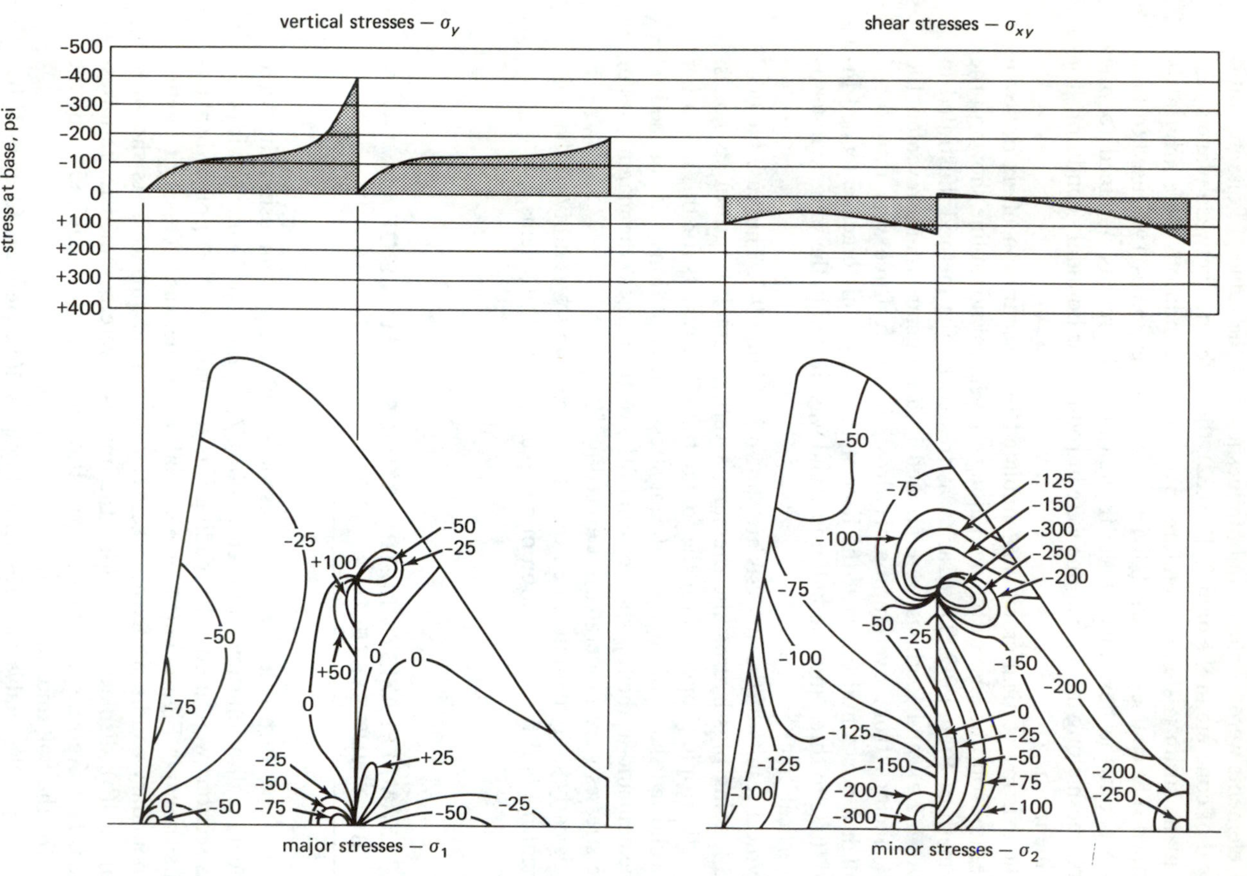

Figure 9-3 Typical results of Norfork Dam analysis (from Reference 11).

9-6 EXAMPLE: ANALYSIS OF PRESTRESSED CONCRETE REACTOR VESSEL

To evaluate the adequacy and relative accuracy of two- and three-dimensional finite element stress analyses for prestressed concrete reactor vessels (PCRV's), Corum and Krishnamurthy[44] compared the two methods with experimental results obtained from a large-scale model. The resulting study is a valuable illustration of the trade-off between the methods in data preparation effort and accuracy of results. A PCRV is a complex, three-dimensional structure; however, by using certain simplifications, two-dimensional methods can be used for approximate analyses.[10]

One of the two-dimensional analyses performed was axisymmetric. The mesh for this approximation is shown in Figure 9-5(a). The idealization involves 831 axisymmetric CST elements with 473 nodes in a typical simplified vertical plane of the vessel. A reinforcing bar cage at the base of the vessel is represented by bar elements. The region at the top with reduced elastic modulus represents a series of penetrations.

In addition, plane stress analyses were performed on two horizontal planes for 30° sectors of the PCRV. The meshes are shown in Figures 9-5(b) and 9-5(c). The first plane stress mesh is designed to study the effect of steel-lined penetrations in the reactor head and consists of 945 CST elements with 537 nodes. The second plane stress mesh is taken through the vessel wall at the level of the circumferential prestressing blocks. It consists of 469 CST elements with 266 nodes.

Figure 9-6 shows the three-dimensional finite element idealization of a 30° sector of the vessel. The primary element used was an octahedron with triangular faces composed of three CSTh. The mesh shown in Figure 9-6 has 12,231 tetrahedra and 2,678 nodes; even with some use of automatic mesh generation, the data preparation for this mesh was very time-consuming. A major portion of the effort was involved in the development of pictorial aids necessary to define the mesh and to check for data errors.

The analyses were carried out by using standardized programs developed for the U.S. Atomic Energy Commission.† The loadings consisted of internal pressure and prestressing forces. The latter were represented by ring loads for the axisymmetric case, and by point loads at the prestressing anchor blocks for the plane stress and three-dimensional cases. These representations are valid for unbonded tendons. The equations were solved by using an accelerated iterative procedure. For each loading, the total central processing time required on an IBM 360/65 for the three-dimensional analysis was about $2\frac{1}{2}$ hours.

† SAFE (Structural Analysis by Finite Elements), Gulf General Atomic, Inc.

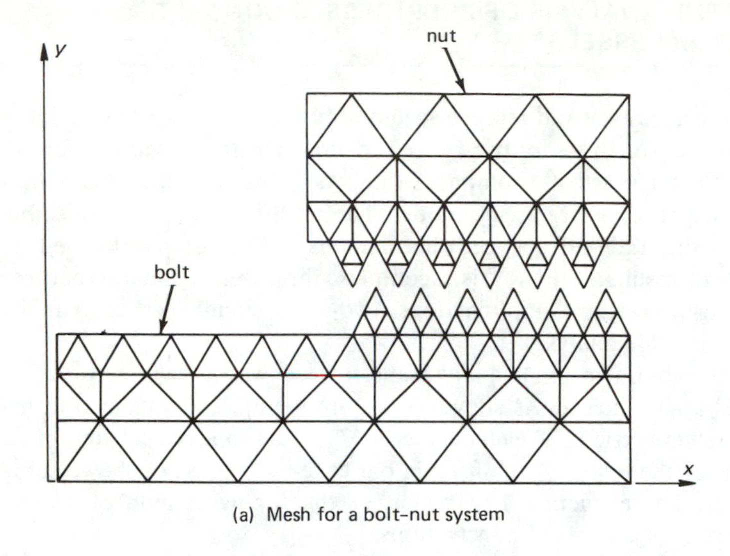

(a) Mesh for a bolt-nut system

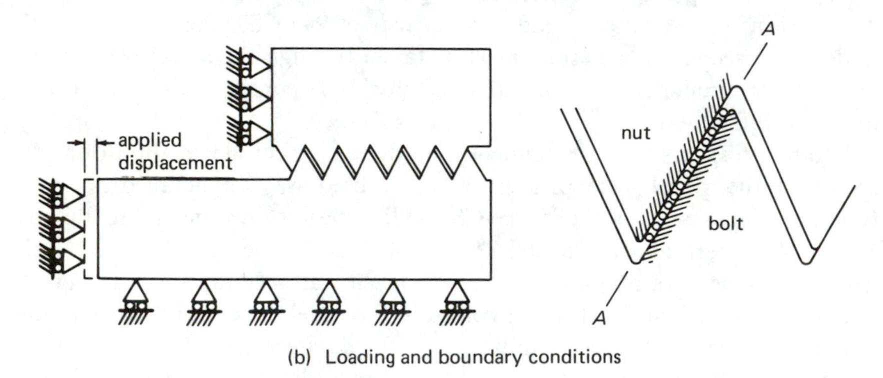

(b) Loading and boundary conditions

Figure 9-4 Analysis of forces in a bolt-nut system (from Reference 43).

Results for the circumferential stresses are shown in Figure 9-7, where they are compared to experimental values. Generally, the two-dimensional analyses overestimate the values of the stresses, whereas the three-dimensional analysis agrees fairly well. General conclusions cannot be drawn from the study of a single structure; nevertheless, this study did point out that some combination of two- and three-dimensional analyses could be used fruitfully for such complex structures. Moreover, the necessity for improved methods of data preparation and output interpretation for three-dimensional studies is clear. The

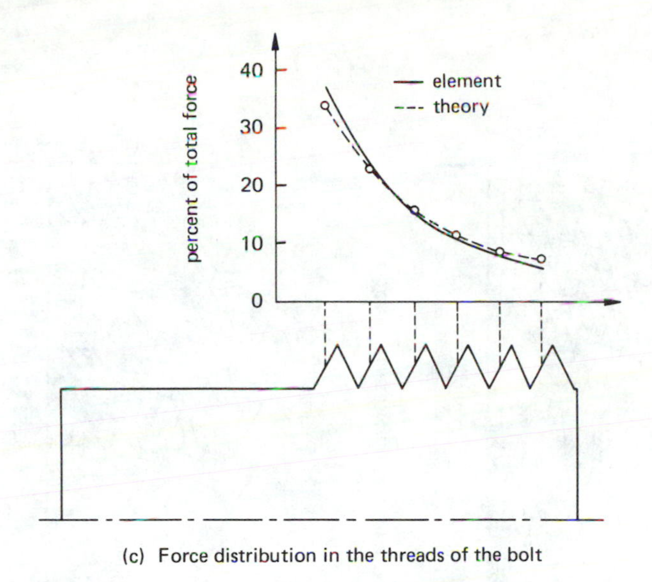

(c) Force distribution in the threads of the bolt

Figure 9-4 (*continued*)

use of higher order three-dimensional elements would require fewer elements to idealize the structure and thus would reduce the necessary input and output.

9-7 EXAMPLE: COMPARISON OF PLATE BENDING FINITE ELEMENTS

The finite element method has been applied successfully to the analysis of practical plate bending problems. However, because there are a large number of different plate-bending element formulations, it is worthwhile to consider a comparison of several different elements, rather than an individual application. Not only will this give us an insight into the behavior of the diverse formulations, but it will permit us to add meaning to theoretical concepts introduced earlier.

Various authors have studied the discretization errors that arise in the analysis of plate bending problems using different finite element formulations. The first of these studies was by Clough and Tocher.[14] As newer elements have been developed, this work has been updated by de Veubeke and Sander,[25] Clough and Felippa[45] and Gallagher.[15] All of these papers compared various elements by depicting the convergence attained with increased subdivision of the plate. The related subject of rounding errors has been treated by Ramstad.[46]

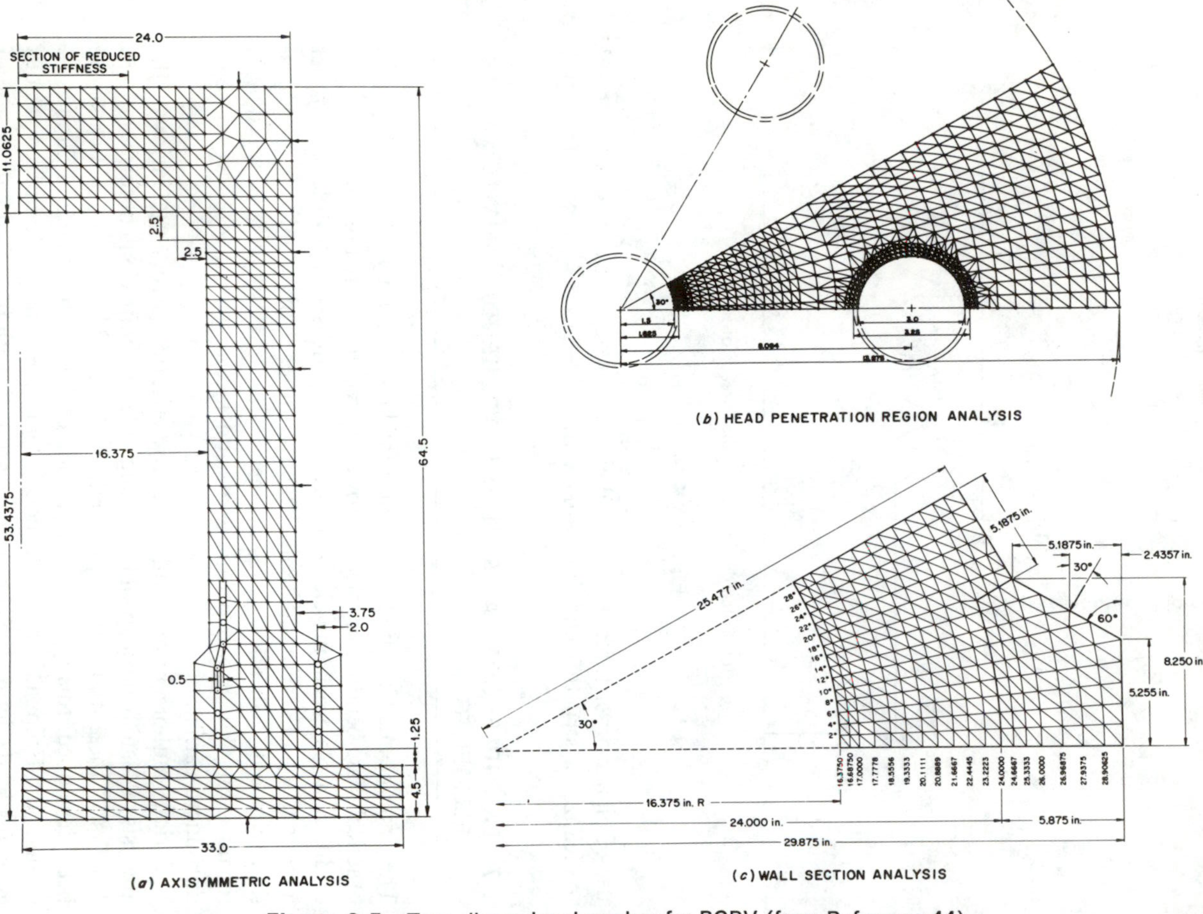

Figure 9-5 Two-dimensional meshes for PCRV (from Reference 44).

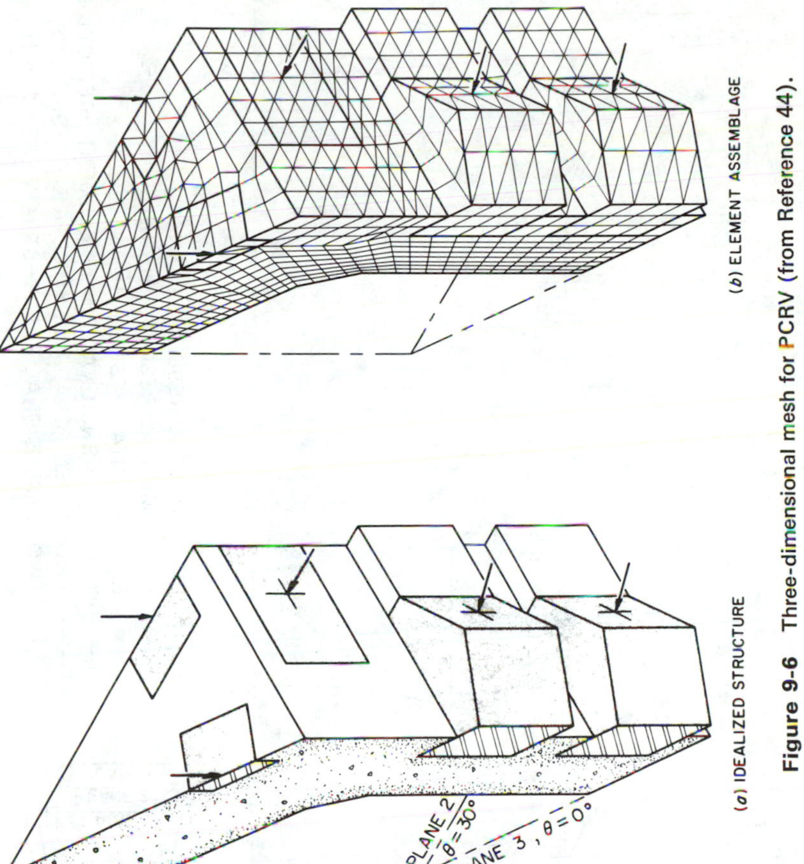

Figure 9-6 Three-dimensional mesh for PCRV (from Reference 44).

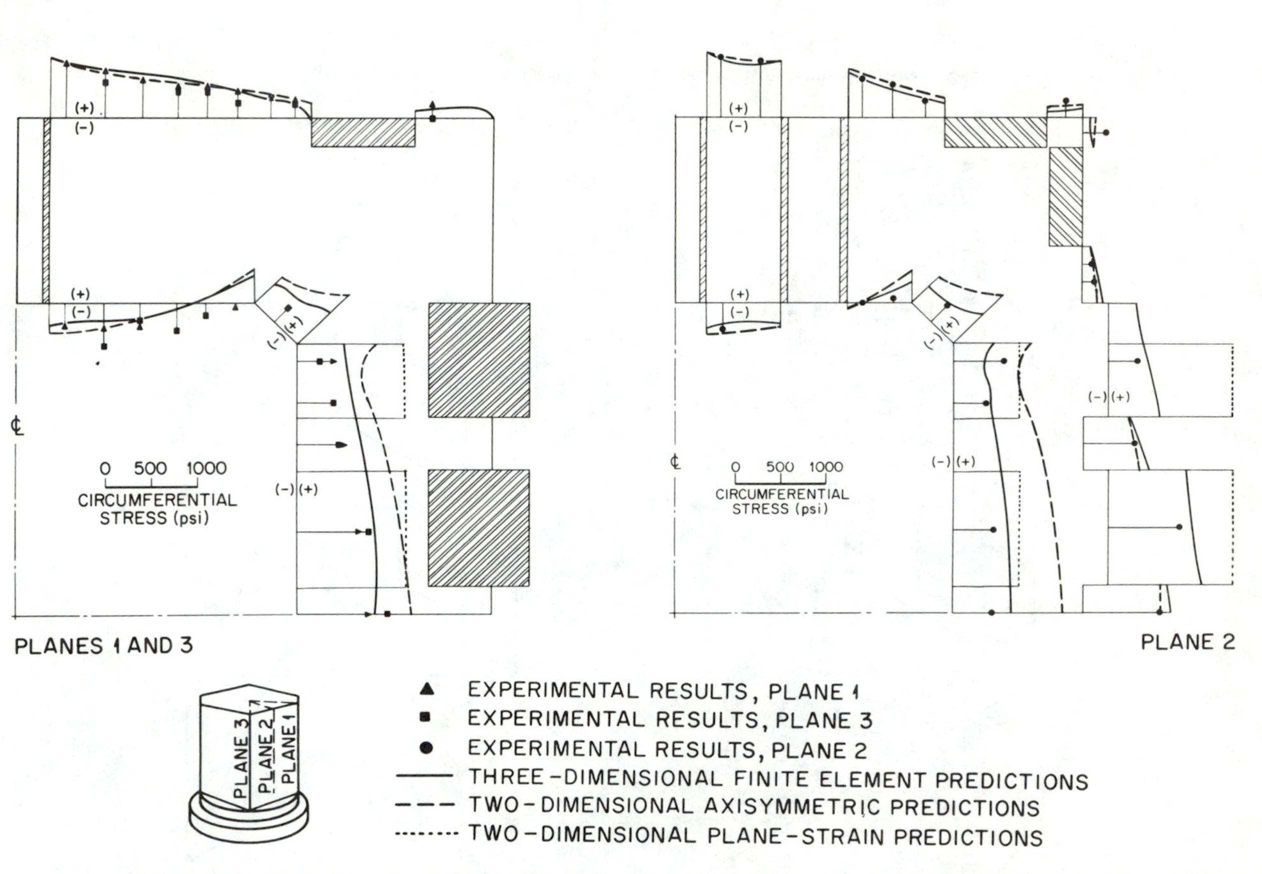

Figure 9-7 Typical results of PRCV analysis compared to experimental results (from Reference 44).

We now consider an abridged version of the convergence comparison for twelve different elements. Table 9-2 contains a brief description of the elements, which include displacement, hybrid, mixed, and equilibrium formulations. The two basic problems used for the comparisons are a square plate with a concentrated load at the center under both simply supported and clamped edge conditions. Because of symmetry about both centerline axes, only a quarter of the plate need be discretized. In all cases, quadrilateral elements or an equivalent assemblage of triangles were used. In Figures 9-8 and 9-9 the percent error in the calculation of the center displacement is shown as a function of NB^2. N is the number of equations used in the discretization, and is given by

$$N = (n + 1)^2 f_p + 2n(n + 1) f_s$$

where n is the number of square elements along one side of the quarter plate, and f_p and f_s are the number of degrees of freedom at primary and secondary external nodes, respectively. B is the semiband width of the equations (Section 6-1). NB^2 is a useful measure of the computer time required to solve banded equations (Section 2-2).

In addition to giving a qualitative comparison of the various element formulations, Figures 9-8 and 9-9 demonstrate several statements made in Part B and Section 9-2. The complete and conforming displacement elements (HCT, LCCT-12, Q-19, BFS, and TUBA-6) provide a lower bound to the displacements, whereas the equilibrium formulations (EQT and E) provide an upper bound. The incomplete displacement model (P) fails to converge to the correct solution. However, the complete but non-conforming elements (ACM and M) do converge. Finally, the mixed method (A) and the hybrid equilibrium formulation (ST) produce results that are generally in between the displacement and equilibrium models of comparable order.

9-8 EXAMPLE: ANALYSIS OF AN ARCH DAM

The stress analysis of arch dams has been studied by several different finite element approaches. The dam may be considered to be a shell structure, or it may be analyzed as a three-dimensional continuum. Although the dimensions of typical arch dams would appear to necessitate a thick-shell analysis, useful solutions have been obtained with thin shell elements which neglect the shearing in the thickness direction.

Here we consider a particular arch dam 120 meters high at the crown section (Figure 9-10). Because of symmetry about the valley axis, only half of the dam was included in the discretization. This structure has been analyzed by Clough and Johnson[18] using thin-shell elements. Their element consisted of four

TABLE 9-2 PLATE BENDING ELEMENTS.

Symbol	Reference	Method	Description of Element	Degrees of Freedom: External	Degrees of Freedom: Internal	f_p	f_s
1. ACM	14, 47, 48	Displacement	Nonconforming but complete rectangle; cubic model.	12	0	3	0
2. M	14, 49	Displacement	Nonconforming but complete rectangle; cubic beam function plus uniform twist.	12	0	3	0
3. P	14, 50	Displacement	Conforming but incomplete rectangle; polynomial terms up to 6th order.	12	0	3	0
4. HCT = LCCT-9	14	Displacement	2 LCCT-9 triangles; cubic model.	12	0	3	0
5. LCCT-12	8, 45	Displacement	2 LCCT-12 triangles; cubic model.	16	1	3	1
6. Q-19	8, 45	Displacement	4 LCCT-11 triangles; cubic model.	12	7	3	0
7. BFS	15, 51	Displacement	Complete and conforming rectangle; cubic Hermitian polynomial model, Example 5-9.	16	0	4	0
8. TUBA-6	52	Displacement	2 TUBA-6 triangles; quintic model.	28	1	6	1
9. ST	27	Hybrid equilibrium	2 triangles; quadratic stresses, cubic interface displacements.	12	0	3	0
10. A	53†	Mixed	4 triangles; cubic displacements; quadratic stress parameters.	16	7	3	1
11. EQT	25	Equilibrium	2 triangles; linear moments	16	3	1	3
12. E	54	Equilibrium	2 triangles; linear stress functions.	8	0	2	0

† ... Numerical Methods in Engineering, Vol. 2, No. 2, 1970 Wiley-Interscience, London.

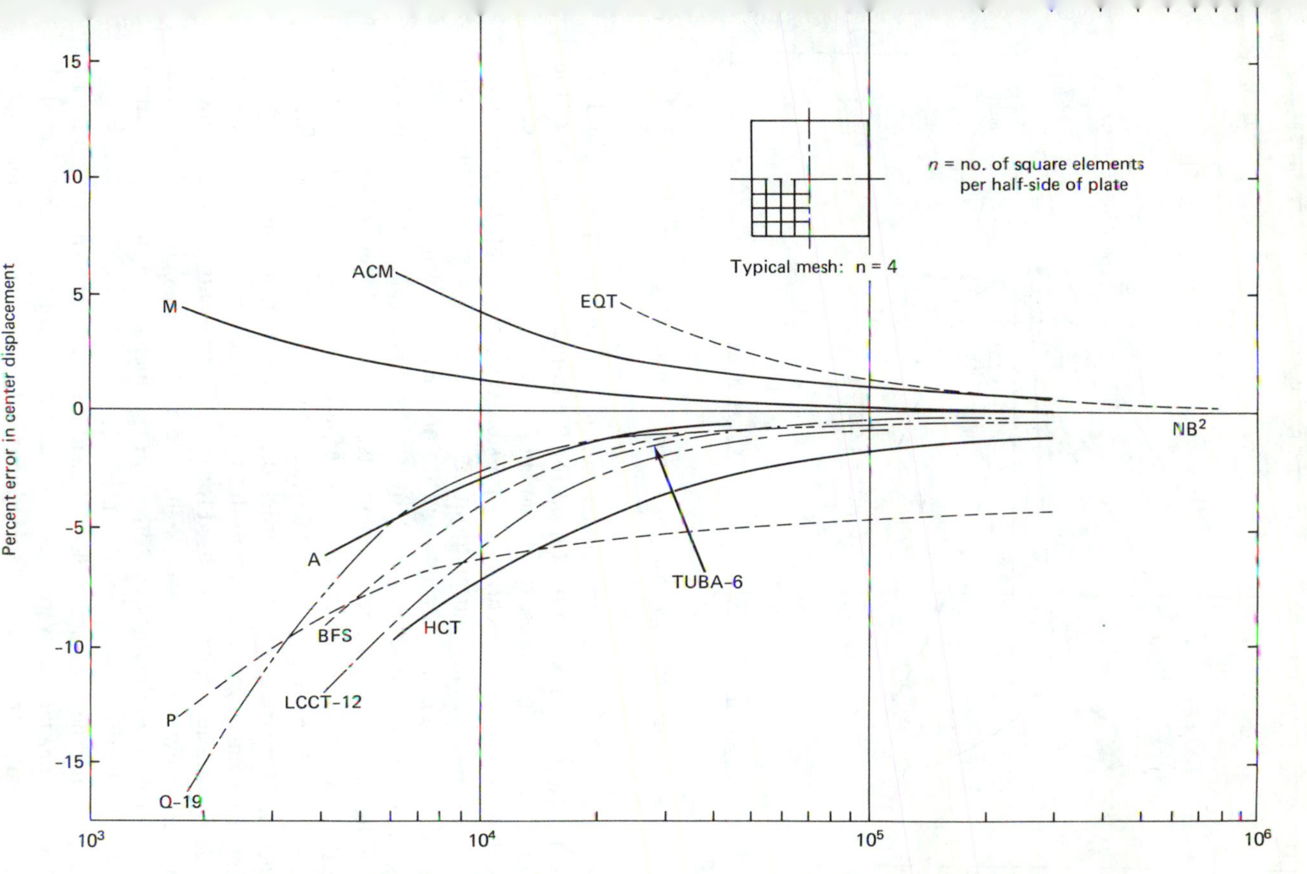

Figure 9-8 Simply supported square plate with concentrated center load.[14,15,25,45]

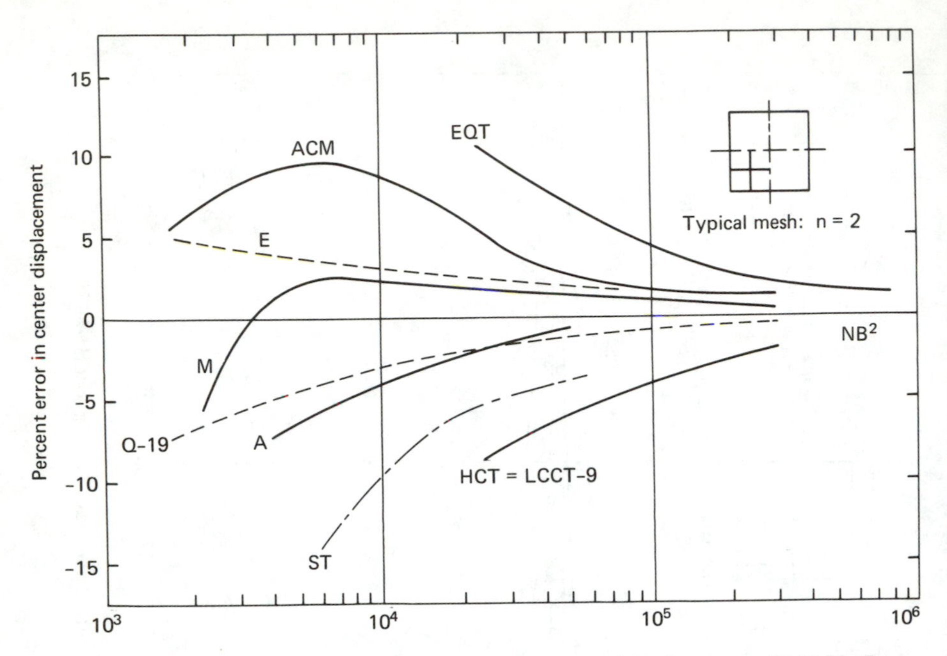

Figure 9-9 Clamped square plate with concentrated load at center.[14,15,25,45]

plane triangular elements combined to form a nonplanar quadrilateral. A linear displacement model was utilized for the in-plane displacements of the triangles, while the transverse displacement function was cubic. The thickness varied linearly in each element. It was possible to approximate the transverse shear effects by assuming that material lines originally straight and normal to the middle surface remain straight but do not remain normal.

The dam has also been analyzed by Ergatoudis *et al.*,[55] who employed three-dimensional isoparametric elements, and by Ahmad *et al.*,[56] who utilized curved, thick-shell elements of superparametric form. *Superparametric elements* are derived from isoparametric forms by adopting a model for the displacements which is of lower order than the shape function used to represent the element geometry. In both the thick-shell and three-dimensional analysis, quadratic models were used for the displacements and the geometry parameters. However, for the thick-shell element the model was two-dimensional, and shear in the thickness direction was included in the same manner as it was in the thin-shell formulation.

The mesh for the thick-shell finite element analysis is shown in Figure 9-10. Similar but finer meshes were used for the three-dimensional and thin-shell analyses. Results for a selected displacement and stress are indicated in Figures 9-11 and 9-12, respectively. Because the three-dimensional discretization is so

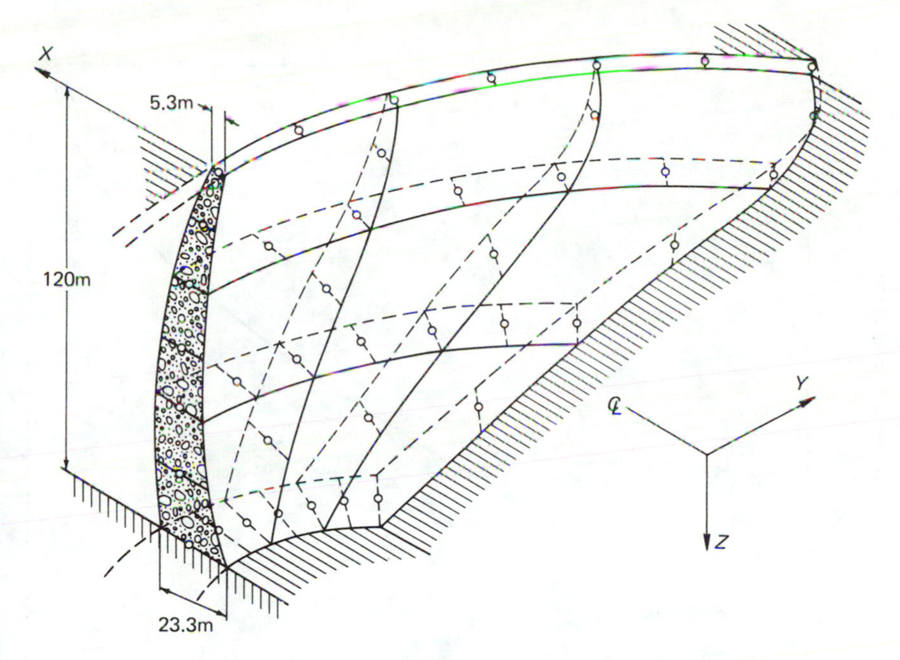

Figure 9-10 Thick shell element mesh for arch dam (from Reference 56).

fine (816 equations), it may be considered a standard of accuracy for the other three analyses (about 200 equations each). A nine-element, three-dimensional analysis provides about the same level of accuracy as the shell solutions. We see that the thick- and thin-shell solutions generally provide the same accuracy of stresses (Figure 9-12), although the thick-shell displacement results are virtually indistinguishable from the three-dimensional results (Figure 9-11).

In this problem the foundation was considered rigid for the three-dimensional analysis. However, one significant advantage of the three-dimensional approach is the possibility of including the foundation in the discretization.[55]

9-9 EXAMPLE: AXISYMMETRIC SHELL APPLICATIONS

The analysis of axisymmetric shells has received considerable attention in the finite element literature because this configuration has a number of important practical applications (Table 9-1). In addition, this class of shell problems is substantially easier to analyze than arbitrary shells because the necessary discretizations are one-dimensional.

A doubly curved axisymmetric shell element was developed by Khojasteh-Bakht and Popov.[17] The meridional cross-section of this element is shown in

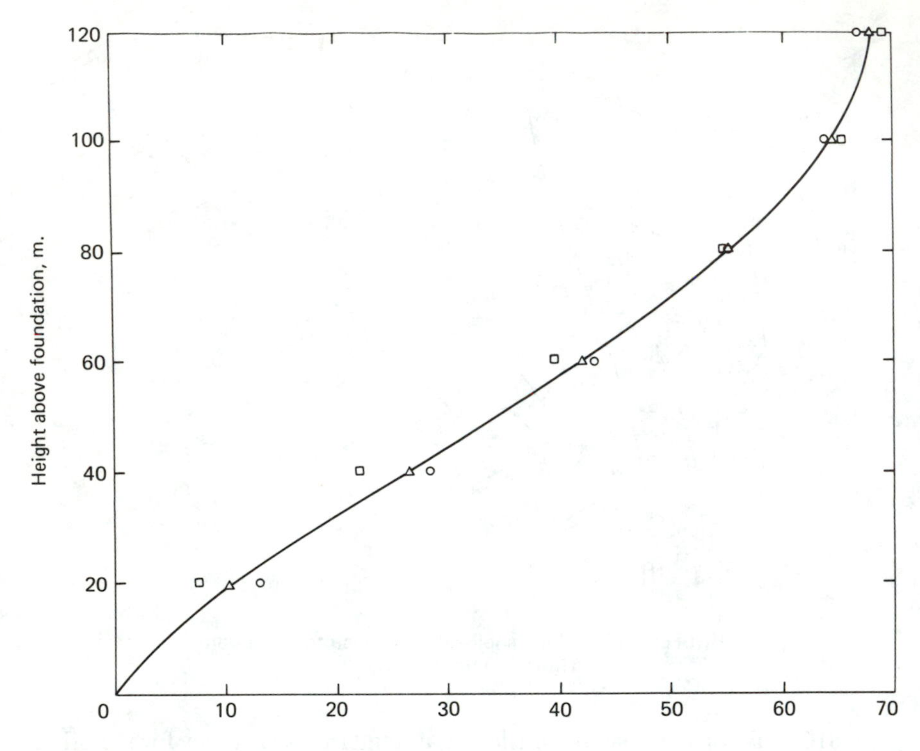

Downstream displacements of air face, on crown section of dam, mm.

Symbol	Type and reference	No. elements	No. nodes	No. equations
——	3D (55, 56)	32	272	816
△	Thick Shell (56)	9	40	200
□	Thin Shell (18)	32	41	205
○	Thin Shell (18) W/Shear	32	41	205

Figure 9-11 Displacements of an arch dam (from References 18 and 56).

Figure 9-13; viewed in plan in the axial direction, the element is annular. The meridian of the element is represented by a fifth order polynomial, which matches the position, slope, and curvature of the actual shell at the two nodes. The displacement model was formulated in the local rectilinear coordinate system ξ-η, with a linear function for the u_1 displacements and a cubic model for the u_2 displacements. For linear elastic problems, this element has provided results which agree well with experiments and alternative solutions.[17]

For elastic-plastic analysis the thin-shell element was further divided into

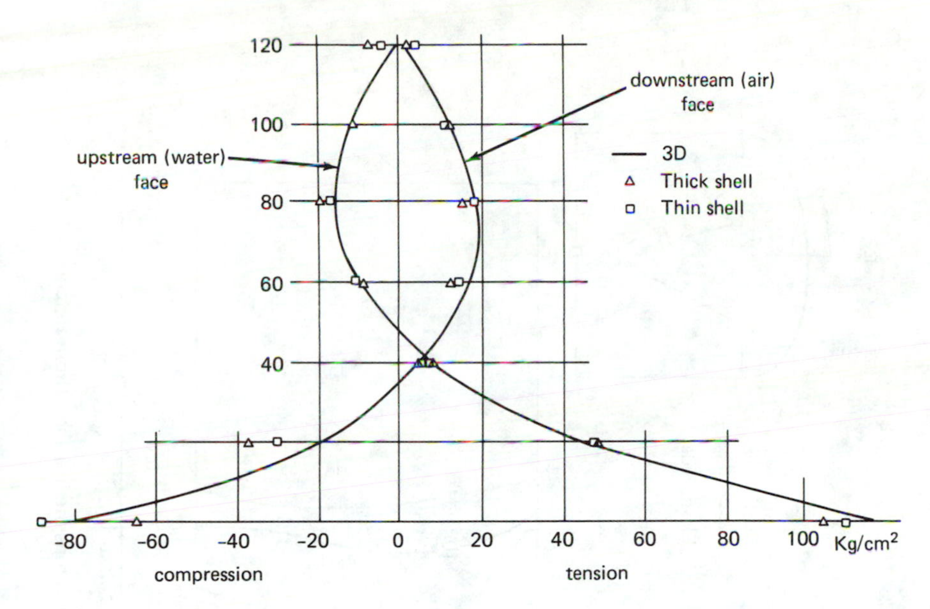

Figure 9-12 Vertical stresses on arch dam at crown section (from References 18 and 56).

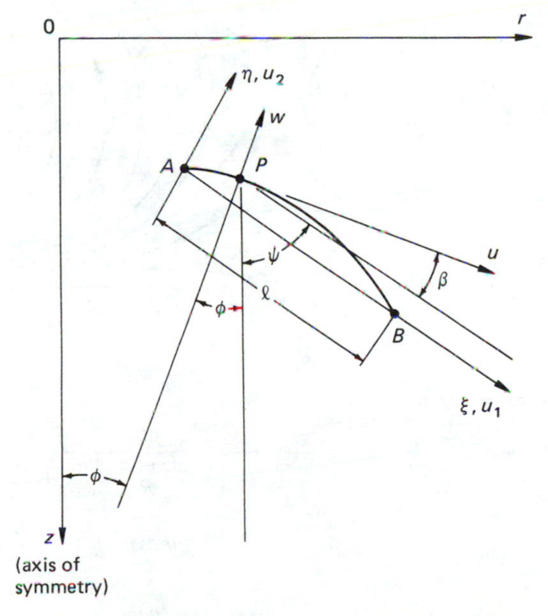

Figure 9-13 Meridional section of doubly curved, axisymmetric shell element (from Reference 17).

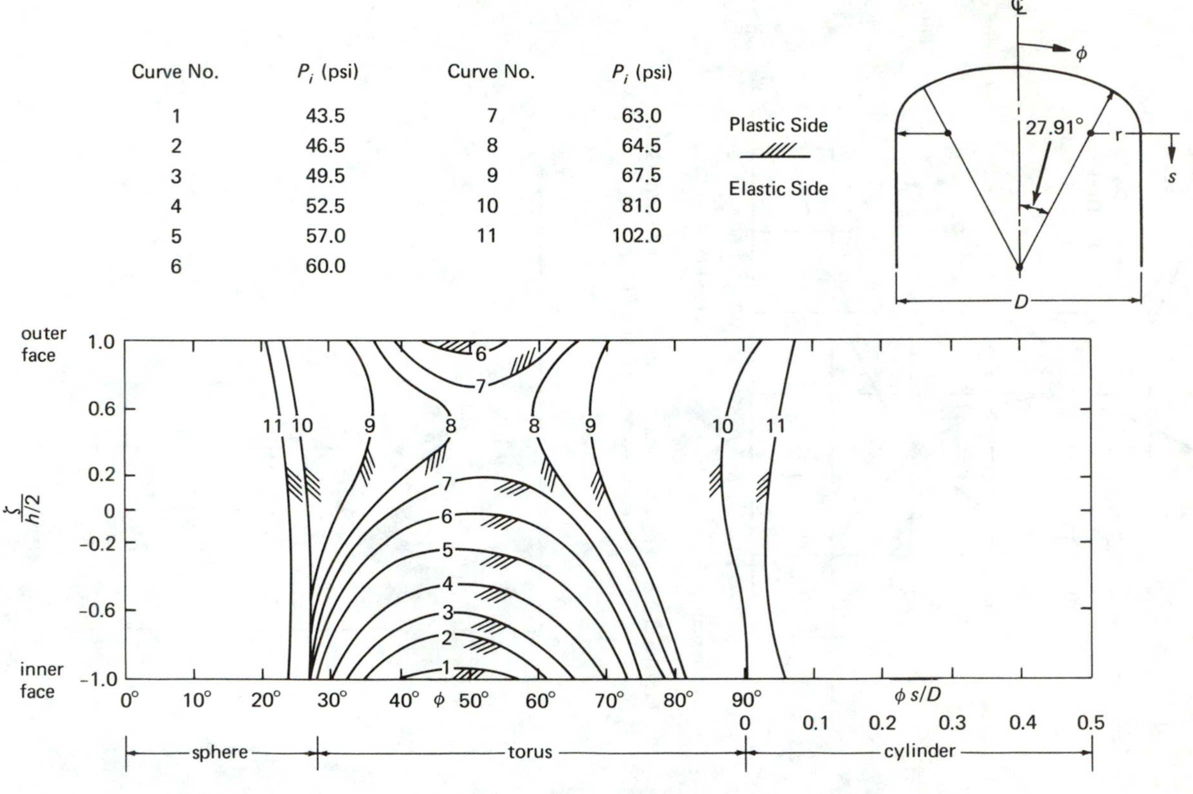

Figure 9-14 Successive elastic-plastic boundaries in a torispherical shell (from Reference 17).

several layers across the thickness of the shell, and the yielding in each lamina was determined according to the von Mises criterion. The flow theory of plasticity was employed in conjunction with the incremental load procedure and the tangent modulus concept. Figure 9-14 shows typical results for the yielding zone of a torispherical pressure vessel head.

Abel and Popov[57] extended the application of the element shown in Figure 9-13 to the elastic analysis of sandwich shells of revolution. The displacement model was augmented to include a linear variation in the ξ-direction of displacement parameters associated with the transverse shear. These parameters were the rotations from the transverse normal of straight lines in the face and core layers of the sandwich. The stresses determined with this element in a shallow spherical cap are shown in Figure 9-15. These results compare well with the analytical solution, despite a very coarse mesh.

The same element configuration has been used by Yaghmai[59] to study the

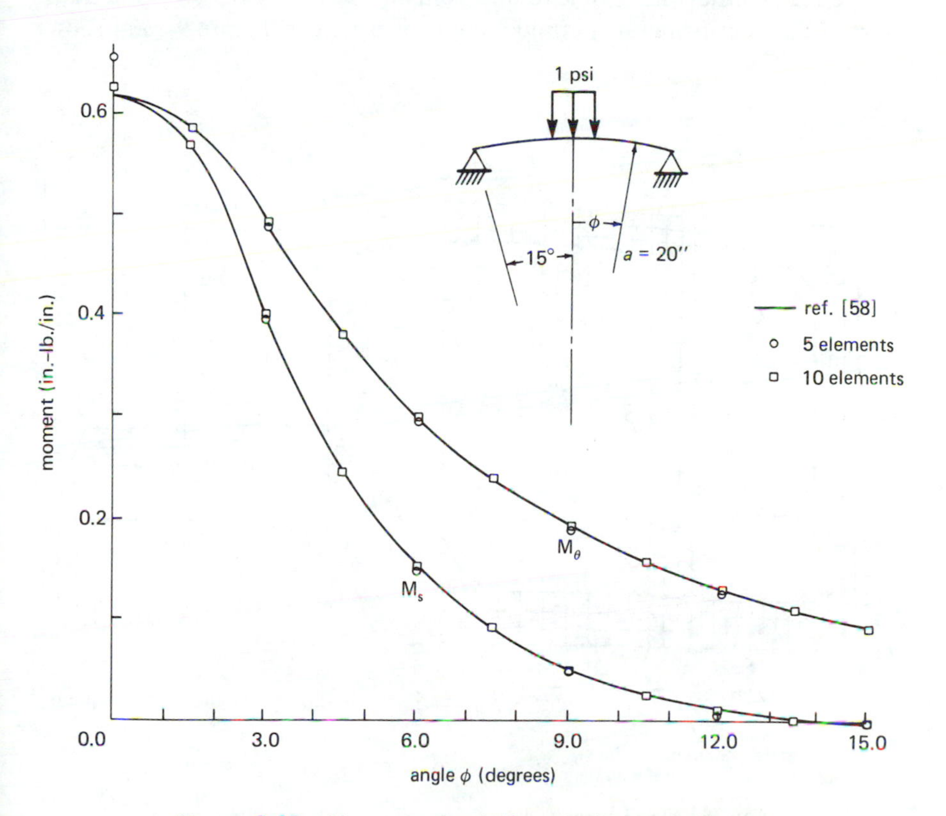

Figure 9-15 Moments of a shallow sherical sandwich cap (from Reference 57).

large inelastic deformations of rotational shells. Sharifi and Popov[60] have further refined the element formulation by utilizing cubic models for the u_1 displacements in the ξ-direction.

9-10 EXAMPLE: SHIP STRUCTURES

Ships are highly complex structural systems subject to static, thermal, and dynamic loads. The dynamic loads include ocean waves, impulses from the propulsion plant, and, in the case of tankers, the sloshing of liquid cargoes. Conventional analysis techniques cannot account for many aspects of a ship's structural behavior, such as the lag problems associated with the interaction of superstructures and hulls, the stress concentrations near holes and irregular stiffeners, and the behavior of deep girders and membrane-like internal structures. The finite element method is being employed increasingly for these difficult problems.[21]

We shall consider here some results from an analysis by Røren[21] of a transverse frame occurring at the mid-section of a tanker. Figure 9-16(a) shows

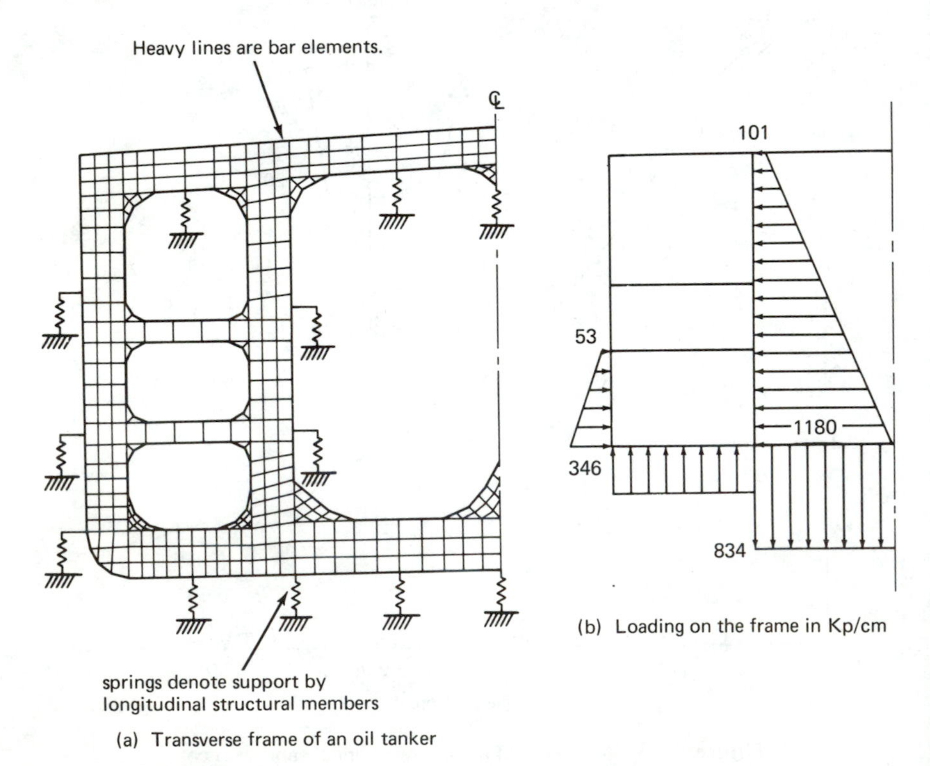

Figure 9-16 Analysis of an oil tanker structure (from Reference 21).

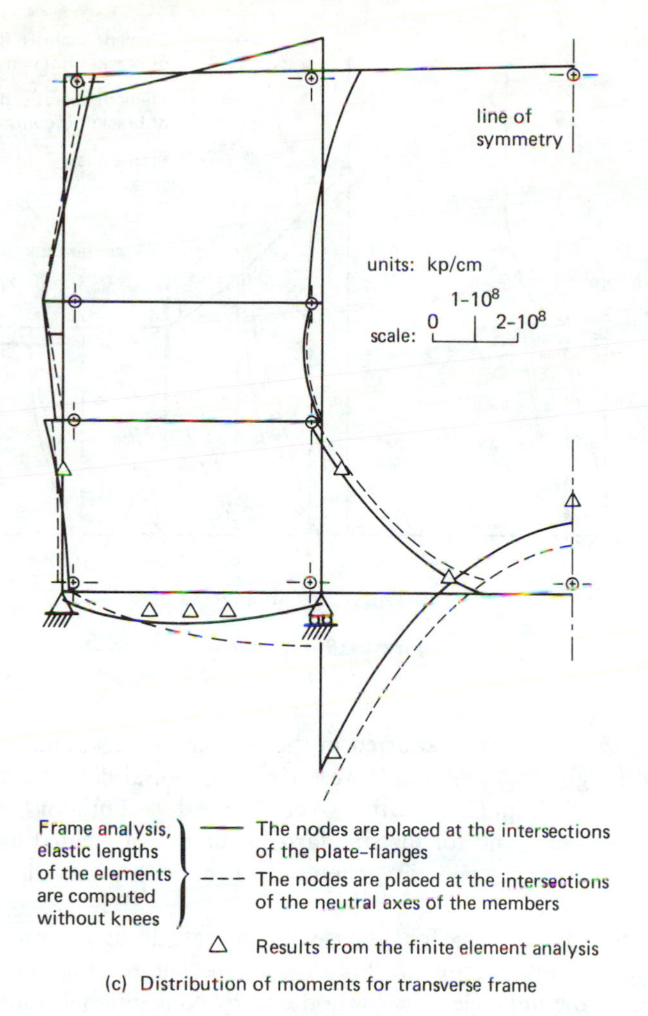

(c) Distribution of moments for transverse frame

Figure 9-16 (*continued*)

the finite element grid for the frame. This is a coarse mesh used for the overall analysis. For detailed study of local regions, the coarse-to-fine and substructure methods described in Section 6-1 were used. The mesh in Figure 9-16(a) contained 343 membrane (plane stress) elements, 188 bar elements, and 406 nodes; and it had a semiband width of 76. Connections with longitudinal girders and the hull were idealized as springs, which were assigned appropriate stiffnesses. (Alternatively, the forces at these connections may be obtained from a substructure analysis.) It was assumed that symmetry existed at the

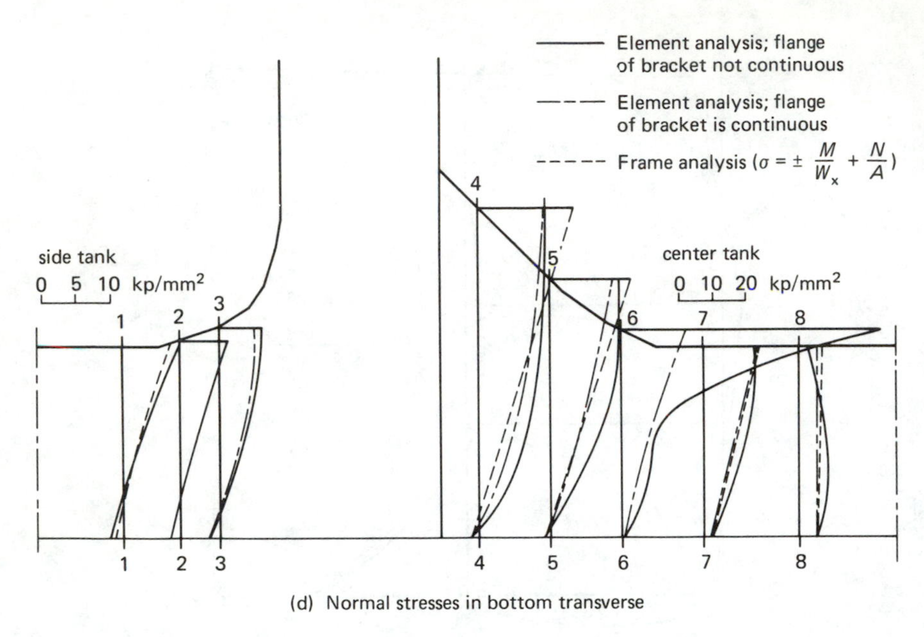

(d) Normal stresses in bottom transverse

Figure 9-16 (*continued*)

centerline, that no girders occurred in the bottom or deck, and that the side shells and longitudinal bulkheads were rigid. A special data-generation program was used for rapid preparation of error-free data, both for the mesh and structural geometry and for the standard loadings. Automated data preparation of this nature is a necessity when only a short time is available for each analysis.

The loading on the frame for the case of moderate draught, empty side-tank, and full center-tank is shown in Figure 9-16(b). The bending moments were computed by the finite element method and by conventional frame analyses, and the results are compared in Figure 9-16(c). In both frame analysis methods, the joint regions were assumed rigid in both bending and shear. The finite element results are the most accurate in this case. For the structure in question, the frame analysis with nodes assumed at flange intersections gave better results than the analysis with nodes taken at the intersection of neutral axes.

Figure 9-16(d) shows the normal stresses in the bottom transverse. The stresses in the bracket region from the finite element analysis differed significantly from the conventional beam hypothesis. This indicates that the bracket flanges transmit much of the total shearing force, especially when the flanges are continuous.

9-11 EXAMPLE: ANALYSIS OF FOLDED PLATE BY SEPARATION OF VARIABLES

Cheung[35] used the finite element method and separation of variables to analyze various folded-plate roof structures. Each plate was divided into a number of strip elements of the type shown in Figures 9-1(a) and 9-17(a). The in-plane displacements u and v and the transverse displacements w were represented as

$$\{u\} = \begin{Bmatrix} u \\ v \\ w \end{Bmatrix}$$

$$= \sum_{n=1}^{N_f} \begin{bmatrix} \{N_1(x)\}^T \cos n\pi y/L & \{0\}^T & \{0\}^T \\ \{0\}^T & \{N_1(x)\}^T \sin n\pi y/L & \{0\}^T \\ \{0\}^T & \{0\}^T & \{N_3(x)\}^T \cos n\pi y/L \end{bmatrix} \{q_n\}$$

where

$$\{q_n\}^T = [u_{1n}\ u_{2n}\ v_{1n}\ v_{2n}\ w_{1n}\ \theta_{1n}\ w_{2n}\ \theta_{2n}]$$

and $\{N_1\}^T$ and $\{N_3\}^T$ are linear and cubic interpolation models, respectively. L is the longitudinal span, and the nodal displacements are taken at the edges of the strip at midspan.

Orthotropic folded-plate roofs symmetric about the crown and with a number of different span lengths were solved by using the above element. The cross section of the roof and the subdivision into elements are shown in Figure 9-17(b). Two different loadings were considered: concentrated ridge loads, and distributed loads, Figure 9-17(b). Results were compared with a direct stiffness solution which also utilized separation of variables, but which employed a classical elasticity stiffness for each plate.[34] Figure 9-17(c) compares the solutions for the longitudinal stresses at midspan, the transverse moments at midspan, and the plate membrane shear at the support for the case of a 70 foot span subject to ridge loads. The agreement between solutions is good.

For conventional, isotropic folded plates, the elasticity method[34] is more accurate and efficient than the finite element method. However, for such complex situations as orthotropic plates, plates with eccentric stiffeners, sandwich plates, and open vierendeel systems, the finite element method is a powerful tool.[35]

9-12 EXAMPLE: NONLINEAR ANALYSIS OF THIN PLATES

An approximate formulation for the large deflection of thin plates has been developed by Murray and Wilson.[61] Triangular finite elements with fifteen degrees of freedom were used; the in-plane behavior was represented by a

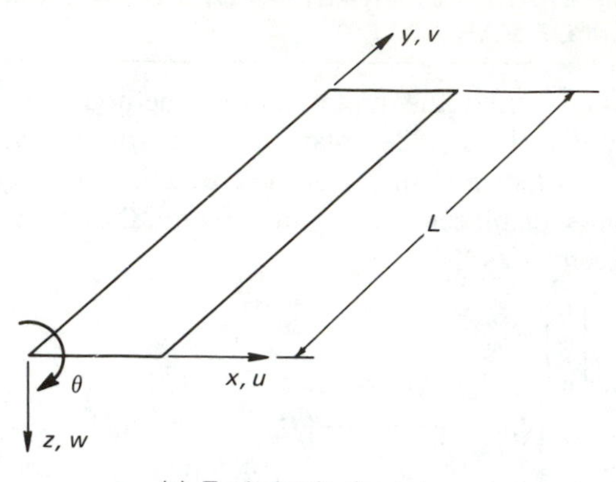

(a) Typical strip element

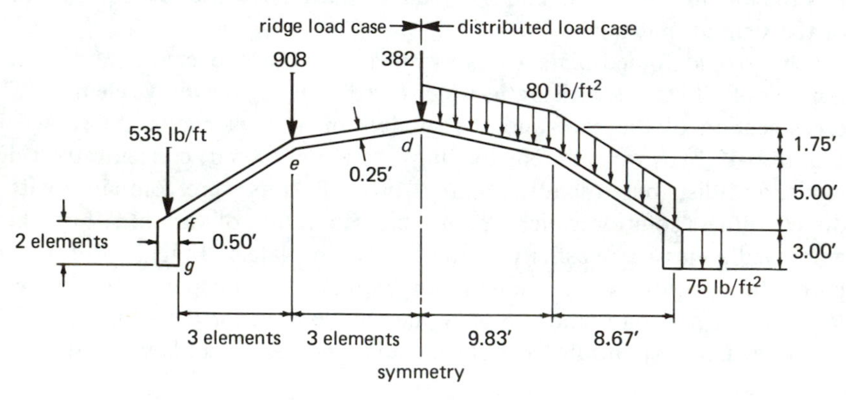

(b) Folded plate section, dimensions and loads

Figure 9-17 Analysis of folded plate by separation of variables (from Reference 35).

CST (linear displacements, six degrees of freedom), and the bending was represented by a LCCT-9[14] (cubic displacements, nine degrees of freedom). Linear elastic material behavior and small strains were assumed, and the geometric nonlinearity was associated with large displacements only.

A mixed solution procedure (Chapter 7) was used for the nonlinear analysis. Because iteration was carried out in each increment until equilibrium was achieved, it was possible to utilize a stiffness matrix that was approximate.

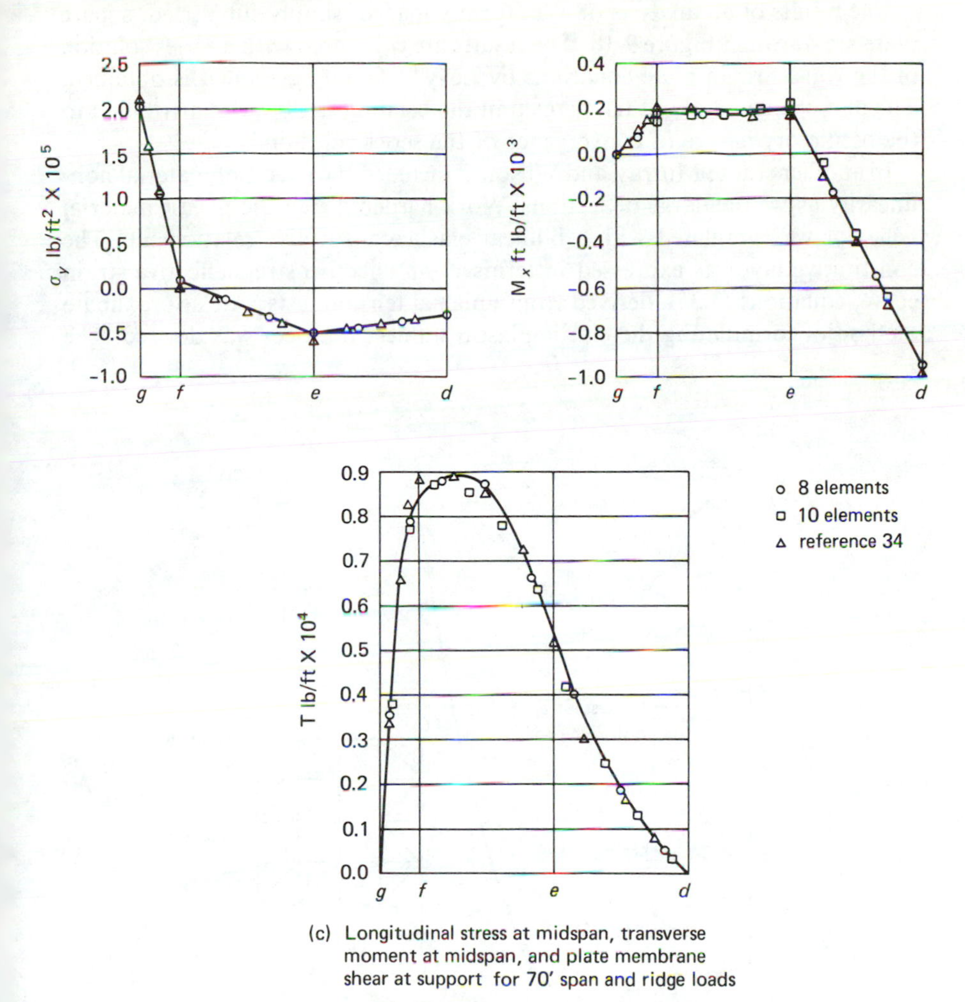

(c) Longitudinal stress at midspan, transverse moment at midspan, and plate membrane shear at support for 70′ span and ridge loads

Figure 9-17 (*continued*)

Hence, the effects of the geometric stiffness were ignored, and only the conventional element stiffnesses were used in assembling the overall stiffness. A local coordinate system which moved with the deforming element was used. With reference to this system, the element displacements at any stage were small; therefore, the linear plate equations were used to formulate the element stiffness in local coordinates. This element stiffness remained constant. The transformation from local to global coordinates was continually updated to account for the changing geometry resulting from large displacements.

The results of an analysis of a uniformly loaded, simply supported, square plate are shown in Figure 9-18. The results are compared with a series solution of the von-Kármán plate equations by Levy.[62] Good agreement is obtained. The discrepancy of about ten percent in the bending stresses was attributed to the oscillatory nature of convergence of the series solution.[61]

In another study, Murray and Wilson[63] included the effects of material non-linearity as well as large deflections. Work-hardening, elastic-plastic material behavior was simulated with a bilinear elastic constitutive relationship. The constitutive law was expressed in terms of an effective stress-effective strain curve, equations (3.35), derived from uniaxial tension tests. An approximate method of formulating the elastic-plastic stiffness matrices was devised. The

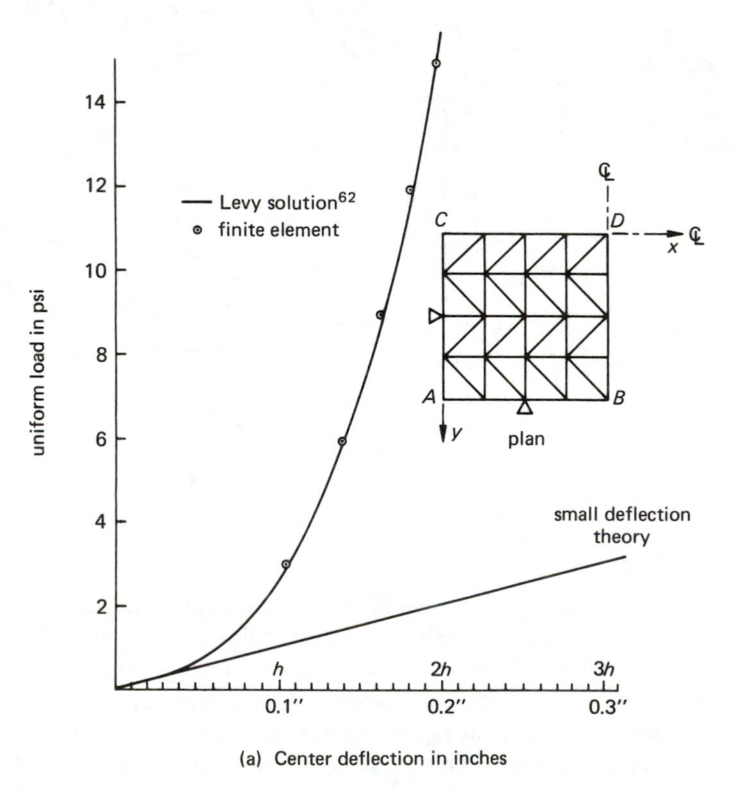

Figure 9-18 Deflections and stresses of a uniformly loaded, simply supported, square plate (from Reference 61).

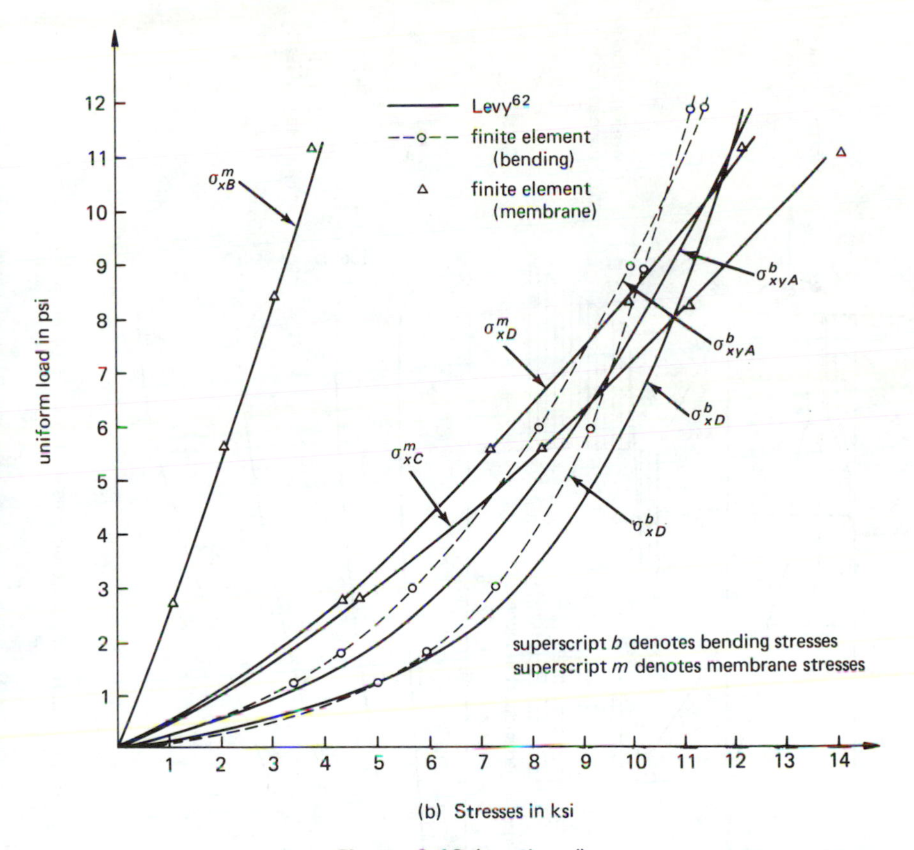

(b) Stresses in ksi

Figure 9-18 (*continued*)

method is applicable to both tangent and secant stiffnesses and gives rise to element stiffnesses which couple the stretching and bending behavior. The mixed method of nonlinear solution was used to study elastic-plastic large-deflection problems. For each increment of loading a tangent stiffness was computed. The same tangent stiffness was used for all iterations within the increment. However, convergence to equilibrium was tested by utilizing a secant stiffness newly computed after each iteration.

Figure 9-19 shows the results for the cylindrical bending of a simply-supported plate subject to a line load along the centerline. The extent of the plastic zones at the end of load increments 6 and 7 are depicted in Figure 9-19(a), whereas Figure 9-19(b) shows the stress distributions across the thickness at selected sections after the seventh increment. Finally, the deflected profiles at different load increments are indicated in Figure 9-19(c). The predominance of membrane behavior at the seventh increment is apparent.

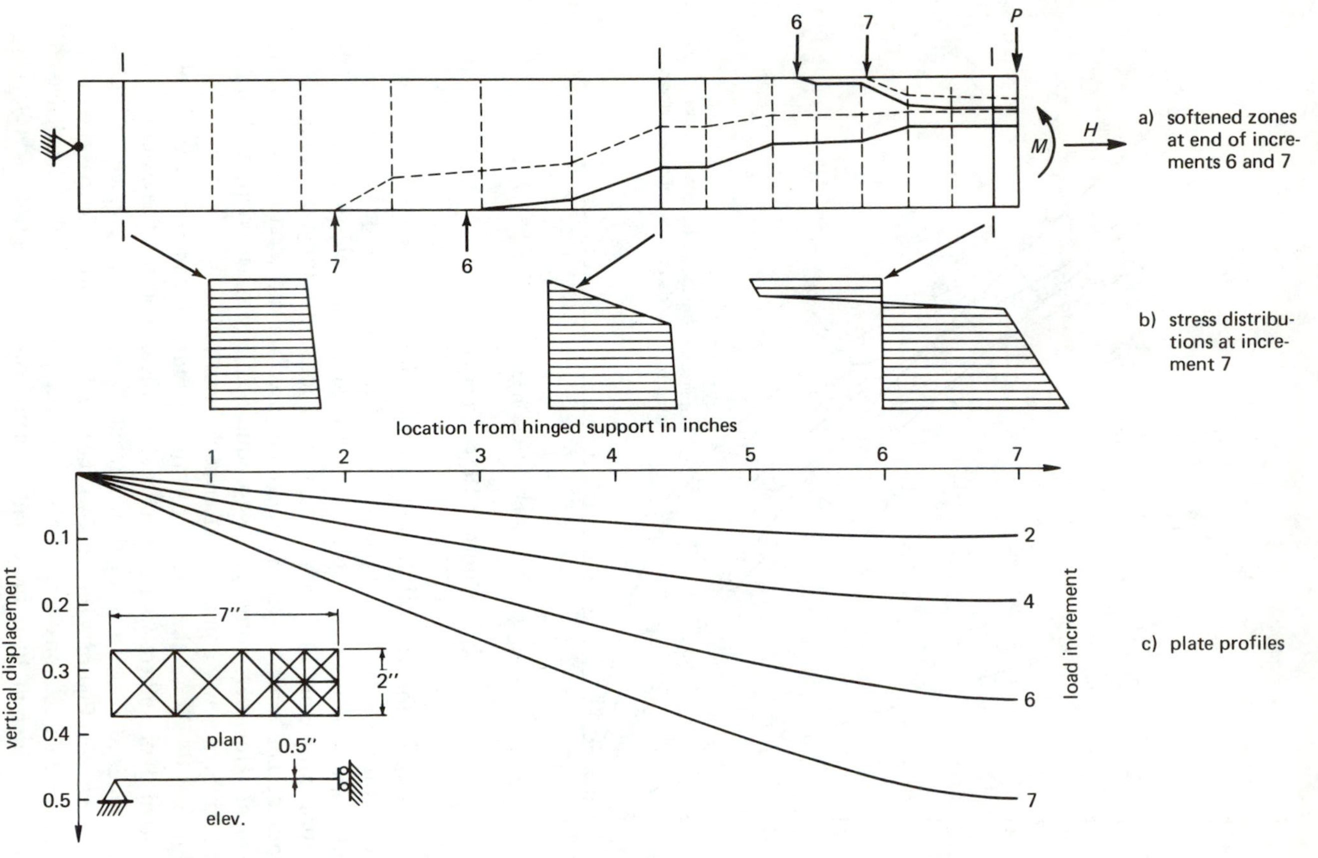

Figure 9-19 Large-deflection cylindrical bending of an elastic-plastic thin plate (from Reference 63).

REFERENCES

(1) See Reference 4, Chapter 1.
(2) *First Conf.*
(3) Zienkiewicz, O. C., and Cheung, Y. K., *The Finite Element Method in Structural and Continuum Mechanics*, London, McGraw-Hill Pub. Co. Ltd, 1967.
(4) See Reference 16, Chapter 8.
(5) *Second Conf.*
(6) *FEM Tapir.*
(7) *Symp. FEM.*
(8) See Reference 13, Chapter 2.
(9) Wilson, E. L., "Structural Analysis of Axisymmetric Solids," *AIAA J.*, Vol. 3, No. 12 (Dec. 1965). Ring elements with triangular cross section are used to analyze complex structures under nonsymmetric loads.
(10) Rashid, Y. R., "Analysis of Axisymmetric Composite Structures by the Finite Element Method," *Nuc. Engr. & Des.*, Vol. 3 (1966). Analysis of prestressed concrete reactor vessel under the influence of body forces, surface loads, thermal loads, residual stresses, and prestressing loads.
(11) See Reference 1, Chapter 6.
(12) Stordahl, H., and Christensen, H., "Experiences from Stress Analysis of Axisymmetric Problems in Machine Design," *FEM Tapir*. Uses CST ring element with triangular cross section to analyze a threaded spindle, a rotor with vanes, and a wheel fixed to a shaft by shrinking.
(13) See Reference 13, Chapter 5.
(14) Clough, R. W., and Tocher, J. L., "Finite Element Stiffness Matrices for Analysis of Plate Bending," *First Conf.* Comparison of the relative accuracy of seven different plate bending elements.
(15) See Reference 6, Chapter 5.
(16) See Reference 14, Chapter 5.
(17) Khojasteh-Bakht, M., and Popov, E. P., "Analysis of Elastic-Plastic Shells of Revolution," *Proc. ASCE, J.EM Dn*, Vol. 96, EM3 (June 1970). Uses doubly curved axisymmetric shell elements for an incremental, tangent stiffness method of elastic-plastic analysis.
(18) Clough, R. W., and Johnson, C. P., "Finite Element Analysis of Arbitrary Thin Shells," ACI Symposium on Concrete Thin Shells, April 1970. Survey and evaluation of various finite elements for thin shell analysis, including planar and curved configurations.
(19) Martin, H. C., "On the Derivation of Stiffness Matrices for the Analysis of Large Deflection and Stability Problems," *First Conf.* A survey of the title problems up to 1965.
(20) Thomas, J. M., "A Finite Element Approach to the Structural Instability of Beam-Columns, Frames, and Arches," U.S. NASA TN D-5782 (May 1970). Formulates the necessary stiffness matrices for a beam-column. Considers both conservative and nonconservative loading behavior.
(21) Røren, E. M. Q., "Finite Element Analysis of Ship Structures," *FEM Tapir*. Briefly surveys applications to naval structures. Includes a short literature

review and some examples. Considers aspects of problems involving transverse frames, bulkheads, and hull sections.
(22) Zienkiewicz, O. C., "Finite Element Procedures in the Solution of Plate and Shell Problems," In *Stress Analysis: Recent Developments in Numerical and Experimental Methods*, edited by Zienkiewicz, O. C., and Holister, G. S., New York, John Wiley & Sons, 1965.
(23) See Reference 8, Chapter 4.
(24) See Reference 11, Chapter 8.
(25) Fraeijs de Veubeke, B., and Sander, G., "An Equilibrium Model for Plate Bending," *Int. J. Solids Structs.*, Vol. 4 (1968). A triangular element is developed for arbitrary oblique coordinates. The convergence rate is compared with that of several other elements.
(26) See Reference 12, Chapter 8.
(27) Severn, R. T., and Taylor, P. R., "The Finite Element Method for Flexure of Slabs When Stress Distributions Are Assumed," *Proc. Inst. Civil Eng.*, Vol. 34 (June 1966). Hybrid equilibrium method for rectangular and triangular elements. Quadratic stress model in element, cubic displacement model over interfaces.
(28) Allwood, R. J., and Cornes, G. M. M., "A Polygonal Finite Element for Plate Bending Problems Using the Assumed Stress Approach," *IJNME*, Vol. 1, No. 2 (1969). Hybrid equilibrium for polygonal elements. Linear, quadratic, and cubic stress models are compared.
(29) See Reference 14, Chapter 8.
(30) See Reference 15, Chapter 8.
(31) See Reference 8, Chapter 1.
(32) See Reference 4, Chapter 5.
(33) Scordelis, A. C., and Lo, K. S., "Computer Analysis of Cylindrical Shells," *J. ACI*, Vol. 61, No. 5 (May 1964). Multiple cylindrical shell roofs are analyzed by separation of variables. Elements are cylindrical segments, and stiffnesses are "exact" from theory of elasticity.
(34) DeFries-Skene, A., and Scordelis, A. C., "Direct Stiffness Solution for Folded Plates," *Proc. ASCE, J.ST Dn.*, Vol. 90, ST4 (Aug. 1964). Similar in concept to Reference 33. Elements are the entire flat plates, and stiffnesses are based on "exact" elasticity solutions, rather than on assumed displacement models.
(35) Cheung, Y. K., "Folded Plate Structures by Finite Strip Method," *Proc. ASCE, J.ST Dn.*, Vol. 95, ST12 (Dec. 1969). Discussion, Vol. 96, ST7 (July 1970).
(36) Percy, J. H., Pian, T. H. H., Klein, S., and Navaratna, D. R., "Application of the Matrix Displacement Method to Linear Elastic Analysis of Shells of Revolution," *AIAA J.*, Vol. 3, No. 11 (Nov. 1965). Applies conical frustum elements to the analysis of axisymmetric shells under symmetric and asymmetric loadings.
(37) Herrmann, L. R., "Elasticity Equations for Incompressible and Nearly Incompressible Materials by a Variational Theorem," *AIAA J.*, Vol. 3, No. 10 (Oct. 1965).
(38) See Reference 3, Chapter 6.
(39) Hughes, T. J. R., and Allik, H., "Finite Elements for Compressible and In-

compressible Continua," *Symp. FEM.* Three-dimensional elements with various order models are compared.

(40) Christian, J. T., "Undrained Stress Distribution by Numerical Methods," *Proc. ASCE, J. SM&F Dn.*, Vol. 94, SM6 (Nov. 1968). Finite difference and finite element solutions for deformations in incompressible elastic soils on the basis of an effective stress concept.

(41) Taylor, R. L., Pister, K. S., and Herrmann, L. R., "On a Variational Theorem for Incompressible and Nearly-Incompressible Orthotropic Elasticity," *Int. J. Solids Structures*, Vol. 4 (1968).

(42) Tong, P., "An Assumed Stress Hybrid Finite Element Method for an Incompressible and Near-Incompressible Material," *Int. J. Solids Structs.*, Vol. 5 (May 1969).

(43) Iversen, P. A., "Some Aspects of the Finite Element Method in Two-Dimensional Problems," *FEM Tapir.* Considers such aspects as computer programs, structural idealization, boundary conditions, and the form of the overall stiffness matrix.

(44) Corum, J. M., and Krishnamurthy, N., "A Three-Dimensional Finite Element Analysis of a Prestressed Concrete Reactor Vessel Model," *Symp. FEM.*

(45) Clough, R. W., and Felippa, C. A., "A Refined Quadrilateral Element for Analysis of Plate Bending," *Second Conf.* A complete and conforming quadrilateral element with twelve external degrees of freedom is developed for the displacement method. Transverse shear is included without increasing external degrees of freedom. Results are presented for static, free vibration, and stability problems.

(46) Ramstad, H., "Convergence and Numerical Accuracy with Special Reference to Plate Bending," *FEM Tapir.* Considers errors due both to discretization and rounding. Shows that although increased refinement of the mesh decreases the discretization error, the corresponding increase in rounding error can be critical. Therefore, double precision solutions or other improved methods of solution may be necessary.

(47) Adini, A., and Clough, R. W., "Analysis of Plate Bending by the Finite Element Method," Rept. to NSF, Grant G7337 (1960).

(48) See Reference 7, Chapter 1.

(49) Melosh, R. J., "A Stiffness Matrix for the Analysis of Thin Plates in Bending," *J. Aero Sci.*, Vol. 28 (1961).

(50) Papenfuss, S. W., "Lateral Plate Deflection by Stiffness Methods with Application to a Marquee," M. S. Thesis, Department of Civil Engineering, University of Washington, Seattle, 1959.

(51) See Reference 21, Chapter 5.

(52) See Reference 15, Chapter 6.

(53) Anderheggen, E., "A Conforming Triangular Finite Element Plate Bending Solution," *IJNME*, Vol. 2, No. 2 (1970).

(54) Elias, Z. M., "Duality in Finite Element Models," *Proc. ASCE, J.EM Dn*, Vol. 94, EM4 (Aug. 1968). Stretching of plates by assumed displacements and bending by assumed stresses.

(55) Ergatoudis, J., Irons, B. M., and Zienkiewicz, O. C., "Three-Dimensional

Analysis of Arch Dams and Their Foundations," *Arch Dams: A Review of British Research & Development*, London, Inst. of Civil Engineers, 1968.
(56) Ahmad, S., Irons, B. M., and Zienkiewicz, O. C., "Curved Thick Shell and Membrane Elements with Particular Reference to Axisymmetric Problems," *Second Conf.* General curved isoparametric elements are degenerated to superparametric elements suitable for shell problems.
(57) See Reference 12, Chapter 6.
(58) Rosettos, J. N., "Deflections of Shallow Spherical Sandwich Shell Under Local Loading," U.S. NASA TN D-3855 (Feb. 1967).
(59) See Reference 24, Chapter 7.
(60) Sharifi, P., and Popov, E. P., "Refined Finite Element Analysis of Elastic-Plastic Thin Shells of Revolution," Structures and Materials Research Rept. SESM 69–8, Berkeley, University of California (Dec. 1969).
(61) Murray, D. W., and Wilson, E. L., "Finite Element Large Deflection Analysis of Plates," *Proc. ASCE, J.EM Dn*, Vol. 95, EM1 (Feb. 1969).
(62) Levy, S., "Bending of Rectangular Plates with Large Deflections," U.S. NACA TN-846 (1942).
(63) Murray, D. W., and Wilson, E. L., "An Approximate Nonlinear Analysis of Thin Plates," *Second Conf.*

FURTHER READING

Abrahamsen, E., "Structural Design Analysis of Large Ships," SNAME Annual Meeting, New York, Nov. 1969.

Abrahamsen, E., Nørdenstrom, N., and Røren, E. M. Q., "Design and Reliability of Ship Structures," SNAME Spring Meeting, Washington, D. C., April, 1970.

Ahmad, S., Irons, B. M., and Zienkiewicz, O. C., "Analysis of Thick and Thin Shell Structures by Curved Finite Elements," *IJNME*, Vol. 2, No. 3 (July–Sept. 1970).

Barsoum, R. S., and Gallagher, R. H., "Finite Element Analysis of Torsional and Torsional-Flexural Stability Problems," *IJNME*, Vol. 2, No. 3 (July–Sept. 1970).

Bouwkamp, J. G., Vaish, A. K., and Terata, H., "A Study of Different Flared Joint Configurations," *Proc. Second Annual Offshore Technology Conference*, April 1970. Thin shell elements are used to study tubular joints. Results are compared to experimental investigations.

Byskov, E., "The Calculation of Stress Intensity Factors Using the Finite Element Method with Cracked Elements," *Intl. J. Fracture Mech.*, Vol. 6, No. 2 (June 1970). A general procedure is developed for computing the stiffness matrix for a cracked plate element.

Davies, J. D., "Analysis of Corner Supported Rectangular Slabs," *J. Inst. of Struct. Engineers*, London, Vol. 48, No. 2 (Feb. 1970). Uses finite element method for deflection and moment analysis of simply supported and clamped slabs. Compares numerical results with experiments on perspex models of slab. Presents curves for design of corner supported flat slabs.

Fjeld, S. A., "Three-Dimensional Theory of Elasticity," *FEM Tapir*. General treatment of various three-dimensional finite elements.

Hodge, P. G., Jr., and Belytschko, T., "Numerical Method for the Limit Analysis of Plates," *J. App. Mech.*, Vol. 35, No. 4 (Dec. 1968). Determines upper and lower bounds on the yield-point loads of plates by using finite element representation of velocity and moment fields.

Iqbal, M. A., and Kokosky, E. M., "Interaction Stresses in Composite Systems," *Proc. ASCE, J.EM Dn*, Vol. 96, EM6 (Dec. 1970). Plane stress finite element analysis of stress distributions generated by inclusions and voids in an otherwise homogeneous matrix.

Kapur, K. K., and Hartz, B. J., "Stability of Plates Using the Finite Element Method," *Proc. ASCE, J.EM Dn*, Vol. 92, EM2 (Apr. 1966).

Lutz, L. A., "Analysis of Stresses in Concrete Near a Reinforcing Bar Due to Bond and Transverse Cracking," *J. ACI*, Vol. 67, No. 10 (Oct. 1970).

Nilson, A. H., "Nonlinear Analysis of Reinforced Concrete by the Finite Element Method," *J. ACI*, Vol. 65, No. 9 (Sept. 1968). Analysis for internal stresses and displacements for reinforced concrete members under steadily increasing loads from zero to ultimate. Accounts for reinforcement, progressive cracking, bond stress transfer, and nonlinear, orthotropic material properties.

Oden, J. T., and Key, J. E., "Numerical Analysis of Finite Axisymmetric Deformations of Incompressible Elastic Solids of Revolution," *Int. J. Solids Structures*, Vol. 6 (1970).

Puppo, A., and Evenson, H., "Calculation and Design of Joints Made from Composite Materials," Tech. Report 70-27, U.S. Army Aviation Material Lab., Fort Eustis, Va., NTIS, Springfield, Va. Finite element analysis of joints made from fiber composites. Different types of joints, such as bolted and adhesive, are considered. A quadratic displacement model is used with a triangular element.

Sawko, F., and Cope, R. J., "Analysis of Spine Beam Bridges Using Finite Elements," *Civil Engg. & Public Works Review* (Feb. 1970). Compares finite element results with experimental results on a model bridge made of perspex.

Smith, P. G., and Wilson, E. L., "Automatic Design of Shell Structures," *Proc. ASCE, J. ST Dn.*, Vol. 97, ST1 (Jan. 1971). Describes a general approach to the selection of optimal shell configurations by using a finite element code for geometrically nonlinear membrane behavior. The goal is to obtain a configuration in which the applied loads are equilibrated by membrane forces.

Various authors, *Recent Advances*.

Various authors, *High Speed Computing of Elastic Structures* (Proc. Symposium of IUTAM, August, 1970), University of Liège, Belgium, 1971.

See Reference 7, Chapter 6.

10
SOIL AND ROCK MECHANICS

The initial development of the finite element method for aerospace and structural engineering was soon followed by application of the method to problems in soil and rock mechanics. The nature of soils and rocks, however, is highly complex and requires different considerations from the materials used in structures. A realistic appraisal of the complexities imposed by such natural causes as joints and other discontinuities would often require that soils and rocks be treated as discontinua. Nevertheless, approximate but acceptable solutions can be obtained by considering them as continuous masses. In most applications of the finite element method, the continuum approach is used; hence, we shall restrict our attention to this idealization. Furthermore, we shall consider only those aspects of soil and rock mechanics directly related to the applications of the finite element method. For fundamentals of these two subjects, the reader may consult standard texts.[1,2,3,4,5,6,7,8]

10-1 SUMMARY OF APPLICATIONS

The uncertainties inherent in the behavior of soils and rocks require experimental or field verification for any analytical technique before it can be used for practical design analysis. For certain problems, such as the analyses of slopes, dams and foundations, and some soil-structure interaction problems, sufficient experimental verifications have been obtained. Additional experimental correlation, however, is desirable for such topics as the complex shapes of dams, jointed rocks, group foundations, multiple wheel loads on pavements, post-cracking phenomena, work softening behavior, and very soft materials.

Tables 10-1 and 10-2 show various topics for which the finite element

TABLE 10-1 TOPICS IN SOIL MECHANICS.

	Topic	*Typical Applications*	*References*
I	Stress-deformation and stability analyses	Stress analysis of slopes, excavations, embankments, dams, riverbanks, etc.	9 to 22
II	Bearing capacity and settlement analyses	Circular and strip footings, deep foundations, piles	17, 18, 23, 24
III	Soil-structure interaction	Pavements, underground structures, lateral pressures, beams on foundations	12, 25 to 31
IV	Constitutive laws	Clays and sands, laboratory correlation	12 to 14, 17, 28, 32 to 36

TABLE 10-2 TOPICS IN ROCK MECHANICS.

	Topic	*Typical Applications*	*References*
I	Stress-deformation and stability analyses	Stress analysis of slopes, excavations	37 to 43
II	Structures	Dams, tunnels, mines, pits, cavities, boreholes	39, 42 to 46
III	Geologic features	Joints, fissures, fractures, layers	39, 47 to 49
IV	Constitutive laws	Intact rocks, discontinuous rocks	7, 8, 17, 35 to 37, 40 to 45, 47 to 49

method has been successfully employed. A few typical applications are described below. Our primary consideration will be the stress-deformation problems, whereas in Chapters 12, 13, and 14 we shall include other topics relevant to soil and rock mechanics, such as seepage, consolidation, and creep effects.

10-2 USE OF THE FINITE ELEMENT METHOD IN SOIL AND ROCK MECHANICS

Deformation behavior of soils and rocks is influenced by a number of factors, such as the physical structure, porosity, density, stress history, loading characteristics, existence and movement of fluids in the pores, and time-dependent or viscous effects in the solid skeleton and the pore fluids. In addition, such geologic features as faults, joints, seams, crushed zones, veins, fissures, folds, and other tectonic effects produce behavior significantly different from that derived on the assumption of continuous mass. These factors render the stress-deformation behavior highly complex and nonlinear. No available analytical solution scheme can handle them all. Numerical methods, such as the finite element method, have proved successful in approximating the effects of many of these factors, and they show promise of incorporating a number of others. Moreover, the initial success of the finite element method has indicated that some of these factors have a more significant effect than had been expected. In this section we shall consider some of the specialized phenomena that can be included in the finite element formulation.

We use the same variational principle, equation (5.43), developed for other solid mechanics problems in applying the finite element method to stress-deformation problems in soil and rock mechanics. Most applications involve use of a linear displacement model such as equation (5.27). Sometimes a higher order model may be warranted. Although any of the common shapes of elements (Figs. 5-7 and 5-8) may be used, usually the quadrilateral with four CST's has been selected.

Loading

Loadings are due chiefly to man-made structures, internal body forces due to gravity, temperature gradients and their seasonal variations, and earthquakes and other dynamic forces. (The last are treated in Chapter 11.) In addition, unloading problems occur as a result of excavation and construction.

Geological factors also contribute loadings and alterations in loading. *Geostatic forces* from the weight of the overburden increase with depth in a manner similar to hydrostatic forces. A side effect of the geostatic forces is the occurrence of horizontal or lateral forces. These forces, which already exist in a soil or rock deposit, are termed *in situ* conditions, also called initial or pre-stress conditions. The *in situ* stresses play a significant role in the subsequent deformation behavior of the mass.

In Fig. 10-1, σ_z denotes the vertical stress resulting from the unit weight γ of a depth of overburden, Z. This vertical stress causes lateral or horizontal

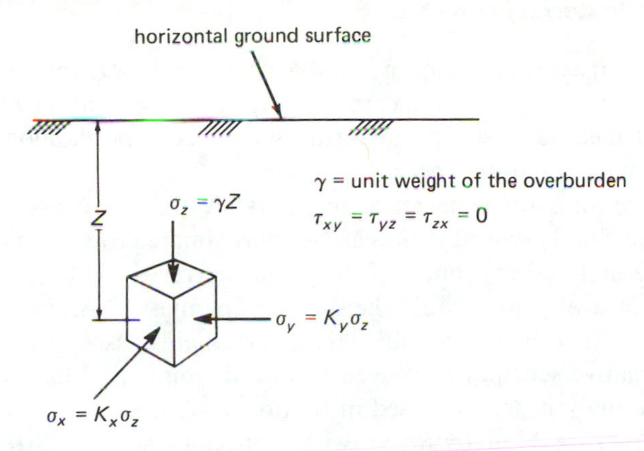

Figure 10-1 *In situ* stress field.

stresses σ_x and σ_y. If the soil deformations have taken place in the vertical direction, only the normal stresses σ_x, σ_y, and σ_z have non-zero values, whereas the shear stress components τ_{xy}, τ_{yz}, and τ_{zx} are zero (Fig. 10-1). The magnitude of the lateral stress is dependent upon the geologic history and is related to σ_z through a factor K (Fig. 10-1) called the *co-efficient of lateral earth pressure*. It can include effects of both the earth and pore fluid (water) pressures. The magnitude of K is governed by such factors as overconsolidation, faults or folds, and tectonic effects. For a normally consolidated soil its value is less than unity, whereas for an overconsolidated soil it may be greater than unity. In the case of a rock with normal folding, its typical value may be 0.5, while for reverse folding or faulting it may be of the order of 2 to 3. The *in situ* stresses may occur as a result of the orogenic movements of the earth's crust and cause the so called *tectonic forces*. Often the *in situ* forces may occur as *residual stresses* resulting from physico-chemical changes. The *in situ* stresses may be computed on the basis of field and/or laboratory experiments.

The vertical and horizontal stresses generally increase with depth of a deposit, and such variations should be included in the analysis. If these variations are determined adequately from field and laboratory observations, they are easily incorporated into the finite element analysis.

The fluid in the pores is generally water, and it causes the so-called *pore pressures*. Pore pressures cause *seepage* and *uplift forces*. In a simple approximate procedure, such forces can be computed from a *flow net* analysis. An alternative procedure for computing approximate pore pressures is to use Skempton's pore pressure parameters.[1,2,4,5] For a more refined analysis, the pore pressure can be assumed as an unknown field variable in addition to the displacements (Chapter 13).

Nonuniform Material Properties

Soils and rock masses are seldom homogeneous and isotropic. If the nonhomogeneity is regular, as in the case of a layered system, application of the finite element method is straightforward. Such abrupt nonhomogeneities as jointing and fissures require special treatment.

Joints are a common occurrence in rocks. A mass traversed by small, numerous, and closely spaced joints can be approximated by a solid continuum with modified material constants. If the joints are relatively large, they must be treated as separate units within the surrounding mass.[49] An important side effect of joints is that they render the parent mass highly susceptible to tension. Various alternative schemes employed to include joints and fissures into the finite element analysis are discussed in Section 10-7.

Soils and rocks exhibit *anistropy* with respect to stresses, strength, and pore pressure characteristics. For instance, a soil or rock deposit is subjected to different *in situ* stresses in the vertical and horizontal directions. This *initial stress anisotropy* produces different strengths in different directions. As an example, the elastic moduli E_v and E_h in the vertical and horizontal directions, respectively, are different. When such an anisotropic mass is loaded, the directions of the principal stresses vary and cause changes in the anisotropy with respect to stresses, strength, and pore pressures. Anisotropy has significant influence on the behavior of soils and rocks.

In their finite element analysis, Dunlop *et al.*[14] account for anisotropy by using the following relation:

$$E_{\beta_i} = E_h - (E_h - E_v) \sin^2 \beta_i \tag{10.1}$$

β_i is the angle between the horizontal and the instantaneous direction of principal compression, and E_{β_i} is the instantaneous modulus value in the β_i direction. Equation (10.1) represents a *sin² variation.* One way to determine E_v and E_h is to conduct vertical and horizontal plane strain tests. A similar relation expressed in terms of undrained shear strength, S_u, was used by D'Appolonia and Lambe.[18]

Tension in Soils and Rocks

Usually soil and rock masses are subjected to compressive loadings, and application of the finite element method is straightforward. However, such factors as geometry, certain types of loading, geologic discontinuities, and the inherent nature of the materials, often cause tensile stresses which must be taken into account. Tensile stresses arise typically at the crest of an excavated slope, the interface between a flexible footing and the soil below, and at various locations within earth and rock fill dams. The most significant effect of excessive tensile stresses is *cracking*, for example the cracking of earth and rockfill dams.[19]

Various approximate approaches are employed to handle tensile stresses in the finite element analysis. For materials like cohesive soils, the elements where tensile stresses occur are assigned a very small tensile strength compared to their compressive strength. However, usually soils and rocks do not possess significant tensile strength. Therefore, these materials are generally assumed to possess zero tensile strength and are called *no tension materials*.

For a realistic assumption of a no tension state, however, the distribution of such stresses in the surrounding elements should be considered. Zienkiewicz *et al.*[41] developed a scheme to handle such a no tension situation. The basic idea in this scheme is similar to the initial stress method described in Chapter 7. The constitutive model assumes linear elastic behavior in the direction of compressive principal stress, and minimal strength in the direction of principal tensile stresses. The procedure involves four steps. First, a standard elastic analysis, including the *in situ* stresses, is performed. Second, elements with tensile stresses are marked, and artificial restraining forces are temporarily applied to nullify these stresses. The restraining forces $\{Q_t\}$ are computed as

$$\{Q_t\} = \iiint_V [B]^{\mathrm{T}}\{\sigma_t\}\, dV \tag{10.2}$$

where $\{\sigma_t\}$ are the element tensile stresses to be supported or restrained by the nodal forces $\{Q_t\}$, and $[B]$ is defined in equation (5.28). Since the forces $\{Q_t\}$ do not really exist, in the third step equal but opposite nodal forces are applied. An elastic analysis is again performed, and the resulting stresses are added to the stresses computed in step 2. In the last step the assemblage is searched for tensile zones; and if they occur, steps 2 and 3 are repeated until such stresses are negligibly small. This iterative procedure is called the *stress transfer method*, and it is analogous to the classical relaxation method. Although it triples or quadruples the time required for a standard elastic analysis, the procedure accounts for no tension materials.

Constitutive Laws

Since soil and rock behavior is usually nonlinear, the linear elastic constitutive law (Chapter 3) can be applied to only a few simple cases. The nonlinear constitutive laws (Chapters 3 and 7) are more commonly employed. They are generally derived from laboratory drained or undrained triaxial, box shear, simple shear, or consolidation tests. The tabular (direct digital) form or the functional form described in Chapter 3 is used to incorporate the constitutive laws into a finite element computer program. Some common forms of material parameters used for nonlinear analysis of soils and rocks are described briefly below.

A digital representation of the tangent Young's modulus, derived from triaxial tests, Figure 3-5(b), can be expressed as [13,17]

$$E_t \simeq \frac{(\sigma_1 - \sigma_3)_i - (\sigma_1 - \sigma_3)_{i-1}}{(\varepsilon_1)_i - (\varepsilon_1)_{i-1}} \tag{10.3a}$$

where i is the stage of incremental loading. An approximate expression for tangent Poisson's ratio is[13]

$$\nu_t \simeq \frac{-1 + \sqrt{1 - 8(E_t/2M_b - 1)}}{4} \tag{10.3b}$$

in which $M_b = E_t/2(1 + \nu)(1 - 2\nu)$. [The symbols in equations (10.3) are explained in Chapter 3.] A more direct procedure would be to compute ν_t from volumetric measurements (Figure 3-5).

In an alternative approach, the test results can be expressed in terms of octahedral and normal stresses and strains. The required material parameters in equation (3.12b) can be expressed as [12,17,33,42]

$$K_t = \frac{\Delta\sigma_{\text{oct}}}{\Delta\varepsilon_{\text{oct}}} \quad \text{and} \quad G_t = \frac{\Delta\tau_{\text{oct}}}{\Delta\gamma_{\text{oct}}} \tag{10.4}$$

Functional forms of the tangent Young's modulus [20,35] and the tangent Poisson's ratio[20] have been used by various investigators who expressed a laboratory stress-strain curve as a hyperbola[50] and formalized its use in the finite element incremental analysis. By using the relations shown in Figures 10-2(a), (b), and (c), and the Mohr-Coulomb yield criterion, equation (3.29), an expression for the tangent modulus was obtained:

$$E_t = \left[1 - \frac{R_f(1 - \sin\phi)(\sigma_1 - \sigma_3)}{2c\cos\phi + 2\sigma_3 \sin\phi}\right]^2 K p_a \left(\frac{\sigma_3}{p_a}\right)^n \tag{10.5a}$$

Equation (10.5a) involves five material parameters: R_f is the *failure ratio* and is always less than unity; c and ϕ are the Mohr-Coulomb strength parameters; and K and n are experimentally determined parameters, Figures 10-2(a), (b), and (c). p_a is the atomospheric pressure and is introduced to make K a dimensionless number. Because in soil and rock mechanics a nonzero initial state of stress is more conveniently introduced than a nonzero initial strain (or displacement), the strain term ε is eliminated in deriving equation (10.5a).[35]

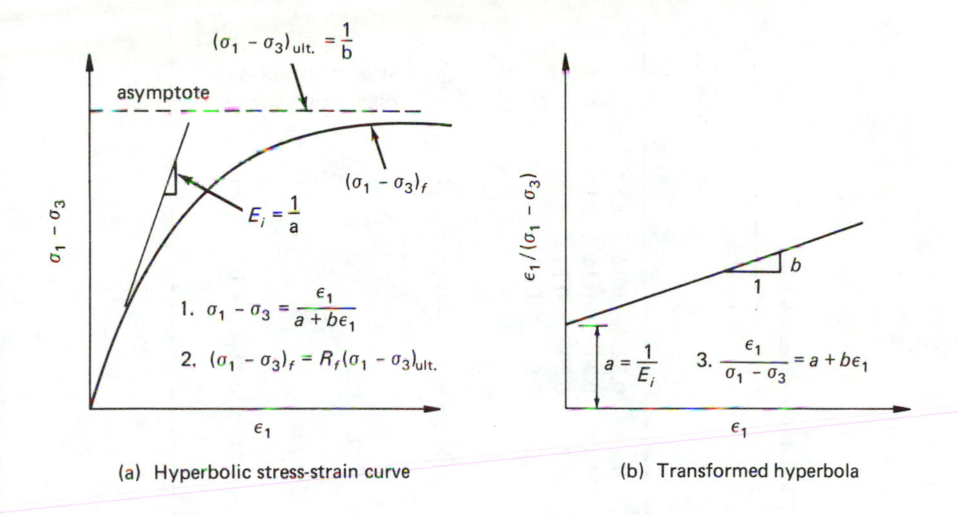

(a) Hyperbolic stress-strain curve (b) Transformed hyperbola

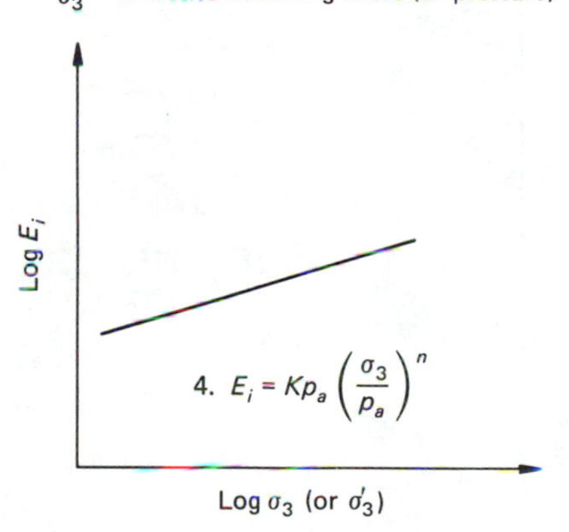

(c) Relation between initial modulus and confining stress

Figure 10-2 Computation of tangent modulus, E_t (from Reference 20).

An expression for the tangent Poisson's ratio is similarly obtained[20] by using the definition in equation (3.22) and the relations in Figures 10.3(a), (b), and (c).

$$\nu_t = \frac{G - F\log(\sigma_3/p_a)}{(1 - A)^2} \tag{10.5b}$$

TABLE 10-3 TYPICAL VALUES OF STRESS-STRAIN PARAMETERS (from Reference 20).

Soil/Group	ϕ' *degrees*		c T/ft^2	ϕ *degrees*	K	n	R_f	G	F	d	*Remarks*
	Low σ_3	*High* σ_3									
GW	47	35			500	0.3	0.7	0.32	0.14	6.4	Drained Tests
GP	46	38			1800	0.3	0.8	0.38	0.15	8.0	
SW	50	35			300	0.5	0.7	0.30	0.13	5.0	
SP	40	30			1200	0.5	0.8	0.54	0.23	4.3	
GC (Oroville Dam Core)			1.32	25.1	345	0.76	0.88	0.30	−0.05	3.8	Unconsolidated, undrained tests
SC (Thomastown Dam)			0.90	17.0	30	0.94	0.61				
CL (Canyon Dam)			2.10	3.0	175	0.41	0.90				
SC (Pittsburgh Sandy Clay)			1.08	14	40	0.48	0.68	0.43	−0.05	0.6	

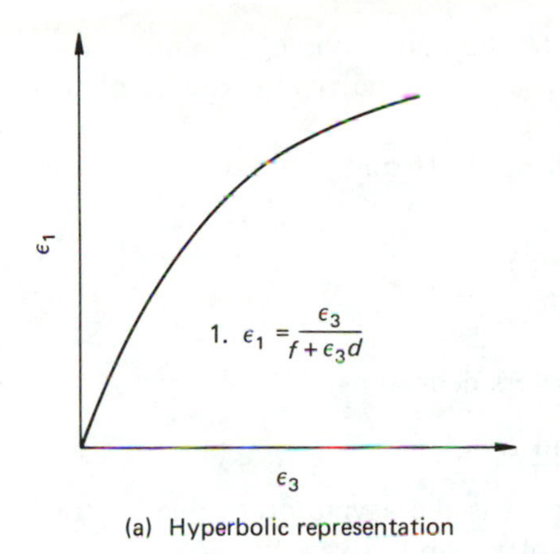

(a) Hyperbolic representation

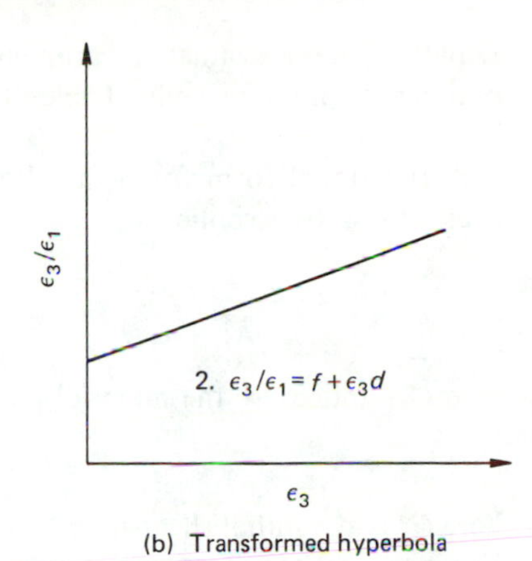

(b) Transformed hyperbola

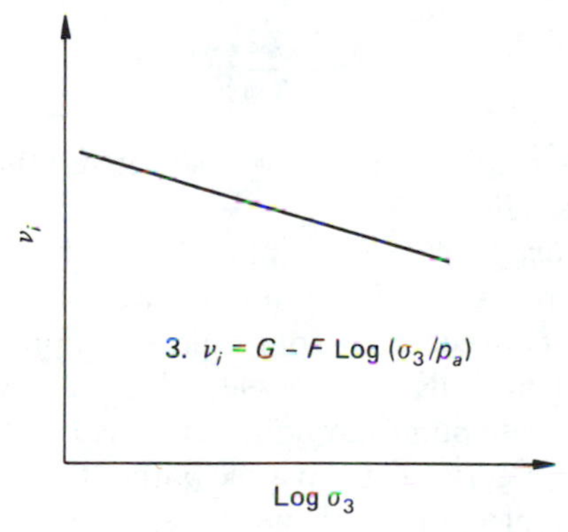

(c) Relation between initial Poisson's ratio and confining pressure

Figure 10-3 Computation of tangent Poisson's ratio, ν (from Reference 20).

where

$$A = \frac{(\sigma_1 - \sigma_3)d}{Kp_a(\sigma_3/p_a)^n\left[1 - \dfrac{R_f(\sigma_1 - \sigma_3)(1 - \sin\phi)}{2c\cos\phi + 2\sigma_3\sin\phi}\right]}$$

In addition to the previous five material parameters, we now need to evaluate two more parameters, G and F, Figure 10-3, from laboratory tests. All parameters can be computed from drained or undrained tests. Kulhawy and

Duncan[20] have evaluated parameters for computations of E_t and v_t for a number of different soils. Table 10-3 presents some typical results of their study.

A functional form for octahedral shear stress and strain may also be expressed in a hyperbolic form as

$$\tau_{oct} = \frac{\gamma_{oct}}{A + B\gamma_{oct}} \tag{10.5c}$$

where A and B are the material parameters defined as

$$A = 1/G_o \text{ and } B = 1/\tau_m$$

here G_o is the initial shear modulus and τ_m is the asymptotic value of stress. From triaxial tests approximate values of τ_{oct} and γ_{oct} are[12,17]

$$\begin{aligned} \tau_{oct} &= \frac{\sqrt{2}}{3}(\sigma_1 - \sigma_3) \\ \gamma_{oct} &= \frac{2\sqrt{2}}{3}(\varepsilon_1 - \varepsilon_3) \end{aligned} \tag{10.5d}$$

Similar functional relationships may also be established for octahedral normal stress and strain.

Mathematical *spline functions* (Chapter 8) provide a satisfactory functional representation of stress-strain curves and of the tangent moduli computed as the first derivative of the curves. A spline function approximates a given curve by a number of polynomials of a preselected degree. Each polynomial spans a limited number of data points, or nodes, such as P in Figure 3-5a. The spline provides as smooth a curve as is possible without resorting to a single polynomial over the complete range of the curve. It provides an analytical curve similar to the fit obtained by employing a mechanical spline or a French curve (References 44 and 45 of Chapter 8).

In terms of the stress difference $\sigma = \sigma_1 - \sigma_3$, and of the axial strain ε, a *cubic spline* can be expressed as (Reference 45, Chapter 8):

$$\sigma(\varepsilon) = \theta_{j-1}\frac{(\varepsilon_j - \varepsilon)^2(\varepsilon - \varepsilon_{j-1})}{h_j^2} - \theta_j\frac{(\varepsilon - \varepsilon_{j-1})^2(\varepsilon_j - \varepsilon)}{h_j^2} \tag{10.5e}$$

$$+ \sigma_{j-1}\frac{(\varepsilon_j - \varepsilon)^2[2(\varepsilon - \varepsilon_{j-1}) + h_j]}{h_j^3} + \sigma_j\frac{(\varepsilon - \varepsilon_{j-1})^2[2(\varepsilon_j - \varepsilon) + h_j]}{h_j^3}$$

for $\varepsilon_{j-1} \le \varepsilon \le \varepsilon_j$. In matrix notation this can be written:

$$\sigma(\varepsilon) = \{N\}^T\{q\} \tag{10.5f}$$

where $\sigma(\varepsilon)$ is the spline representing stress, σ_j and ε_j are the data points, θ is the first derivative of $\sigma(\varepsilon)$, $n + 1$ is the number of data points, h_j is $\varepsilon_j - \varepsilon_{j-1}$, $\{N\}$ is the vector of interpolation functions, and $\{q\}$ is the vector of nodal amplitudes θ_j, σ_j, etc. A set of simultaneous equations results from equation (10.5e) when continuity of second derivatives at all data points is imposed. Solution of these equations with the proper boundary conditions determines the nodal amplitudes and this uniquely defines the spline. Desai[53] compared the cubic spline representation with the hyperbolic (Figure 10-2a) and the "parabolic" representations. The spline was introduced in a finite element analysis to compute load-settlement behavior of a laboratory-test footing on a cohesive subsoil.[17] Comparisons were obtained between the load-settlement curves from the experiment, the spline formulation, and the hyperbolic representation. The spline provided a better simulation of the stress-strain curves, the tangent Young's moduli, and the load-settlement behavior of the footing.

For behavior expressed by a number of curves, Figure 3-5b, the spline formulation involves three measured quantities. Such a spline represents a surface in a three-dimensional space. A *bicubic spline* can be used for this behavior. For example the spline in terms of the stress difference, axial strain, and confining pressure can be expressed as [54,55, Ref. 45 of Chap. 8]

$$\sigma(p, \varepsilon) = \sum_{i=0}^{3} \sum_{j=0}^{3} \alpha_{ij} p^i \varepsilon^j \tag{10.5g}$$

or in matrix notation

$$\sigma(p, \varepsilon) = [\phi]\{\alpha\} \tag{10.5h}$$

Here $p = \sigma_3$ is the confining pressure; $\{\alpha\}$ is the 16×1 vector of generalized stresses; $[\phi]$ is the 16×16 matrix of coordinates; and the values of σ are given at all interior and boundary mesh points of a small rectangular region in the $p - \varepsilon$ space. The bicubic spline can be established by solving for a number of cubic splines. Desai[55] used the bicubic spline to represent laboratory curves expressed as stress difference-axial strain-confining pressure, radial strain-axial strain-confining pressure, deviatoric stress-deviatoric strain-mean normal stress, and volumetric strain-mean normal stress-relative density. These four splines permitted computation of the tangent Young's modulus, tangent Poisson's ratio, tangent shear modulus, and tangent bulk modulus, respectively, These tangent moduli, used in nonlinear analysis, were computed as the first derivatives of the bicubic splines.

Elastic-plastic Behavior

The above nonlinear formulations, equations (10.3) and (10.5), assume elastic behavior for the complete range of loading (Section 3-5). However, in certain situations soils and rocks are elastic-plastic. Often a simple approximate elastic-plastic representation is obtained by treating the nonlinear behavior as bilinear. In that case a set of constant elastic moduli, such as E and ν, are assumed in the elastic range; and their reduced values are adopted in the plastic range after a specified yield stress is reached.[14,18]

For general elastic-plastic behavior the incremental theory of plasticity (Section 3-5) is often employed. A criterion such as the extended Mohr-Coulomb rule, equation (3.29d), accounts for both cohesive and frictional strengths. Together they provide a rational consideration of soils and rocks as elastic-plastic work hardening materials.[36,40,43]

Hysteritic Behavior

The incremental procedures based on equations (10.3) and (10.5) may also be applied to hysteritic behavior. A constant modulus, often equal to the initial elastic modulus, E_i, Figure 10-2(a), is employed for an approximate analysis of the unloading during hysterisis.[14,22]

Time Effects

Time effects play an important role in the stress distributions and deformations, particularly when we are interested in long-term behavior. Typical examples that involve significant time or viscous effects are the creep behavior of cohesive soils, the consolidation of clays, and the stress relaxations in and around underground cavities or tunnels. Brief consideration to these effects in soils is given in Chapters 13 and 14.

Sequential Construction

Engineering structures are usually constructed in a definite sequence of operations. A conventional linear analysis of such structures is performed by assuming that the entire construction takes place in a single operation. In other words, stresses and deformations are computed by considering loads on completed structures. However, for the nonlinear problems typical in soil and rock mechanics and foundation engineering, the behavior at a particular stage of loading is dependent upon the state of stress and stress history. Thus the stresses in the final configuration are dependent upon the sequence of intermediate configurations and loadings.

In soil and rock engineering, sequential construction is encountered during *excavations* and *embankments*. Goodman and Brown[10] have developed an analytical simulation procedure that is commonly employed in finite element applications.[11,14,16,20,22,25,28] This procedure is illustrated in Figures 10-4

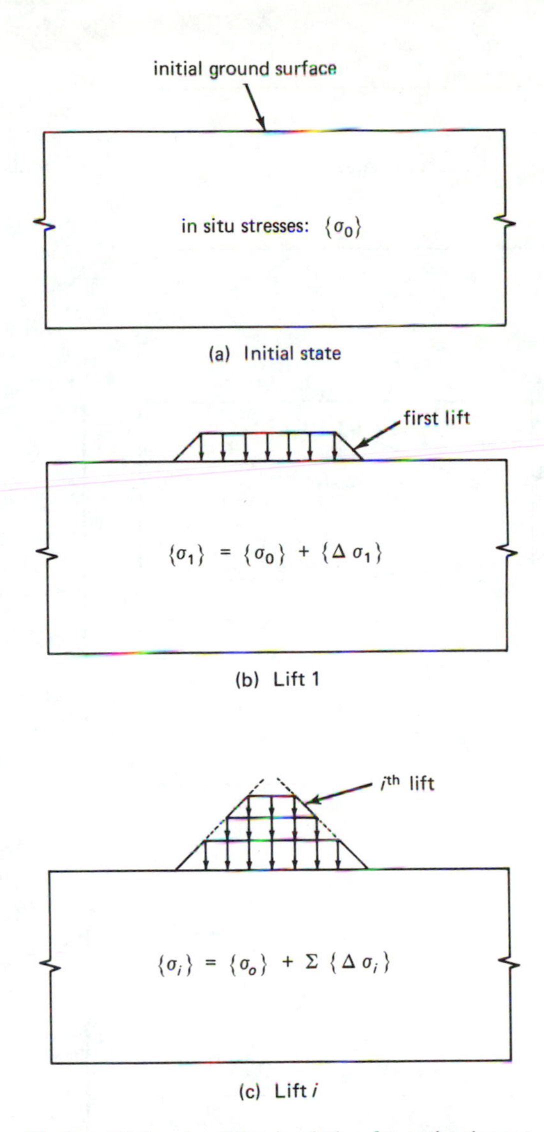

Figure 10-4 Analytic simulation for embankment.

and 10-5. The soil or rock mass under consideration is first divided into a finite element mesh with side and bottom boundaries located at appropriate distances (Section 6-2). Before the construction starts, the mass is subjected to *in situ* stresses (Fig. 10-1). A cycle of finite element analysis is performed for the *in situ* loads in which the resulting stresses, $\{\sigma_o\}$, are computed.

The first lift is now laid or excavated, and the increment of stress $\{\Delta\sigma_1\}$ is

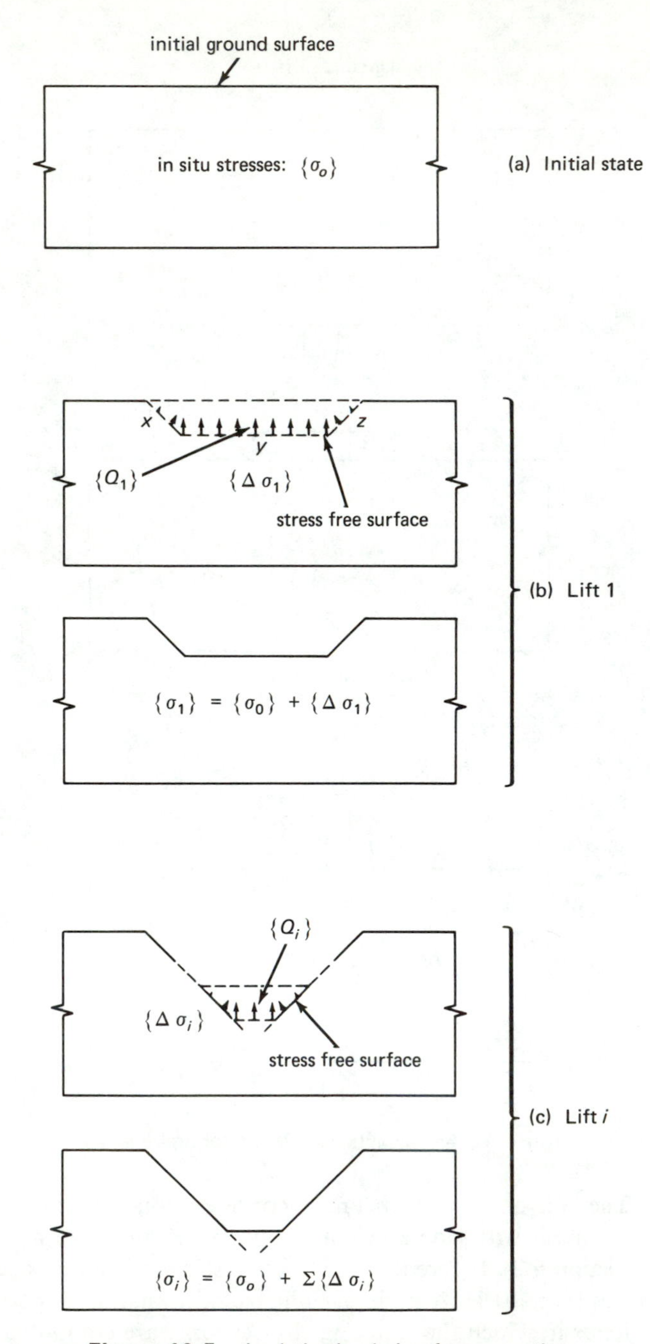

Figure 10-5 Analytic simulation for excavation.

computed for the modified assemblage by performing the next cycle of analysis. For an embankment the modified assemblage is obtained by adding elements in the first lift to the initial assemblage, whereas for excavation it is obtained by removing the elements contained in the lift. The total stresses at the end of the first embankment lift are obtained by adding $\{\Delta\sigma_1\}$ to $\{\sigma_o\}$ to get $\{\sigma_1\}$.

Simulation of excavation needs an additional step for the computation of $\{\sigma_1\}$. After the lift is removed, the surface *x-y-z* at the base has no stresses on it; that is, it is a "stress free" surface. In the additional step equivalent forces, $\{Q_1\}$, at the nodes on *x-y-z* are computed, Figure 10-5(b). Forces equal but opposite in sign are now applied at these nodes, which approximately create the required stress free surface. Finally, a cycle of finite element analysis under the loading, $\{Q_1\}$, gives $\{\Delta\sigma_1\}$, which is added to $\{\sigma_o\}$ to give $\{\sigma_1\}$.

A general recursive form of the incremental simulation is

$$\{\sigma_i\} = \{\sigma_o\} + \sum_i \{\Delta\sigma_i\} \tag{10.6}$$

Displacements are computed by a similar formula.

An alternative procedure for simulating excavation or embankment is called *gravity turn-on* analysis and was used by Dunlop *et al.*[14, 11] It is relatively simple and uses the standard finite element analysis for the *final* configuration of the slope under the effect of gravity, and it involves no sequential steps. It gives stresses identical to those from the analytical simulation described above, if the *in situ* stress field satisfies certain conditions. In the plane strain case, this condition is that the lateral earth pressure coefficient, K, be equal to $\nu/(1 - \nu)$.

10-3 EXAMPLE: ANALYSIS OF SLOPES IN SOILS

The conventional analysis of slopes, based on the plastic equilibrium concept permits an acceptable estimation of the stability of slopes and their behavior at failure. It is difficult, however, to obtain information regarding displacements, strains, and regions of local failures from such conventional analyses. Dunlop *et al.*[14] performed a detailed finite element analysis of excavated slopes. In this analysis the 4-CST quadrilateral element was used in a plane strain idealization of the slope. The study provided computations of displacements, strains, and stress distributions, and it enabled the location of failure zones. The effects of such factors as *in situ* stresses, anisotropy, variation of strength within the soil deposit, pore pressure distributions, and sequential construction were included. Various constitutive laws, such as linear, bilinear, and multilinear (piecewise linear) were compared.

A typical finite element mesh used is shown in Figure 10-6(a). The

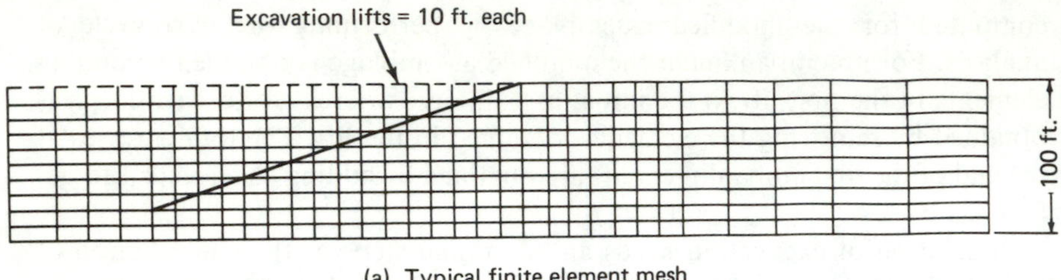

(a) Typical finite element mesh

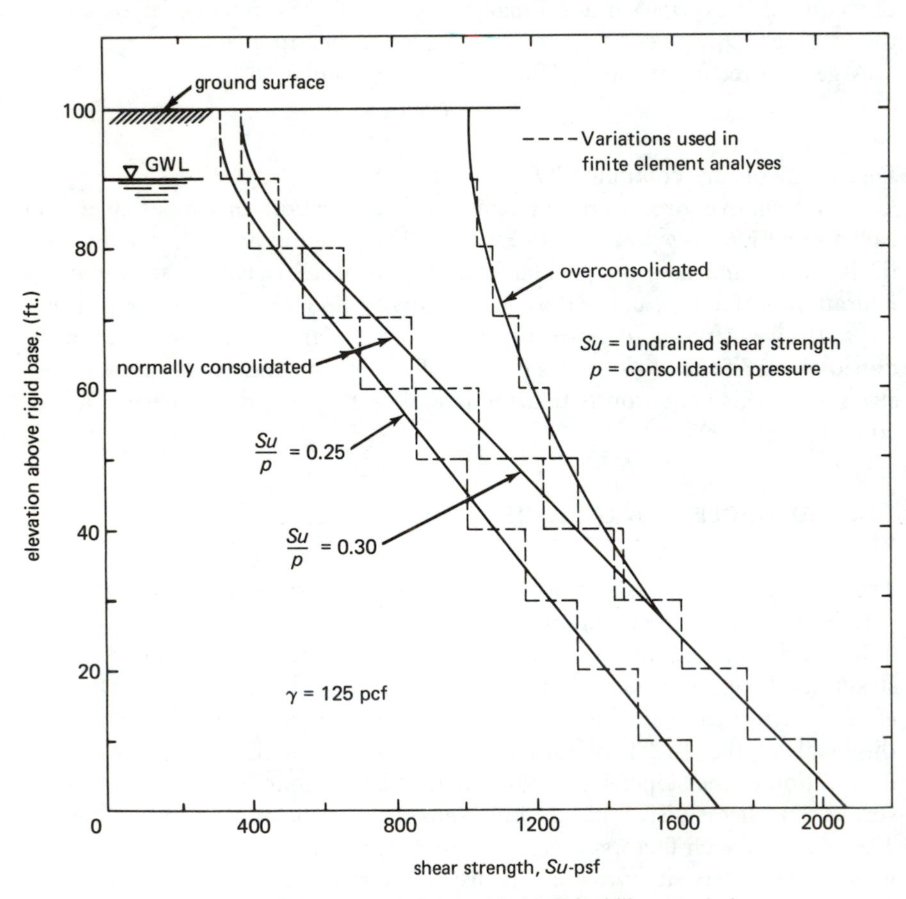

(b) Strength variation with depth for bilinear analysis

Figure 10-6 Finite element mesh and variation of strength (from Reference 14).

excavation was simulated by using the analytic procedure (Fig. 10-5) and was assumed to be performed under undrained conditions. Hence, the total stresses were used to express stress-strain relations and the failure criterion. In the bilinear analysis, the elastic modulus in the initial part of the stress-strain curve was adopted as 100 times the undrained shear strength of the soil. For instance, the typical undrained strength of 130 psf near the ground surface gave an elastic modulus of 13,000 psf. The Poisson's ratio was adopted as 0.475.

Failure zones were located by examining maximum shear stress induced in each element after each step of excavation. If this stress was equal to or greater than the shear strength of the soil at that location, the modulus value for the element was reduced to a small magnitude. For example, the reduced modulus for failed elements was typically taken as 10 psf.

We shall include results here only from a part of the study concerning the failure behavior of slopes in normally consolidated and overconsolidated soils. Figure 10-6(b) shows typical variations of strength with depth for the two soils. Figures 10-7(a) and (b) show zones of failure, critical failure surfaces, and the factors of safety at different stages of excavation. The critical surface was determined from a number of trial failure surfaces, and the factor of safety was obtained on the basis of mobilized shear stresses.

The failure of a slope in a normally consolidated clay, Figure 10-7(a), starts *beneath the slope* and extends upward toward the slope crest and downward toward the toe. This trend agrees with the development of curved sliding surfaces in actual slopes. In contrast, the failure in the overconsolidated soil, Figure 10-7(b), initiates *beneath the base* of the excavation, extends upward toward the bottom of the slope and downward toward the base of the layer or the lift being excavated. This behavior conforms with field observations of a phenomenon known as *loss of ground*,[1] Figure 10-7(c).

10-4 EXAMPLE: BEARING CAPACITY AND SETTLEMENT ANALYSIS

Constitutive laws for immediate stress and settlement and bearing capacity analyses can be derived from undrained laboratory tests. Desai and Reese[17] conducted detailed stress, deformation, contact pressure, and settlement studies for a circular footing in layered cohesive subsoils. Results from the incremental procedure, Chapter 7 and equation (10.3a), were compared with laboratory tests on a steel footing 3.0″ diameter and 0.5″ thick resting on two layers, each consisting of a different cohesive soil. Initial moduli and Poisson's ratios of the clays are shown in Figure 10-8, which also shows a typical mesh for the two-layered system. The constitutive law was derived from undrained triaxial tests under various confining pressures, Figure 3-5(b). Load-settlement behavior was obtained by applying either increments of pressure or

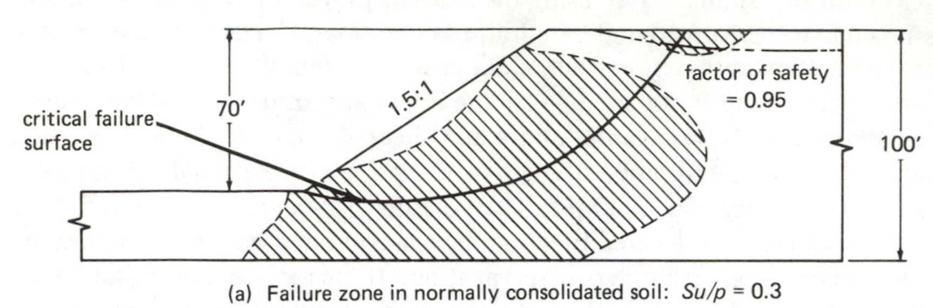

(a) Failure zone in normally consolidated soil: $Su/p = 0.3$

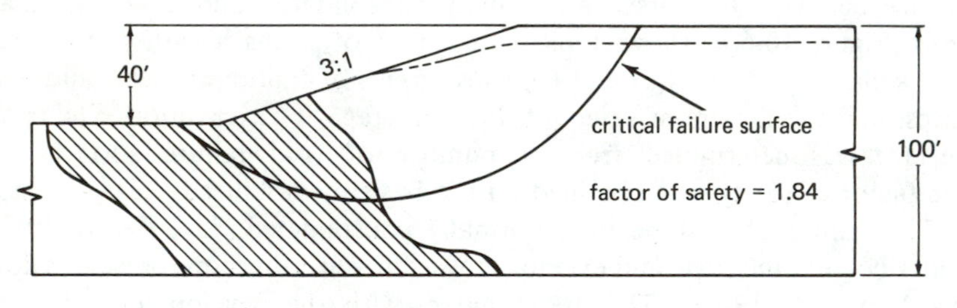

(b) Failure zone in overconsolidated soil ($k = 1.25$)

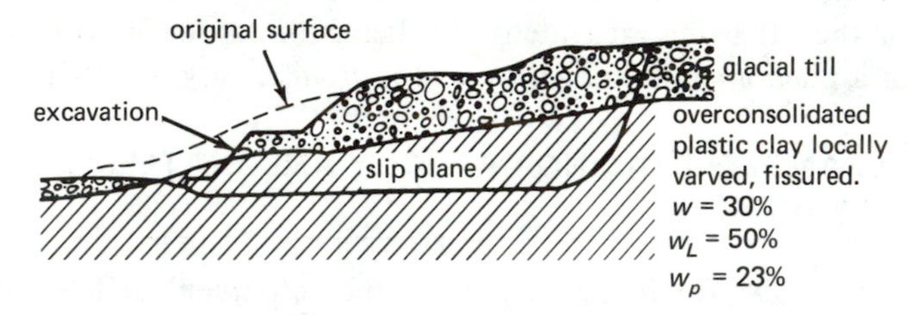

(c) Observed slope failure in overconsolidated clay: excavation for Seattle freeway (from Reference 51)

Figure 10-7 Behavior of excavated slopes in soil (from Reference 14).

increments of rigid displacements, which respectively simulated flexible or rigid footings. A typical comparison of load-settlement curves obtained from the finite element analysis and the experiments for the layered systems are shown in Figure 10-9, and it indicates good correlation. From the observations of the experimental behavior of the footing, a central displacement of about

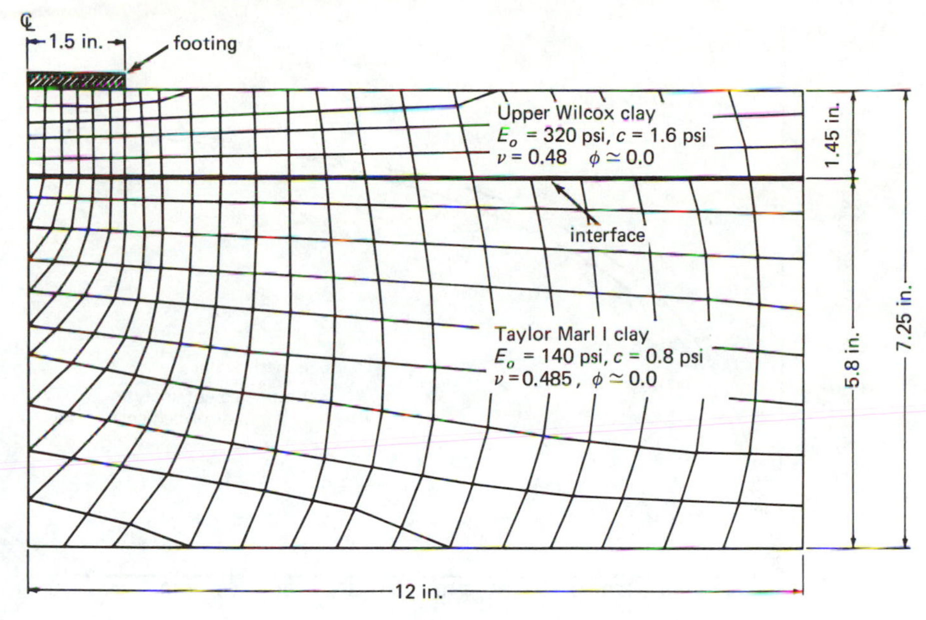

Figure 10-8 Mesh layout for two-layer soil system (from Reference 17).

TABLE 10-4 BEARING CAPACITIES (IN PSI) BY VARIOUS METHODS (from Reference 17).

System	*Experi-mental*	*Finite Element Analysis*	*Terzaghi*†	*Skempton*†	*Meyerhof*†	*Button*†	*Remarks*
Two layers	7.15	7.10				7.8	Footing on surface
Single layer	9.9	10.0	7.95	9.92	9.90		Footing on surface
		10.92			11.2		Footing at 1.2 in. below surface

† Details of the formulae for computations of bearing capacities may be found in References 1, 3, and 17.

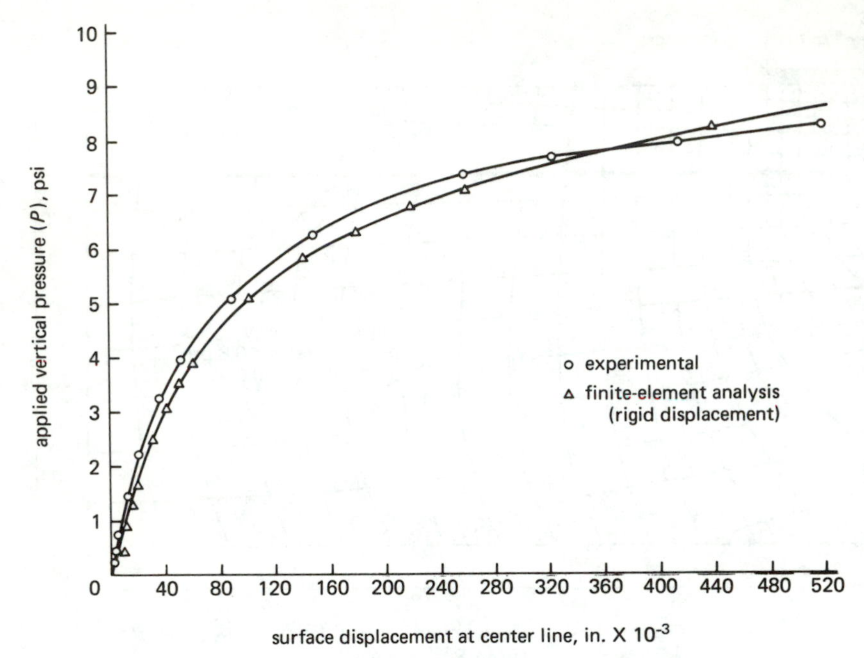

Figure 10-9 Load-settlement curves: comparisons (from Reference 17).

7.5 percent of the diameter of the footing was adopted as the critical displacement. Ultimate capacities corresponding to this critical displacement were derived from the experiments, from the finite element results, and from conventional methods, Table 10-4. This table also includes ultimate capacities for a single layer system of the cohesive soils loaded by the same footing. It can be seen that the values from finite element analysis compare well with experimental and conventional results.

The finite element analyses indicated that changes in the flexibility of the footing altered stress and contact pressure distributions, but did not appreciably change the load-settlement behavior.[17] Moreover, the bearing capacity of a circular footing was found to be about 1.3 times that of an equivalent strip footing.[17,18]

10-5 EXAMPLE: STATIC ANALYSIS OF AN EARTH DAM

Clough and Woodward[13] analyzed Otter Brook Dam by using the incremental construction procedure, equation (10.3). Recently Kulhawy *et al.*[20] performed a detailed study of the same dam. In their finite element analysis the

2-LST quadrilateral element was employed. The dam construction was simulated as explained in Figure 10-4. In addition to homogeneous dams, zoned embankments were also considered.

The nonlinear analysis was performed by using the incremental procedure and the concept of tangent moduli, equations (10.5a) and (10.5b). Various parameters used in the analysis are shown in the last row of Table 10-3. Because permeability of the soil was low, conditions of no drainage were assumed, and the moduli were derived from unconsolidated-undrained tests.

Figure 10-10(a) shows a cross-section of the Otter Brook dam. Data from the instruments installed in the section provided the histories of displacement at various locations. Figure 10-10(b) shows a finite element mesh for the

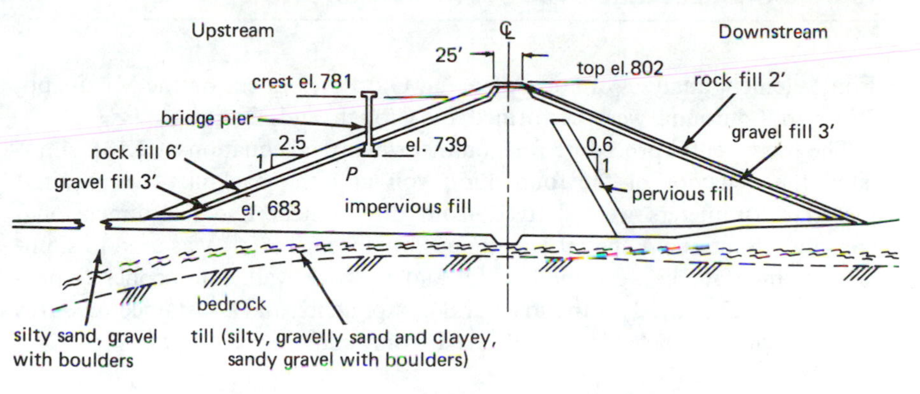

(a) A section of Otter Brook Dam (from Reference 52)

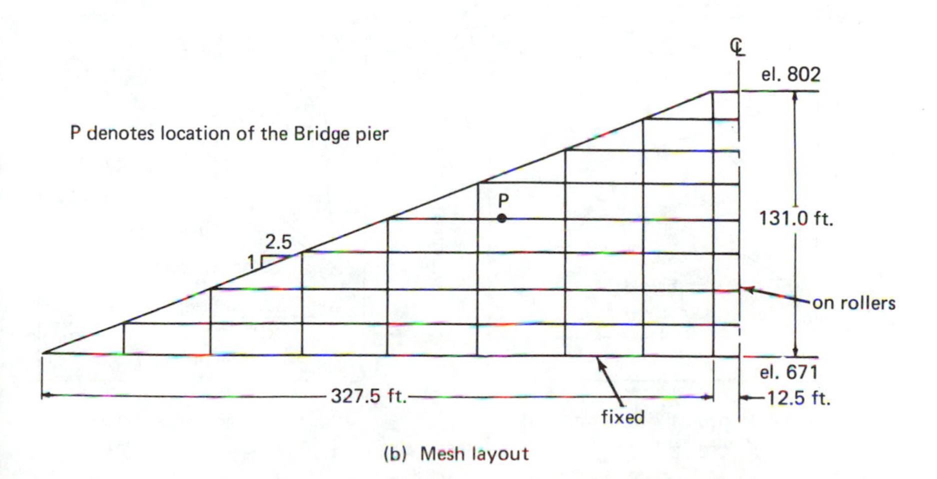

(b) Mesh layout

Figure 10-10 Otter Brook Dam and finite element mesh.

homogeneous idealization of the left half of the dam. Comparison between the finite element solutions for the horizontal displacements of the upstream face and the measured movements is shown in Figure 10-11(a).

Figure 10-11(b) shows the vertical and horizontal displacements of the base of the bridge pier in comparison with field movements as the construction proceeded from El. 744 to the top of the dam. Agreement between the observed results and the finite element solutions is very close.

10-6 EXAMPLE: SOIL-STRUCTURE INTERACTION

Finite element analyses of Port Allen and Old River Locks on the Mississippi River in Louisiana were performed by Clough and Duncan.[28]

The incremental procedure for nonlinear analysis, equations (7.10), simulated the behavior of the foundation soil and the backfill. The required material parameters were derived from direct shear and one-dimensional consolidation tests. A special one-dimensional element[47,49] was used to simulate the interface between the backfill and the lock wall. Mass concrete properties were employed for the structural components, and a sustained modulus for the long-term behavior was derived from uniaxial creep tests.

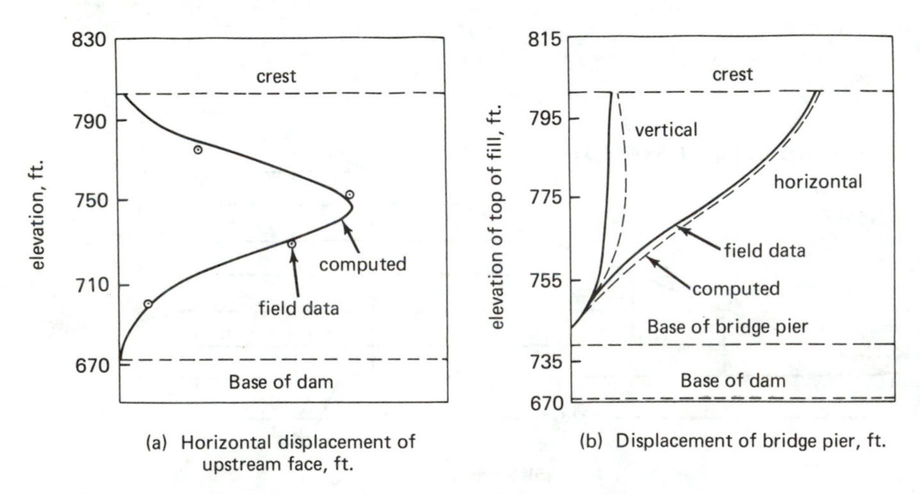

(a) Horizontal displacement of upstream face, ft.

(b) Displacement of bridge pier, ft.

Figure 10-11 Finite element analysis of Otter Brook Dam: comparisons (from Reference 20).

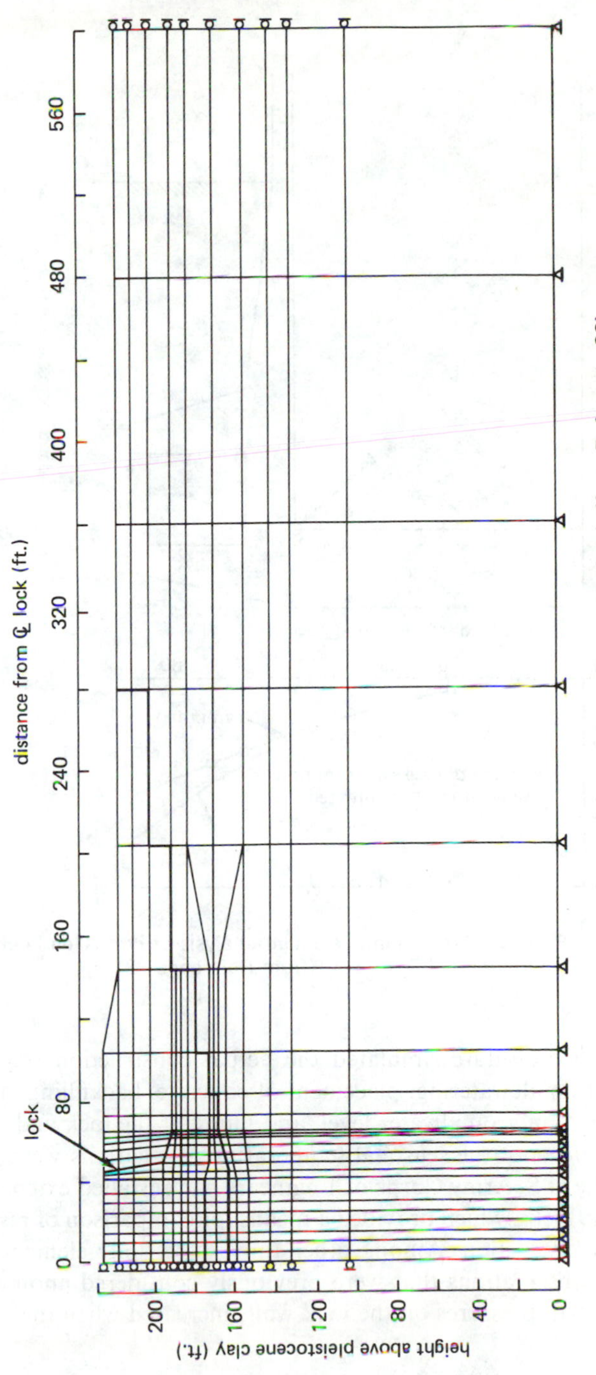

Figure 10-12 Typical mesh for Port Allen Lock (from Reference 28).

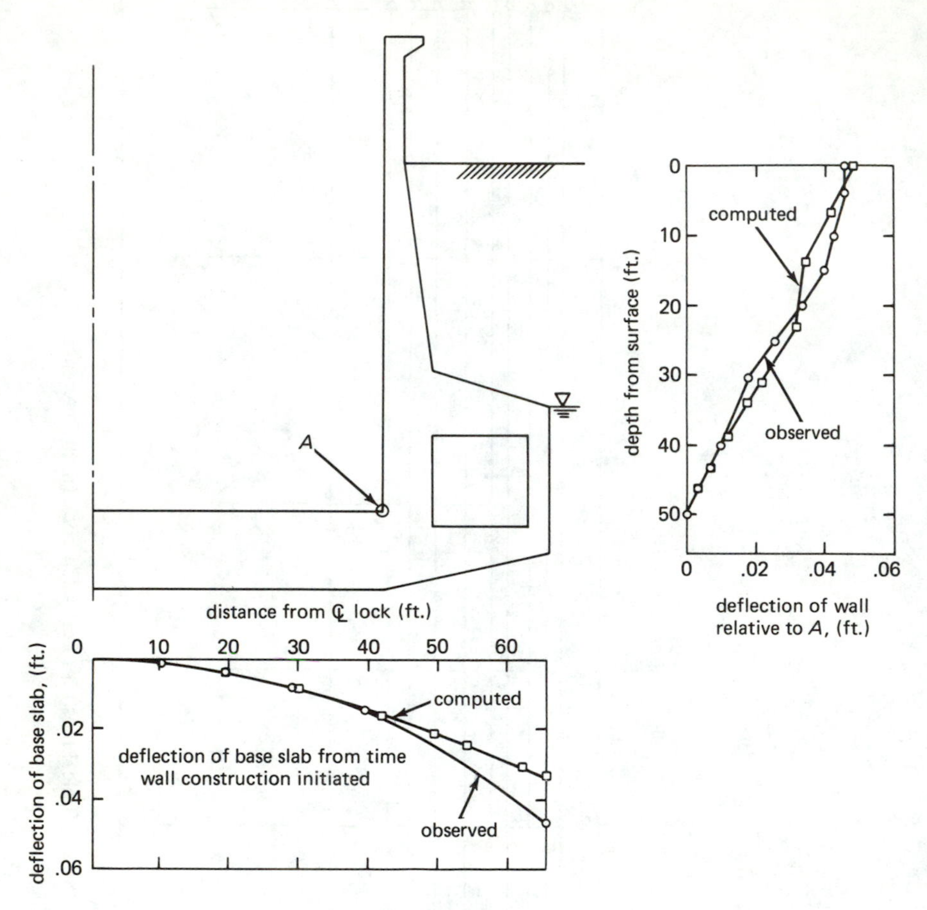

Figure 10-13 Finite element analysis of Port Allen Lock: comparisons (from Reference 28).

The procedure simulated the actual construction sequence including excavation, dewatering, placement of concrete, backfilling at non-uniform rates, changes in groundwater level, and filling of the lock and culvert. In addition, the effects of the seasonal temperature fluctuations were taken into account.

The U.S. Army Corps of Engineers had installed extensive instrumentation in the locks, which provided a means for comparison of results at each stage of the construction. Among other things, the finite element results verified the field observations that were previously considered anomalous. For instance, the earth pressures on the lock walls increased when the walls moved inward.

This comprehensive study of the Port Allen and Old River Locks has proved significant to the understanding of the behavior of lock systems and has provided a valuable design-analysis basis for future locks.

A typical finite element discretization for Port Allen Lock is shown in Figure 10-12. Figure 10-13 shows a comparison between the measured and numerically computed values of the deflections of the base and walls of the lock.

10-7 EXAMPLE: SLOPES IN JOINTED ROCKS

The behavior of a jointed rock mass is influenced by such factors as the sizes and locations of the joints and the properties of the materials filling the joints. Typical fillers are clays and crushed rock.

Duncan and Goodman[47] used three different procedures for incorporating joints in rock masses. In the *ubiquitous joint* analysis, the material was assumed homogenous and isotropic, but a large number of hypothetical joint orientations were studied for possible slips or openings. In the *orthotropic continuum* analysis, the mass was treated as orthotropic (or stratified), and the corresponding stress-strain relations, equations (3.13), were employed. In the *single joint* studies where the joint had predominant dimensions, special two-dimensional and one-dimensional formulations for the individual joint were developed. (For details of these formulations the reader may consult References 47, 48, and 49).

The finite element analysis for the above three procedures was performed by using linear elastic properties for both the parent rock and the joints. Effects of *in situ* stresses were included, and the excavation was simulated as explained in Figure 10-5. A typical finite element mesh for a joint that makes a 30° angle with the horizontal is shown in Figure 10-14(a). Figures 10-14(b) and (c) show displacements and shear stress distributions obtained on the basis of single joint analysis for a typical K value equal to 1, and the ratio of Young's moduli for the parent rock and the joint, R, equal to 100. K is defined in Figure 10-1.

For realistic analyses, reliable techniques for the determination of *in situ* stresses should be developed. In addition, adequate laboratory and/or field techniques should be developed for sampling of joints in the field and for determination of the properties of the joints. Finally, comparisons between the finite element method and field or laboratory observations should provide confidence for using the method for practical design analysis.[47]

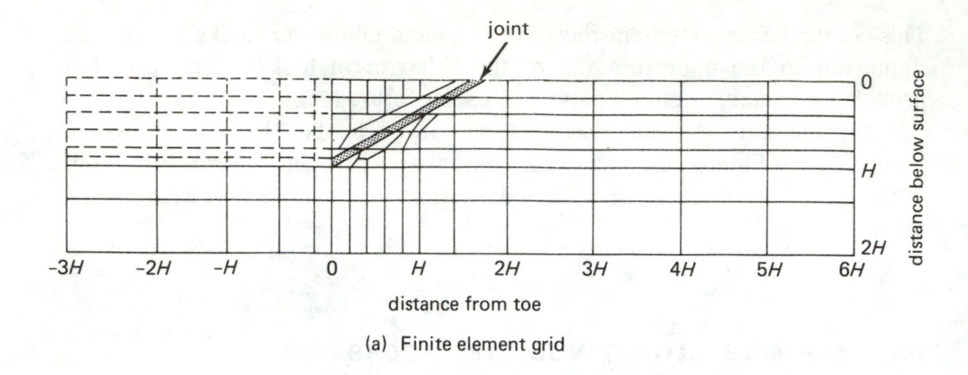

(a) Finite element grid

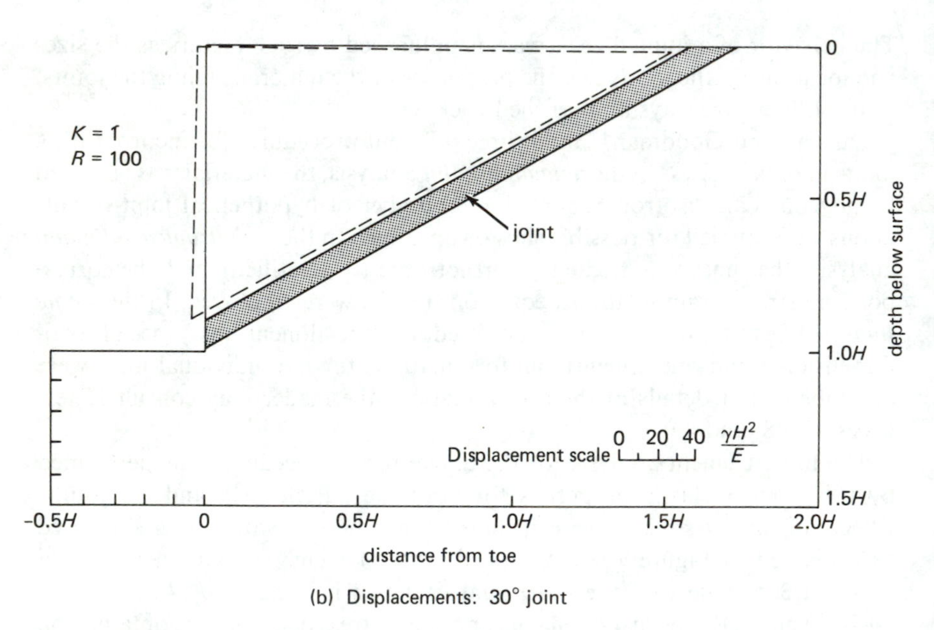

(b) Displacements: 30° joint

Figure 10-14 Analysis of slopes in jointed rock (from Reference 47).

10-8 EXAMPLE: DEEP BOREHOLES

Desai and Reese[42] examined the stability of an open borehole drilled in a Green River shale, whose constitutive law was derived from undrained triaxial tests. The stress-strain relation was expressed as in Figure 3-5(b). A small region of the bottom of the borehole was discretized as shown in Figure 10-15, and an indirect use of the incremental load procedure approximately

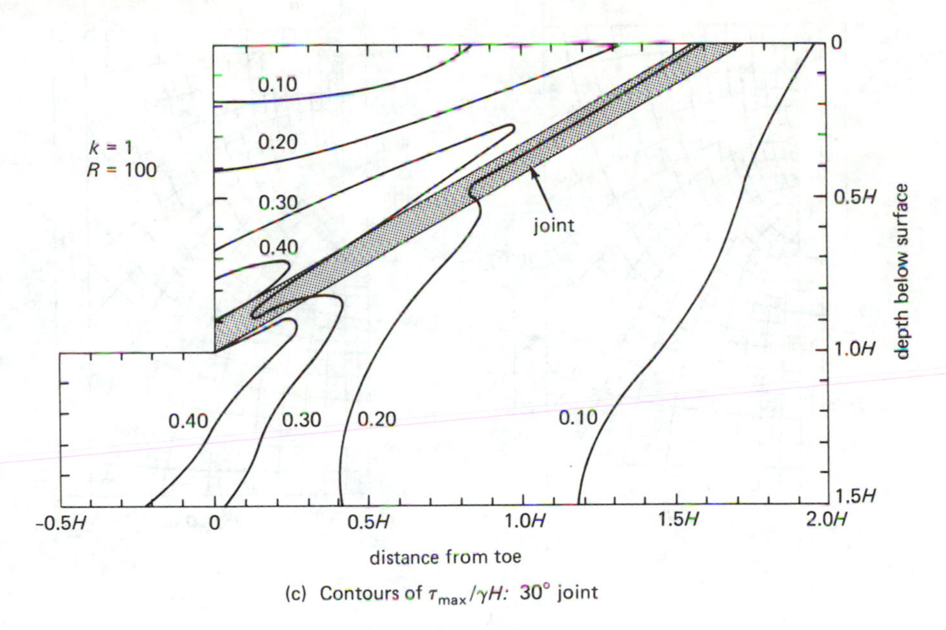

(c) Contours of $\tau_{max}/\gamma H$: 30° joint

Figure 10-14 (*continued*)

simulated the progress of drilling. The load was applied in increments of 1000 or 4000 psi at the top nodes. This loading can be related to the depth of drilling if the local geology is known.

After an increment of loading was applied each element in the assemblage was examined; and the maximum stress developed in it was compared with the limiting stress determined from the Mohr-Coulomb criterion, equation (3.29). The extent of the plastic zone was then established on the basis of the elements in which the induced stress equalled or exceeded the limiting stress. In the borehole, Figure 10-15, the first plastic zone was initiated at a vertical loading of about 22,000 psi. At about 28,000 psi the entire borehole was in a state of plastic failure.

On the assumption that the drilling bits jammed at the critical displacement corresponding to a decrease of borehole radius of 0.5″, the critical vertical collapse load can be read from the computed load-displacement curve, Figure 10-16.

The study also compared linear elastic finite element results with photoelastic experiments on an epoxy resin model. The model simulated the bottom of a borehole subjected to hydrostatic pressure. The computed and the photoelastic stress distributions agreed well.[42]

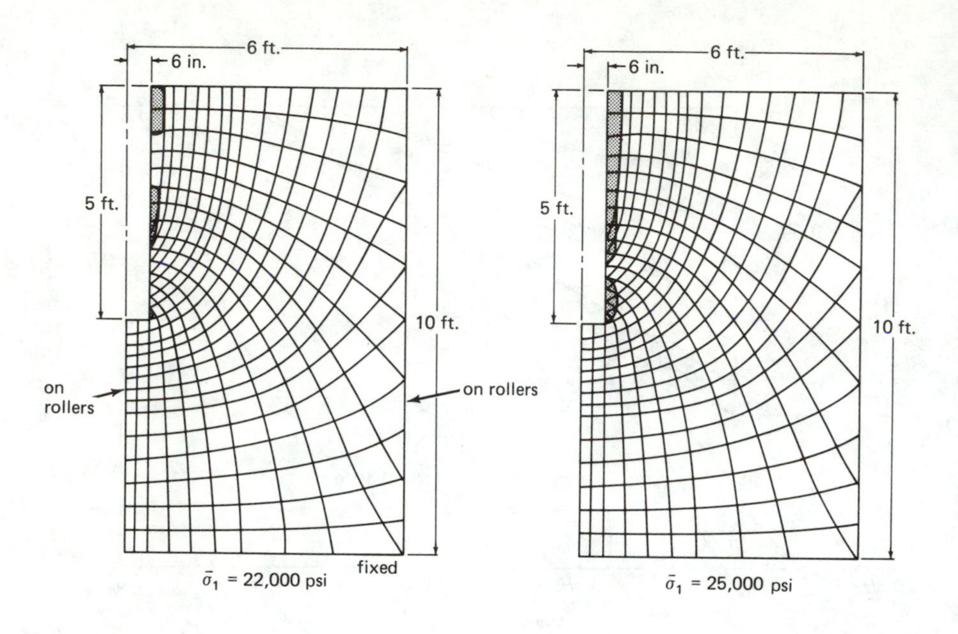

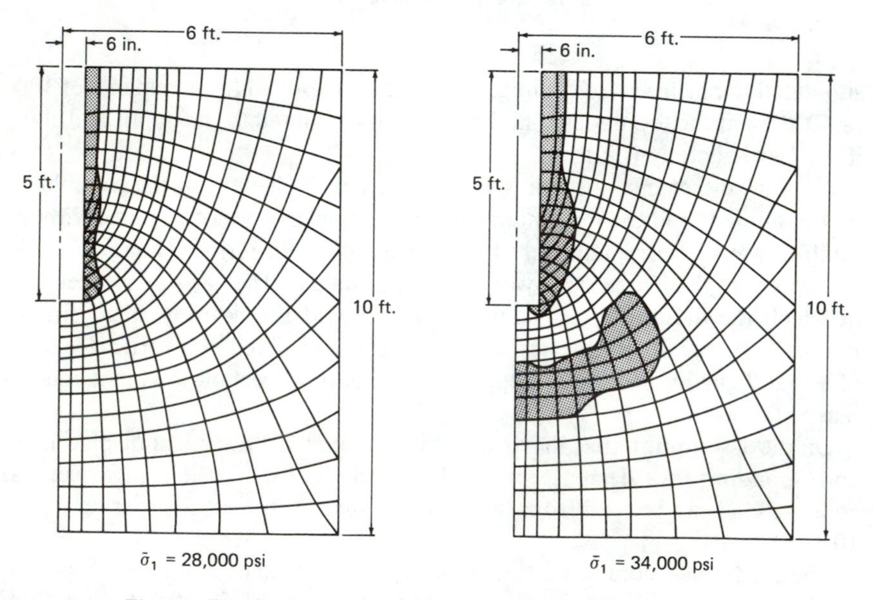

Figure 10-15 Discretization for deep borehole and spread of plastic zones (from Reference 42).

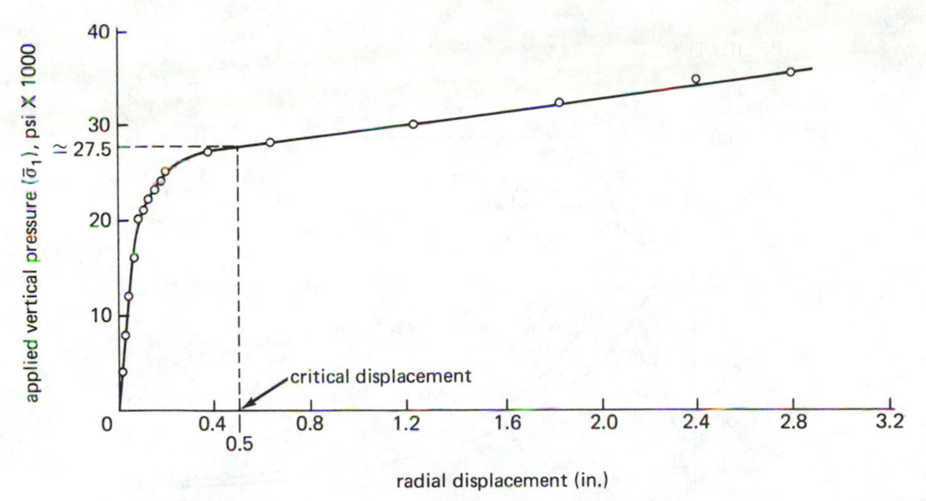

Figure 10-16 Load-radial displacement curve: deep borehole (from Reference 42).

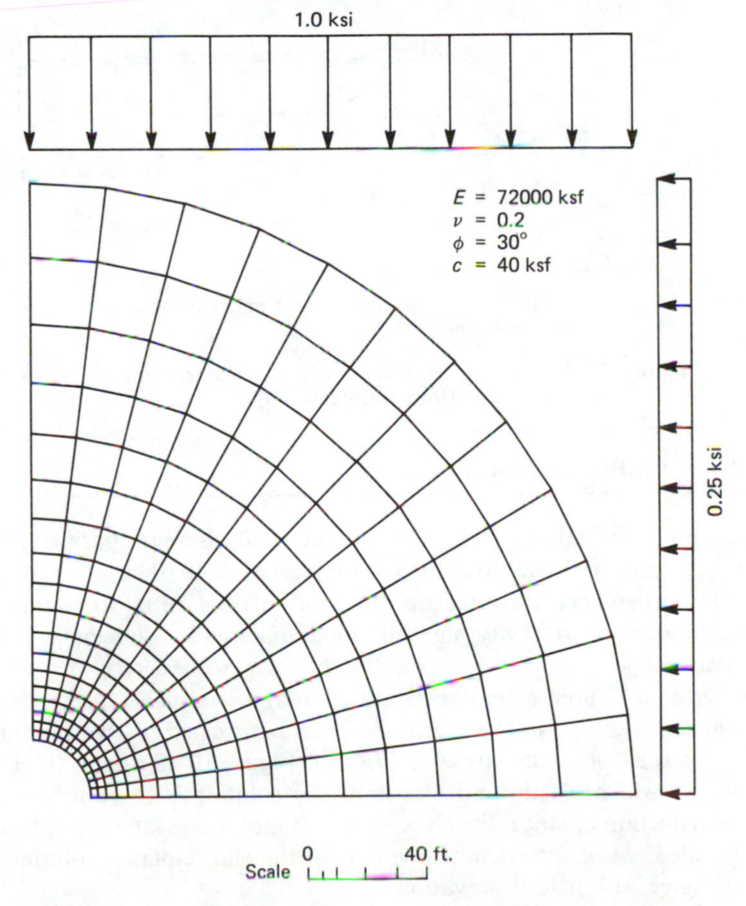

Figure 10-17 Mesh and loading for tunnel (from Reference 43).

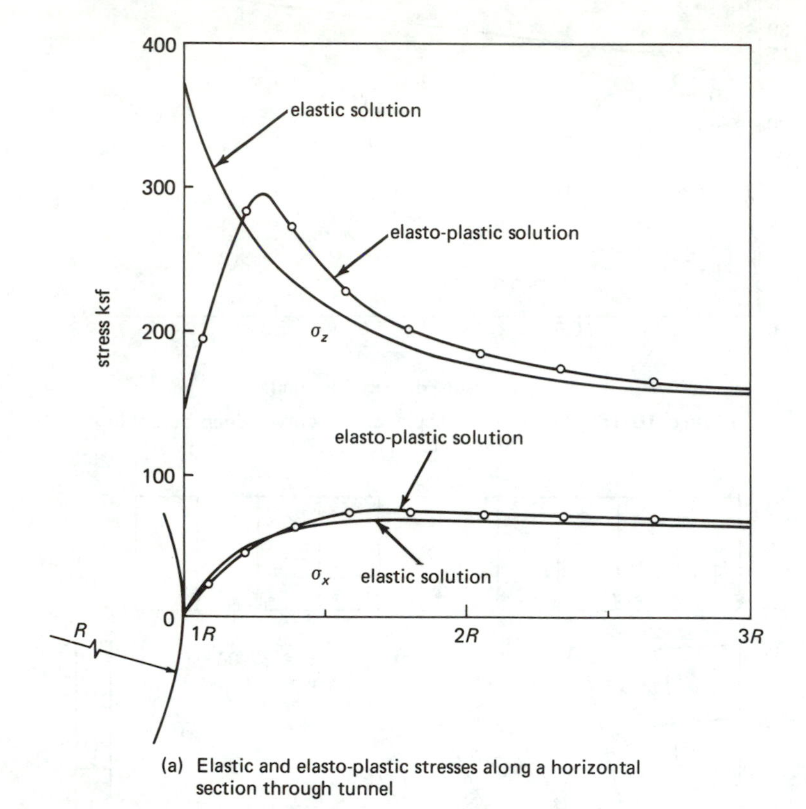

(a) Elastic and elasto-plastic stresses along a horizontal section through tunnel

Figure 10-18 Comparisons for stresses and displacements: tunnel (from Reference 43).

10-9 EXAMPLE: TUNNELS

Reyes and Deere[40] developed an incremental analysis on the basis of equations (3.29d), (3.36), and the flow theory of plasticity. In their study the stress distributions and extents of plastic zones for different initial stress ratio values for an underground cavity showed good qualitative agreement with field observations.

Baker *et al.*[43] used a similar formulation and analyzed a deep tunnel in a rock mass (Fig. 10-17). The *in situ* stress field consisted of a vertical stress of 144 ksf and a horizontal stress of 57.6 ksf. Figure 10-18(a) and (b) show the results, in which elastic and elastic-plastic solutions are compared. There occurs reduction in tangential stresses and an increase in lateral displacements, often called *plastic intrusion*, in the case of the elastic-plastic solution. These trends agree with field observations.

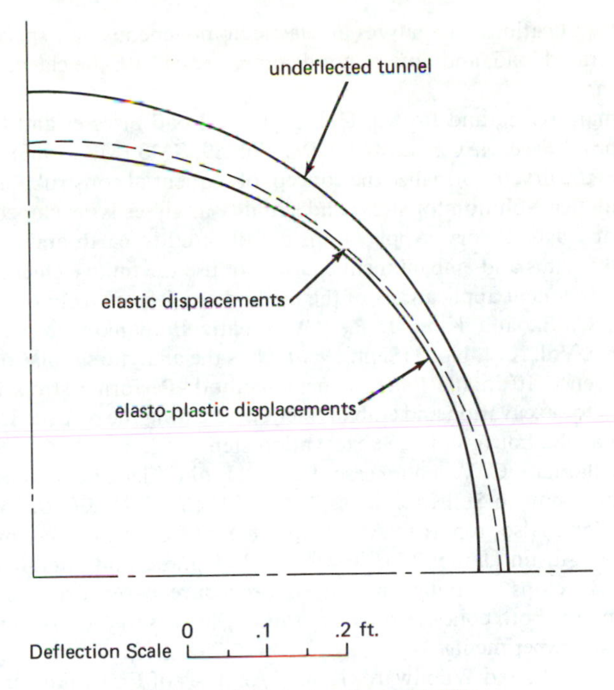

(b) Tunnel deformation as computed by elastic and elasto-plastic analysis

Figure 10-18 (*continued*)

REFERENCES

(1) Terzaghi, K., and Peck, R. B., *Soil Mechanics in Engineering Practice*, New York, John Wiley & Sons, 1967.
(2) Taylor, D. W., *Fundamentals of Soil Mechanics*, New York, John Wiley & Sons, 1964.
(3) Scott, R. F., *Principles of Soil Mechanics*, Reading, Mass., Addison-Wesley, 1963.
(4) Harr, M. E., *Foundations of Theoretical Soil Mechanics*, New York, McGraw-Hill Book Co., 1966.
(5) Jumikis, A. R., *Soil Mechanics*, New York, Van Nostrand Reinhold Co., 1962.
(6) Obert, L., and Duvall, W. I., *Rock Mechanics and the Design of Structures in Rock*, New York, John Wiley & Sons, 1967.
(7) Stagg, K. G., and Zienkiewicz, O. C. (Ed.), *Rock Mechanics in Engineering Practice*, London, John Wiley & Sons, 1968.
(8) Jaeger, J. C., and Cook, N. G. W., *Fundamentals of Rock Mechanics*, London, Methuen and Co., 1968.
(9) Clough, R. W., and Rashid, Y., "Finite Element Analysis of Axi-symmetric Solids," *Proc. ASCE, J. EM Dn*, Vol. 91, EM1 (Feb. 1965). This is one of the

earlier applications. It analyzes an elastic homogeneous half-space loaded by a concentrated load, and obtains good comparisons with the classical Boussinesq solutions.

(10) Goodman, R. E., and Brown, C. B., "Dead Load Stresses and the Instability of Slopes," *Proc. ASCE, J.SM&F Dn*, Vol. 89, SM3 (May 1963). This work is perhaps the first to formalize the concept of sequential construction procedures. An analytical solution for stress and stability analyses is developed on the basis of linear elastic theory. Applications are directed to earth and concrete structures like dams and embankments. Does not use the finite element method, but many subsequent applications of the method are derived from this formulation.

(11) Brown, C. B., and King, I. P., "Automatic Embankment Analysis," *Geotechnique*, Vol. XVI, No. 3 (Sept. 1966). Uses the analytic simulation procedures of Reference 10 in the finite element method. Performs stress and stability analyses for excavation and embankment by creating the so-called "stress free" surface at the bottom of each excavation step.

(12) Girijavallabhan, C. V., and Reese, L. C., "Finite Element Method Applied to Some Problems in Soil Mechanics," *Proc. ASCE, J.SM&F Dn*, Vol. 94, SM2 (Mar. 1968). Also, report to API, Department of Civil Engineering, University of Texas, Austin (Jan. 1967). Considers Boussinesq and lateral pressure problems. Develops a nonlinear analysis procedure based on octahedral stress approach for both cohesive soils and sands. Shows satisfactory agreement with laboratory experiments.

(13) Clough, R. W., and Woodward, R. J., "Analysis of Embankment Stresses and Deformations," *Proc. ASCE, J.SM&F Dn*, Vol. 93, SM4 (July 1967). Uses nonlinear incremental procedure and simulates construction sequence in dams. Derives constitutive law from laboratory triaxial tests. Obtains satisfactory correlation with field displacements in the Otter Brook Dam.

(14) See Reference 9, Chapter 6.

(15) Banks, D. C., and Palmerton, J. B., "Application of Finite Element Method in Determining Stability of Crater Slopes," Misc Paper S-68-3, USAEWES (May 1968). Studies stability of irregularly shaped slopes caused by nuclear excavations.

(16) Duncan, J. M., and Dunlop, P., "Slopes in Stiff-Fissured Clays and Shales," Contract Report No. S-68-4, USAEWES (June 1968). Also in *Proc. ASCE, J. SM&F Dn*, Vol. 95, SM2 (Mar. 1969). Reviews behavioral aspects of slopes in stiff-fissured clays and shales. Studies effects of nonlinear stress-strain behavior, fissures, and *in situ* stress fields. Derives analytic simulation procedure for excavation.

(17) Desai, C. S., "Solution of Stress-Deformation Problems in Soil and Rock Mechanics Using the Finite Element Method," Ph. D. Dissertation, The University of Texas, Austin, Aug. 1968. Also with L. C. Reese in (1) *Proc. ASCE, J.SM&F Dn*, Vol. 96, SM4 (July 1970); (2) *J. Indian National Soc. SM&FE*, Vol. 9, No. 1 (Jan. 1970).

(18) D'Appolonia, D. J., "Prediction of Stress and Deformation for Undrained Loading Conditions," Ph. D. Dissertation, MIT, Cambridge, Sept. 1968. Also, with Lambe, T. W., in *Proc. ASCE, J. SM&F Dn*, Vol. 96, SM2 (Mar. 1970). Studies effects of *in situ* stress fields, initial stress anisotropy, and deviations in

initial elastic modulus, yield stress, and undrained shear strength on instantaneous or initial settlement prediction of a strip footing on Boston Blue clay.

(19) Covarrubias, S. W., "Cracking of Earth and Rockfill Dams," Ph. D. Dissertation, Harvard University, Cambridge, April 1969. In addition to a number of hypothetical dam-valley-abutment combinations, examines actual cracking behavior of four dams (Gespatch in Austria, Infiernillo in Mexico, Hyttejuvet in Norway, and Dam X) and obtains good correlation. Uses linear elastic analysis and suggests that nonlinear and viscous effects should be included.

(20) Kulhawy, F. H., Duncan, J. M., and Seed, H. B., "Finite Element Analysis of Stresses and Movements in Embankments During Construction," Contract Report No. S-69-8, USAEWES (Nov. 1969).

(21) Thoms, R. L., and Arman, A., "Analysis of Stress Distribution Under Embankments on Soft Foundations," Bulletin No. 99, Louisiana State University, Baton Rouge, 1969. Compares finite element results with experiments on gelatin simulating soft muck in embankment foundations. Considers large displacements.

(22) Palmerton, J. B., "Preliminary Finite Element Alalysis, Atchafalaya Basin Protective Levees," Misc Paper S-69-53, USAEWES (Dec. 1969). Uses nonlinear incremental procedure for settlement analysis of soft foundations, settlements being of the order of 10–20 feet over a period of 10–15 years. Obtains agreement with field data to the same order of magnitude. Envisages inclusion of creep and large displacements in the analysis.

(23) Radhakrishnan, N., and Reese, L. C., "Behavior of Strip Footings on Layered Cohesive Soils," *Symp. FEM*. Uses the procedure and the cohesive soil systems of Reference 17 to examine behavior of a strip footing.

(24) Ellison, R. D., "An Analytical Study of the Mechanics of Single Pile Foundations," Ph. D. Dissertation, Carnegie-Mellon University, 1969. Uses the finite element method for load-settlement and stress-distribution analyses and for evaluation of failure mechanics of isolated piles in soil foundations. Uses incremental procedure for nonlinear elastic analysis with the concept of octahedral stresses and strains. Obtains comparisons between results from finite element analysis and from tests on large model piles buried in a sand and a stiff clay.

(25) Brown, C. B., Green, D. R., and Pawsey, S., "Flexible Culverts Under High Fills," *Proc. ASCE, J.ST Dn*, Vol. 94, ST4 (April 1968). On the basis of the procedure in Reference 10, examines pressure and stresses in flexible culverts under effects of high fill, organic fill and its rotting, various boundary conditions and arching. In a previous paper—Vol. 93, ST5 (Oct 1967)—similar analysis is done for rigid culverts.

(26) Cheung, Y. K., and Nag, D. K., "Plates and Beams on Elastic Foundations: Linear and Nonlinear Behavior," *Geotechnique*, Vol. XVIII, No. 2 (July 1968). The diagonal foundation flexibility matrix is inverted to obtain the stiffness matrix, which is added to the plate stiffness to form the composite stiffness matrix. Tensile or zero contact pressure at the interface is included by appropriately modifying the foundation matrix.

(27) See Reference 8, Chapter 6.

(28) Clough, G. W., and Duncan, J. M., "Finite Element Analysis of Port Allen and Old River Locks," Contract Report No. TE 69-3, USAEWES (Sept. 1969).

(29) Kirwan, R. W., and Glynn, T. E., "Experimental and Theoretical Investigation of Pavement Deflections," Tech. Report, University of Dublin, Ireland, Nov. 1969. Viscoelastic response of bituminous concrete is represented by creep compliance, and granular and cohesive subgrade response by resilient modulus. Stress transfer approach (Reference 41) is used for tensile stresses. Comparisons between laboratory, field, and finite element results are shown.
(30) Hoyaux, B., and Ladanyi, B., "Gravitational Stress Field Around a Tunnel in Soft Ground," *Canadian Geotechnical Journal*, Vol. 7, No. 1 (Feb. 1970). Nonlinear iterative procedure with secant modulus is used to analyze stress and displacements around shallow and deep unlined tunnels in both sensitive and insensitive clays.
(31) Haddadin, M. J., "Discussion: Photoelastic Study of Beams on Elastic Foundations," *Proc. ASCE, J.ST Dn*, Vol. 96, ST4 (April 1970). Compares finite element results for finite and infinite beams on elastic foundations with experiments from photoelastic studies.
(32) Chang, T. Y., Ko, N. Y., and Scott, R. F., "An Integrated Approach to the Stress Analysis of Granular Materials," Report to Soil Mech. Lab., California Institute of Technology, Pasadena, 1967. Establishes a second order stress-strain law. Employs deformation theory of plasticity (Chapter 3) and examines qualitatively the influence of hydrostatic and deviatoric components in tests such as pure shear, one-dimensional compression, and triaxial shear. Suggests second order strain-stress law to obviate instability caused by work softening.
(33) Girijavallabhan, C. V., and Mehta, K. C., "Stress-Strain Relationship from Compression Tests on Nonlinear Materials," *Symp. FEM.* Uses octahedral stress approach for nonlinear analysis of triaxial test specimens of soils and concrete.
(34) Perloff, W. H., and Pombo, L. E., "End Restraint Effects in the Triaxial Test," *Proc. 7th Int. Conf. on SM&FE*, Mexico City, Aug. 1969. Uses octahedral stress approach and extended Von Mises flow rule for incremental elastic-plastic analysis of triaxial test specimen.
(35) Duncan, J. M., and Chang, C. Y., "Nonlinear Analysis of Stress and Strain in Soils," *Proc. ASCE, J. SM&F Dn*, Vol. 96, SM5 (Sept. 1970).
(36) See Reference 25, Chapter 3.
(37) Blake, W., "Application of the Finite Element Method of Analysis in Solving Boundary Value Problems in Rock Mechanics," *Int. J. Rock Mech. and Min. Sci.*, Vol. 3, No. 3 (1966). Compares results from linear finite element analysis with classical Kirsch and photoelastic solutions for a plate with circular hole under compressive load. Effects of faults and local geology are considered.
(38) Finn, W. D. L., "Static and Dynamic Stresses in Slopes," *Proc., 1st. Intl. Cong. Rock Mech.*, Lisbon, 1966. Changes in the *in situ* stress fields in rock slopes subjected to geostatic forces, earthquake forces, and excavations are analyzed.
(39) Goodman, R. E., "On the Distribution of Stresses Around Circular Tunnels in Non-homogeneous Rocks," *Proc. 1st. Int. Cong. Rock Mech.*, Lisbon, 1966. Effects of heterogeneity or planar layering and *in situ* stress fields around circular tunnels in multilayered rock systems.

(40) See Reference 24, Chapter 3.
(41) Zienkiewicz, O. C., Valliapan, S., and King, I. P., "Stress Analysis of Rock as a 'No Tension' Material," *Geotechnique*, Vol. XVIII, No. 1 (Mar. 1968)
(42) Desai, C. S., and Reese, L. C., "Stress-Deformation and Stability Analyses of Deep Boreholes," *Proc. 2nd Int. Cong. Rock Mech.*, Belgrade, 1970.
(43) Baker, L. E., Sandhu, R. S., and Shieh, W. Y., "Application of Elasto-Plastic Analysis in Rock Mechanics by Finite Element Method," *Proc. 11th Symp. on Rock Mech.*, Berkeley, June 1969. Computes factor of safety for a rock slope using procedure of Reference 36. Good comparison with results from conventional analysis. Examples of an elastic-plastic wedge and a deep tunnel are also considered.
(44) Zienkiewicz, O. C., and Cheung, Y. K., "Stresses in Buttress Dams," *Water Power*, Vol. 17 (Feb. 1965). Considers effects of pore water pressure, temperature gradients, and development of tension zones in dam-rock foundation system. In a previous paper—*Water Power*, Vol. 15 (May 1964)—effects of gravity, hydrostatic pressure, and jointing were considered.
(45) Nair, K., Sandhu, R. S., and Wilson, E. L., "Time-Dependent Analysis of Underground Cavities under Arbitrary Initial Stress Field," *Proc. 10th Symp. on Rock Mech.*, Austin, Texas, 1968. An approximate power law is adopted for time dependent and nonlinear analysis. Different shapes of axisymmetric cavities are considered and effects of *in situ* stresses are examined under three different initial stress ratios.
(46) McMahon, B. K., and Kendrick, R. F., "Predicting the Block Caving Behavior of Orebodies," *J. Soc. of Min. Engrs, AMIE*, No. 69-AU-51 (1969). Block caving behavior of some sections of Urad mines in Colorado is examined and the results are found in qualitative agreement with field behavior.
(47) See Reference 10, Chapter 6.
(48) Goodman, R. E., "Effects of Joints on the Strength of Tunnels," Tech. Report No. 5, Omaha Dist., U.S. Army CE, Omaha, 1968. Field and laboratory methods of simulating joints are presented. Two methods called "ubiquitous joint analysis" and "joint stiffness analysis" are proposed and used. A power law connecting the octahedral shear and normal stresses is used as failure criterion. The analysis agreed well with field behavior of the particular sections of the Piledriver Drifts.
(49) Goodman, R. E., Taylor, R. L., and Brekke, T. L., "A Model for the Mechanics of Jointed Rock," *Proc. ASCE, J.SM&F Dn*, Vol. 94, SM3 (May 1968). A special joint stiffness matrix is developed in which the joint is characterized by normal and tangential stiffness and shear strength. These properties can be derived from a box shear test. The approach permits consideration of failure in tension or shear, rotation of blocks, arching phenomenon, and collapse patterns.
(50) Kondner, R. L., and Zelasko, J. S., "A Hyperbolic Stress-Strain Formulation for Sands," *Proc. Second Pan-Am Conf., Soil Mech. & Found. Eng.*, Vol. I, Brazil, 1963.
(51) Bjerrum, L., "Progressive Failure in Slopes of Overconsolidated Plastic Clay and Clay Shales," *Proc. ASCE, J.SM&F Dn*, Vol. 93, SM5 (Sept. 1967).

(52) Linell, K. A., and Shea, H. F., "Strength and Deformation Characteristics of Various Glacial Tills in New England," *Proc. Res. Conf. on Shear Strength of Cohesive Soils, ASCE SM&F Dn*, Boulder, Colo., June 1960.

(53) Desai, C. S., "Nonlinear Analysis Using Spline Functions," *Proc. ASCE, J. SM&F Dn.*, Vol. 97, SM 10 (Oct. 1971).

(54) Birkhoff, G., and DeBoor, C. R., "Piecewise Polgynomial Interpolation and Approximation," *Approximations of Functions*, H. L. Garabedian (Editor), Amsterdam, Elsevier Publishing Co., 1965.

(55) Desai, C. S., "Bicubic Spline Simulation of Stress-Strain Curves," published as closure to Ref. 53, *Proc. ASCE, J. SM&F Dn.*, Vol. 98, No. SM9, Sept. 1972.

FURTHER READING

Agarwal, R. K., and Boskhov, S. H., "Stresses and Displacements Around Circular Tunnel in a Three-Layer Medium—I," *Int. J. Rock Mech. Min. Sci.*, Vol. 6 (1969). A tunnel in soft rock surrounded by layers of hard materials, with different values of ratio of height of soft layer to tunnel diameter.

Boughton, N. O.," Elastic Analysis for Behavior of Rockfill," *Proc. ASCE, J.SM & F Dn*, Vol. 96, SM5 (Sept. 1970). Derives constitutive relations from special triaxial and direct shear tests. Establishes functional relationships for the tangent Young's modulus and tangent Poisson's ratio for the nonlinear elastic analysis.

Douglas, A., "Finite Elements for Geological Modelling," *Nature*, Vol. 226, May 16, 1970. Simulates earth's crust, 800 km long and 25 km thick. Computes displacements resulting from deposition of a lens of sediment, 1 km thick, at the center of the trough between two mountains. The computed displacement patterns agree with classical solutions. Proposes use of the method for studying effects of erosion and deposition leading to formation of such geological features as rift valleys.

Hayashi, M., "Numerical Analysis of Progressive Failure of Sandy Slope—Comparison with Model Test," Tech. Report C 68004, Tech. Lab., Central Res. Inst. Elect. Power Industry, Tokyo, Japan (Feb. 1970). Accounts for nonlinear constitutive law, strain hardening resulting from consolidation of soils, variations in Poisson's ratio, redistribution of stresses due to local failures, and partial restraint of viscoelastic strain increment. Obtains good comparisons with experiments on model tests.

Malina, H., "The Numerical Determination of Stresses and Deformations in Rock Taking into Account Discontinuities," *Rock Mechanics*, Vol. 2 (1970). Presents the finite element method. Solves a problem of a continuous footing on rock with a joint system.

Smith, I. M., "A Finite Element Approach to Elastic Soil-Structure Interaction," *Canadian Geotechnical Journal*, Vol. 7. No. 2 (March 1970). Compares results from finite element analyses with tests on flexible perspex plates bearing on an overconsolidated sand foundation and with field observations of deformations in cylindrical storage tanks.

Voight, B., and Dahl, H. D., "Numerical Continuum Approaches to Analysis of Nonlinear Rock Deformation," *Canadian Journal of Earth Sci.*, Vol. 7, No. 3 (June 1970). Reviews finite element method for problems in geological engineering. Discusses elastic, elastic-plastic and viscous behavior.

Zienkiewicz, O. C., "Problems in Rock Mechanics," *The Engineer*, No. 5690, Feb. 12, 1965. Considers basic aspects of rock mechanics and presents applications of the finite element method.

Proc., Seventh Int. Conf. on Soil Mech. & Found. Engg., Mexico City, Aug. 1969.

Proc., First Int. Congress on Rock Mechanics, Lisbon, 1966.

Proc., Second Int. Congress on Rock Mechanics, Belgrade, 1970.

See Reference 12, Chapter 3.

11

DYNAMICS, INCLUDING EARTHQUAKE ANALYSIS

Dynamics is a rapidly expanding area of application of the finite element method in the fields of structural and soils engineering. Indeed, many problems in dynamics cannot be solved effectively by analytical methods, and the evolution of both the digital computer and various associated numerical techniques of analysis has significantly enhanced solution capability. Therefore, the finite element method is at least as valuable a tool in dynamics as it is in the static problems of soils and structures.[1] For the fundamentals and background of structural and soil dynamics, the reader is referred to textbooks concerned with these general topics.[2,3,4,5,6,7]

11-1 SUMMARY OF APPLICATIONS IN DYNAMICS

Basically, four different types of problems can be distinguished in the field of dynamics: free vibration, steady-state response, transient response to known excitations, and response to random excitation. However, as indicated in Table 11-1, we shall consider only the free vibration and transient response aspects in this chapter. Once we formulate these problems for the finite element method, we can readily extend our formulations to the remaining two categories.[2,3,4,7]

Solution methods for eigenvalue problems are discussed in Section 2-3. In addition, a general step-by-step method for transient problems is presented

TABLE 11-1 TOPICS IN DYNAMICS AND EARTHQUAKE ANALYSES

	Topic	*Typical Applications*	*References*
I	Eigenvalue problems	Free vibrations. Prerequisite to mode superposition method. Linear viscoelastic damping (structural damping).	Section 2-3; References 1, 5, 8 to 17
II	Transient problems	Response to arbitrary dynamic loadings, such as wind, explosions, earthquakes, water waves. Propagation of stress or displacement waves.	Sections 2-4 and 11-4; References 1, 5, 8 to 10, 14, 18 to 26

in Section 2-4. However, another important technique for transient problems is the *mode superposition method.* This approach will be detailed subsequently.

Later in this chapter we shall consider several examples of the applications listed in Table 11-1. All the material in this chapter is restricted to problems in solid mechanics. In Chapters 13 and 14 some dynamic problems related to the topics of hydroelasticity and fluid mechanics will be treated.

11-2 FORMULATION OF INERTIAL PROPERTIES

In addition to the stiffness of the body or structure, a property essential in dynamic behavior is the inertia or mass, which is represented by a *mass matrix*. We shall consider two basic types of mass matrices, the *consistent mass matrix* and the *lumped mass matrix*. The former may be derived by the application of Hamilton's principle, which is the variational theorem associated with the displacement finite element method of dynamic analysis. The lumped mass is assembled in a direct method by assuming that all the mass tributary to a node is concentrated at that node. There is an obvious analogy between the two mass matrices and the two types of load vectors, consistent and lumped (Sections 5-6 through 5-8).

Consistent Mass Matrix

In dynamic problems the displacements, velocities, strains, stresses, and loadings are all time dependent. When interpolation displacement models are used, we may insert equations (5.8), (5.28b), and (5.29b) into equation (4.20) to obtain

$$L = \frac{1}{2}\iiint\limits_V (\{q\}^{\mathrm{T}}[B]^{\mathrm{T}}[C][B]\{q\} - \rho\{\dot{q}\}^{\mathrm{T}}[N]^{\mathrm{T}}[N]\{\dot{q}\} - 2\{q\}^{\mathrm{T}}[N]^{\mathrm{T}}\{\bar{X}\})\,dV - \iint\limits_{S_1} \{q\}^{\mathrm{T}}[N]^{\mathrm{T}}\{\bar{T}\}\,dS_1 \tag{11.1}$$

Applying the variational principle, equation (4.21), we obtain

$$\int_{t_1}^{t_2}\left(\{\delta q\}^{\mathrm{T}}\iiint\limits_V [B]^{\mathrm{T}}[C][B]\,dV\{q\} - \{\delta\dot{q}\}^{\mathrm{T}}\iiint\limits_V \rho[N]^{\mathrm{T}}[N]\,dV\,\{\dot{q}\} - \{\delta q\}^{\mathrm{T}}\iiint\limits_V [N]^{\mathrm{T}}\{\bar{X}\}\,dV - \{\delta q\}^{\mathrm{T}}\iint\limits_{S_1} [N]^{\mathrm{T}}\{\bar{T}\}\,dS_1\right) dt = 0 \tag{11.2}$$

Integration of the second term by parts with respect to time gives

$$\int_{t_1}^{t_2}\{\delta\dot{q}\}^{\mathrm{T}}\iiint\limits_V \rho[N]^{\mathrm{T}}[N]\,dV\,\{\dot{q}\}\,dt = \left[\{\delta q\}^{\mathrm{T}}\iiint\limits_V \rho[N]^{\mathrm{T}}[N]\,dV\,\{\dot{q}\}\right]_{t_1}^{t_2} - \int_{t_1}^{t_2}\{\delta q\}^{\mathrm{T}}\iiint\limits_V \rho[N]^{\mathrm{T}}[N]\,dV\,\{\ddot{q}\}\,dt \tag{11.3}$$

According to Hamilton's principle, the tentative displacement configuration must satisfy given conditions at times t_1 and t_2. Hence, $\{\delta q(t_1)\} = \{\delta q(t_2)\} = \{0\}$, so the first term on the right-hand side of equation (11.3) vanishes. Substituting the remaining term into equation (11.2), we find

$$\int_{t_1}^{t_2}\{\delta q\}^{\mathrm{T}}\left(\iiint\limits_V \rho[N]^{\mathrm{T}}[N]\,dV\,\{\ddot{q}\} + \iiint\limits_V [B]^{\mathrm{T}}[C][B]\,dV\{q\} - \iiint\limits_V [N]^{\mathrm{T}}\{\bar{X}\}\,dV - \iint\limits_{S_1} [N]^{\mathrm{T}}\{\bar{T}\}\,dS_1\right) dt = 0 \tag{11.4}$$

Since the variations of the nodal displacements, $\{\delta q\}$, are arbitrary, the expression in parentheses must vanish. Therefore we obtain the *equations of motion for the element*:

$$[m]\{\ddot{q}\} + [k]\{q\} = \{Q\} \tag{11.5}$$

Here $[k]$ and $\{Q\}$ are the element stiffness matrix and load vector defined in the usual manner, equations (5.45a) and (5.45b), and $[m]$ is the consistent mass matrix defined by

$$[m] = \iiint\limits_V \rho[N]^{\mathrm{T}}[N]\,dV \tag{11.6}$$

If we perform the element formulation in terms of generalized coordinate displacement models, the use of equations (5.5), (5.28a), and (5.29a) in equation (11.1) will give the results stated in equations (5.46), plus the mass matrix

$$[m_\alpha] = \iiint_V \rho[\phi]^T[\phi]\,dV \tag{11.7}$$

As is done in equations (5.47), $[m_\alpha]$ can also be transformed to $[m]$ by

$$[m] = [A^{-1}]^T[m_\alpha][A^{-1}] \tag{11.8}$$

Hence the dynamic equilibrium relation for the element, equation (11.5), is applicable to both the interpolation model and generalized coordinate cases.

The mass matrices given by equations (11.6) and (11.8) are termed *consistent* because the same displacement functions are used to formulate the stiffness and mass matrices. Hence the potential and kinetic energies have the same basis. If the displacement models used are complete and conforming, the frequencies obtained from a consistent formulation can be shown to be upper bounds of the true frequencies, and the displacements are lower bounds.[1,27]

Lumped Mass Matrix

In the direct method of formulation, the mass of the length, area, or volume tributary to a particular node is considered to be concentrated at the node. Both rotary and translational inertia may be treated in the same way. The resulting lumped mass matrix is diagonal.

We may write an expression analogous to equations (11.6) and (11.7) for the lumped mass matrix as

$$[m] = \iiint_V \rho[\psi]^T[\psi]\,dV \tag{11.9}$$

where $[\psi]$ is the matrix of functions ψ_i which have unit value over the region tributary to node i and zero value elsewhere.[1] As long as the tributary regions do not overlap, the lumped mass matrix calculated from equation (11.9) is diagonal.

Consistent Masses vs. Lumped Masses

Lumped masses provide some significant economies compared to consistent masses. The diagonal lumped mass matrix for the assemblage requires less storage space than the banded consistent mass matrix. Moreover, the diagonal lumped form greatly facilitates matrix calculations. In addition, for some types of problems, such as beam, plate, and shell analyses, lumped masses permit a marked reduction of the number of equations occurring in the dynamic problem. This occurs when rotary inertia is neglected, because in that case rotational degrees of freedom have no associated lumped masses. A lumped mass representation having the same number of equations for the eigenvalue problem as the consistent mass method will often give more accurate natural frequencies.[12] However, the mode shapes are usually less reliable for the lumped mass case. If upper-bound natural frequencies and more accurate mode shapes are desired, the results from a preliminary calculation using a lumped mass approach can be employed in conjunction with consistent mass equations and a Rayleigh-Ritz technique.[12] Finally, the lumped mass formulation must be used in all wave propagation problems if reliable results are to be obtained.[1]

The principal advantages of the consistent mass approach are the more accurate mode shapes and the frequencies which are proven upper bounds. In most applications, nevertheless, the computational advantages of lumped masses are the overriding consideration.

Condensation and Assembly of Mass Matrices

At the element level we can eliminate the internal degrees of freedom from the equations of motion in the lumped mass case only. The appropriate partitioned form of equation (11.5) is

$$\begin{bmatrix} [\bar{m}] & [0] \\ [0] & [0] \end{bmatrix} \begin{Bmatrix} \{\ddot{q}_1\} \\ \{\ddot{q}_2\} \end{Bmatrix} + \begin{bmatrix} [k_{11}] & [k_{12}] \\ [k_{21}] & [k_{22}] \end{bmatrix} \begin{Bmatrix} \{q_1\} \\ \{q_2\} \end{Bmatrix} = \begin{Bmatrix} \{Q_1\} \\ \{Q_2\} \end{Bmatrix} \tag{11.10}$$

Here $\{q_2\}$ is the vector of internal degrees of freedom which have corresponding lumped masses of zero; $[\bar{m}]$ is the diagonal condensed element mass matrix. Using a procedure identical to that of Section 5-9, we obtain the condensed equilibrium equations for the element

$$[\bar{m}]\{\ddot{q}_1\} + [\bar{k}]\{q_1\} = \{\bar{Q}\} \tag{11.11}$$

where the definitions of $[\bar{k}]$ and $[\bar{Q}]$ are given by equations (5.63c) and (5.63d), respectively.

The mass matrix for the assemblage of elements is constructed by a direct stiffness procedure exactly analogous to that for the stiffness matrix (Chapter 6). The assembly rule for the inertial properties is

$$[M] = \sum_{e=1}^{E} [M_e] \tag{11.12}$$

where e denotes the element number, E is the total number of elements in the assemblage, and $[M_e]$ is the $N \times N$ expanded element matrix obtained from the $n \times n$ element mass matrix in global coordinates, $[m_g]$. The transformation of coordinates that may be necessary to obtain $[m_g]$ is the same as that given in Section 6-5.

$$[m_g] = [T]^{\mathrm{T}}[m_\ell][T] \tag{11.13}$$

However, for the lumped mass case the transformation is not necessary, since the mass matrix is invariant with respect to the transformation of equation (11.13).

The assembled lumped mass matrix remains diagonal, while the assembled consistent mass matrix has the same banded form as the stiffness matrix.

The procedure used to eliminate degrees of freedom corresponding to zero lumped masses from the assemblage equations for the eigenvalue problem is given in Section 2-3. This process is accomplished simultaneously with the reduction of the eigenvalue problem to standard form.

Example: Sample Problem 1. We shall now derive the lumped and consistent mass matrices for the two elements used as sample problems in Chapter 5. The beam element with uniformly distributed mass in shown in Figures 5-4 and 5-18. The tributary length for each node of the beam element is $\ell/2$. If we neglect rotary inertia, we may write the element lumped mass matrix directly as

$$[m] = \frac{\rho b h \ell}{2}\begin{bmatrix} 1 & 0 & 0 & 0 \\ 0 & 0 & 0 & 0 \\ 0 & 0 & 1 & 0 \\ 0 & 0 & 0 & 0 \end{bmatrix}$$

Note that the total mass of the element is equally divided between the nodes.

In order to compute the consistent mass matrix for the element, we employ equation (11.6). The interpolation displacement models are given by equation (5.26)

$$\underset{1\times 4}{[N]} = [L_1^2(3-2L_1),\ L_1^2L_2\ell,\ L_2^2(3-2L_2),\ -L_1L_2^2\ell] \tag{5.26}$$

Substituting this into equation (11.6) and using the integration formula given by equation (5.17e), we obtain the consistent mass matrix

$$[m] = bh\rho\int_\ell \begin{bmatrix} L_1^4(3-2L_1)^2 & L_1^4L_2(3-2L_1)\ell & L_1^2L_2^2(3-2L_1)(3-2L_2) & -L_1^3L_2^2(3-2L_1)\ell \\ & L_1^4L_2^2\ell^2 & L_1^2L_2^3(3-2L_2)\ell & -L_1^3L_2^3\ell^2 \\ & & L_2^4(3-2L_2)^2 & -L_1L_2^4(3-2L_2)\ell \\ \text{Symmetrical} & & & L_1^2L_2^4\ell^2 \end{bmatrix} d\ell$$

$$= \frac{bh\rho\ell}{420}\begin{bmatrix} 156 & -22\ell & 54 & 13\ell \\ & 4\ell^2 & -13\ell & -3\ell^2 \\ & & 156 & 22\ell \\ \text{Symmetrical} & & & 4\ell^2 \end{bmatrix}$$

where rotary inertia is neglected.

Example: Sample Problem 2. The constant strain triangle (CST) of Sample Problem 2 is shown in Figure 5-5. The tributary area for each node is taken as one-third of the element area. Hence the lumped mass matrix can be written as

$$[m] = \frac{\rho h A}{3} \begin{bmatrix} 1 & 0 & 0 & 0 & 0 & 0 \\ 0 & 1 & 0 & 0 & 0 & 0 \\ 0 & 0 & 1 & 0 & 0 & 0 \\ 0 & 0 & 0 & 1 & 0 & 0 \\ 0 & 0 & 0 & 0 & 1 & 0 \\ 0 & 0 & 0 & 0 & 0 & 1 \end{bmatrix}$$

where ρ is the uniform mass density of the triangle, and h is the constant thickness. Note that the total mass of the element is equally divided among the nodes for *each* basic type of translation, u and v.

The interpolation displacement model used for the triangular element is given by equation (5.27)

$$[N] = \begin{bmatrix} L_1 & L_2 & L_3 & 0 & 0 & 0 \\ 0 & 0 & 0 & L_1 & L_2 & L_3 \end{bmatrix} \tag{5.27}$$

Substituting this displacement model into equation (11.6) and applying the integration formula equation (5.18f), we obtain the consistent mass matrix

$$[m] = \rho h \iint_A [N]^{\mathrm{T}}[N]\, dA = \frac{\rho h A}{12} \begin{bmatrix} 2 & 1 & 1 & 0 & 0 & 0 \\ 1 & 2 & 1 & 0 & 0 & 0 \\ 1 & 1 & 2 & 0 & 0 & 0 \\ 0 & 0 & 0 & 2 & 1 & 1 \\ 0 & 0 & 0 & 1 & 2 & 1 \\ 0 & 0 & 0 & 1 & 1 & 2 \end{bmatrix}$$

Recall that the corresponding arrangement of the vector $\{q\}$ is

$$\{q\}^{\mathrm{T}} = [u_1 \; u_2 \; u_3 \; v_1 \; v_2 \; v_3]$$

11-3 FORMULATION OF DAMPING PROPERTIES

The complete equations of motion for a discretized body or structure often must include a term to account for energy dissipation. Usually this mechanism takes the form of viscous damping that is linearly proportional to the velocity. Hence the equations of motion for a damped system are

$$[M]\{\ddot{r}\} + [C]\{\dot{r}\} + [K]\{r\} = \{R(t)\} \tag{11.14}$$

where $[C]$ is the *viscous damping matrix*. Unfortunately, little is known about the evaluation of the damping coefficients that are the elements of the matrix $[C]$. However, since the effect of damping is generally less than the inertial and stiffness effects, it is reasonable to represent $[C]$ by a simplified approximation.

Wilson[19] employs a viscous damping matrix of the form

$$[C] = \alpha[M] + \beta[K] \tag{11.15}$$

Here α and β are proportionality constants which relate damping to the velocity of the nodes and the strain velocities, respectively. Most of the experimental data on damping properties of structures consist of *modal damping ratios*, that is, ratios of actual damping to critical damping for particular natural modes of vibration. Therefore, it is useful if values of α and β in equation (11.15) can be assigned from known values of modal damping. To do this, Wilson[19] utilizes the following relation between α and β and the damping ratio for the ith mode, c_i:

$$c_i = \alpha/2\omega_i + \beta\omega_i/2 \tag{11.16}$$

where ω_i is the natural frequency of the ith mode. Equation (11.16) can be rewritten as

$$c_i = (\bar{\omega}/\omega_i + \omega_i/\bar{\omega})\bar{c}/2 \tag{11.17}$$

where $\bar{c}$ is the minimum damping ratio and $\bar{\omega}$ is the frequency corresponding to $\bar{c}$. Equation (11.17) can be used to give curves of c_i vs. $\bar{\omega}/\omega_i$ for various values of $\bar{c}$. Hence, if a frequency range and a modal damping ratio are selected, α and β may be calculated from

$$\alpha = \bar{c}\bar{\omega} \tag{11.18a}$$

$$\beta = \bar{c}/\bar{\omega} \tag{11.18b}$$

Wilson has developed an alternative method of obtaining the matrix $[C]$ from the modal damping ratios.[1] This technique takes advantage of the orthogonality of the modal matrix, $[X]$. Some of the properties of the modal matrix are given in Section 2-3. In addition, the matrix is orthogonal with respect to $[M]$ and $[K]$ as weighting functions:

$$[X]^{\mathrm{T}}[M][X] = [D] \tag{11.19a}$$

$$[X]^{\mathrm{T}}[K][X] = [\Lambda][D] \tag{11.19b}$$

As a consequence of orthogonality, $[D]$ is a diagonal matrix. $[\Lambda]$ is the spectral matrix, which is also diagonal with elements ω_i^2. Wilson assumes that the modal matrix is also orthogonal with respect to the damping matrix $[C]$ as a weighting function:

$$[X]^{\mathrm{T}}[C][X] = 2[c][\Lambda^{1/2}][D] \tag{11.19c}$$

Here $[c]$ is the diagonal matrix of modal damping ratios. The fact that the right hand side of equation (11.19c) is diagonal implies that the damped and undamped assemblages have the same natural mode shapes.[18] If equation (11.19c) is premultiplied by $[X^{-1}]^{\mathrm{T}}$ and postmultiplied by $[X^{-1}]$, we obtain

$$[C] = 2[X^{-1}]^{\mathrm{T}}[c][\Lambda^{1/2}][D][X^{-1}]$$

However, if the eigenvectors are normalized so that $[X]$ is orthonormal with respect to the identity matrix as a weight function (Section 2-3), $[X]^{\mathrm{T}} = [X^{-1}]$; and the last equation becomes

$$[C] = 2[X][c][\Lambda^{1/2}][D][X]^{\mathrm{T}} \tag{11.20}$$

Equation (11.20) can be used to compute the damping matrix $[C]$ when propagation problems are solved by direct integration, such as the step-by-step method outlined in Section 2-4. The disadvantage of the approach in equation (11.20) is that the eigenvalue problem must be solved in order to obtain $[\Lambda]$ and $[X]$. However, only a few of the fundamental modes and frequencies are needed to assemble an approximation for $[C]$.

11-4 SOLUTION OF TRANSIENT PROBLEMS BY MODE SUPERPOSITION

One method of solving linear transient problems is the *mode superposition* approach.[1,14,18] This technique is also known as the *normal mode method*[4] or as *modal analysis*.[3] The basis of this method is that the modal matrix can be used to diagonalize the mass, damping, and stiffness matrices, and thus uncouple the multi-degree of freedom problem to give several one-degree of freedom problems.

To formulate the mode superposition equations, the orthogonality properties of the modal matrix given in equations (11.19) are used in conjunction with the following transformation of coordinates from $\{r\}$ to $\{p\}$

$$\underset{N\times 1}{\{r\}} = \underset{N\times M}{[X]}\ \underset{M\times 1}{\{p\}} \tag{11.21}$$

where $M \leq N$ is the total number of normal modes, and N is the number of degrees of freedom in the original equations of motion, equation (11.14). By substituting equation (11.21) into equation (11.14) and premultiplying by $[X]^{\mathrm{T}}$, we obtain

$$[X]^{\mathrm{T}}[M][X]\{\ddot{p}\} + [X]^{\mathrm{T}}[C][X]\{\dot{p}\} + [X]^{\mathrm{T}}[K][X]\{p\} = [X]^{\mathrm{T}}\{R(t)\}$$

However, from equations (11.19) we observe that this equation reduces to

$$[D]\{\ddot{p}\} + 2[c][\Lambda^{1/2}][D]\{\dot{p}\} + [\Lambda][D]\{p\} = \{P(t)\} \tag{11.22a}$$

where

$$\{P(t)\} = [X]^{\mathrm{T}}\{R(t)\} \tag{11.22b}$$

Since all the matrices in equation (11.22a) are diagonal, we may rewrite the uncoupled equation for each normal mode

$$\ddot{p}_i + 2c_i\omega_i\dot{p}_i + \omega_i^2 p_i = P_i(t)/D_i,\ i = 1,2,\ldots, M \tag{11.23}$$

where D_i is the ith element of the diagonal matrix $[D]$.

The individual uncoupled equations can be solved by any of the standard methods developed for single degree of freedom structures, such as the Duhamel integral.[1,4,5]

$$p_i(t) = \frac{1}{D_i\,\omega_{Di}}\int_0^t P_i(t')e^{-c_i\omega_{Di}(t-t')}\sin\omega_{Di}(t-t')\,dt' \tag{11.24a}$$

where

$$\omega_{Di} = \omega_i\sqrt{1-c_i^2} \tag{11.24b}$$

Alternatively, any of the methods mentioned in Section 2-4, including the step-by-step method detailed for coupled equations, may be used to solve these single-degree of freedom equations. Once the individual normal mode responses $\{p\}$ are calculated for a particular instant of time, equation (11.21) can be employed to recover the desired solution $\{r\}$.

One advantage of the mode superposition method over the direct integration methods is that the former permits a marked reduction in the number of equations to be solved. Because the more fundamental normal modes play a more significant role in the response than the higher modes, only the lower modes need to be used in the transformation of coordinates, equation (11.21). Another advantage of the normal mode method is that we need not formulate a damping matrix $[C]$; rather, we work directly with the modal damping ratios c_i. If anything is known about the damping of the assemblage, it will probably be the modal damping ratios.

Mode superposition has the disadvantage of requiring that the eigenvalue problem be solved before the transient problem can be attempted. Furthermore, if the number of degrees of freedom of the assemblage is extremely large, it may not be feasible to solve the eigenvalue problem. Finally, the superposition method is applicable only to linear problems in which all the loads have the same variation with respect to time. Therefore, modal analysis is less general than the direct integration which can be used for very large systems of equations, for arbitrary loading, and for nonlinear problems.

11-5 EXAMPLE: FREE VIBRATION OF THIN PLATES

Felippa[12] used both lumped mass (LM) and consistent mass (CM) formulations to compute the natural frequencies and mode shapes of thin plates. The basic element utilized was the Q-19 quadrilateral, Section 9-7. For both stiffness and mass matrices, the equations for the internal degrees of freedom at the centroidal node were retained. Consistent and lumped masses were formulated only at the corner nodes and the centroidal node of each element.

The eigenvalue equations were reduced to standard form by using the methods explained in Section 2-3. For the lumped mass case, the rotational degrees of freedom were eliminated to reduce the order of the eigenvalue problem. We shall consider the results for a simply supported square plate. A quadrant of this plate was discretized into three different meshes with 4, 16, and 36 elements (Fig. 11-1). Table 11-2 gives the number of equations for the quarter plate for the various analyses. The results of the analysis are summarized in Table 11-3 and are compared to other element formulations in

TABLE 11-2 DEGREES OF FREEDOM FOR PLATE QUADRANT (from Reference 12).

Type of Analysis	*Type of Mode*	*Degrees of Freedom for Mesh*		
		2×2	4×4	6×6
LM	S-S	13	41	83
	A-S	8	32	72
CM	S-S	24	96	216
	A-S	22	92	210

LM = Lumped mass
CM = Consistent mass

S = Symmetrical mode
A = Asymmetrical mode

TABLE 11-3 REDUCED FREQUENCIES FOR A SIMPLY SUPPORTED SQUARE PLATE (from Reference 12).

Reduced Fre-quency	*Type of Mode*	*Theore-tical Frequency*	2×2		4×4		6×6	
			LM	CM	LM	CM	LM	CM
λ_{11}	S-S	2.000	2.034	2.031	2.007	2.007	2.001	2.001
λ_{12}	A-S	5.000	5.142	5.135	5.032	5.031	5.013	5.012
λ_{13}	S-S	10.000	10.185	10.390	10.076	10.078	10.032	10.032
λ_{23}	A-S	13.000	13.499	14.028	13.317	13.289	13.131	13.125
λ_{14}	A-S	17.000	—	19.014	17.136	17.176	17.059	17.065
λ_{33}	S-S	18.000	17.117	19.747	18.733	18.647	18.305	18.283
λ_{34}	A-S	25.000	23.672	29.880	26.283	26.153	25.557	25.509
λ_{15}	S-S	26.000	30.267	28.954	26.186	26.390	26.095	26.127

LM = Lumped mass
CM = Consistent mass

S = Symmetrical mode
A = Asymmetrical mode

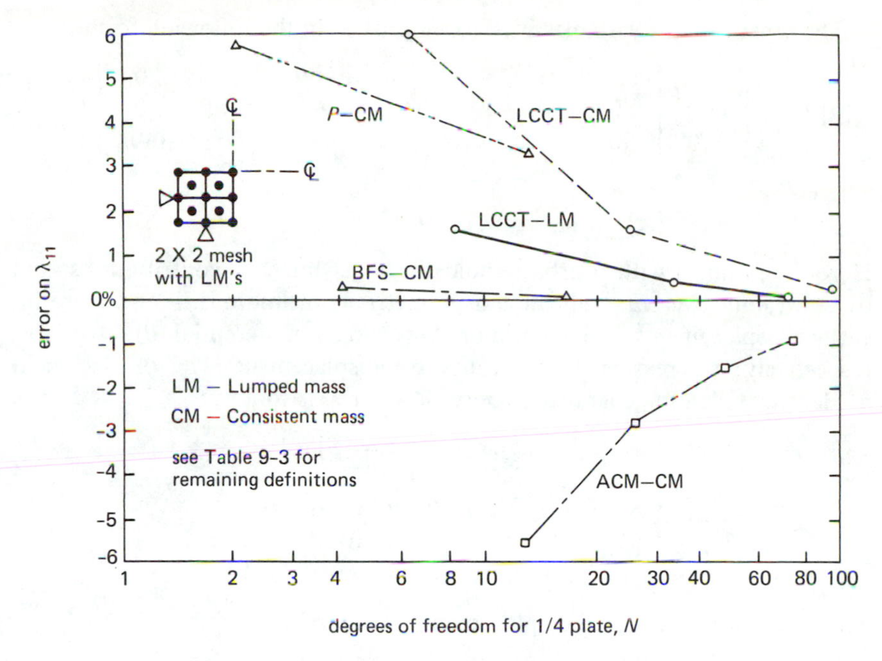

Figure 11-1 Predictions of lowest frequency λ_{11} for square, simply supported plates by various elements (from Reference 13).

Figure 11-1. Comparisons like the one in Figure 11-1 indicate clearly that the lumped mass approach provides a more satisfactory solution for the frequencies in plate problems. For example, for the same number of equations, the LCCT–LM approach gives a more accurate frequency than the LCCT–CM method, Figure 11-1. Moreover, for the same mesh size, the LM frequencies are about as accurate as the CM frequencies, despite the fact that there are two to three times more equations in the CM idealization (Tables 11-2 and 11-3). In fact, Felippa[12] found that the LM analysis was approximately fifteen times faster than the CM analysis for the same mesh. Similar results are reported in Reference 13.

11-6 EXAMPLE: FREE VIBRATION OF CYLINDRICAL SHELLS

The natural frequencies of axisymmetric and asymmetric vibration of cylindrical shells were computed by Ghosh and Wilson,[14] who used the method of separation of variables (Section 9-3). The shell was represented by an assemblage of conical frustum elements (Fig. 6-15).

The separation of variables was accomplished in the following form

$$\{u\} = \begin{Bmatrix} u \\ v \\ w \end{Bmatrix} = \sum_{n=0}^{N_f} \begin{bmatrix} \cos n\theta & 0 & 0 & \sin n\theta & 0 & 0 \\ 0 & \sin n\theta & 0 & 0 & \cos n\theta & 0 \\ 0 & 0 & \cos n\theta & 0 & 0 & \sin n\theta \end{bmatrix} \begin{Bmatrix} \{u_n\} \\ \{\bar{u}_n\} \end{Bmatrix}$$

where

$$\{u_n\}^T = [u_n \; v_n \; w_n], \quad \{\bar{u}_n\}^T = [\bar{u}_n \; \bar{v}_n \; \bar{w}_n]$$

Here the symbols with overbars indicate the asymmetric contributions, and those without overbars indicate axisymmetric contributions. u and v are in-surface displacements in the meridional (s) and circumferential (θ) directions, respectively; whereas w is the transverse displacement. The displacement model was taken in generalized coordinate form as follows:

$$\{u_n\} = [\phi]\{\alpha_n\}, \quad \{\bar{u}_n\} = [\phi]\{\bar{\alpha}_n\}$$

where

$$[\phi] = \begin{bmatrix} 1 & s & 0 & 0 & 0 & 0 & 0 & 0 \\ 0 & 0 & 1 & s & 0 & 0 & 0 & 0 \\ 0 & 0 & 0 & 0 & 1 & s & s^2 & s^3 \end{bmatrix}$$

The vectors of local nodal displacements were

$$\{q_n\}^T = [u_{n1} \; v_{n1} \; w_{n1} \; \chi_{n1} \; u_{n2} \; v_{n2} \; w_{n2} \; \chi_{n2}]$$
$$\{\bar{q}_n\}^T = [\bar{u}_{n1} \; \bar{v}_{n1} \; \bar{w}_{n1} \; \bar{\chi}_{n1} \; \bar{u}_{n2} \; \bar{v}_{n2} \; \bar{w}_{n2} \; \bar{\chi}_{n2}]$$

where χ is defined as dw/ds, and 1 and 2 are the nodes of the conical element. The element stiffnesses for each harmonic were formulated in the standard manner (Section 5-7). Lumped masses were employed.

The cylindrical shell, Figure 11-2, was considered clamped at the ends and the natural frequencies were computed for different numbers of axial nodes, m, and of circumferential nodes, $2n$. In Table 11-4 the results from the finite element analysis are compared with those obtained from Flügge's shell equation[28] and from experiments.[29] Satisfactory agreement was attained.

11-7 EXAMPLE: RESPONSE OF EARTH DAM BY MODE SUPERPOSITION

One of the earliest applications of the finite element method to earthquake analysis was by Clough and Chopra.[18] They used the mode superposition method (Section 11-5) to perform a plane strain analysis of a typical earth dam. By assuming that the foundation of the dam was rigid, they were able to utilize the following load vector:

$$\{R(t)\} = -[M]\{\ddot{u}_g(t)\}$$

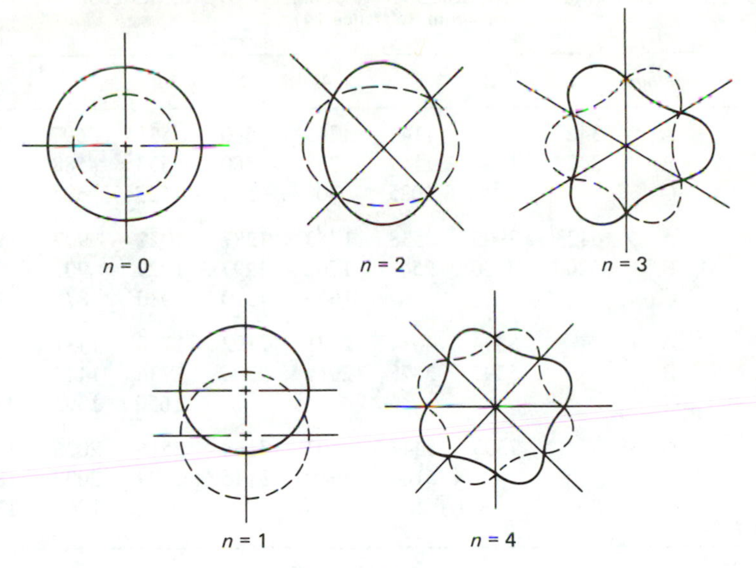

(a) Circumferential nodal pattern

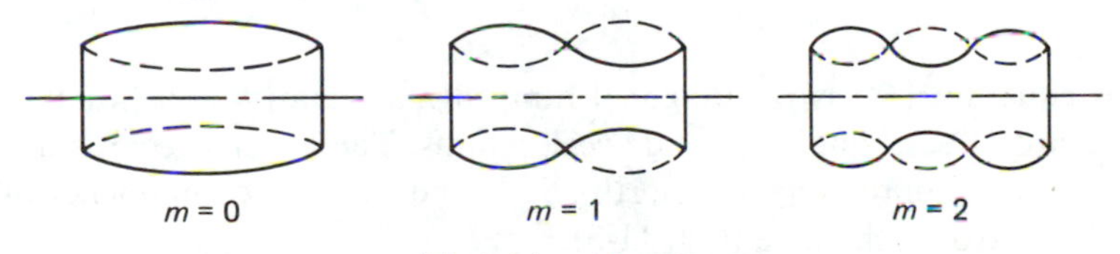

(b) Axial nodal pattern

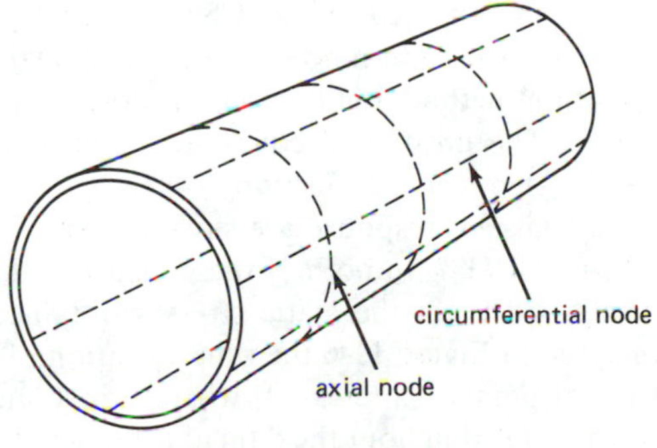

(c) Nodal arrangement for $n = 3$, $m = 3$

Figure 11-2 Nodal patterns for free vibration of cylindrical shells (from Reference 14).

TABLE 11-4 NATURAL FREQUENCIES OF A CLAMPED CYLINDRICAL SHELL, CPS (from Reference 14).

$n\rightarrow$ / $\downarrow m$	Method	1	2	3	4	5	6	7	8
	A	3427	1918	1145	765	580	538	597	721
0	B	3422	1920	1159	768	580	533	588	710
	C			1025	700	559	525	587	720
	A	6423	3905	2538	1753	1287	1022	907	911
1	B	5200	3730	2541	1762	1297	1029	907	902
	C				1620	1210	980	875	900
	A		5844	4054	2921	2192	1720	1431	1287
2	B		5292	3875	2917	2204	1736	1443	1289
	C						1650	1395	1350
	A		7303	5447	4104	3168	2516	2076	1797
3	B			5160	3940	3146	2529	2090	1810
	C							1960	1765

m = number of axial nodes.
$2n$ = number of circumferential nodes

A: Flügge's shell equation, Reference 28.
B: Finite element method.
C: Experimental values, Reference 29.

In this equation $[M]$ is the diagonal lumped mass matrix and $\{\ddot{u}_g(t)\}$ is the time record of earthquake ground accelerations. The ground records used by Clough and Chopra were the north-south and vertical components of the 1940 El Centro earthquake (Fig. 11-3).

The cross section of the idealized earth dam is shown in Figure 11-4. The dam material was assumed to be linearly elastic, homogeneous, and isotropic. The finite element mesh consisted of 100 CST elements with 66 nodes. The first eight natural frequencies and mode shapes are shown in Figure 11-5. In the mode superposition method, the damping ratios, c_i, were assumed to be 20% for each mode. The uncoupled equation for each normal mode was solved by the step-by-step method (Section 2-4).

Solutions for the transient response are shown in the computer-generated plots of Figures 11-6 and 11-7. In both of these displays, the dynamic stresses have been superimposed upon the static stresses obtained from a separate finite element analysis. In Figure 11-6 the time variation of computed stresses at two selected nodal points is indicated, whereas in Figure 11-7 the major principal stress is shown throughout the dam at three selected instants of time.

Because of the limiting assumptions, the example presented here is only a demonstration of the feasibility of applying the finite element method to the earthquake analysis of dams. In particular, material nonlinearity must be

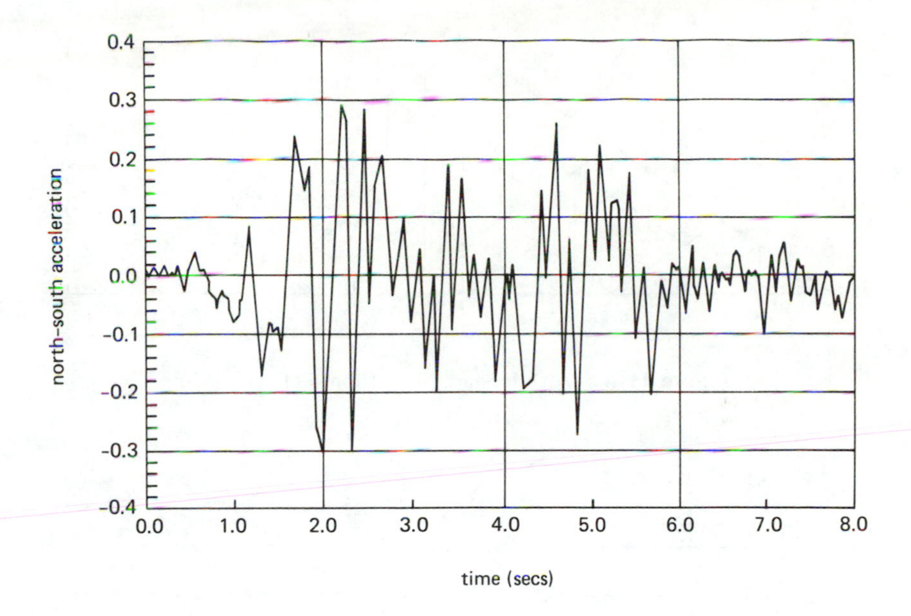

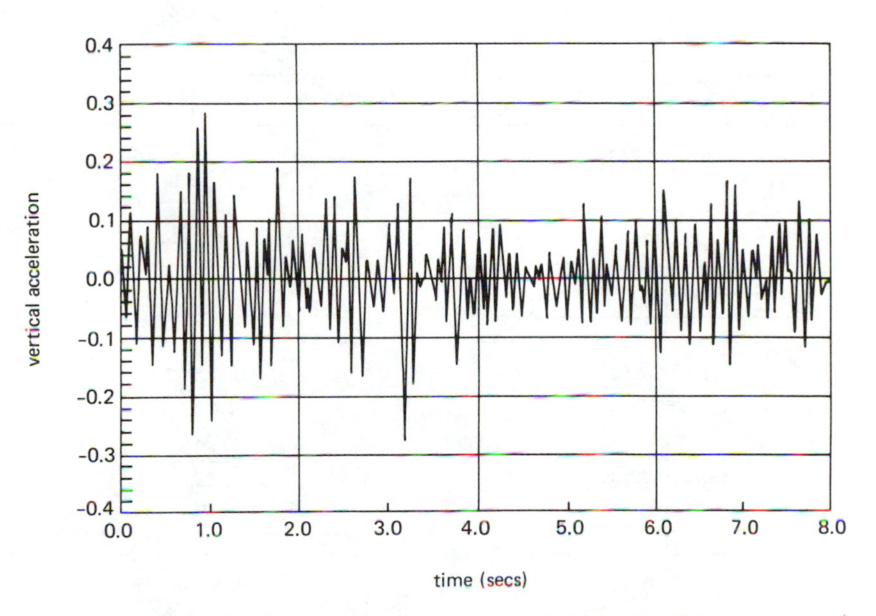

Figure 11-3 Acceleration records for 1940 El Centro earthquake (from Reference 18).

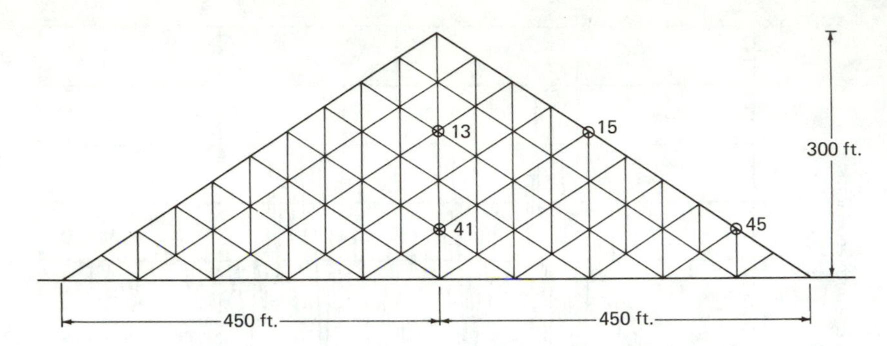

Figure 11-4 Mesh for earth dam (from Reference 18).

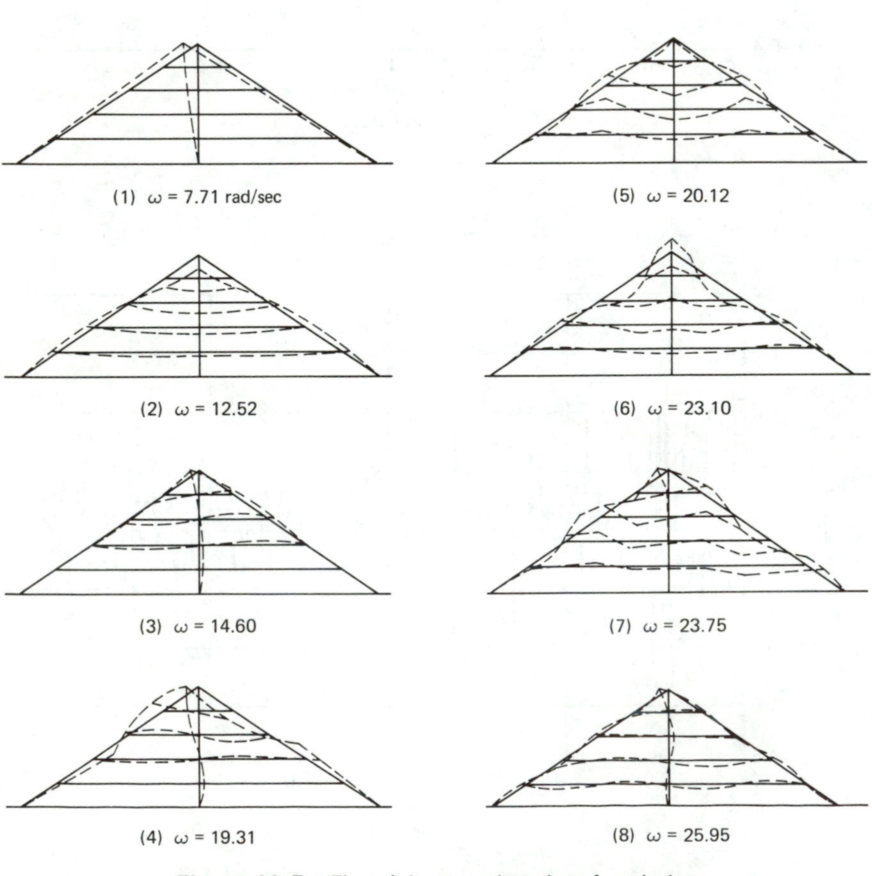

Figure 11-5 First eight natural modes of earth dam (from Reference 18).

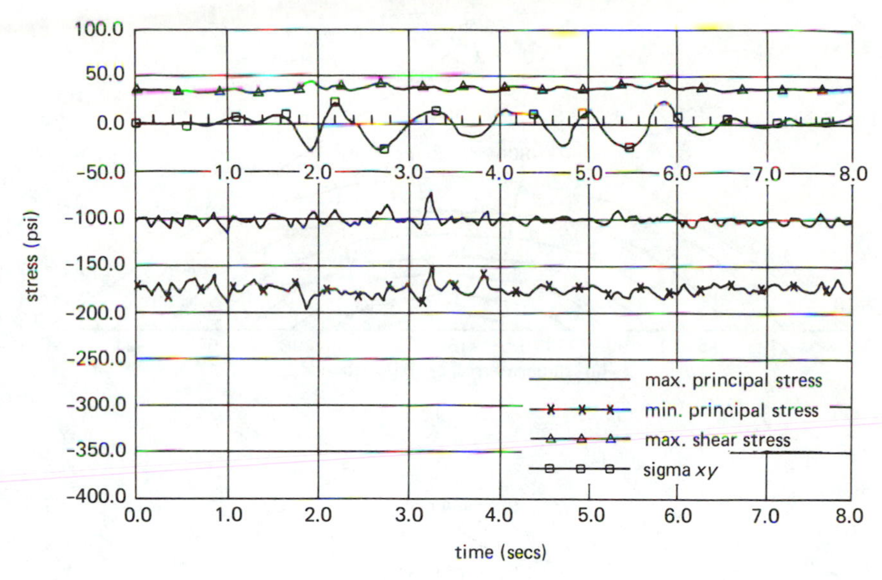

(a) Centerline: nodal point 41

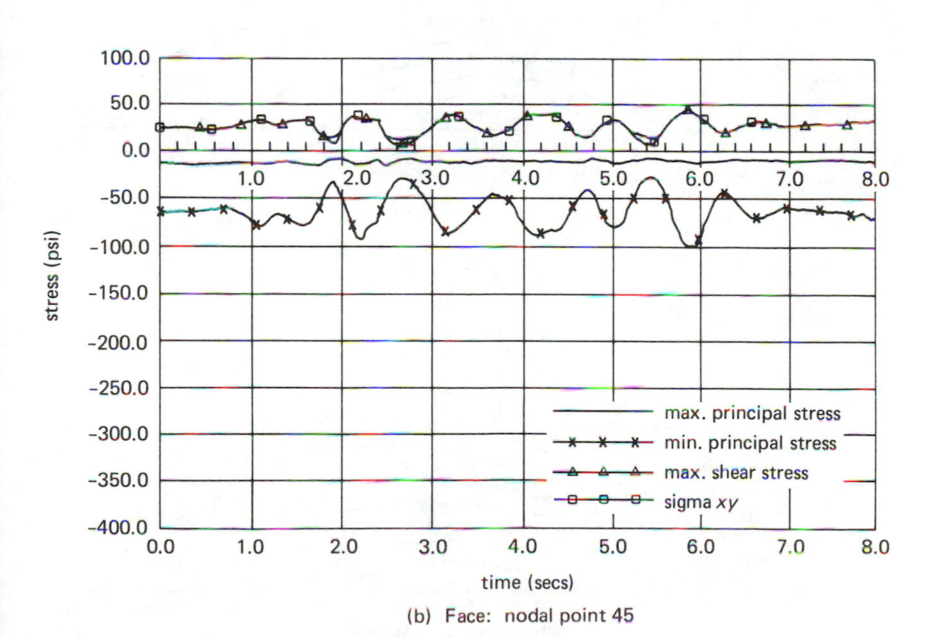

(b) Face: nodal point 45

Figure 11-6 Calculated stress records at 60 ft. level (from Reference 18).

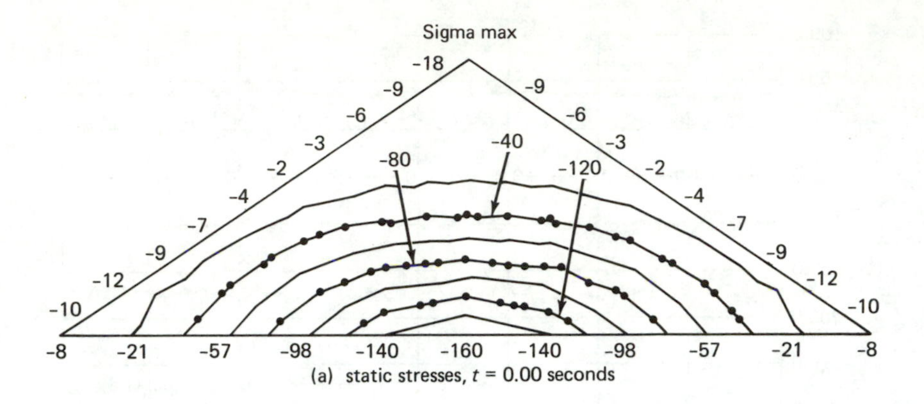

(a) static stresses, t = 0.00 seconds

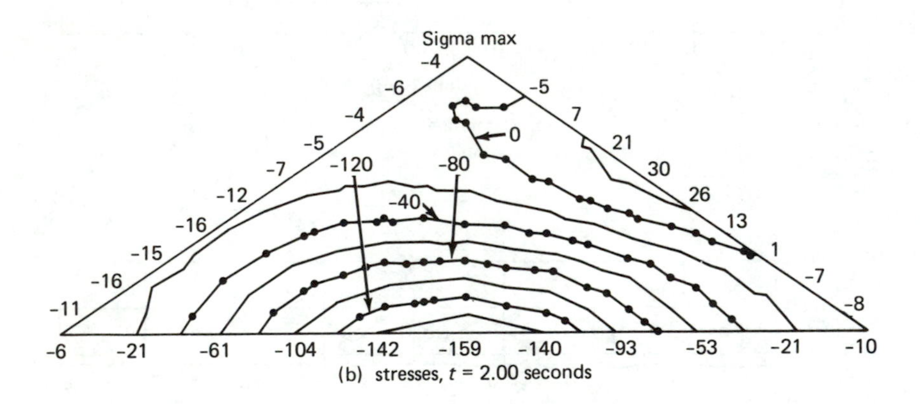

(b) stresses, t = 2.00 seconds

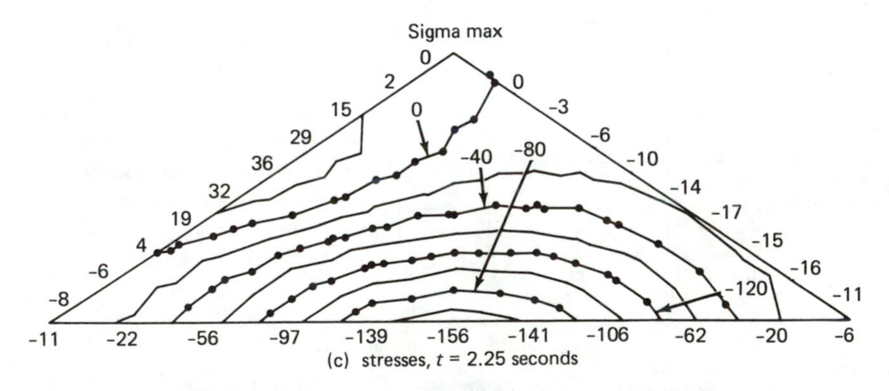

(c) stresses, t = 2.25 seconds

Figure 11-7 Major principal stress contours (from Reference 18).

included to obtain realistic results. Subsequent developments have extended the finite element method to the elastic-plastic analysis of nonhomogeneous dams with flexible bases.[10,23] A flexible base makes it possible to account for different earthquake ground motions at different locations on the base of the dam.

11-8 EXAMPLE: NONLINEAR DYNAMIC ANALYSIS OF UNDERGROUND STRUCTURES

The nonlinear analysis of underground structures is a typical complex structure-medium interaction problem. The nonlinear finite element methods described in Chapter 7 can be applied to such problems, provided realistic constitutive models are available that include such effects as hysteresis. These stress-strain laws should be derived from dynamic field or laboratory tests that simulate the prototype conditions. One possible approach is to utilize an incremental method of nonlinear analysis in conjunction with a stress-strain relation in terms of bulk and shear moduli, equation (3.12g). Figure 11-8 shows typical stress-strain curves that may be obtained from dynamic uniaxial or triaxial tests. As indicated in the figure, the tangent or instantaneous values of K and G can be obtained for an incremental procedure.

Confined uniaxial, dynamic laboratory tests have been conducted at the U. S. Army Engineer Waterways Experiment Station and reported by Jackson.[30] These results have been used by Farhoomand and Wilson[20] in

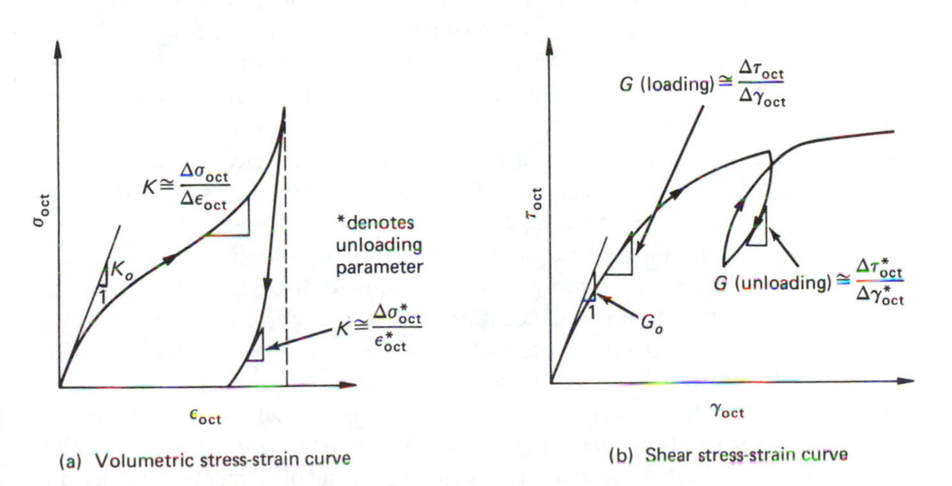

Figure 11-8 Dynamic stress-strain relations.

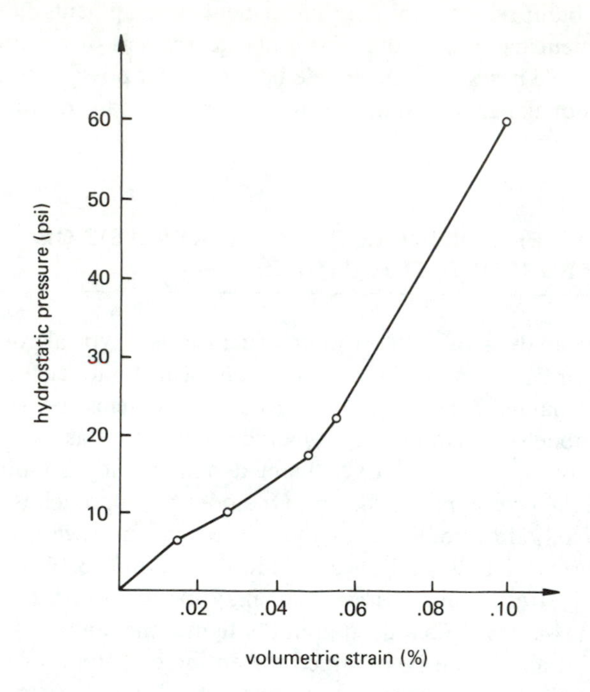

Figure 11-9 Volumetric stress-strain test for granular soil (from Reference 20).

nonlinear finite element analyses of blast-loaded underground structures. One problem studied was the response of an axisymmetric structure buried in granular soil. The volumetric stress-strain curve for this material is shown in Figure 11-9. In Figure 11-10 the finite element mesh and the time variation of the blast loading are shown. A linear isoparametric formulation was used for the axisymmetric elements with quadrilateral cross section. The soil was idealized by 47 such elements, while the structure was represented by 3 axisymmetric membrane elements. The total number of nodes was 63. The step-by-step integration scheme of Section 2-4 was utilized to compute the displacements. Figure 11-11 compares the computed displacements of a point in the soil medium with a linear solution and with experimental results. The experiments were conducted in a blast load generator at the U. S. Army Engineer Waterways Experiment Station. The nonlinear analysis agrees reasonably well with the test data, including the permanent set; on the other hand, the linear solution correlates poorly. The most severe drawback of the linear analysis is that it predicts a return to zero displacements, that is, no permanent set.

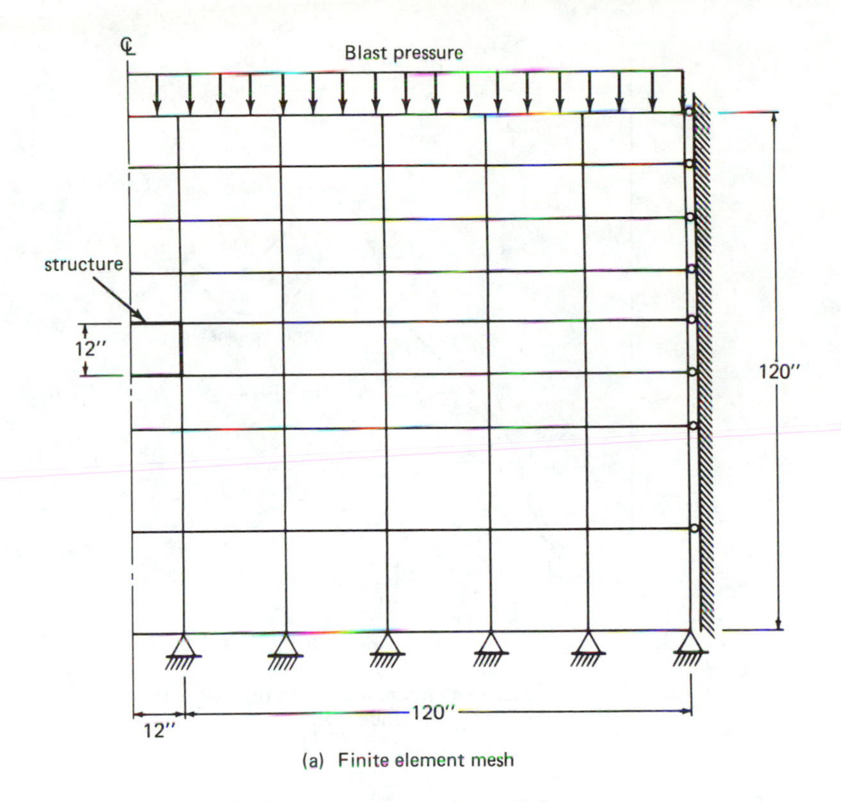

(a) Finite element mesh

pressure (psi)

40

0 10 20 30

time (msec)

(b) Time variation of blast pressure

Figure 11-10 Axisymmetric structure buried in granular soil and subjected to blast loading (from Reference 20).

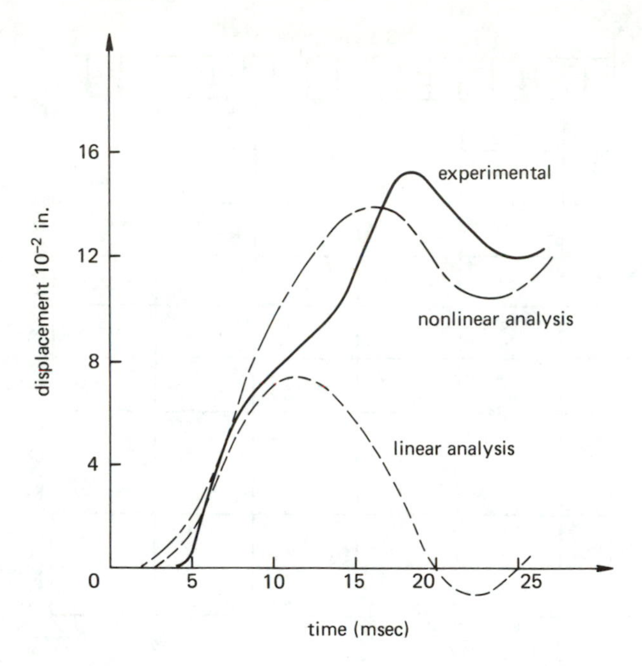

Figure 11-11 Displacements at a point in the soil medium (from Reference 20).

11-9 EXAMPLE: ANALYSIS OF AN EARTH DAM FAILURE

An application of the finite element method to the dynamic analysis of an earth dam failure was reported by Seed *et al.*[21] The Sheffield Dam, Santa Barbara, California, was constructed in 1917 and failed as a result of the 1925 Santa Barbara earthquake. Available field data indicated that the slide failure of the downstream face was triggered by liquefaction of a stratum of loose, saturated silty sand near the base of the dam. In order to assess the factors that affected the failure and to evaluate the adequacy of current analytical and laboratory procedures, Seed *et al.* applied a number of different analysis techniques to the problem, including the finite element method.

The finite element mesh used for the problem is shown in Figure 11-12, along with an estimated horizontal base acceleration. This acceleration was obtained by scaling the maximum ground acceleration, the duration, and the predominant acceleration frequency of the 1940 El Centro record so as to correspond with the available observations of these parameters for the 1925 Santa Barbara earthquake.

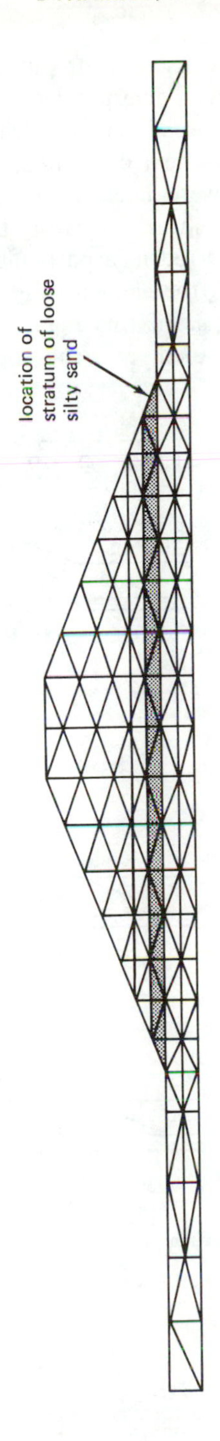

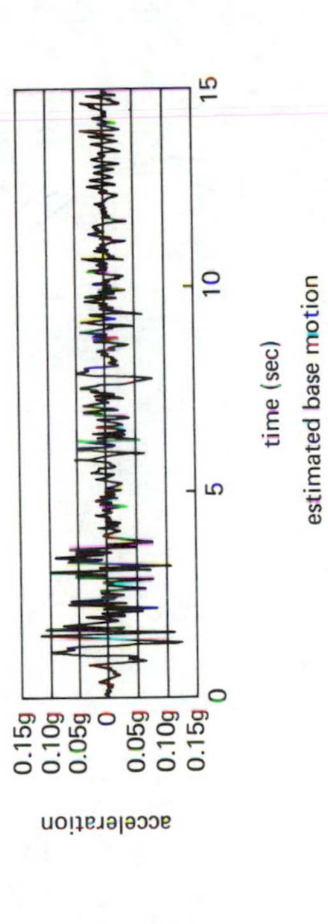

Figure 11-12 Finite element mesh and base accelerations for Sheffield Dam (from Reference 21).

First, a static finite element analysis was used to compute the initial stresses in the dam for a hydrostatic loading which corresponded to a water depth of 15 feet, the approximate depth at the time of the earthquake. The dynamic response to the ground acceleration was next computed, and the dynamic shear stresses along the base of the dam were determined. These stresses were compared to the results of cyclic, simple shear laboratory tests to determine whether failure or liquefaction had occurred in a particular element in the base layer. When liquefaction did occur, the element was assumed to carry zero shear for all subsequent time steps. In this manner the progressive liquefaction along the base was determined, and the shear stresses were

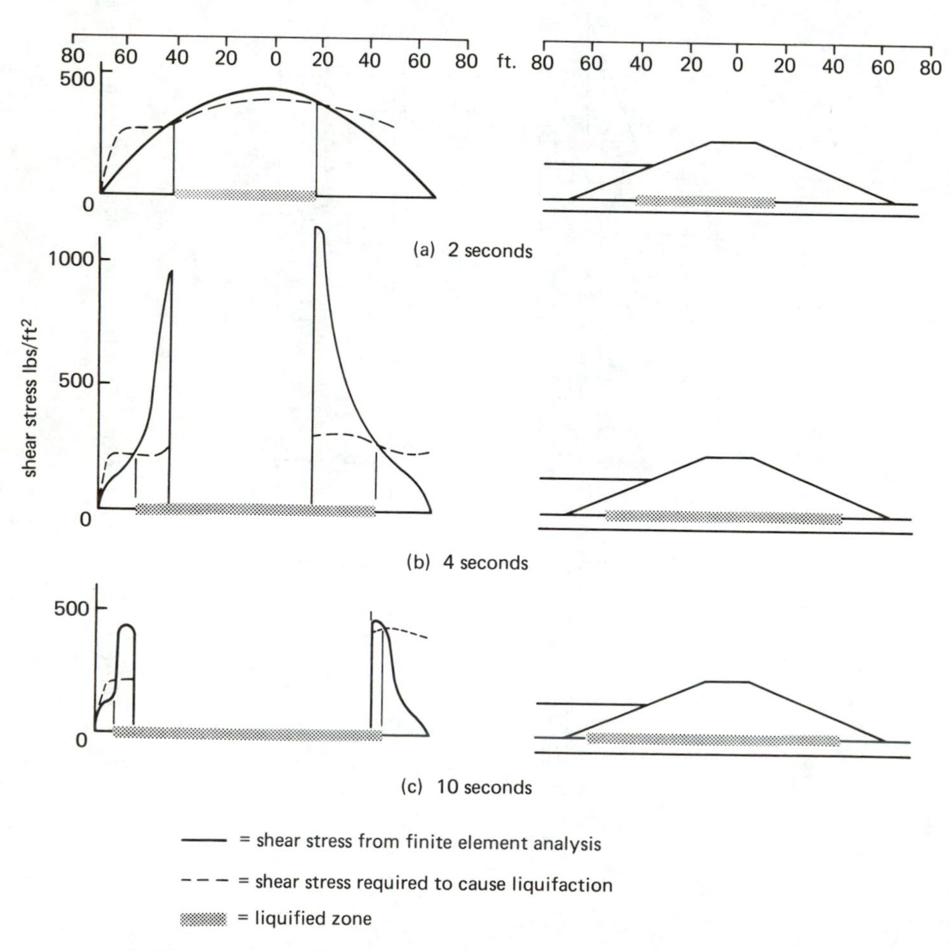

Figure 11-13 Progressive liquefaction at base of Sheffield Dam (from Reference 21).

redistributed to unfailed portions of the base. The shear stresses and zones of liquefaction for 2, 4, and 10 seconds of ground motion are shown in Figure 11-13.

The analysis indicated that the liquefaction started near the centerline and extended upstream faster than it did downstream. However, sliding in the upstream direction was prevented by the hydrostatic pressure. When liquefaction reached the upstream toe, the small remaining unfailed section at the downstream toe was insufficient to resist the horizontal water pressure, and a slide movement of the entire embankment occurred. After sufficient loss of water, the embankment would tend to slump laterally in both directions. All of these deductions were found to be in accord with the observations of the actual failure, thus confirming the efficacy of the finite element method for practical dynamic analyses of earth dams.

REFERENCES

(1) Clough, R. W., "Analysis of Structural Vibrations and Dynamic Response," *Recent Advances.* Summarizes analytical procedures used in finite element applications to dynamics.
(2) Timoshenko, S., and Young, D. H., *Vibration Problems in Engineering*, New York, Van Nostrand Reinhold Co., 1961.
(3) Biggs, J. M., *Introduction to Structural Dynamics*, New York, McGraw-Hill Book Co., 1964.
(4) Hurty, W. C., and Rubinstein, M. F., *Dynamics of Structures*, Englewood Cliffs, Prentice-Hall, Inc., 1964.
(5) See Reference 16, Chapter 8.
(6) Barkan, D. D., *Dynamics of Bases and Foundations*, New York, McGraw-Hill Book Co., 1962.
(7) Lin, Y. K., *Probabilistic Theory of Structural Dynamics*, New York, McGraw-Hill Book Co., 1967.
(8) *First Conf.*
(9) *Second Conf.*
(10) See Reference 17, Chapter 8.
(11) See Reference 3, Chapter 9.
(12) See Reference 13, Chapter 2.
(13) See Reference 45, Chapter 9.
(14) Ghosh, S., and Wilson, E. L., "Dynamic Stress Analysis of Axisymmetric Structures Under Arbitrary Loading," Earthquake Eng. Res. Ctr. Rept. EERC 69-10, University of California, Berkeley, Sept. 1969, NTIS, Springfield, Va. Axisymmetric shell and axisymmetric solid elements used in mode superposition, direct integration, and response spectrum techniques.
(15) Zienkiewicz, O. C., Irons, B. M., and Nath, B., "Natural Frequencies of Complex, Free or Submerged Structures by the Finite Element Method," *Proc. Symp. Vibration in Civil Eng.*, London, Butterworth's, 1966.

(16) Anderson, R. G., Irons, B. M., and Zienkiewicz, O. C., "Vibration and Stability of Plates Using Finite Elements," *Int. J. Solids Structures*, Vol. 4 (1968). Displacement method applied to the title problems. Presents an eigenvalue economizer which permits reduction of order of eigenvalue problem for fine subdivisions.
(17) Abel, J. F., "Static and Dynamic Analysis of Sandwich Shells with Viscoelastic Damping," Ph. D. Dissertation, University of California, Berkeley, 1969. Includes complex eigenvalue problem to compute effective structural damping at natural frequencies.
(18) Clough, R. W., and Chopra, A. K., "Earthquake Stress Analysis in Earth Dams," *Proc. ASCE, J. EM Dn.*, Vol. 92, EM2 (Apr. 1966).
(19) See Reference 14, Chapter 2.
(20) Farhoomand, I., and Wilson, E. L., "A Nonlinear Finite Element Code for Analyzing the Blast Response of Underground Structures," Contract Rept. N-70-1, USAEWES, Vicksburg, Miss., Jan. 1970.
(21) Seed, H. B., Lee, K. L., and Idriss, I. M., "Analysis of Sheffield Dam Failure," *Proc. ASCE, J. SM&F Dn.*, Vol. 95, SM6 (Nov. 1969).
(22) Costantino, C. J., "Finite Element Approach to Stress Wave Problems," *Proc. ASCE, J. EM Dn.*, Vol. 93, EM2 (Apr. 1967). Two-dimensional linear elastic problems.
(23) Dibaj, M., and Penzien, J., "Nonlinear Seismic Response of Earth Structures," Earthquake Eng. Res. Ctr. Rept. No. EERC 69-2, University of California, Berkeley, Jan. 1969, NTIS, Springfield, Va. Employs incremental theory of plasticity for earthquake analysis of earth dams. Considers hysteretic behavior.
(24) Koenig, H. A., and Davids, N., "Dynamical Finite Element Analysis for Elastic Waves in Beams and Plates," *Int. J. Solids Structures*, Vol. 4 (1968). Traveling elastic waves in finite beams and plates, including the effect of shear and rotary inertia.
(25) Koenig, H. A., and Davids, N., "The Damped Transient Behavior of Finite Beams and Plates," *IJNME*, Vol. 1, No. 2 (1969). Extension of reference 24 to damped behavior.
(26) Lysmer, J., "Lumped Mass Method for Rayleigh Waves," *Bull. Seis. Soc. Amer.*, Vol. 60, No. 1 (Feb. 1970). Analysis of generalized Rayleigh waves in multilayered elastic media.
(27) Archer, J. S., "Consistent Mass Matrix for Distributed Mass Systems," *Proc. ASCE, J. ST Dn*, Vol. 39, ST4 (Aug. 1963). The basic reference on consistent mass formulation.
(28) Smith, B. L., and Haft, E. E., "Natural Frequencies of Clamped Cylindrical Shells," *AIAA J.*, Vol. 6, No. 4 (Apr. 1968).
(29) Koval, L. R., and Cranch, E. T., "On the Free Vibrations of Thin Cylindrical Shells Subjected to Initial Torque," *Proc. U.S. Nat. Cong. App. Mechs.*, 1962.
(30) Jackson, J. G., "Factors that Influence the Development of Soil Constitutive Relations," Misc. Paper No. 4-980, USAEWES, Vicksburg, Miss., July 1968. Presents data from dynamic uniaxial and triaxial compression tests conducted in the laboratory. Discusses influence of factors such as rate of loading, hysteresis, saturation, and *in situ* stresses on the constitutive laws.

FURTHER READING

Dezfulian, H., and Seed, H. B., "Seismic Response of Soil Deposits Underlain by Sloping Rock Boundaries," *Proc. ASCE, J. SM&F Dn.*, Vol. 96, SM6 (Nov. 1970).

Nickell, R. E., "On Stability of Approximation Operators in Problems of Structural Dynamics," *Int. J. Solids Structures*, Vol. 7 (1971). Investigates numerical stability of three direct integration procedures: Newmark's acceleration operator, Wilson's averaging variant of the linear acceleration operator (Reference 14, Chapter 2) and the averaging scheme from the variational principle derived by Gurtin (Reference 27, Chapter 8).

Richart, F. E., Hall, J. R., Jr., and Wood, R. E., *Vibrations of Soils and Foundations*, Englewood Cliffs, Prentice-Hall, 1970.

Tahbildar, U. C., and Tottenham, H., "Earthquake Response of Arch Dams," *Proc. ASCE, J. ST Dn.*, Vol. 96, ST 11 (Nov. 1970). Shell elements are used to idealize the dam. Viscous damping and added mass are included. Step-by-step solution method.

Various authors, *Recent Advances.*

12

TORSION, HEAT CONDUCTION, SEEPAGE

Torsion, steady and transient temperature distributions, and steady and transient fluid flow in porous media are significant problems belonging to the class known as *field problems* (Section 8-3). The phenomenon of diffusion and convection is governed by similar equations, and therefore is also included in this chapter. Among other problems in this class are potential flow of fluids and surface waves on a fluid, which will be grouped with problems in hydraulics in Chapter 14. Table 12-1 lists various topics covered in this chapter with a number of references.

12-1 USE OF THE FINITE ELEMENT METHOD

Although the problems covered in this chapter usually involve variational principles different from those in solid mechanics (Chapter 4), there does exist a close analogy between the two finite element formulations.

Many applications of the finite element method employ linear field variable models, particularly the 4-CST quadrilateral. Nevertheless, higher order models are being utilized increasingly, and the development of isoparametric elements has added impetus to this trend.[19]

TABLE 12-1 TOPICS COVERED IN CHAPTER 12.

Topics	*Typical Applications*	*References*
I Torsion	Torsion of irregularly shaped, prismatic solids. Torsion of nonhomogeneous prismatic solids.	1 to 4
II Heat conduction	Steady-state and transient temperature distributions in solids.	5 to 12
III Seepage		
A Steady, confined	Seepage through a confined, permeable foundation under constant head.	13 to 16, 19, 22
B Steady, unconfined	Seepage through a permeable embankment under constant head.	14 to 16, 19
C Unsteady, confined	Flow toward wells in saturated confined aquifers.	17 to 19
D Unsteady, unconfined	Flow through embankments under rising or falling head.	20, 21
IV Diffusion-convection	Studies of nonreactive pollutant concentration.	8, 23

Governing Equations

The basic differential equation governing torsion, heat conduction, and seepage may be expressed in general form as

$$\frac{\partial}{\partial x}\left(k_x \frac{\partial \psi}{\partial x}\right) + \frac{\partial}{\partial y}\left(k_y \frac{\partial \psi}{\partial y}\right) + \frac{\partial}{\partial z}\left(k_z \frac{\partial \psi}{\partial z}\right) + \bar{Q} = c\,\frac{\partial \psi}{\partial t} \tag{12.1a}$$

For heat conduction problems ψ is the unknown temperature, k_x, k_y, and k_z are the thermal conductivities in the three directions, $\bar{Q}$ is the externally applied heat flux, and c is the specific heat.

The problems represented by equation (12.1a) involve time as one of the parameters and hence are referred to as *propagation*, *transient*, or *time dependent* problems. If the time dependent term on the right hand side vanishes, we are left with the Poisson equation, equation (8.6), which represents a *steady state* problem. In this case, $\bar{Q}$ corresponds to C in equation (8.6).

The boundary conditions associated with equation (12.1a) are[19]

$$\psi = \bar{\psi}(t) \text{ on } S_1 \tag{12.1b}$$

$$k_x \frac{\partial \psi}{\partial x} \ell_x + k_y \frac{\partial \psi}{\partial y} \ell_y + k_z \frac{\partial \psi}{\partial z} \ell_z + \alpha(\psi - \psi_0) + \bar{q}(t) = 0 \text{ on } S_2 \text{ and } S_3 \tag{12.1c}$$

S_1 is the part of the boundary on which ψ is prescribed; S_2 and S_3 are the parts on which $\bar{q}$, the intensity of heat input, and $\alpha(\psi - \psi_0)$ are prescribed, respectively. α is the heat transfer co-efficient, and $(\psi - \psi_0)$ denotes the difference between the temperature within the medium, ψ, and the temperature of the surroundings, ψ_0. ℓ_x, ℓ_y, and ℓ_z are the direction cosines of the outward normal to the boundary.

Associated Functional

A functional corresponding to equation (12.1) is expressed as[5,13]

$$A = \iiint\limits_V \frac{1}{2}\left[k_x\left(\frac{\partial \psi}{\partial x}\right)^2 + k_y\left(\frac{\partial \psi}{\partial y}\right)^2 + k_z\left(\frac{\partial \psi}{\partial z}\right)^2 - 2\left(\bar{Q} - c\frac{\partial \psi}{\partial t}\right)\psi\right] dV$$
$$- \iint\limits_{S_2} \bar{q}\psi \, dS_2 + \iint\limits_{S_3} \frac{1}{2}\alpha(\psi - \psi_0)^2 \, dS_3 \tag{12.2}$$

In order to use equation (12.2), the boundary condition of equation (12.1b) must be satisfied. Also, $\partial\psi/\partial t$ must be considered fixed in the calculus of variations formulation. An alternative functional proposed by Gurtin is often employed.[6,17,18] A residual approach based on Galerkin's principle has also been used.[19,21]

Whichever functional is used, we obtain essentially the same forms for the finite element equations. Since the functional in equation (12.2) is relatively easy to handle, we shall adopt it as our basis.

Analogy Between Solid Mechanics and Field Problem Formulations

We write equation (12.2) in matrix notation as

$$A = \iiint\limits_V \left(\tfrac{1}{2}\{g\}^{\mathrm{T}}[R]\{g\} - \{\psi\}^{\mathrm{T}}\{\bar{Q}\} + c\{\psi\}^{\mathrm{T}}\{\dot{\psi}\}\right) dV$$
$$- \iint\limits_{S_2} \{\psi\}^{\mathrm{T}}\{\bar{q}\}\, dS_2 + \iint\limits_{S_3} \tfrac{1}{2}\alpha(\{\psi\} - \{\psi_0\})^{\mathrm{T}}(\{\psi\} - \{\psi_0\})\, dS_3 \tag{12.3}$$

where $\{g\}$ is the vector of gradients of the field variable,

$$\{g\}^{\mathrm{T}} = \left[\frac{\partial \psi}{\partial x}\ \frac{\partial \psi}{\partial y}\ \frac{\partial \psi}{\partial z}\right],$$

$\{\dot{\psi}\}$ denotes the invariant time rate of change, and $[R]$ is the material property matrix.

Equation (12.3) may be considered analogous to equation (4.13) and equation (11.1).[5] We describe this analogy by listing various terms of the two formulations in Table 12-2. The terms containing α can be grouped with the

TABLE 12-2 ANALOGY BETWEEN SOLID MECHANICS AND FIELD PROBLEM FORMULATIONS[5]

Equations (4.13) *and* (11.1)	*Equation* (12.3)
$\{\varepsilon\}^{\mathrm{T}}[C]\{\varepsilon\}$	$\{g\}^{\mathrm{T}}[R]\{g\}$
$\{u\}^{\mathrm{T}}\{\bar{X}\}$	$\{\psi\}^{\mathrm{T}}\{\bar{Q}\}$
$\{u\}^{\mathrm{T}}\{\bar{T}\}$	$\{\psi\}^{\mathrm{T}}\{\bar{q}\}$
$\rho\{\dot{u}\}^{\mathrm{T}}\{\dot{u}\}$	$c\{\psi\}^{\mathrm{T}}\{\dot{\psi}\}$

first term, $\{g\}^{\mathrm{T}}[R]\{g\}$, in equation (12.3) and may be interpreted as an addition to be made to the material property matrix to account for external effects.

Finite Element Equations

By finding the stationary condition of the functional in equation (12.3), we obtain the following finite element equations

$$[p]\{\dot{\psi}_n\} + [k]\{\psi_n\} = \{f\} \tag{12.4a}$$

where

$$[p] = \iiint_V c[N]^{\mathrm{T}}[N]\,dV \tag{12-4b}$$

$$[k] = \iiint_V [B]^{\mathrm{T}}[R][B]\,dV + \iint_{S_3} \alpha[N]^{\mathrm{T}}[N]\,dS_3 \tag{12.4c}$$

$$\{f\} = \iiint_V [N]^{\mathrm{T}}\{\bar{Q}\}\,dV + \iint_{S_2} [N]^{\mathrm{T}}\{\bar{q}\}\,dS_2 + \iint_{S_3} \alpha[N]^{\mathrm{T}}\{\psi_0\}\,dS_3 \tag{12.4d}$$

and where $[N]$ is the field variable interpolation model, and $[B]$ is the corresponding field variable gradient model. The subscript n denotes nodal values.

The equations for the various elements, equation (12.4a), are assembled by the direct stiffness method (Chapter 6). The resulting assemblage equations are

$$[P]\{\dot{\Psi}_n\} + [K]\{\Psi_n\} = \{F\} \tag{12.5}$$

Introduction of the Boundary Conditions

Equation (12.1c) represents the natural boundary conditions implicit in the variational formulation (Chapter 4). Equation (12.1b) gives the geometric boundary conditions introduced into the assemblage equations in the same manner as specified displacements (Section 6-6).

12-2 EXAMPLE OF A FORMULATION: SEEPAGE

We shall consider the problem of seepage to illustrate a finite element formulation. Moreover, we shall restrict our attention to the two-dimensional seepage through a slice of unit thickness. Extension to three-dimensional problems is not difficult.

Inertia effects are neglected in transient seepage problems because the accelerations of the fluid are very small. Hence these problems may be classified as *quasi-static*.

Fluid potential or total head is generally adopted as the unknown field variable in seepage problems. The potential may be defined as the sum of the pressure head, p/γ, and the elevation head, H. Here γ is the fluid density and p is the pressure. A model for the potential may be written as

$$\psi = \{N\}^{\mathrm{T}}\{\psi_n\} \tag{12.6a}$$

where ψ is the potential at any point in the element, $\{\psi_n\}$ is the vector of nodal potentials, and $\{N\}^{\mathrm{T}}$ is the matrix of interpolation functions, equation (5.23). If we adopt a linear model for a triangular element, there will be three nodal potentials, and hence the vector $\{\psi_n\}$ will be 3×1.

Differentiation of ψ in equation (12.6a) with respect to x and y coordinates gives the vector of gradients, $\{g\}$:

$$\underset{2\times 1}{\{g\}} = \begin{Bmatrix} \dfrac{\partial \psi}{\partial x} \\ \dfrac{\partial \psi}{\partial y} \end{Bmatrix} = \frac{1}{2A}\begin{bmatrix} b_1 & b_2 & b_3 \\ a_1 & a_2 & a_3 \end{bmatrix}\begin{Bmatrix} \psi_1 \\ \psi_2 \\ \psi_3 \end{Bmatrix} = [B]\{\psi_n\} \tag{12.6b}$$

a_i and b_i are defined in equation (5.18c), and 1, 2, and 3 denote the nodes of the triangular element (Fig. 5-5). The matrix of permeabilities in the principal directions, may be expressed as

$$[R] = \begin{bmatrix} k_x & 0 & 0 \\ 0 & k_y & 0 \\ 0 & 0 & k_z \end{bmatrix} \tag{12.6c}$$

$\bar{Q}$ is now the applied fluid flux per unit volume, c is the porosity and $\bar{q}$ is the applied flow intensity on the surface of the element. The potential difference at the surface, $\psi - \psi_0$, is assumed to vanish.

When we substitute equations (12.6) into equations (12.3) and (12.4), we obtain the following element property matrices and element action vector:

$$[p] = c \iiint_V \{N\}^{\mathrm{T}}\{N\}\, dV \tag{12.7a}$$

$$[k] = \iiint_V [B]^{\mathrm{T}}[R][B]\, dV \tag{12.7b}$$

$$\{f\} = \iiint_V \{N\}^{\mathrm{T}}\{\bar{Q}\}\, dV + \iint_{S_2} \{N\}^{\mathrm{T}}\{\bar{q}\}\, dS_2 \tag{12.7c}$$

These are called the element porosity matrix, the element permeability matrix, and the element nodal vector of applied flow, respectively.

The direct stiffness method is used to obtain the equations for the assemblage, equation (12.5). The applied potentials are introduced in these assemblage relations, and the time-dependent equations are solved by the techniques presented in Section 2-4.

Element Resultants

Solutions of equation (12.5) give the values of nodal fluid heads as primary quantities. Once these heads are known, fluid velocities and quantities of flow may be evaluated as element resultants. For instance, in seepage, Darcy's law is often used to compute the velocities, v, and flows, Q:

$$v = -ki \quad \text{and} \quad Q = vA \tag{12.8a}$$

where i is the hydraulic gradient. In matrix notation, the velocity components are

$$\{v\} = \begin{Bmatrix} v_x \\ v_y \end{Bmatrix} = -[R]\begin{Bmatrix} \dfrac{\partial \psi}{\partial x} \\ \dfrac{\partial \psi}{\partial y} \end{Bmatrix} = -[R][B]\{\psi_n\} \tag{12.8b}$$

For the linear model of the potential, the velocity components v_x and v_y in an element are constant. The quantity of flow can be related to the seepage velocities by computing equivalent flows at the nodes. For this purpose, the net flows in the x and y directions at the nodes are evaluated as the uniform flow tributary to the node

$$\{Q\} = \begin{Bmatrix} Q_{x1} \\ Q_{x2} \\ Q_{x3} \\ Q_{y1} \\ Q_{y2} \\ Q_{y3} \end{Bmatrix} = -\frac{1}{2} \begin{bmatrix} b_1 & 0 \\ b_2 & 0 \\ b_3 & 0 \\ 0 & a_1 \\ 0 & a_2 \\ 0 & a_3 \end{bmatrix} \{v\} \tag{12.8c}$$

Categories of Seepage Problems

The case of steady state seepage in which the fluid potentials at all pervious external boundaries are known and constant is called *steady confined* seepage. For this class of problems, $\{\dot{\psi}\} = 0$. An example is the flow in the foundation of an impervious dam.

Problems involving transient or unsteady seepage under confined conditions are referred to as *transient confined.* Flow of fluid to a well in a saturated aquifer during pumping is a typical example of transient confined seepage.

The class of seepage problems that involves a free, or phreatic, surface is called *unconfined* seepage. Here the free boundary is not known in advance, and its location is governed by a nonlinear law. Such a free surface can occur in both steady and transient categories, and it requires that the following additional boundary condition be fulfilled:

$$\psi = H(x, y, t) \text{ at the free surface} \tag{12.1d}$$

Here ψ may be defined as the total head, and H represents the elevation head.

In the case of *steady unconfined* seepage, the free surface is unique and does not involve variations with time. An example is seepage through a pervious earth dam on an impervious foundation subjected to a reservoir head on the upstream side and a tailwater head on the downstream side.

In the case of *transient* or *unsteady unconfined* seepage, the free surface changes continuously with time as the externally applied fluid head varies. This phenomenon is relatively complicated, and its solution involves modifications of the flow region with time and satisfaction of the boundary conditions, equations (12.1b, c, and d). Flows of water through a riverbank under varying floods and through a dam under drawdown conditions are some examples of this category. Later in this chapter we consider some applications to the various categories described above.

12-3 NUMERICAL EXAMPLE: STEADY CONFINED SEEPAGE

We shall consider the problem of steady-state confined seepage through a two-dimensional soil mass which is subjected to externally applied fluid pressures. The equation (12.1a) for this problem reduces to

$$k_x \frac{\partial^2 \psi}{\partial x^2} + k_y \frac{\partial^2 \psi}{\partial y^2} + \bar{Q} = 0$$

with head $\bar{\psi}$ applied on a part of the boundary. The corresponding variational functional, equation (12.2), is

$$A = \iiint_V \frac{1}{2} \left[k_x \left(\frac{\partial \psi}{\partial x} \right)^2 + k_y \left(\frac{\partial \psi}{\partial y} \right)^2 - \bar{Q}\psi \right] dV$$

For a simple illustration, consider the soil mass in Figure 12-1, which is enclosed between two impervious surfaces. This problem can represent a permeameter used to determine soil permeability in the laboratory. The soil

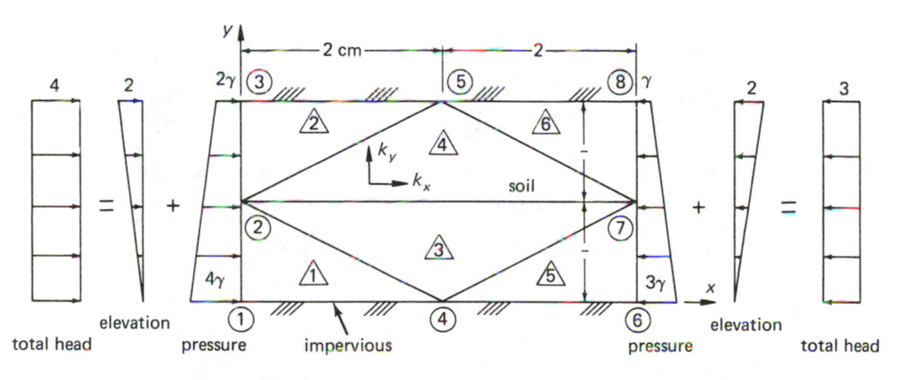

Total head on the left $= p/\gamma + z = 4$ cm
Total head on the right $= p/\gamma + z = 3$ cm
Applied nodal heads:

$\psi_1 = \psi_2 = \psi_3 = 4$ cm
$\psi_6 = \psi_7 = \psi_8 = 3$ cm
ψ_4 and ψ_6 to be determined

γ = density of fluid = 1 gm/cm^3
$k_x = k_y$ = 1 cm/sec = horizontal and vertical permeabilities

Figure 12-1 Example of steady, confined seepage through soil.[22]

mass is subjected to the fluid pressure distribution shown in the figure. As seen from the definition

$$\psi = p/\gamma + H$$

the total head distribution at each end of the soil sample is uniform (Fig. 12-1). No flow occurs across the impervious top and bottom boundaries, and the sample is assumed to have a unit thickness with no flow normal to the x–y plane.

By using the linear potential model, equations (12.6a) and equation (12.6b), we obtain the element equations for a triangular element as

$$\frac{1}{4A}\begin{bmatrix} k_x b_1^2 + k_y a_1^2 & k_x b_1 b_2 + k_y a_1 a_2 & k_x b_1 b_3 + k_y a_1 a_3 \\ & k_x b_2^2 + k_y a_2^2 & k_x b_2 b_3 + k_y a_2 a_3 \\ \text{Symmetrical} & & k_x b_3^2 + k_y a_3^2 \end{bmatrix} \begin{Bmatrix} \psi_1 \\ \psi_2 \\ \psi_3 \end{Bmatrix} = \begin{Bmatrix} f_1 \\ f_2 \\ f_3 \end{Bmatrix}$$

where A is the area of the element of unit thickness. By using the values of k_x and k_y and the dimensions of the elements as shown in figure 12-1, we can compute element characteristic (permeability) matrices as

$$[k_1] = \tfrac{1}{4}\begin{bmatrix} 5 & -1 & -4 \\ -1 & 1 & 0 \\ -4 & 0 & 4 \end{bmatrix},\ [k_2] = \tfrac{1}{4}\begin{bmatrix} 4 & 0 & -4 \\ 0 & 1 & -1 \\ -4 & -1 & 5 \end{bmatrix}$$

$$[k_3] = \tfrac{1}{4}\begin{bmatrix} \frac{5}{2} & -4 & \frac{3}{2} \\ -4 & 8 & -4 \\ \frac{3}{2} & -4 & \frac{5}{2} \end{bmatrix},\ [k_4] = \tfrac{1}{4}\begin{bmatrix} \frac{5}{2} & \frac{3}{2} & -4 \\ \frac{3}{2} & \frac{5}{2} & -4 \\ -4 & -4 & 8 \end{bmatrix}$$

$$[k_5] = \tfrac{1}{4}\begin{bmatrix} 1 & -1 & 0 \\ -1 & 5 & -4 \\ 0 & -4 & 4 \end{bmatrix},\ [k_6] = \tfrac{1}{4}\begin{bmatrix} 1 & 0 & -1 \\ 0 & 4 & -4 \\ -1 & -4 & 5 \end{bmatrix}$$

We now use the direct stiffness method to assemble the element stiffnesses and obtain the assemblage equations as

$$\begin{bmatrix} 5 & -4 & 0 & -1 & 0 & 0 & 0 & 0 \\ & 13 & -4 & -4 & -4 & 0 & 3 & 0 \\ & & 5 & 0 & -1 & 0 & 0 & 0 \\ & & & 10 & 0 & -1 & -4 & 0 \\ \text{Symmetrical} & & & & 10 & 0 & -4 & -1 \\ & & & & & 5 & -4 & 0 \\ & & & & & & 13 & -4 \\ & & & & & & & 5 \end{bmatrix} \begin{Bmatrix} 4 \\ 4 \\ 4 \\ \psi_4 \\ \psi_5 \\ 3 \\ 3 \\ 3 \end{Bmatrix} = \begin{Bmatrix} F_1 \\ F_2 \\ F_3 \\ F_4 = 0 \\ F_5 = 0 \\ F_6 \\ F_7 \\ F_8 \end{Bmatrix}$$

In this simple example, all nodal potentials are known except those at nodes 4 and 5. Since we know that the flows added at nodes 4 and 5, F_4 and F_5, are zero, we can solve for ψ_4 and ψ_5 directly.

$$-4 - 16 + 10\psi_4 - 3 - 12 = 0, \qquad \psi_4 = 3.5$$

and

$$-16 - 4 + 10\psi_5 - 12 - 3 = 0, \qquad \psi_5 = 3.5$$

Hence, pressures at points 4 and 5 are

$$p_4 = (3.5 - 0)\gamma = 3.5\gamma$$
$$p_5 = (3.5 - 2)\gamma = 1.5\gamma$$

which indicates that the pressure variation in the x direction is linear.

Using Darcy's law, equation (12.8), we can compute the element velocities from

$$\begin{Bmatrix} v_x \\ v_y \end{Bmatrix} = -\frac{1}{2A}\begin{bmatrix} k_x & 0 \\ 0 & k_y \end{bmatrix}\begin{bmatrix} b_1 & b_2 & b_3 \\ a_1 & a_2 & a_3 \end{bmatrix}\begin{Bmatrix} \psi_1 \\ \psi_2 \\ \psi_3 \end{Bmatrix}$$

For example, in element 3: $\psi_1 = 4$, $\psi_2 = 3.5$, $\psi_3 = 3$, $b_1 = -1$, $b_2 = 0$, $b_3 = 1$, $a_1 = 2$, $a_2 = -4$, $a_3 = 2$, and the area $A = 2\text{cm}^2$. Hence,

$$\begin{Bmatrix} v_x \\ v_y \end{Bmatrix} = -\tfrac{1}{4}\begin{bmatrix} 1 & 0 \\ 0 & 1 \end{bmatrix}\begin{bmatrix} -1 & 0 & 1 \\ 2 & -4 & 2 \end{bmatrix}\begin{Bmatrix} 4 \\ 3.5 \\ 3 \end{Bmatrix}$$

$$\begin{Bmatrix} v_x \\ v_y \end{Bmatrix} = \begin{Bmatrix} 0.25 \\ 0 \end{Bmatrix} \text{ cm/sec}$$

and the equivalent fluid flows at the nodes of element 3 are, equation (12.8c):

$$\begin{Bmatrix} Q_{x1} \\ Q_{x2} \\ Q_{x3} \\ Q_{y1} \\ Q_{y2} \\ Q_{y3} \end{Bmatrix} = -\tfrac{1}{2}\begin{bmatrix} -1 & 0 \\ 0 & 0 \\ 1 & 0 \\ 0 & 2 \\ 0 & -4 \\ 0 & 2 \end{bmatrix}\begin{Bmatrix} 0.25 \\ 0 \end{Bmatrix} = \begin{Bmatrix} \frac{1}{8} \\ 0 \\ -\frac{1}{8} \\ 0 \\ 0 \\ 0 \end{Bmatrix}$$

12-4 TORSION OF PRISMATIC SHAFTS

St. Venant Torsion

In the St. Venant formulation of the problem of torsion of prismatic shafts, Figure 12-2(a), all cross sections are assumed to warp in the same manner.[24] The displacements are given by

$$u = -\theta z y, \qquad v = \theta z x, \qquad w = \theta\psi(x, y) \qquad (12.9a)$$

where θ is the angle of twist per unit length of shaft, and $\psi(x, y)$ is the *warping function*. The governing differential equation for St. Venant torsion is the Laplace equation

$$\frac{\partial^2 \psi}{\partial x^2} + \frac{\partial^2 \psi}{\partial y^2} = 0 \tag{12.9b}$$

which is a special case of equation (12.1). The boundaries of the cross-section of the shaft are stress-free; hence, the boundary conditions on the entire lateral surface are[2]

$$\left(\frac{\partial \psi}{\partial x} - y\right) \ell_x + \left(\frac{\partial \psi}{\partial y} + x\right) \ell_y = 0 \tag{12.9c}$$

where ℓ_x and ℓ_y are the direction cosines of the outward normal. The relations giving the non-zero stresses, τ_{xz} and τ_{yz}, and the applied moment, M_z, are

$$\tau_{xz} = G\theta\left(\frac{\partial \psi}{\partial x} - y\right), \qquad \tau_{yz} = G\theta\left(\frac{\partial \psi}{\partial y} + x\right) \tag{12.9d}$$

$$M_z = G\theta \iint_A \left(x \frac{\partial \psi}{\partial y} - y \frac{\partial \psi}{\partial x} + x^2 + y^2\right) dx\, dy \tag{12.9e}$$

where G is the shear modulus and A is the cross-sectional area of the shaft.

In the finite element analysis, we can adopt the warping function, ψ, as our field variable and assume a warping function model as

$$\psi = \{N\}^{\mathrm{T}}\{\psi_n\} \tag{12.9f}$$

The associated functional for St. Venant torsion is[2]

$$\Pi = L \iint_A \frac{G}{2} \theta^2 \left[\left(\frac{\partial \psi}{\partial x} - y\right)^2 + \left(\frac{\partial \psi}{\partial y} + x\right)^2\right] dx\, dy \tag{12.9g}$$

which is a special case of equation (12.2). L is the length of the shaft. Further details of the finite element formulation for torsion are analogous to the seepage problem (Sections 12-2 and 12-3). Equations (12.9c) are the natural boundary conditions corresponding to the stationary value of Π.

Prandtl Torsion

An alternative approach to the torsion problem is Prandtl's theory.[24] In this approach the components of stress are derived from a stress function, ϕ.

The governing equation in terms of stress function is the Poisson equation.

$$\frac{\partial}{\partial x}\left(\frac{1}{G}\frac{\partial \phi}{\partial x}\right) + \frac{\partial}{\partial y}\left(\frac{1}{G}\frac{\partial \phi}{\partial y}\right) = -2\theta \tag{12.10}$$

In the finite element formulation of Prandtl torsion, we adopt the stress function as the field variable.

Hybrid Formulation

The warping function and the stress function approaches described above represent the displacement and the equilibrium methods, respectively (Chapter 8). It is also possible to adopt a hybrid approach (Chapter 9), in which the stress function ϕ is assumed within the element, and the warping function ψ is expressed so that the displacements are compatible along the interelement boundaries.[4]

Example: *Torsion of an Elliptic Shaft*. We shall now consider application of the finite element method to torsion of an elliptic shaft. Yamada *et al.*[3] used both the stress function and the warping function as the field variables and compared the finite element results with exact solutions. In their elastic analysis, linear models were adopted for ϕ and ψ, and the elliptic bar was discretized as shown in Figure 12-2(b).

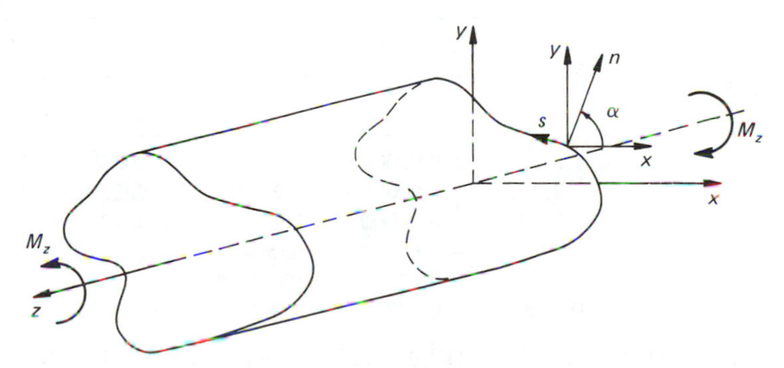

(a) Torsion of a non-circular shaft

Figure 12-2 Torsion of prismatic bars.

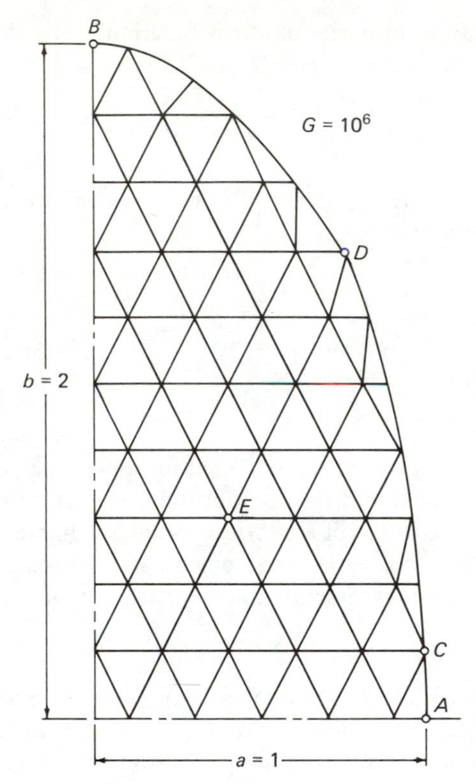

(b) Discretization of elliptic bar (from Reference 3)

Figure 12-2 (*continued*)

TABLE 12-3 COMPARISONS OF SOLUTIONS FOR TORSION OF ELLIPTIC SHAFT (from Reference 3).

		Points in Figure 12-2(b)				
		A	*B*	*C*	*D*	*E*
τ_{yz}	I	1.27324	0.00000	1.26686	0.90928	0.50930
	II	1.23100	0.00000	1.22450	0.90560	0.50860
	III	1.27135	0.00000	1.24450	0.90474	0.50926
τ_{xz}	I	0.00000	−0.63662	−0.06366	−0.44563	−0.19099
	II	0.00000	−0.62570	−0.06173	−0.44845	−0.19340
	III	0.00000	−0.64781	−0.06471	−0.44861	−0.19226

θ (rad.) (I) 0.795775×10^{-6}, (II) 0.80500×10^{-6}, (III) 0.796953×10^{-6}

I Exact solution
II Stress function (force method)
III Warping function (displacement method)

Table 12-3 compares the stresses and the angles of twist computed from the finite element analyses and the exact solution. The three results compare satisfactorily.

12-5 EXAMPLE: TRANSIENT HEAT CONDUCTION

Brisbane *et al.*[8] solved the problem of transient heat flow in a cylinder with a star-shaped perforation, Figure 12-3(a). The cylinder was subjected to severe convective cooling conditions. In the finite element analysis the variational functional given by equation (12.2) was employed. The four-CST quadrilateral was adopted and a linear model was used for the unknown temperatures. A finite difference scheme (Section 2-4) was introduced for solutions in the time domain.

Figure 12-3(a) shows the finite element mesh. The computed distributions of temperatures with respect to radius of the cylinder for various time levels are compared with analytical solutions in Figure 12-3(b). The analytical solutions were obtained by using a combination of the finite difference method and conformal mapping.[25]

12-6 EXAMPLE: STEADY CONFINED SEEPAGE UNDER A SHEET-PILE WALL

Figure 12-4(a) shows an impervious, sheet-pile wall driven into a porous, anisotropic soil foundation. The soil has different permeabilities in different directions, as shown in the figure. The heads at the upstream and downstream sides are known, and the flow occurs only through the foundation soil. This problem typifies steady confined seepage.

A finite element solution for the above sheet-pile foundation problem was obtained by Banks,[22] who used a computer program developed by Taylor and Brown.[14] Solutions were obtained in terms of fluid pressures (potentials), element velocities, and fluid flows. Figure 12-4(b) shows the finite element grid for the foundation.

Figure 12-5(a) shows plots of equipotential lines and the directions of fluid flow as computed from the analysis. For design, it is necessary to compute the seepage forces induced in the foundation. Such computations are performed on the basis of a flow net analysis. The flow net shown in Figure 12-5(b) was derived from the equipotential lines, Figure 12-5(a). As explained in Section 12-2, one can also compute the flow rates.

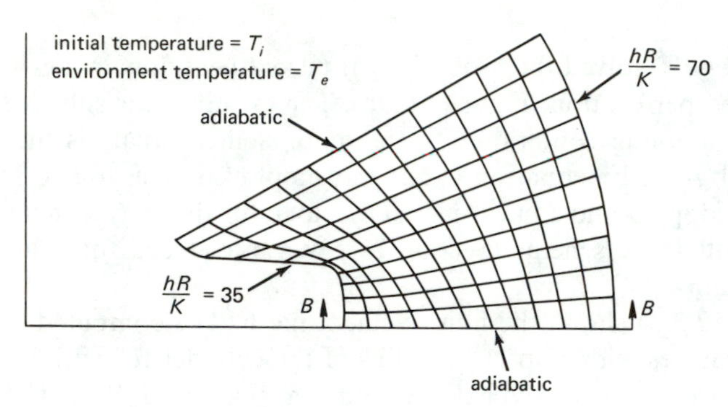

(a) Finite element mesh

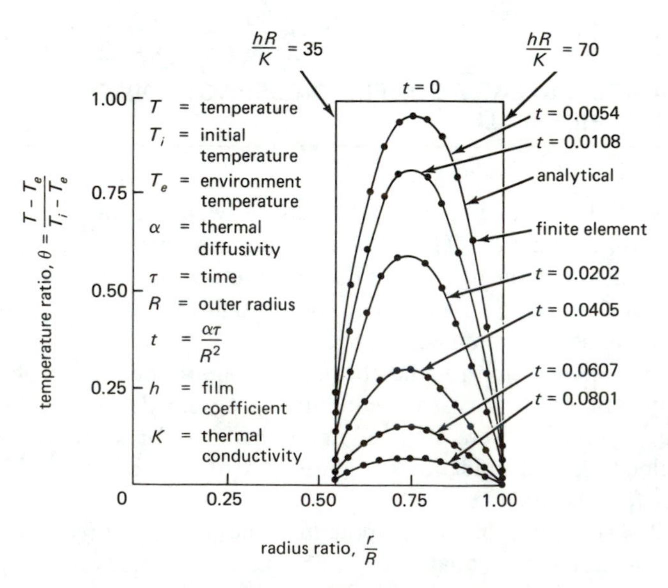

(b) Thermal history at section B-B

Figure 12-3 Thermal analysis of star-shaped cylinder (from Reference 8).

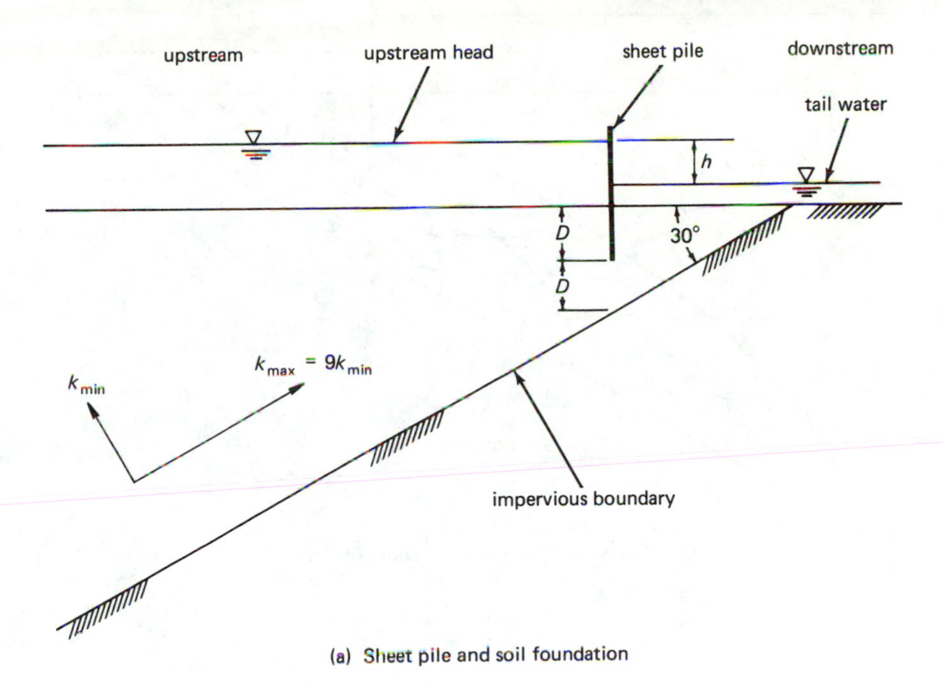

(a) Sheet pile and soil foundation

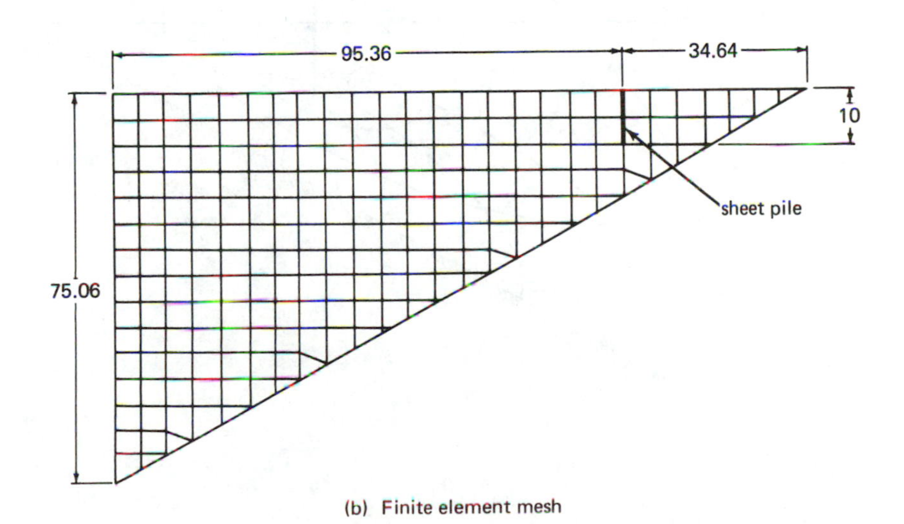

(b) Finite element mesh

Figure 12-4 Steady, confined seepage through sheet-pile foundation (from Reference 22).

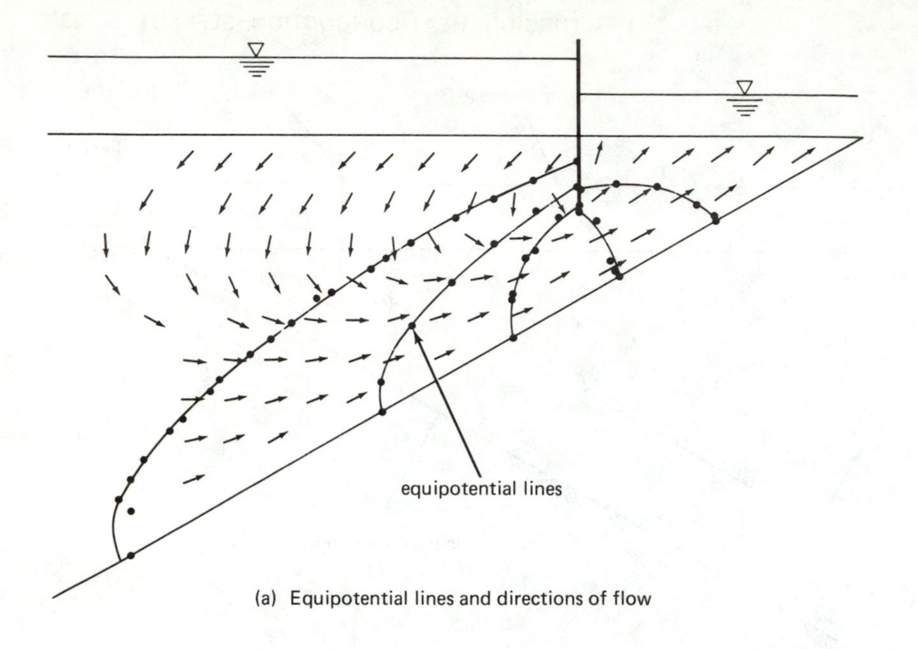

(a) Equipotential lines and directions of flow

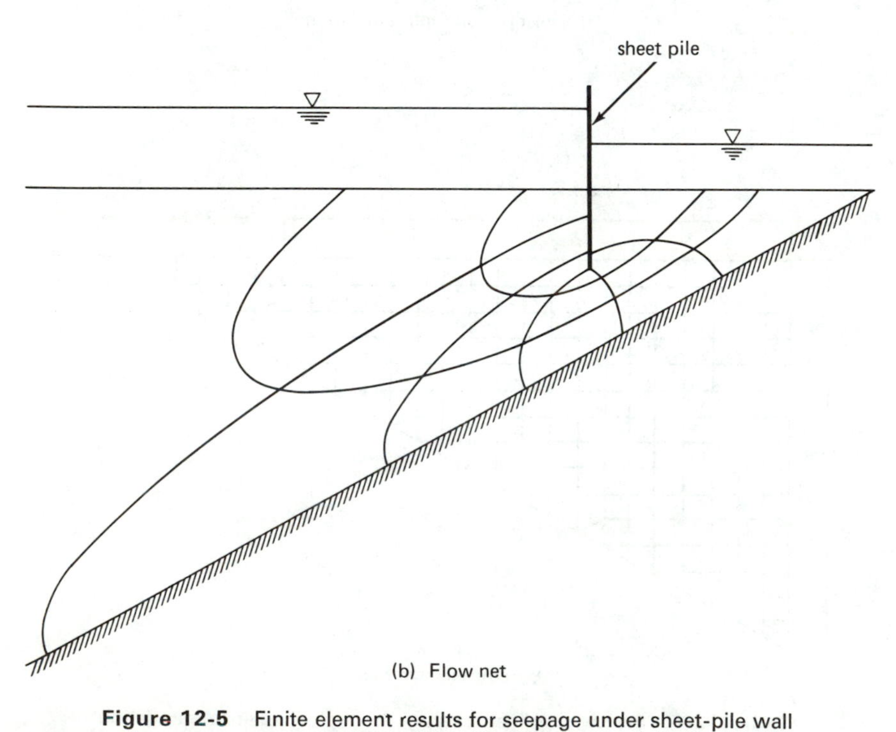

(b) Flow net

Figure 12-5 Finite element results for seepage under sheet-pile wall (from Reference 22).

12-7 EXAMPLE: TRANSIENT CONFINED SEEPAGE IN LAYERED AQUIFERS

The problem of transient confined flow of fluids toward wells in multilayer aquifers occurs both in petroleum production and in utilization of groundwater resources. By using Gurtin's variational principle (Section 8-3), Javandel and Witherspoon[17] developed a finite element procedure to study this problem.

In the analysis, 4-CST quadrilaterals were used. Two different types of boundary conditions were considered for the well. In the constant potential boundary condition, the fluid level in the well was suddenly lowered, and hence the potential along the well bore was constant. This condition is often referred to as *operation at constant terminal pressure*. In the second condition, the rate of flow of pumping from the well was kept constant. This case is called the *constant flow*, or *constant terminal rate* boundary condition. Solutions were obtained for a number of situations by assuming that the aquifer was entirely saturated and that the piezometric surface remained above the top of the aquifer.

A typical solution for the constant terminal rate boundary in a two-layer aquifer is depicted in Figure 12-6. Figure 12-6(a) shows the finite element mesh, and Figure 12-6(b) compares the finite element solutions with results from both a finite difference analysis[26] and an analytical solution for a corresponding homogeneous system.[27]

12-8 EXAMPLE: STEADY UNCONFINED SEEPAGE THROUGH AN EARTH DAM

Taylor and Brown[14] used a finite element formulation based on Darcy's law to solve problems of flow toward an axisymmetric well and of seepage in earth dams. They adopted fluid pressure as the primary field variable and used a linear pressure model. The 4-CST quadrilateral was the basic finite element.

In the earth dam problem, both the upstream and downstream boundaries had specified constant heads. The foundation was assumed impervious. The water seeping through the pervious dam had a free (phreatic) surface. This problem is a typical example of steady unconfined seepage.

The location of the phreatic surface was obtained by approximately satisfying the conditions of zero (atmospheric) pressure and of zero normal flow across the free surface. In the iterative procedure, an initial location of the free surface was estimated, and the condition of zero normal flow was enforced on this surface. Then, results from successive finite element analyses

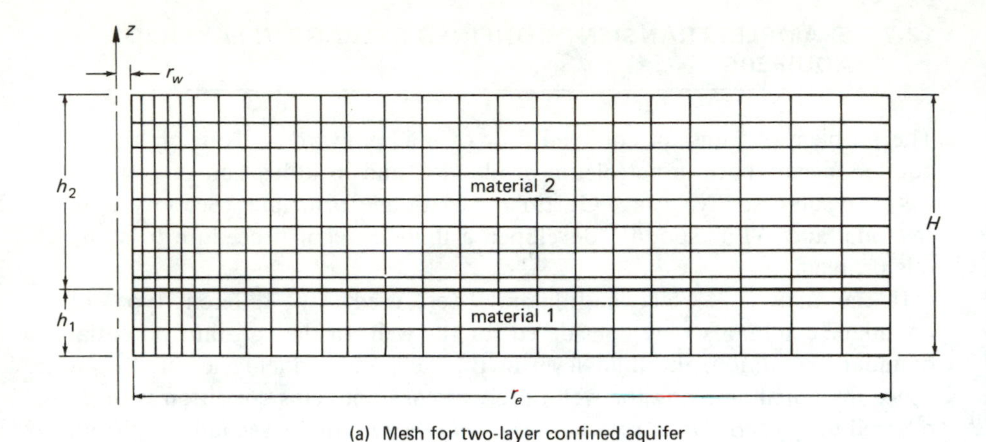

(a) Mesh for two-layer confined aquifer

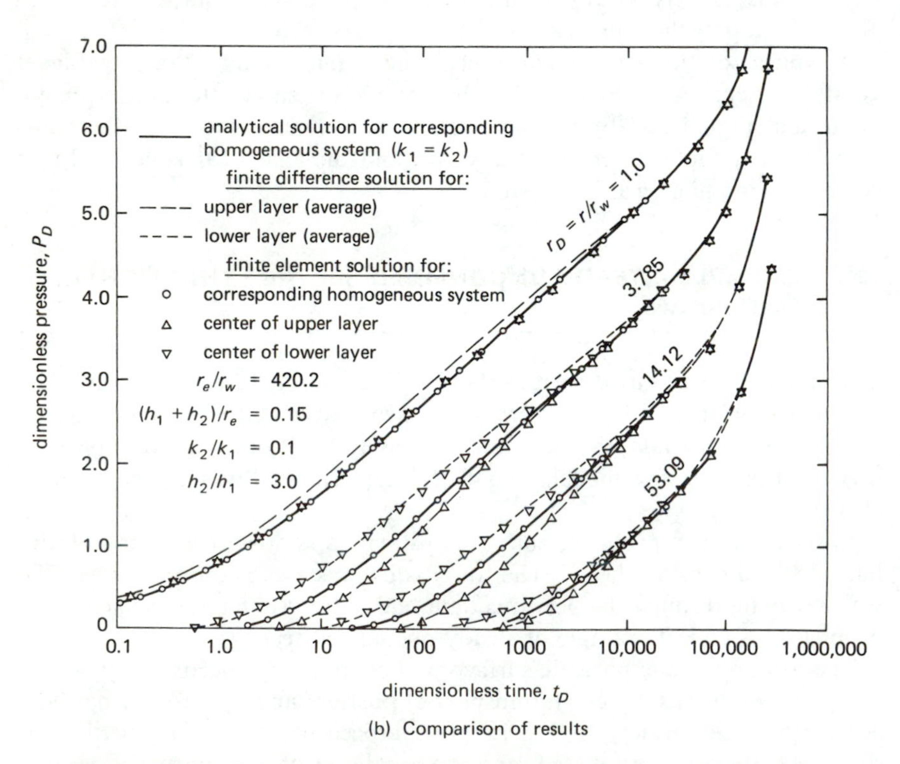

(b) Comparison of results

Figure 12-6 Solutions for transient flow toward a well in layered aquifer (from Reference 17).

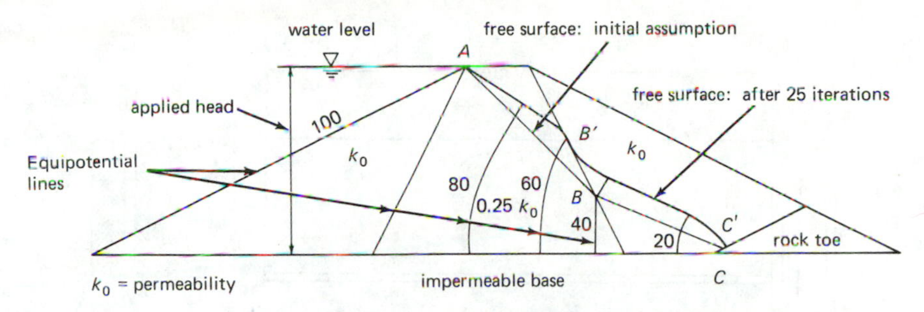

Figure 12-7 Steady, unconfined seepage with free surface (from Reference 14).

were used to adjust the surface until the zero pressure condition was satisfied. This procedure usually required about 20 cycles of iteration. After each iteration, the flow region was modified by adding (or subtracting) new elements, depending upon the adjusted location of the free surface.

An earth dam problem solved by Taylor and Brown is shown in Figure 12-7. The dam consisted of three zones with two different materials having different permeabilities. The rock toe was a highly pervious material. The upstream head was 100 units, and zero downstream head was assumed. Line *A-B-C* was the initial estimate for the phreatic surface. The final location of the free surface was obtained as the curve $A\text{-}B^1\text{-}C^1$ after 25 cycles of iteration. Figure 12-7 also shows equipotential lines determined from the finite element analysis.

12-9 EXAMPLE: TRANSIENT UNCONFINED SEEPAGE

Desai[20, 21] adopted the following form of the Boussinesq equation for an approximate finite difference solution of transient unconfined seepage:[28]

$$\frac{k_x}{2}\frac{\partial^2 h^2}{\partial x^2} + \frac{k_y}{2}\frac{\partial^2 h^2}{\partial y^2} = c\,\frac{\partial h}{\partial t} \qquad (12.11a)$$

Here h denotes height of the free surface above the horizontal impervious base, and c is the porosity of the soil. A parallel plate viscous flow model, Figure 12-8(a), was employed for laboratory tests that provided comparisons with numerical solutions. The changing location of the free surface was obtained at each time step by satisfying the condition that the head at a point on the free surface must be equal to the elevation of that point. Before applying the method for the drawdown case, the extent of the surface of seepage, Figure 12-9, had to be determined. This was accomplished by using Pavlovsky's

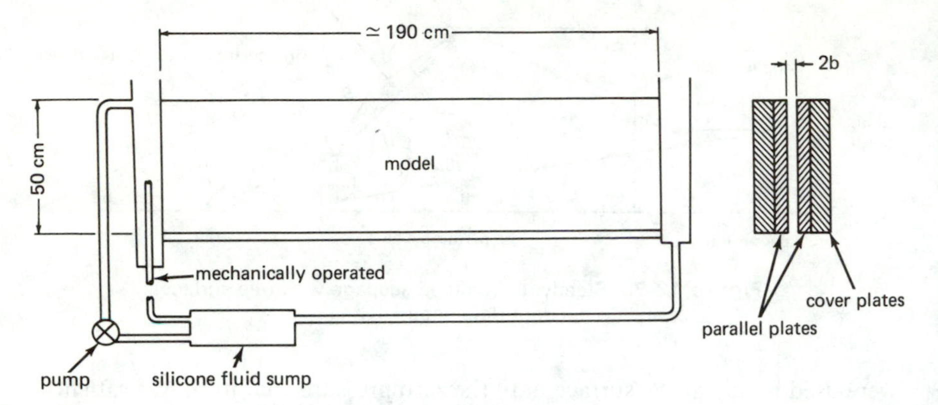

(a) View of parallel plate model

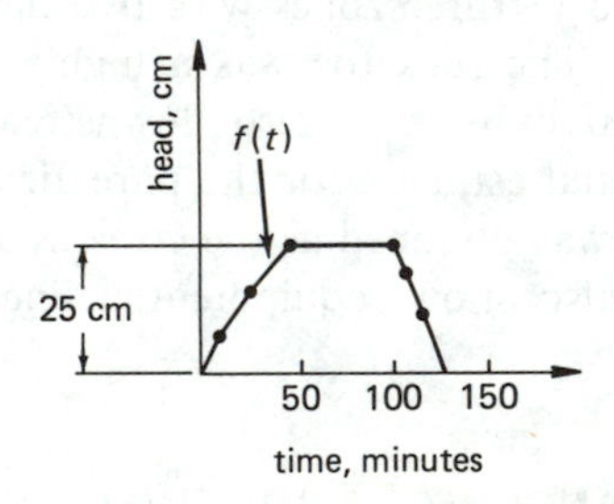

(b) Variations in external head

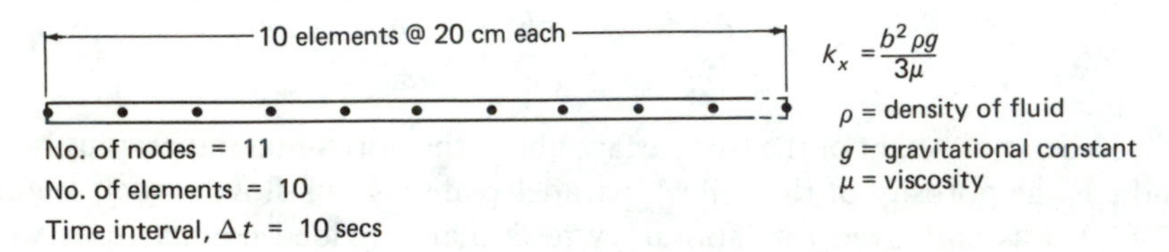

(c) Finite element mesh: one-dimensional elements

Figure 12-8 Apparatus and finite element mesh for transient, unconfined seepage (from Reference 21).

method of fragments with an iterative procedure that equated the outflow to the amount of fluid contained between two consecutive locations of the free surface.[28, 29]

We shall consider here some typical preliminary finite element results for a similar problem. For one-dimensional flow involving small variations in head, a linearized form of equation (12-11a) was

$$k_x \bar{h} \frac{\partial^2 h}{\partial x^2} = c \frac{\partial h}{\partial t} \tag{12.11b}$$

Here $\bar{h}$ denotes the mean value of entrance head, $h(0, t)$ and is given by

$$\bar{h} = \{h(0, t) + h(0, t + \Delta t)\}/2 \text{ for rise of head}$$
$$\bar{h} = \{h_e(0, t) + h_e(0, t + \Delta t)\}/2 \text{ for drawdown}$$

where Δt is the time interval, and subscript e denotes elevation of the point of exit. $\Delta t = 10$ sec was used in this study.

A linear potential model, equation (5.17), was used, and the finite element equations were obtained by using Galerkin's residual method (Section 8-4). The numerical results were compared with tests on the rectangular parallel plate viscous flow model, Figure 12-8(a). The entrance head was varied as shown in Figure 12-8(b).

Figure 12-8(c) shows the one-dimensional finite element mesh. Here the entrance head at node 1 was given by the variation depicted in Figure 12-8(b):

$$h(1, t) = f(t)$$

An approximation was made to handle the fluid head conditions at the tail face. An additional point was introduced outside the model, and the head there was assumed as zero.

Figure 12-9(a) shows a comparison between the finite element solutions and the test results for three typical time levels, and Figure 12-9(b) shows a typical flow net during drawdown. Agreement is good in the rise case. For steady state and drawdown cases the correlation between analysis and experiments is found to be good near the entrance face and is considered satisfactory in other zones. The results can be improved by employing the nonlinear form, equation (12-11a). In this problem the nonlinearity is analogous to geometric nonlinearity (Chapter 7), and an iterative, incremental or mixed analysis technique must be used. A method for improving both linear and nonlinear solutions is the adoption of higher order models in which additional degrees of freedom (unknowns) at the external nodes are included.[21] Similar finite element formulations can be developed for the two-dimensional equation, equation (12.11a).

An extension of the iterative procedure employed by Taylor and Brown[14] can be developed for transient unconfined flow. For each time step the known

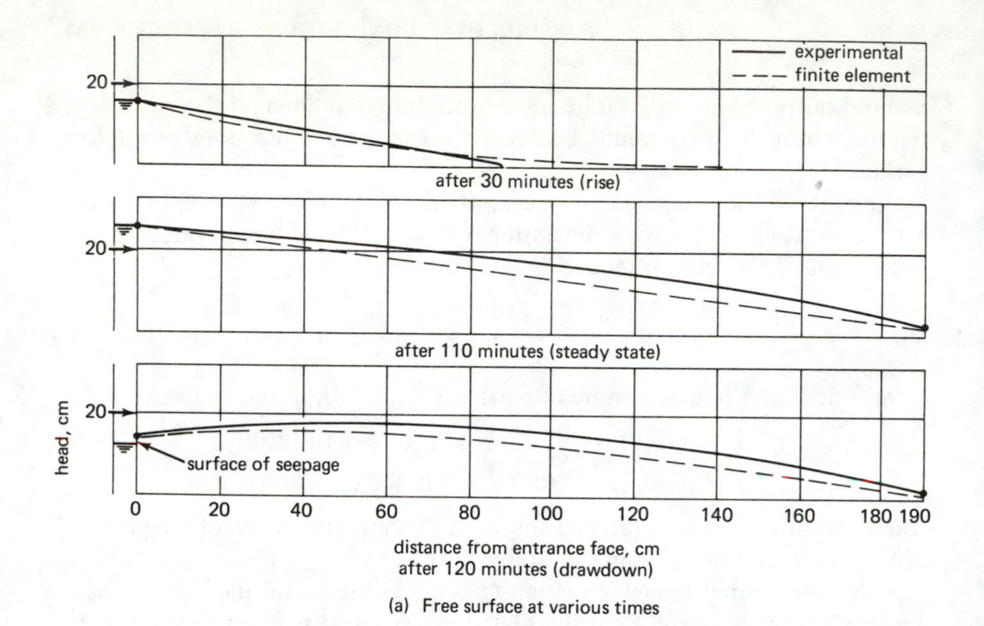

(a) Free surface at various times

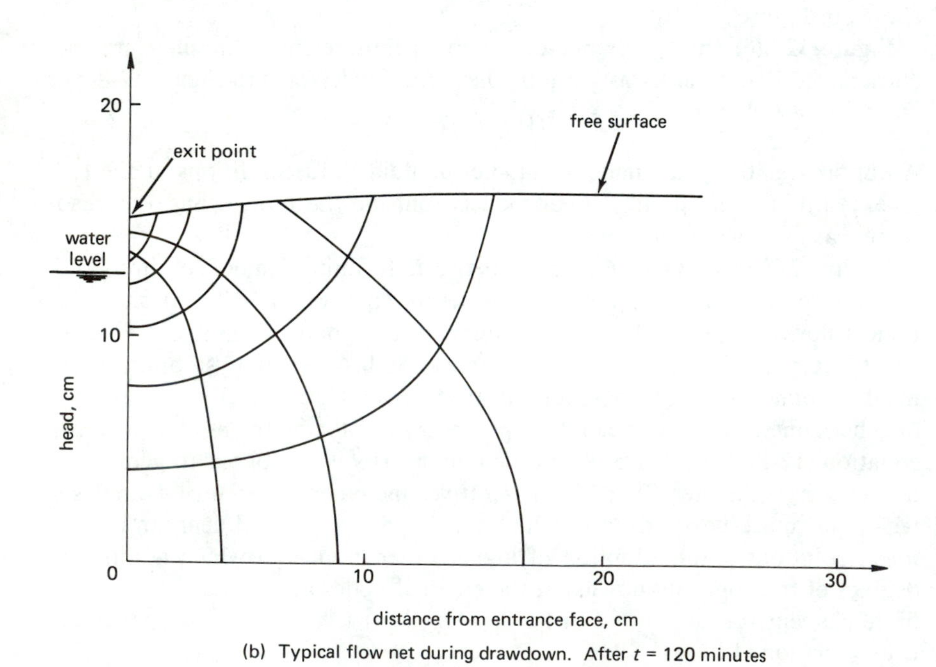

(b) Typical flow net during drawdown. After t = 120 minutes

Figure 12-9 Free surface at various times and typical flow net for transient, unconfined seepage (from Reference 21).

change in boundary heads can be applied to perform a finite element analysis of the flow within the domain. From this solution the velocities of the nodes on the free surface are computed. Hence, the mesh can be modified by computing the distance travelled by the free-surface nodes during the time step. A number of iterations may be required to satisfy adequately the phreatic-surface conditions at the end of the time step.[21]

REFERENCES

(1) Zienkiewicz, O. C., and Cheung, Y. K., "Finite Elements in the Solution of Field Problems," *The Engineer*, Vol. 220, 24 Sept. 1965. Solves problems of torsion of an equilateral triangle, a rectangle, and a bimetallic bar with a central circular hole. Obtains good comparisons with previous finite difference solutions. This work is one of the earlier applications of the finite element method to field problems.

(2) Herrmann, L. R., "Elastic Torsional Analysis of Irregular Shapes," *Proc. ASCE, J.EM Dn*, Vol. 91, EM6 (Dec. 1965). Finite element solution is obtained on the basis of minium potential energy and St. Venant's warping function, Compares numerical results with exact solutions for various torsional constants defined as M_z/G.

(3) See Reference 20, Chapter 3.

(4) Yamada, Y., Nakagiri, S., and Takatsuka, K., "Analysis of Saint-Venant Torsion Problem by a Hybrid Stress Model," *Seisan-Kenkyu* (Monthly Journal of Institute of Industrial Science), Vol. 21, No. 11 (1969).

(5) Visser, W., "A Finite Element Method for the Determination of Non-Stationary Temperature Distribution and Thermal Deformations," *First Conf.* This work presents the parallel between the formulations for transient heat flow problems and the displacement method. Solves two problems: a cylindrical slab with zero initial temperature subjected to constant temperature field on its surfaces; and the same slab with constant rate of change of temperature.

(6) Wilson, E. L., and Nickell, R. E., "Application of the Finite Element Method to Heat Conduction Analysis," *Nuclear Engg. and Design*, Amsterdam, North Holland Publ. Co., 1966. First use of Gurtin's variational principle for transient heat flow problem.

(7) Sandhu, R. S., Wilson, E. L., and Raphael, J. M., "Two-Dimensional Stress Analysis with Incremental Construction and Creep," Rept. SESM 67-34, Department of Civil Engineering, University of California, Berkeley, Dec. 1967. Description in Chapter 14.

(8) Brisbane, J. J., Becker, E. B., and Parr, C. H., "The Application of Finite Element Methods To the Solution of Stress and Diffusion Problems of Continua," *J. of Alabama Academy of Sci.*, Vol. 39, April 1968. Develops the method on the basis of Herrmann's variational principle for thermoelastic stress analysis.

(9) Wilson, E. L., "The Determination of Temperatures Within Mass Concrete Structures," Contract Report No. 68-17, USAE Walla Walla Dist., Dec. 1968.

Uses a direct formulation for transient heat conduction to determine temperature history in mass concrete dams due to temperature fluctuations caused by hydrating cement and cooling pipes. Computes approximate stresses on the basis of the temperature history determined from the finite element analysis.

(10) Fujino, T., and Ohsaka, K., "The Heat Conduction and Thermal Stress Analysis By the Finite Element Method," *Second Conf.* Studies accuracy of the analysis for various element shapes and displacement models. Uses a thin plate problem as the basis of application.

(11) Zienkiewicz, O. C., Bahrani, A. K., and Arlett, R. L., "Solution of Three-Dimensional Field Problems by the Finite Element Method," *The Engineer*, Oct. 27, 1967. Employs three-dimensional elements such as a brick composed of five tetrahedra. Solves example problems: distribution of electrical potential, temperature, and hydrodynamic pressures on a wall moving in an incompressible fluid.

(12) Rybicki, E. F., and Hopper, A. T., "Higher Order Finite Element Method for Transient Temperature Analysis of Inhomogeneous Materials," Proc. Winter Annual Meeting, ASME, Los Angeles, Nov. 1969. Higher order models with 16 and 36 element degrees of freedom are used. Stability and convergence for different magnitudes of temporal steps are studied. Spatial variations of material properties such as k and c are included.

(13) See Reference 9, Chapter 8.

(14) Taylor, R. L., and Brown, C. B., "Darcy Flow Solutions with A Free Surface," *Proc. ASCE, J. HY Dn*, Vol. 93, HY2 (Mar. 1967).

(15) Finn, W. D. L., "Finite Element Analysis of Seepage Through Dams," *Proc. ASCE, J. SM&F Dn*, Vol. 93, SM6 (Nov. 1967). Determines free surface for seepage in earth dams and around cofferdam walls by employing iterative schemes similar to Reference 14.

(16) Fenton, J. D., "Hydraulic and Stability Analyses of Rockfill Dams," DR15, Department of Civil Engineering, University of Melbourne, July 1968. Uses the finite difference and the finite element methods for steady flow through dams with a free surface. Also examines stability of such dams.

(17) Javandel, I., and Witherspoon, P. A., "Application of the Finite Element Method to Transient Flow in Porous Media," *Trans. Soc. of Pet. Engrs.*, Vol. 243 (1968).

(18) Neuman, S. P., and Witherspoon, P. A., "Transient Flow of Ground Water To Wells in Multiple-Aquifer Systems," Publ. No. 69-1, Department of Civil Engineering, University of California, Berkeley, Jan. 1969. Uses Gurtin's variational principle and solves example problems for transient fluid flow to a well in a multiple-aquifer system.

(19) See Reference 19, Chapter 8.

(20) Desai, C. S., "Analysis of Transient Seepage Using Viscous Model and Numerical Methods," Misc. Paper S-70-3, USAEWES, Jan. 1970. Obtains finite difference and finite element solutions for transient unconfined seepage with a free surface under rise in external water levels. Compares finite difference results with experiments from a parallel plate model.

(21) See Reference 56, Chapter 8.

(22) Courtesy of Mr. D. C. Banks, USAE Waterways Expt. Sta., Vicksburg, Miss.

(23) Guymon, G. L., "A Finite Element Solution of the One-dimensional Diffusion-Convection Equation," *J. Water Resources Research*, Vol. 6, No. 1 (Feb. 1970). A variational principle is presented for the one-dimensional diffusion equation. The matrix equations from the finite element formulation are solved in exact form by using eigenfunctions of the associated matrix. Comparisons with exact solutions are obtained.

(24) See Reference 7, Chapter 3.

(25) Willoughby, D. A., "Heat Conduction in Star Perforated Solid Propellant Grains," *AIAA J.*, Vol. 2 (1965).

(26) Vacher, J. P., and Cazabat, V., "Ecoulement des Fluides dans les milieux poreux Statifies Resultat Obtenus sur le Models du Biconche avec Communication," *Revue Institute Francais de Petrole*, Vol. XVI, No. 10 (1961).

(27) van Everdingen, A. F., and Hurst, W., "The Application of the Laplace Transformation to Flow Problems in Reservoirs," *Trans. of AIME*, Vol. 186, (1949).

(28) Desai, C. S., and Sherman, W. C., "Unconfined Transient Seepage in Sloping Banks," *Proc. ASCE, J. SM&F Dn*, Vol. 97, SM2 (Feb. 1971). Considers seepage in river banks resulting from both rise and drawdown conditions caused by time dependent floods in rivers. Uses finite difference method with the method of fragments for the location of the point of exit of the free surface along the bank. Compares numerical results to laboratory experiments with a parallel plate model.

(29) See Reference 4, Chapter 10.

FURTHER READING

McCorquodale, J. A., "Variational Approach to Non-Darcy Flow," *Proc. ASCE, J. HY Dn.* Vol. 96, HY11, Nov. 1970. Presents a basic solution procedure for transient confined and unconfined seepage. On the basis of Lagrangian coordinates modifies the coordinates of nodal points after each time interval.

Visser, W., "The Finite Element Method in Deformation and Heat Conduction Problems," Proefschrift, Technical University of Delft, Netherlands, 1968.

Volker, R. E., "Nonlinear Flow in Porous Media by Finite Elements," *Proc. ASCE, J. HY Dn*, Vol. 95, HY6 (Nov. 1969). Develops finite element equations for nonlinear steady unconfined two-dimensional flow. Solves problems of flow through dams and compares finite element results with free surface experiments and Darcy free surface solutions.

13

THERMOELASTICITY, CONSOLIDATION, HYDROELASTICITY

This chapter deals with the applications of the finite element method to problems in linear and coupled thermoelasticity, linear fluid stress effects, and coupled phenomena of consolidation and hydroelasticity. In Chapter 8 we discussed some basic properties of these problems. However, since the finite element formulations are relatively involved and often necessitate tensor notation for concise presentation, we shall not detail the derivations.

TABLE 13-1 TOPICS COVERED IN CHAPTER 13.

	Topics	*Typical Applications*	*References*
I	Thermoelasticity		
A	Uncoupled	Stress analysis of bodies with known temperature distributions.	1 to 6
B	Coupled	Stress-temperature response of solids subjected to thermal loads.	7, 8
II	Flow through deformable media	Consolidation. Land subsidence.	9 to 12
III	Hydroelasticity	Sloshing of liquids in flexible containers. Reservoir-dam interaction.	13 to 16

Instead, we shall only present a few brief examples to illustrate the application of the finite element method to coupled problems. Table 13-1 lists some significant applications of the method to thermoelasticity, consolidation, and hydroelasticity.

13-1 EXAMPLE: STRESS ANALYSIS FOR THERMAL AND PORE PRESSURE EFFECTS

In Section 7-1 we considered some details of the linear effects of prescribed temperature and fluid stress distributions. On the basis of Biot's theory, Crose[2] developed a finite element formulation to include both these effects; and he performed stress analyses for a solid cylinder subjected to axially symmetric mechanical loads, temperature gradients, and pore pressures (Fig. 13-1). The cylinder was idealized as an assemblage of axisymmetric elements (Example 5-6).

The elastic material properties of the cylinder were $E = 10^6$ psi and $\nu = 0.49$. The applied axial load was 1.0 lb. The thermal effects were given by $\alpha T = 1.5 \times 10^{-4}$, where α is the coefficient of linear thermal expansion and T is the temperature difference. The applied fluid stresses, which were caused by a fluid diffusing in the porous medium, are indicated in Figure 13-1(b).

Figure 13-1(c) shows a simple finite element grid for a typical 0.1-in. height of the cylinder, which has a radius of 1.0 in. Figures 13-2(a) and 13-2(b) compare the stresses computed from the finite element analysis with the exact solution:

$$\sigma_r = 100r(1 - r)$$
$$\sigma_z = \sigma_r - 17 = 100r(1 - r) - 17$$

where r is the radius of the cylinder.[17]

13-2 EXAMPLE: HEATING OF LINEAR THERMOELASTIC HALF-SPACE

The first finite element formulation for linear coupled thermoelasticity was obtained by Nickell and Sackman[7] on the basis of Gurtin's variational principle. Oden and Kross[8] used energy balance equations to derive the finite element equations for the general problem of dynamic thermoelasticity. They considered the problem of ramp heating on the stress-free bounding surface of a linear thermoelastic half-space, shown schematically in Figure 13-3(a).

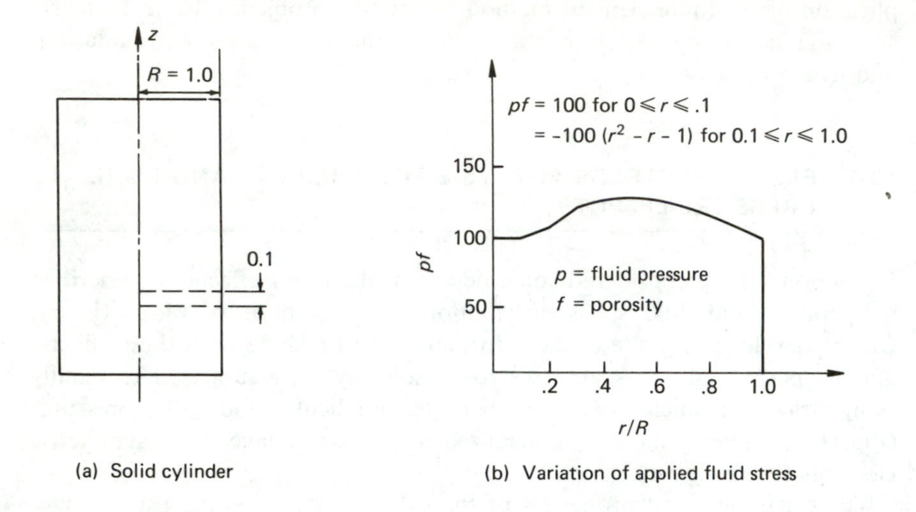

(a) Solid cylinder

(b) Variation of applied fluid stress

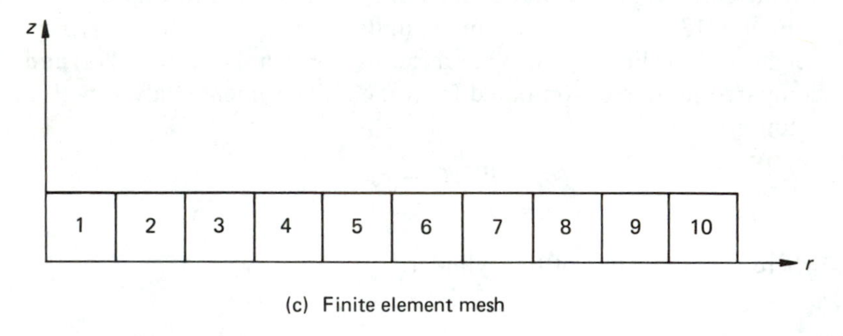

(c) Finite element mesh

Figure 13-1 Solid cylinder and mesh layout (from Reference 2).

Displacements and temperatures are the two field variables in this coupled problem. The field variable models for the displacement, $\{u\}$, and the temperature, T, may be expressed as

$$\{u(x,y,z,t)\} = [N_1(x,y,z)]\{q(t)\} \tag{13.1a}$$

$$T(x,y,z,t) = \{N_2(x,y,z)\}^{\mathrm{T}}\{T_n(t)\} \tag{13.1b}$$

$\{q\}$ and $\{T_n\}$ are the nodal displacement and nodal temperature vectors, respectively; $[N_1]$ and $\{N_2\}$ are the matrices of interpolation functions, and

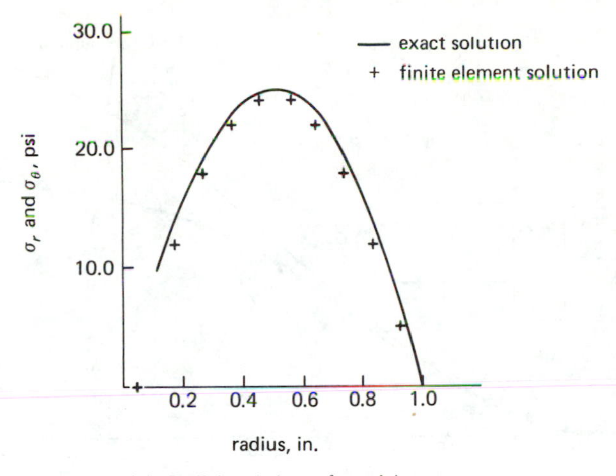

(a) Radial and circumferential stresses

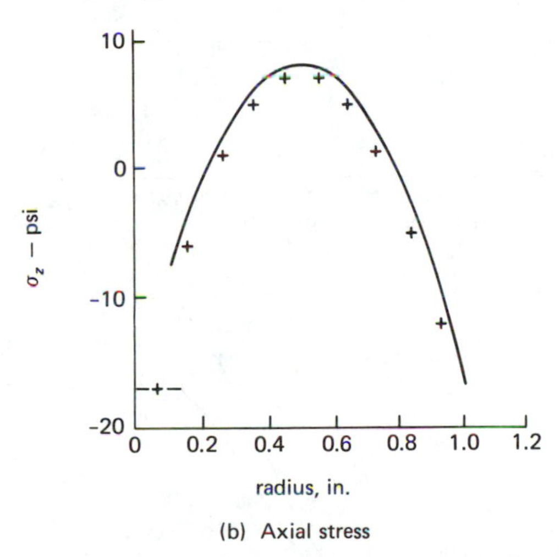

(b) Axial stress

Figure 13-2 Comparisons between exact and finite element solutions for problem in Figure 13-1 (from Reference 2).

x, y, z and t are space and time coordinates. In this problem, Oden and Kross used linear interpolation models for both field variables.

Comparisons of the finite element solutions with an exact solution are shown in Figures 13-3(a) and 13-3(b). The exact solutions were obtained by using a numerical scheme to invert Laplace transforms. In these figures, $\bar{\theta}$, $\bar{u}$, τ and ξ represent the dimensionless temperature, displacement, time, and distance from the free surface, respectively. The results were obtained for

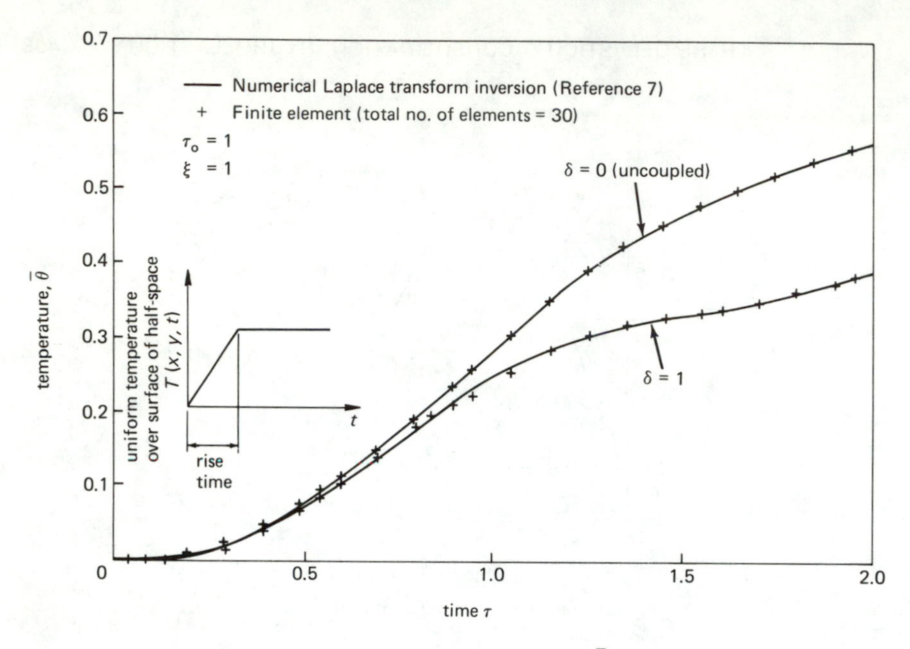

(a) Comparison for temperature, $\bar{\theta}$

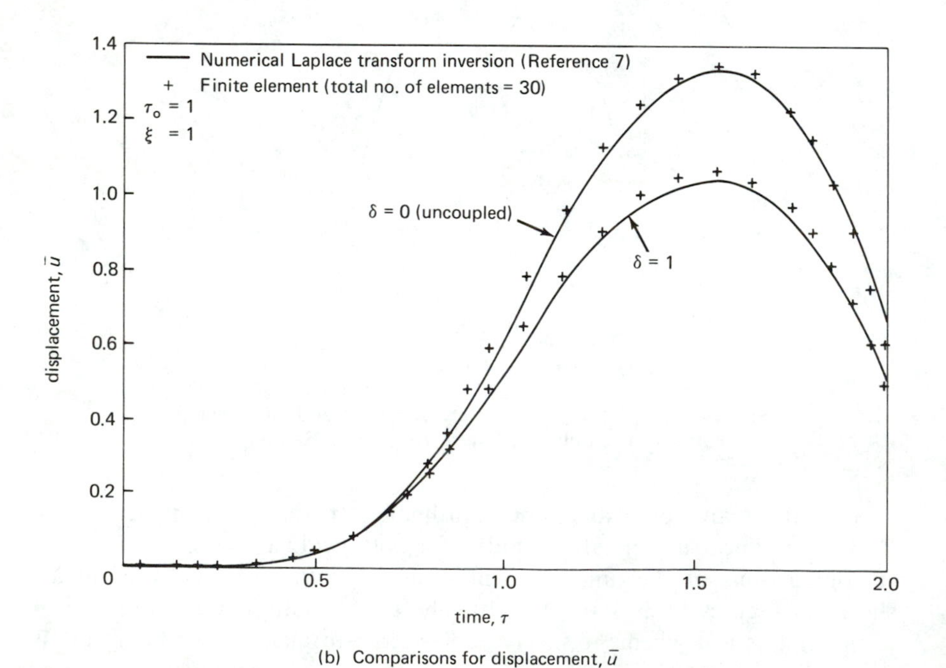

(b) Comparisons for displacement, $\bar{u}$

Figure 13-3 Ramp heating of linear thermoelastic half-space (from Reference 8).

various dimensionless rise-times, τ_0. Here we have included only the results for $\tau_0 = 1$ and $\xi = 1$.

Oden and Kross also performed analyses for various mesh layouts and for convergence with refined mesh. The reader may consult the original paper and other works by Oden (Further Reading) for further information on coupled thermoelasticity.

13-3 EXAMPLE: TERZAGHI'S CONSOLIDATION

Sandhu and Wilson[9] used Gurtin's principle to solve the linear coupled problem of seepage through elastic porous media. They assumed a quadratic interpolation model, equation (5.23), for displacements, and a linear model, equation (5.27), for the isotropic fluid stress. Two classical problems were solved by using the finite element method: Terzaghi's one-dimensional consolidation, and Biot's settlement of porous elastic half-space ($\nu = 0$) subjected to a strip load.

Figure 13-4 compares the settlement-time curves from the finite element solutions with the analytical solutions obtained from Terzaghi's one-dimensional theory. The problem involved a bounded soil layer with a uniformly distributed load. The results for both settlements and pore fluid stresses (pressures) from the two analyses agree completely.

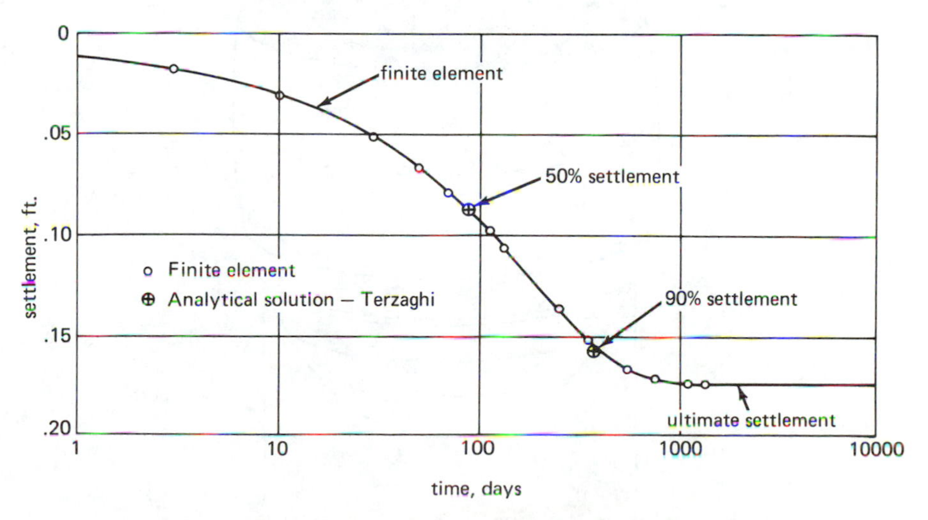

Figure 13-4 Comparison: finite element and Terzaghi solutions for consolidation (from Reference 9).

13-4 EXAMPLE: MOTION OF FLUID IN FLEXIBLE CONTAINER

The analysis of the sloshing of liquid propellants in elastic containers of various shapes has received considerable attention in aerospace structural dynamics. Interaction occurs between the deformations of the container walls and the motion of the liquid, and such coupling must be considered in the analysis.

Tong,[13] Guyan *et al.*,[14] and Luk[15] have used the finite element method for the solution of the coupled sloshing problem. Figure 13-5 shows a schematic diagram of a partially filled, axisymmetric container.[14]

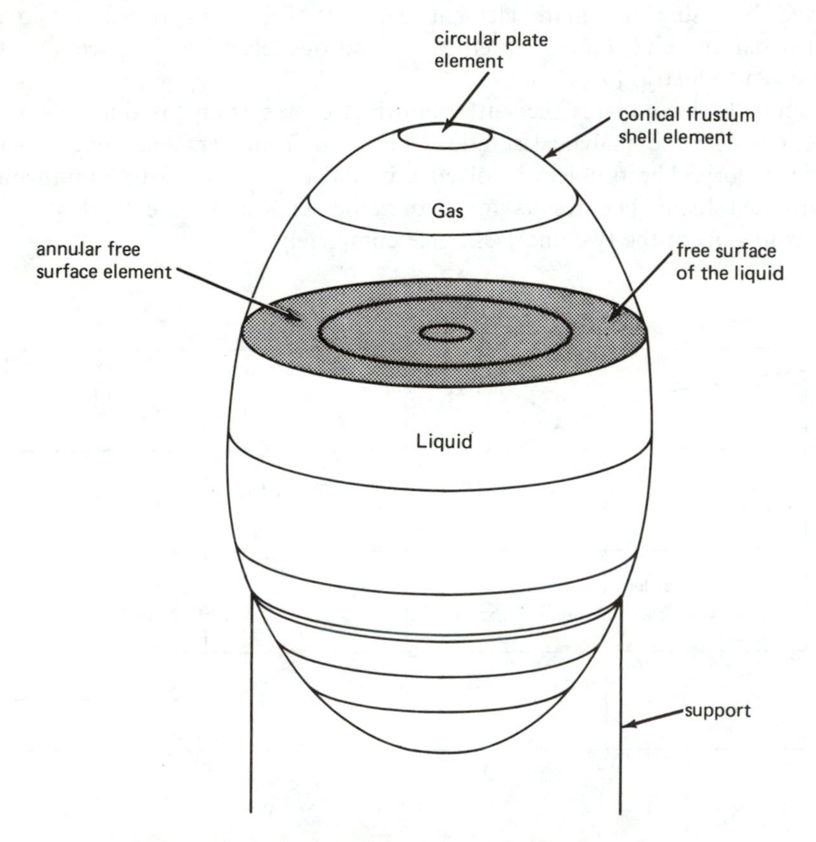

Figure 13-5 Liquid in a flexible, axisymmetric structure (from Reference 14).

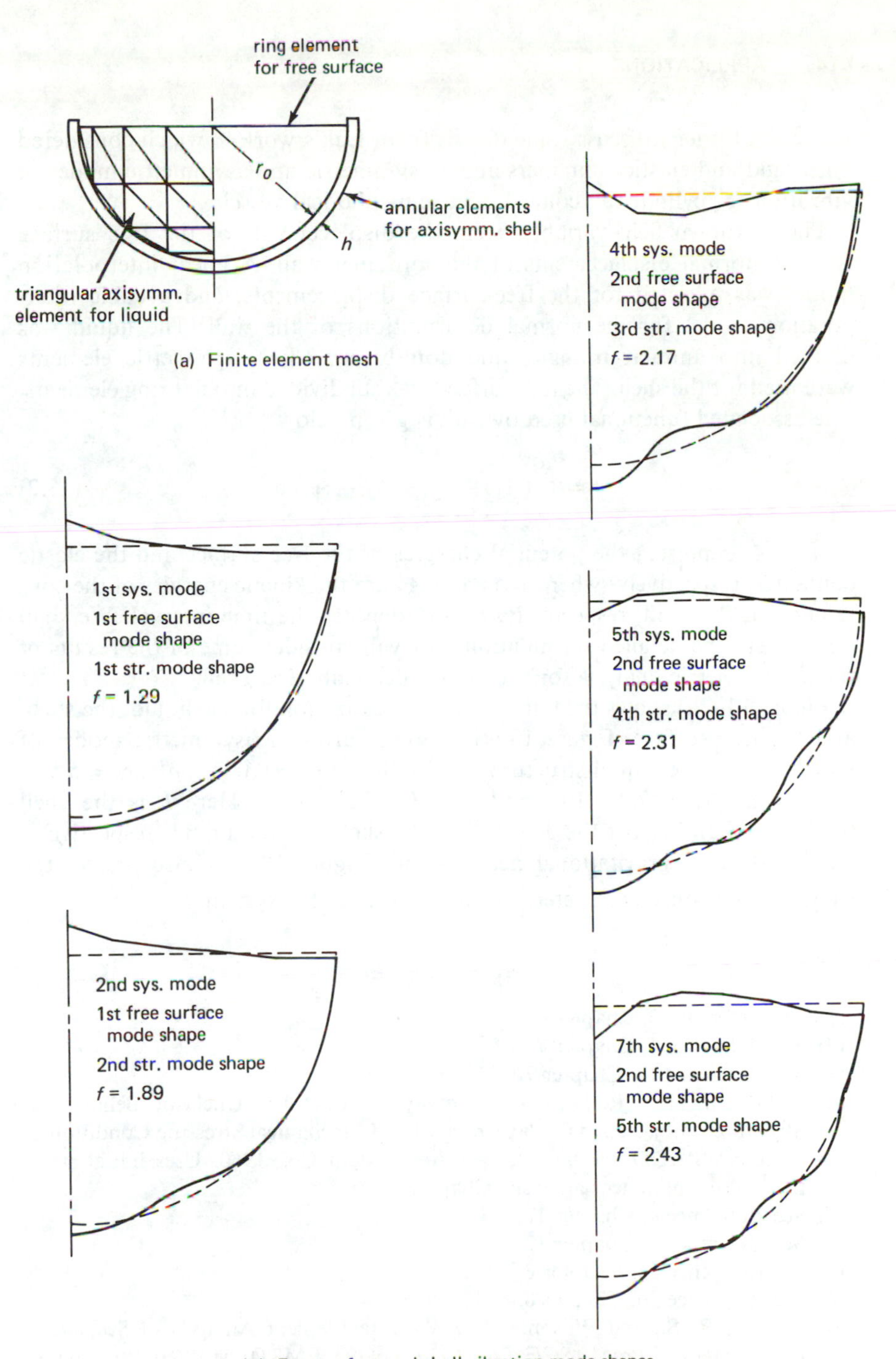

Figure 13-6 Analysis of liquid sloshing (from Reference 15).

We shall briefly discuss some details from Luk's work,[15] which considered both rigid and elastic containers and axisymmetric and asymmetric modes of vibration for cylindrical, conical, and semispherical vessels.

The unknown field variables were the displacements of the free surface and the normal displacements of the container wall. A linear interpolation model was adopted for the free-surface displacements, and a cubic interpolation model for the normal deformations of the wall. The liquid was divided into annular triangles and doubly curved, axisymmetric elements were used for the shell. The free surface was subdivided into flat ring elements. The associated functional used by Luk is given below:[13,15]

$$A = \int_{t_1}^{t_2} (A_1 + A_2 - A_3 - A_4)\, dt \tag{13.2}$$

A_1 and A_2 represent the potential energies of the free surface and the elastic container, respectively; whereas A_3 and A_4 are the kinetic energies of the container and the fluid, respectively. $t_1 - t_2$ denotes the time interval. We shall not present the detailed formulation, but will consider some of the results of the axisymmetric analysis for the elastic hemispherical shell.

Figure 13-6(a) shows the finite element idealization for the liquid, the shell, and the free surface. Figure 13-6(b) shows various axisymmetric modes of vibration for the liquid-structure system for $h/r_0 = 0.002$, $\rho h/\rho_0 r_0 = 5.5 \times 10^{-3}$, $\rho_0 g_0 r_0^4/D = 7.9 \times 10^4$, and $D = Eh^3/12(1 - \nu^2)$. Here h is the shell thickness; ρ and ρ_0 are the densities of the shell and the liquid, respectively; and g_0 denotes gravitational acceleration. Figure 13-6(b) also shows the frequency response parameter, $f = \omega\sqrt{r_0/g_0}$, for the system.

REFERENCES

(1) See Reference 3, Chapter 6.
(2) See Reference 1, Chapter 7.
(3) See Reference 17, Chapter 7.
(4) Padlog, J., Huff, R. D., and Holloway, G. F., "The Unelastic Behavior of Structures Subjected to Cyclic, Thermal and Mechanical Stressing Conditions," Report WPADD TR 60-271, Bell Aerosystem Co., 1960. Uses initial strain method for nonlinear analysis (Chapter 7).
(5) See Reference 4, Chapter 12.
(6) See Reference 9, Chapter 12.
(7) See Reference 25, Chapter 8.
(8) See Reference 26, Chapter 8.
(9) Sandhu, R. S., and Wilson, E. L., "Finite Element Analysis of Seepage in Elastic Media," *Proc. ASCE, J. EM Dn*, Vol. 95, EM3 (June 1969). By using the theory of References 23 and 25 of Chapter 8, develops a formulation for the consolidation problem.

(10) Sandhu, R. S., and Wilson, E. L., "Finite Element Analysis of Land Subsidence," *Proc. Int. Symp. on Land Subsidence*, Tokyo, Japan, September 1969.
(11) See Reference 40, Chapter 9.
(12) Christian, J. T., and Boehmer, J. W., "Plane Strain Consolidation by Finite Elements," *Proc. ASCE, J. SM&F Dn*, SM4 (July 1970). Uses the concept in Reference 11 for one- and two-dimensional (plane strain) consolidation of linear elastic soils. Adopts a quadratic model for unknown pore pressures. Obtains good comparison between finite element solutions and results from Terzaghi's one-dimensional theory. Solves for consolidation in (radial) triaxial and plane strain shear specimens and for consolidation of semi-infinite half space.
(13) Tong, P., "Liquid Sloshing in an Elastic Container," AFOSR 66-0943 (June 1966). The fluid is divided into toroidal elements. Fluid velocity is adopted as the field variable for these elements, and the fluid mass system is coupled with the elastic structure. Employs displacement method.
(14) See Reference 28, Chapter 8.
(15) See Reference 29, Chapter 8.
(16) Chopra, A. K., Wilson, E. L., and Farhoomand, I., "Earthquake Analysis of Reservoir-Dam Systems," *Proc. 4th World Conf. on Earthquake Eng.*, Santiago, Chile, 1969. Also published in "Earthquake Engineering Research at Berkeley," Earthquake Eng. Res. Ctr. Rept. No. 69-1, University of California, Berkeley, 1969, NTIS, Springfield, Va. Compares a finite element solution with an exact solution for a simple case of coupled hydroelastic problem.
(17) Evans, S., McDonough, T., and Menkes, E., "A Computer Program for the Thermoelastic Analyses of an Inhomogeneous, Composite, Thick Cylinder, Including the Effect of Diffusing Gas in Porous Materials," BSK-TR-65-473, General Electric, Re-entry Syst. Dept., Philadelphia, Pa., December 1965. Available—Defense Documentation Center.

FURTHER READING

Oden, J. T., "Finite Element Analysis of Nonlinear Problems in the Dynamical Theory of Coupled Thermoelasticity," *Nuclear Engrg. and Design*, Amsterdam, North-Holland Publishing Co., Vol. 10, 1969. Presents formulation for nonlinear analysis of problems in dynamical theory of coupled thermoelasticity. Compares results from linear and nonlinear (small strains) analysis for a slab subjected to ramp temperature rise.

Oden, J. T., and Poe, J., "On the Numerical Solution of a Class of Problems in Dynamic Coupled Thermoelasticity," UARI Res. Report No. 72, University of Alabama at Huntsville, October 1969. Develops the finite element method for transient coupled thermoelasticity. Includes material nonlinearities and temperature-dependent variations of thermal conductivity and specific heat.

14

MISCELLANEOUS APPLICATIONS

In the foregoing chapters of Part C we have considered logical groupings of applications of the finite element method. However, these groupings have been discerned more on the basis of disciplines than of mathematical similarities. Some of the problems considered in this chapter bear mathematical resemblance to those presented earlier. Nevertheless, here we shall consider primarily the subjects of viscoelasticity and fluid mechanics. Some additional topics in engineering are also mentioned in the summary of miscellaneous applications given in Table 14-1.

TABLE 14-1 MISCELLANEOUS APPLICATIONS OF THE FINITE ELEMENT METHOD

	Topic	*Typical Applications*	*References*
I	Viscoelasticity	Analysis of solids with rate-dependent material behavior.	1 to 5
		Viscous behavior of clays.	6
		Creep of concrete.	7
II	Fluid mechanics and hydraulics	Irrotational flow of incompressible fluids.	8, 9
		Viscous flow of fluids.	10 to 12
		Seiche of lakes and harbors.	13 to 15
		Open channel flow.	16
III	Other miscellaneous	Flow of compressible fluids.	17
		Electrical engineering.	14, 18, 19
		Dynamic behavior of gases.	20, 21
		General applications in mathematical physics.	21
		Water resources.	Further Reading

14-1 EXAMPLE: VISCOUS BEHAVIOR OF CLAY SOILS

Viscous effects play a significant role in the time-dependent deformations of some soft organic clays. An example of pronounced creep behavior is the deformations of the order of 10–20 feet which occurred in the Atchafalaya levee on the Mississippi River in Louisiana, over a period of 10 to 15 years. Field observations of this levee showed that the deformations were mainly due to creep, since no significant pore pressure dissipations took place.[22,23] Comparisons of the field data with nonlinear elastic and elastic-plastic finite element analyses indicated that creep effects must be included in the analysis.[23,24]

Recently Watt[6] performed a detailed analysis of linear viscous effects in soft incompressible clays, such as San Francisco Bay mud and London clay. Finite element formations were obtained for both the displacement and mixed procedures. The constitutive laws for characterizing the viscous behavior were derived from laboratory creep tests. The study considered a number of factors, such as normal and overconsolidation conditions; drained and undrained cases; elastic, elastic-plastic, and viscous behavior; and different

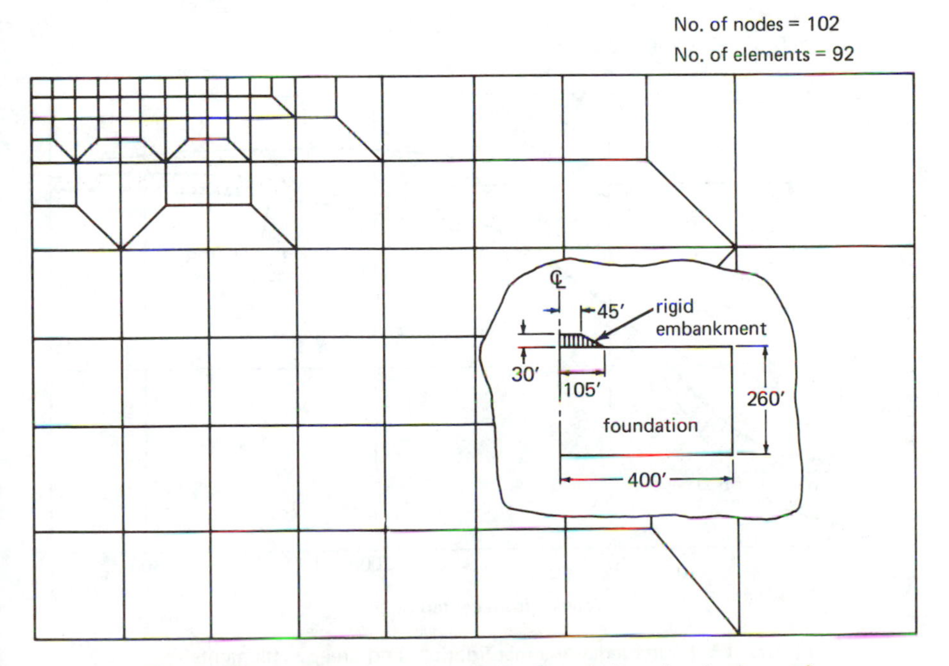

Figure 14-1 Finite element mesh for foundation of embankment.[6]

types and rates of loading. The procedures were applied to the analysis of laboratory plane strain and triaxial test specimens and to a number of practical situations. We shall include here only the analysis of the foundation of an embankment.

Figure 14-1 shows the finite element mesh layout for the embankment foundation consisting of highly overconsolidated (O.C. ratio = 20) London Clay. The embankment configuration has some similarities with one constructed for Interstate 95 in Massachusetts. The embankment was assumed to be constructed at uniform rates over periods of ten and forty days.

Among the many conclusions drawn from the computed stress and displacement distributions was that the deviatoric creep caused significant deformations and redistribution of stresses. Pronounced differences in the stress and deformation patterns occurred for drained and undrained conditions. Figure 14-2 shows the total vertical creep settlement after 500 days, plotted in comparison with the immediate and the consolidation settlements. The creep deformations form about twenty percent of the total settlement.

Figure 14-3(a) shows horizontal displacements due to immediate, creep, and consolidation effects. These analytical results show qualitative agreement with the field behavior of the similar embankment on Interstate 95, Figure 14-3(b).

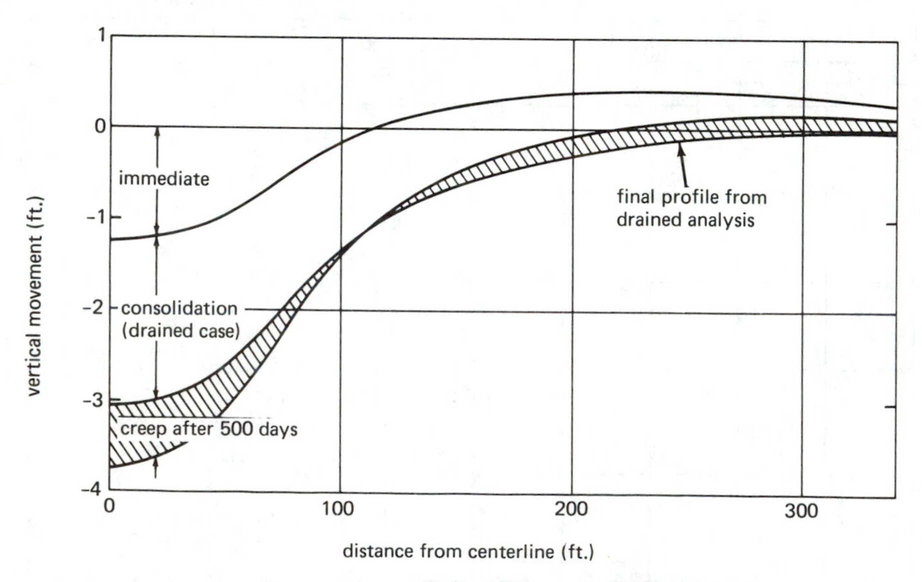

Figure 14-2 Immediate, consolidation, and creep settlements at surface of embankment foundation (from Reference 6).

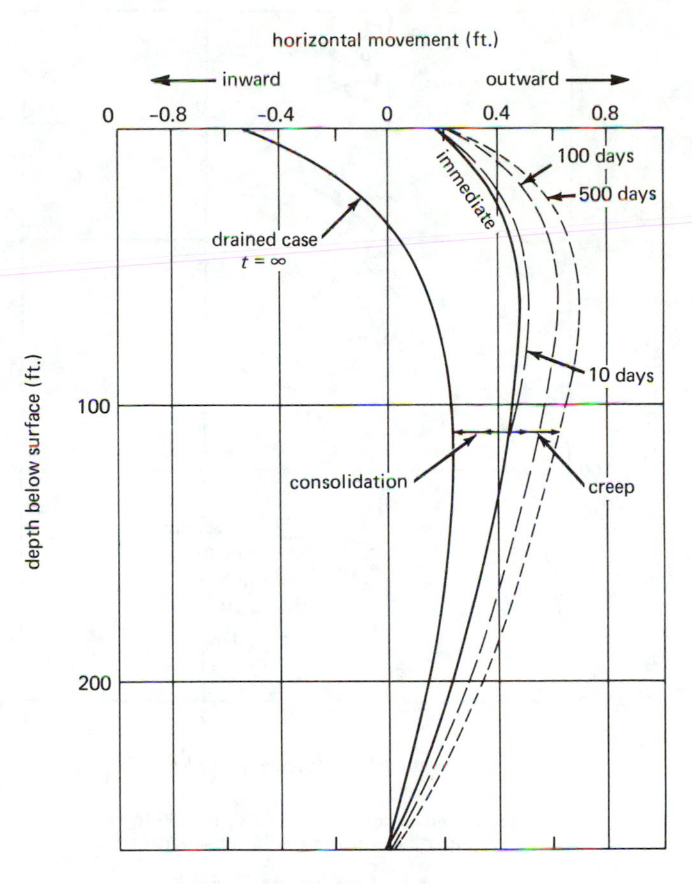

(a) Horizontal displacements: 80 ft. off centerline of embankment. Computed from finite element analysis.

Figure 14-3 Comparisons for horizontal displacements (from Reference 6).

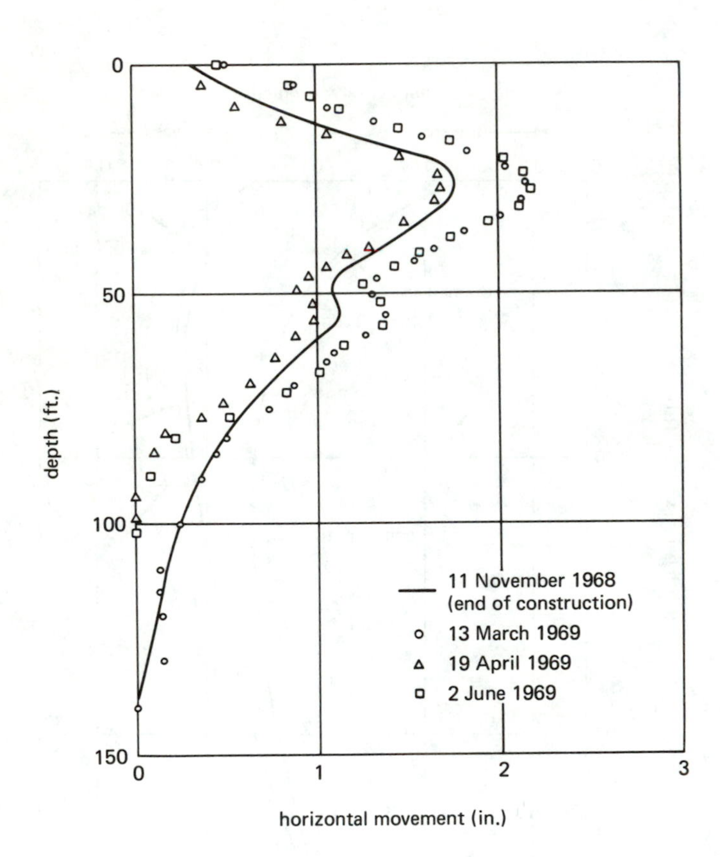

(b) Observed horizontal movements: 90 ft. off centerline of Interstate 95 embankment.[25]

Figure 14-3 (*continued*)

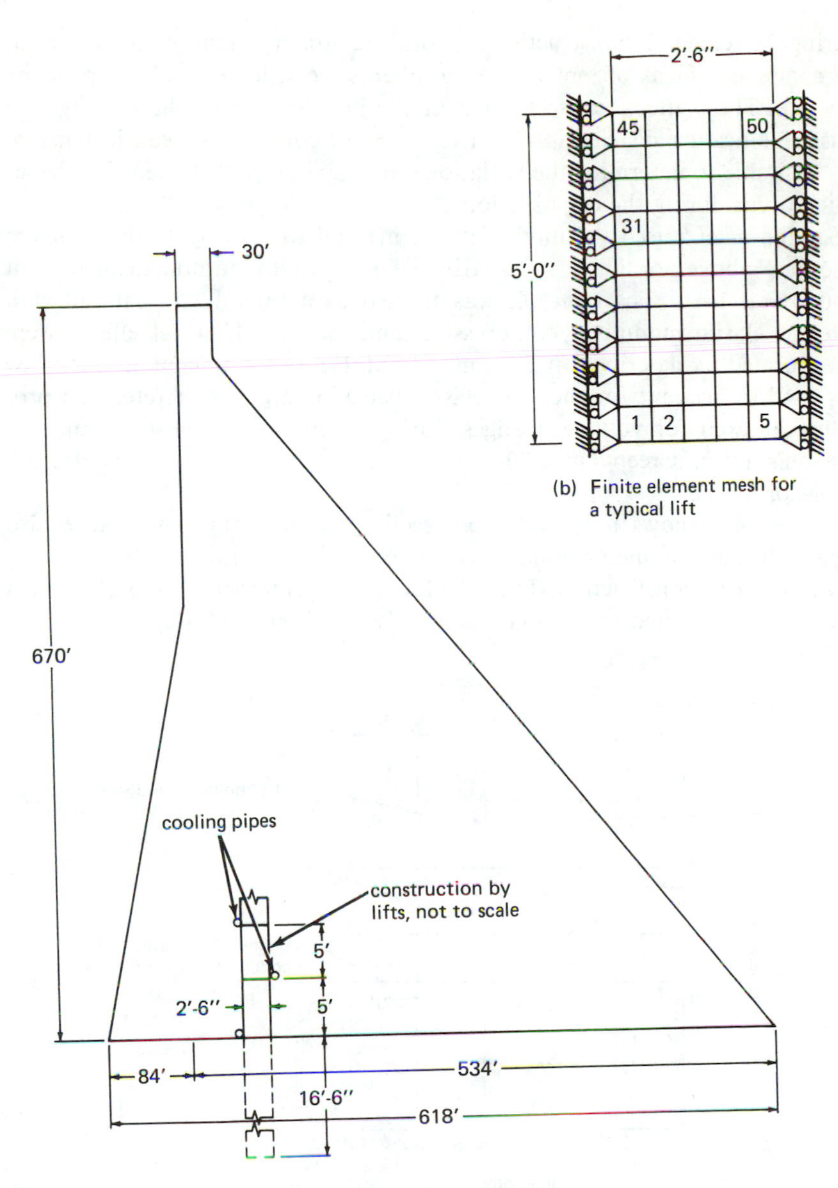

(b) Finite element mesh for a typical lift

(a) Cross section of concrete gravity dam

Figure 14-4 Incremental construction of concrete gravity dam (from Reference 7).

14-2 EXAMPLE: CREEP IN CONCRETE DAMS

During incremental construction a concrete gravity dam is subjected to the combined effects of temperature gradients, dead loads, and creep of the concrete. These effects cause tension at the junctions of the lifts. They also cause significant redistribution of stresses. Since concrete is weak in tension, it is desirable to determine the variations in magnitudes of the tensile stresses which occur during the construction.

Sandhu *et al.*[7] used the finite element method to investigate the problem described above. A linear strain triangle element (quadratic displacement model) was used. The concrete was treated as a bimodular material with different elastic moduli in compression and tension. Thermal effects were incorporated as described in Section 7-1. McHenry's exponential creep law was used to account for the viscoelastic behavior of the concrete. The procedure allowed for volume changes during creep with a constant value of Poisson's ratio. Creep curves for use in the analysis were derived from laboratory tests.

Figure 14-4 shows a typical cross section of the dam. This figure also depicts the steps in incremental construction and the finite element mesh for a typical lift of construction. The schedule of air temperature and placement of concrete simulated in the analysis is shown in Figure 14-4(c).

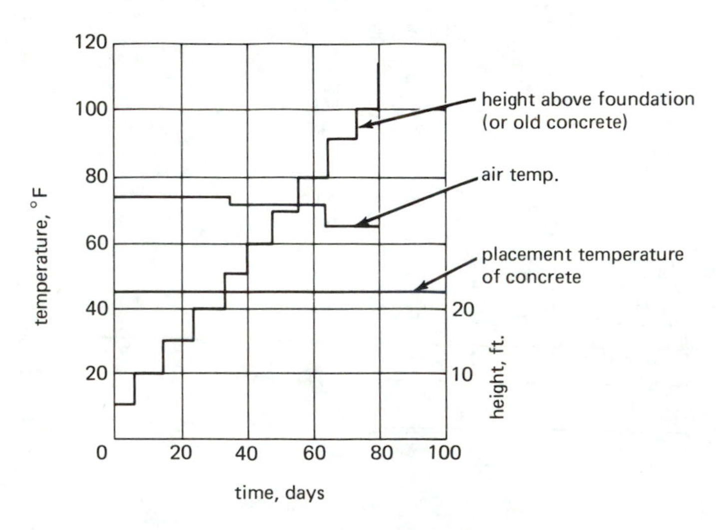

(c) Air temperature and placement schedule

Figure 14-4 (*continued*)

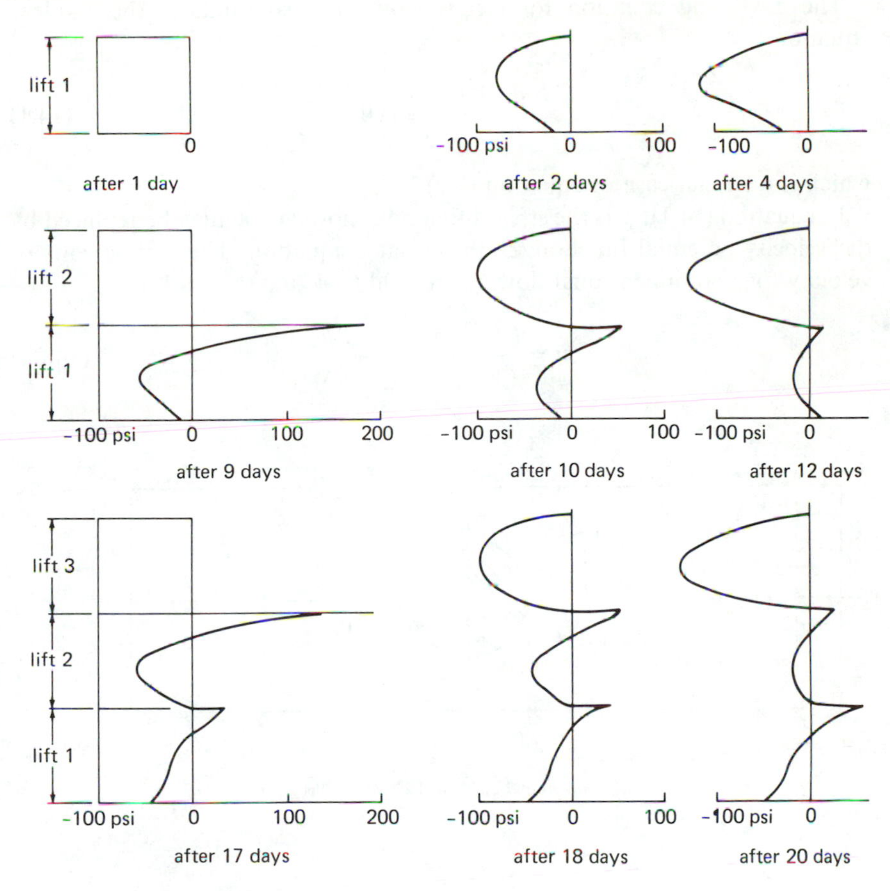

Figure 14-5 Typical distributions of average horizontal stress on vertical section of concrete gravity dam (from Reference 7).

Figure 14-5 depicts the computed time history of the horizontal stress distribution as the lifts are added. Significant tensile stresses are induced at the lift interfaces, particularly at the top of the lift below the one most recently laid. However, as time elapses, redistribution of stresses occurs due to creep, and the tensile stresses are progressively reduced.

14-3 EXAMPLE: FLOW OF IDEAL FLUIDS

Martin[8] showed the analogy between the structural and non-structural formulations of the finite element method, and he applied available structural finite element codes to the solution of inviscid, incompressible fluid flow.

The governing equation for steady flow of ideal fluids is the Laplace equation

$$\frac{\partial^2\psi}{\partial x^2} + \frac{\partial^2\psi}{\partial y^2} = 0 \tag{14.1}$$

which is a special case of equation (8.5).

In equation (14.1), ψ is the stream function; however, ψ may be replaced by the velocity potential function, ϕ, in the same equation. The components of velocity for two-dimensional flow are related to ψ and ϕ as follows:

$$u = \frac{\partial\psi}{\partial y}, \qquad v = -\frac{\partial\psi}{\partial x} \tag{14.2a}$$

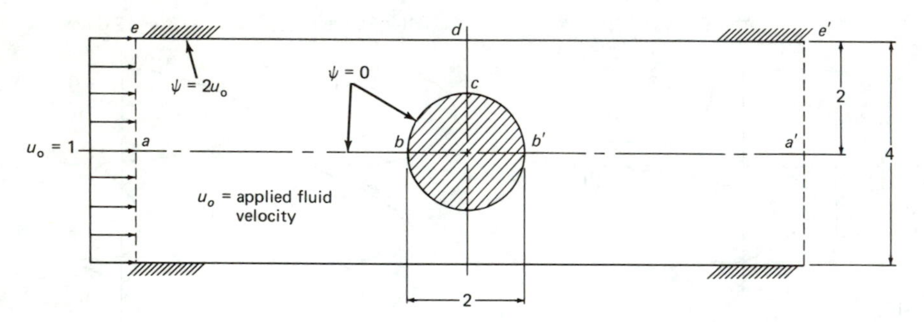

(a) Flow around a cylinder between parallel walls

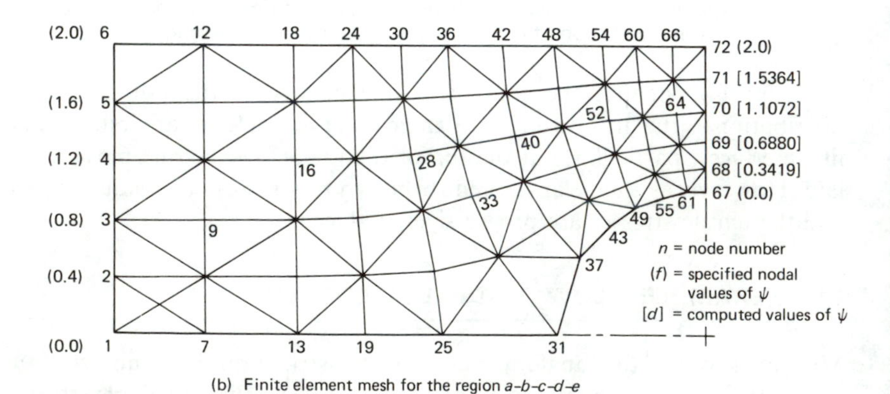

(b) Finite element mesh for the region *a-b-c-d-e*

Figure 14-6 Problem in the flow of ideal fluids (from Reference 8).

$$u = \frac{\partial \phi}{\partial x}, \qquad v = \frac{\partial \phi}{\partial y} \tag{14.2b}$$

The boundary conditions may occur as *Dirichlet conditions*, in which ψ is prescribed on the boundary. Martin also considered *Neumann type boundary conditions*, in which the normal gradient, $\partial\psi/\partial n$, is specified. He recommended a procedure to handle the latter condition by treating the specified normal velocity as a constraint on the assemblage equations.

In addition to solving a flow problem illustrating St. Venant's principle in fluid flow, Martin considered the problem of flow around a cylinder between parallel walls, Figure 14-6(a). The upper half of the flow region was discretized into finite elements. The mesh for the quadrant *a-b-c-d-e* is shown in Figure 14-6(b); the other half is obtained as the mirror image about line *c-d*. Because of symmetry, $\psi = 0$ was specified along *a-b-c-b′-a′*, and from the definition of the stream function, $\psi = 2$ was found on *e-d-e′*. Further Dirichlet conditions were specified along *a-e* and *a′-e′* from the assumption that the flow was uniform across these sections.

The nodal values of ψ were computed as the primary field variables, and the velocity components u and v were evaluated as element resultants by using equations (14.2a). The computed nodal values of ψ along *c-d* are shown in Figure 14-6(b). Figure 14-7 shows the computed distribution of velocity on *c-d*. The figures in parentheses show solutions for u at c and d obtained from an approximate theoretical solution based on a vertical distribution of

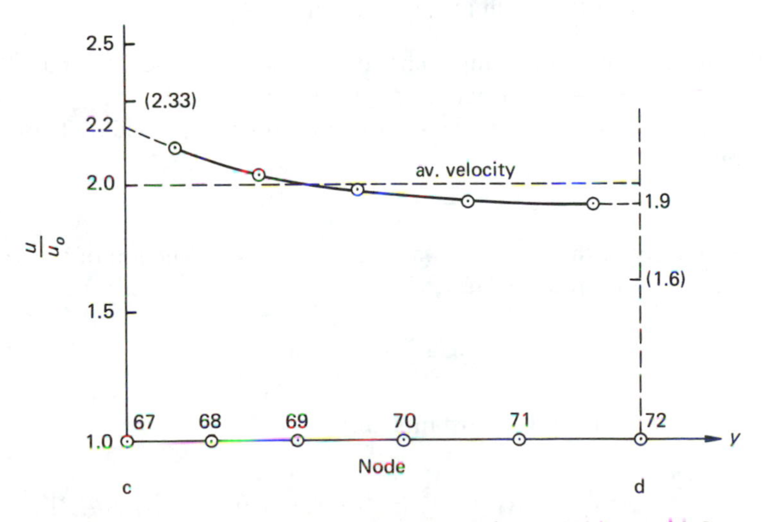

Figure 14-7 Horizontal velocities on the section *c-d* of Figure 14-6 (from Reference 8).

doublets in a horizontal flow. The two results compare well at point c but not at point d; nevertheless, the finite element results are closer to the exact solution, since the computed profile will provide continuity of flow (Fig. 14-7).

The finite element formulation used by Martin[8] is almost identical to the formulation for solid mechanics (Chapters 5 and 6). In fact, the discretization in Figure 14-6(b) is the same as that used for stress analysis of a flat sheet with a circular cut-out.

14-4 EXAMPLE: FLOW OF VISCOUS FLUIDS

Thompson *et al.*[10] considered the flow of a viscous incompressible non-Newtonian fluid for the case of flow sufficiently slow to neglect inertia effects. The variational functional for this case may be written in matrix form as

$$A = \iiint_V (\{\sigma_D\}^T\{\dot{\varepsilon}_D\} - \{u\}^T\{\bar{X}\} - \{\varepsilon_v\}^T\{p\})\, dV - \iint_{S_1} \{u\}^T\{\bar{T}\}\, dS_1 \quad (14.3)$$

where

$\{\dot{\varepsilon}_D\}^T = [\dot{\varepsilon}_{xD}\ \dot{\varepsilon}_{yD}\ \dot{\varepsilon}_{zD}\ \dot{\gamma}_{xy}\ \dot{\gamma}_{yz}\ \dot{\gamma}_{zx}]$

$\{\sigma_D\}$ = deviatoric stress vector, equation (3.4)

$\dot{\varepsilon}_{xD} = \dot{\varepsilon}_x - \dot{\varepsilon}_v = \dot{\varepsilon}_x - (\dot{\varepsilon}_x + \dot{\varepsilon}_y + \dot{\varepsilon}_z)/3$, and so on,

$\{u\}$ = fluid velocity vector

$\{p\}^T = [p\ p\ p\ 0\ 0\ 0]$ = fluid stress vector

and the overdot on strain components denotes strain rates. $\{\bar{X}\}$ and $\{\bar{T}\}$ are the body force and surface traction intensities, respectively.

The constitutive law relating the stresses and strain rates used by Thompson *et al.* was

$$\{\sigma\} = -\{p\} + 2\mu\{\dot{\varepsilon}\} \quad (14.4)$$

μ is the viscosity of the fluid and was assumed to be a function of the second invariant, the strain-rate deviator, $\dot{\varepsilon}'$

$$\mu = \frac{B}{2}\dot{\varepsilon}'^{(1-n)/n} \quad (14.5)$$

where B and n are material constants, and

$$\dot{\varepsilon}' = \frac{\sqrt{2}}{3}[(\dot{\varepsilon}_x - \dot{\varepsilon}_y)^2 + (\dot{\varepsilon}_y - \dot{\varepsilon}_z)^2 + (\dot{\varepsilon}_z - \dot{\varepsilon}_x)^2 + 6(\dot{\gamma}_{xy}^2 + \dot{\gamma}_{yz}^2 + \dot{\gamma}_{zx}^2)]^{1/2}$$

The finite element equations were obtained by extremizing equation (14.3).

$$\begin{bmatrix} [k] & [G]^{\mathrm{T}} \\ [G] & [0] \end{bmatrix} \begin{Bmatrix} \{q\} \\ -\{P\} \end{Bmatrix} = \begin{Bmatrix} \{Q\} \\ \{0\} \end{Bmatrix}$$

Here $[k]$ is the viscous property matrix, $[G]$ is the incompressibility matrix, $\{q\}$ is the vector of nodal fluid velocities, $\{P\}$ is the vector of nodal fluid pressures, and $\{Q\}$ is the nodal force vector.

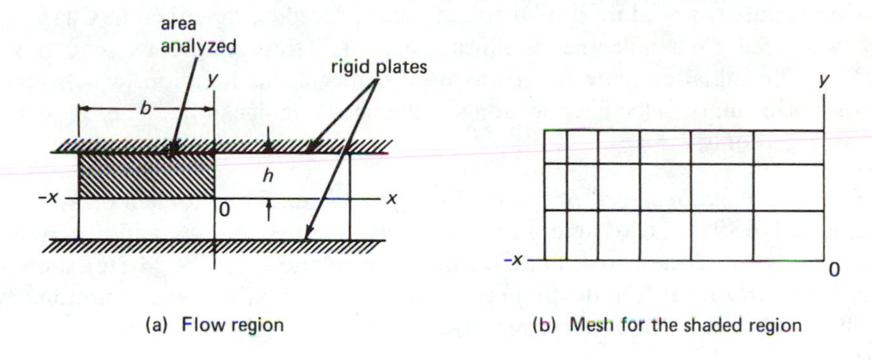

(a) Flow region

(b) Mesh for the shaded region

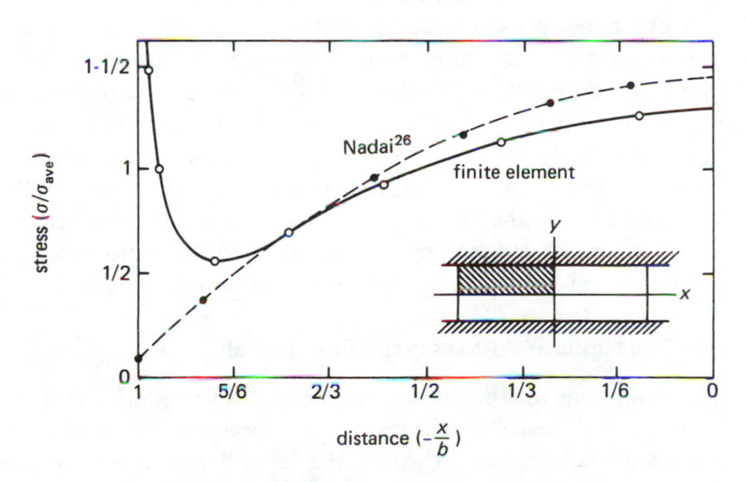

(c) Comparison of the finite element and theoretical results

Figure 14-8 Viscous flow of incompressible non-Newtonian fluid (from Reference 10).

Thompson *et al.* found that triangular finite elements were less satisfactory than rectangular elements in fulfilling the conditions of incompressibility. The following bilinear velocity model was used for the rectangular elements:

$$\begin{Bmatrix} u_x \\ u_y \end{Bmatrix} = \begin{bmatrix} 1 & x & y & xy & 0 & 0 & 0 & 0 \\ 0 & 0 & 0 & 0 & 1 & x & y & xy \end{bmatrix} \begin{Bmatrix} \alpha_1 \\ \alpha_2 \\ \vdots \\ \alpha_8 \end{Bmatrix} \tag{14.6}$$

An iterative procedure similar to the basic procedure described in Chapter 7 was used to handle the nonlinear viscosity stress-strain law, equation (14.5). An initial estimate for μ was used to begin this iteration, which was continued until velocities no longer changed significantly. The process converged only for $n > 1$.

Example: Fluid Squeezed Between Two Rigid Plates. The problem of a fluid squeezed between two rigid parallel plates, Figures 14-8(a) and (b), was solved by using the above finite element formulation. Figure 14-8(c) shows the comparison between the finite element results and a theoretical solution.[26] The agreement is good except near the free surface.

14-5 EXAMPLE: OSCILLATIONS OF HARBORS AND LAKES

If long-period waves excite a resonance condition in a harbor, moored vessels may experience considerable horizontal movements that interfere with cargo operations or damage ships. Some harbors on the Pacific Coast of the United States have persistent problems of this nature. The natural periods and mode shapes of such oscillations are useful information in the design of both new harbors and modifications to existing harbors. In such designs we should avoid configurations with natural periods that match either the predominant excitations or the resonance of the moored ships. Moreover, the locations of wave nodes are regions of maximal horizontal movement; and if possible, mooring positions should not be located at such nodes.

Governing Equations and Associated Functional

The usual assumptions for the linear, long-period wave equations are that the friction can be neglected, the convective accelerations are small, and the velocities are essentially constant through the depth. Under these assumptions, the governing differential equation for the free harmonic oscillations of a body of water is[13]

$$\frac{\partial}{\partial x}\left(h\frac{\partial \eta}{\partial x}\right) + \frac{\partial}{\partial y}\left(h\frac{\partial \eta}{\partial y}\right) + \frac{4\pi^2}{gT^2}\eta = 0 \tag{14.7a}$$

h is the depth, η is the water surface elevation with respect to the mean water level, and T is the period of oscillation. The boundary conditions at the solid boundaries are that the normal component of velocity is zero

$$\frac{\partial \eta}{\partial n} = 0 \tag{14.7b}$$

Equation (14.7a) is a special form of the Helmholz equation.[14]

The associated functional for equation (14.7a) is[13]

$$\Pi = \iint_A \frac{1}{2}\left[h\left(\frac{\partial \eta}{\partial x}\right)^2 + h\left(\frac{\partial \eta}{\partial y}\right)^2 - \lambda^2 \eta^2\right] dA \tag{14.8a}$$

The Euler equation for the stationary value of Π is equation (14.7a), and the natural boundary conditions corresponding to equation (14.7b) are

$$h\frac{\partial \eta}{\partial n} = 0 \tag{14.8b}$$

Finite Element Formulation

Taylor *et al.*[13] have used the finite element method to determine the natural periods and modes of harbors that may be approximated as completely enclosed bodies of water. This problem is identical to the seiche problem of lakes. Triangular elements were used and a linear model of water surface elevation was assumed.

$$\eta = \underset{1\times 3}{\{N\}^{\mathrm{T}}}\underset{3\times 1}{\{\eta\}} = [L_1 \;\; L_2 \;\; L_3]\{\eta\}$$

The gradients of η were given by

$$\begin{Bmatrix} \dfrac{\partial \eta}{\partial x} \\[2ex] \dfrac{\partial \eta}{\partial y} \end{Bmatrix} = \frac{1}{2A}\begin{bmatrix} b_1 & b_2 & b_3 \\ a_1 & a_2 & a_3 \end{bmatrix}\{\eta\} = [B]\{\eta\}$$

By substituting into the functional and obtaining the stationary value, Taylor *et al.* obtained the following element equations

$$[k]\{\eta\} = \lambda^2[m]\{\eta\}$$

where

$$\lambda^2 = 4\pi^2/gT^2$$

$$[k] = \bar{h}A[B]^{\mathrm{T}}[B]$$

$$[m] = \iint_A \{N\}\{N\}^{\mathrm{T}}\, dA$$

and where $\bar{h}$ is the average depth of the element. The assemblage equations obtained from the direct stiffness method were

$$[K]\{\eta\} = \lambda^2[M]\{\eta\}$$

This eigenvalue problem was converted to standard form by using the methods described in Section 2-3. However, to make $[K]$ nonsingular, Taylor *et al.* had to apply a small perturbation to a single element of the matrix. This is the equivalent of weakly restraining the pumping mode of the enclosed harbor, which would otherwise have an infinite period.

It is difficult to obtain a satisfactory idealization of a totally enclosed harbor. An alternative approximation is to assume a fixed location of a wave node across the entrance to the harbor. At this location, the geometric boundary conditions $\eta = 0$ can be applied. This renders the $[K]$ matrix nonsingular and avoids the necessity for utilizing the device of perturbation.

Example: Oscillations of Rectangular, Constant-Depth Harbor. Abel[15] developed a finite element code on the basis of the work of Taylor *et al.*[13] He used 4-CST elements and a lumped approximation to the matrix $[M]$ (Section 11-2). The element resultants computed for each mode shape were the horizontal velocities and the horizontal displacements.

The periods computed for rectangular, constant-depth harbors with open and closed entrances (Figure 14-9) are given in Tables 14-2 and 14-3, respectively. Good agreement with the theoretical values is obtained for the most fundamental modes.

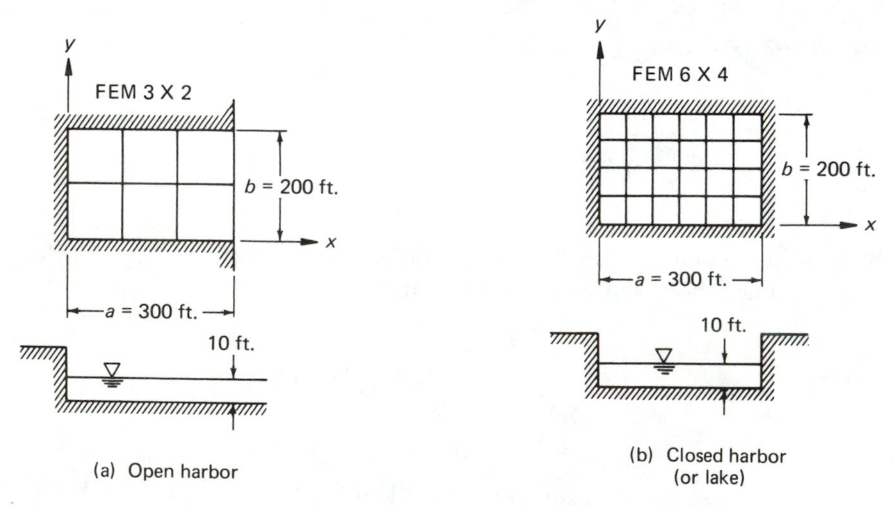

Figure 14-9 Open and closed rectangular harbors.

TABLE 14-2 COMPARISON OF PERIODS: OPEN HARBOR.

				FEM 3 × 2		FEM 6 × 4		FEM 12 × 8	
Mode No.	*m*	*n*	*T*†	*T*	% *Error*	*T*	% *Error*	*T*	% *Error*
1	1	0	66.90	67.67	1.15	67.09	0.29	66.95	0.075
2	3	0	22.30	24.77	11.1	22.88	2.7	22.44	0.63
3	1	1	21.16	23.98	13.3	21.83	3.2	21.32	0.76
4	3	1	15.77	20.22	28.2	16.81	6.6	16.02	1.58
5	5	0	13.38	18.13	35.5	14.39	7.5	13.62	1.80
6	5	1	11.47	17.82	55	12.87	13.3	11.81	2.97
7	1	2	11.00			12.28	11.7	11.30	2.73
8	3	2	9.97			11.57	16.0	10.35	3.81
9	7	0	9.56			11.04	15.5	9.90	3.55
10	7	1	8.78			10.59	20.6	9.20	4.78
11	5	2	8.57			10.58	23.5	9.04	5.50

† Theoretical solution, $\eta = \cos \dfrac{m\pi x}{2a} \cos \dfrac{n\pi y}{b}$

TABLE 14-3 COMPARISON OF PERIODS: CLOSED HARBOR.

				FEM 3 × 2		FEM 6 × 4		FEM 12 × 8	
Mode No.	*m*	*n*	*T*†	*T*	% *Error*	*T*	% *Error*	*T*	% *Error*
1	1	0	33.45	33.99	1.61	33.52	0.21	33.46	0.0003
2	0	1	22.30	24.34	9.15	22.70	1.80	22.39	0.40
3	1	1	18.55	21.40	15.36	19.11	3.02	18.68	0.70
4	2	0	16.73	19.87	18.77	17.36	3.77	16.88	0.90
5	2	1	13.38	18.30	36.80	14.41	7.70	13.62	1.80
6	0	2	11.15			12.38	11.66	11.44	2.60
7	3	0	11.15			12.30	10.31	11.40	2.24
8	1	2	10.58			11.88	12.29	10.87	2.74
9	3	1	9.97			11.44	14.74	10.31	3.41
10	2	2	9.28			10.94	17.88	9.66	4.10
11	4	0	8.36			10.11	20.93	8.74	4.55
12	3	2	7.88			10.03	27.28	8.38	6.35

† Theoretical Solution, $\eta = \cos \dfrac{m\pi x}{a} \cos \dfrac{n\pi y}{b}$

REFERENCES

(1) See Reference 32, Chapter 8. Applications to viscoelastic behavior of thick-wall cylinder encased in an elastic shell, and a two-dimensional analysis of star grain of solid propellants.

(2) Nickell, R. E., "Stress Wave Analysis in Layered Thermoviscoelastic Materials by the Extended Ritz Method," Tech. Report S-175, Vol. II, Rohm and Haas Co., Huntsville, Ala., Oct. 1968, NTIS, Springfield, Va., AD 841 224. Develops solutions for stress wave propagation in thermoviscoelastic bodies by employing the extended Ritz procedure.

(3) See Reference 33, Chapter 8. Viscoelastic analyses of solid propellant grains, involving a cylinder with steel casing and a deep tunnel in rock with concrete liner.

(4) Oden, J. T., "A Generalization of the Finite Element Concept and its Application to a Class of Problems in Nonlinear Viscoelasticity," *Proc. Developments in Theoretical and Applied Mechanics*, Vol IV, IV SECTAM, London, Pergamon Press, 1968. The finite element equations are derived from energy balances associated with isothermal deformations. Behavior of a viscoelastic cuboid under simple shear conditions is studied.

(5) White, J. L., "Finite Elements in Linear Viscoelasticity," *Second Conf.* Presents a direct finite element formulation for plane strain analysis of thermoviscoelastic problems.

(6) See Reference 34, Chapter 8.

(7) See Reference 7, Chapter 12.

(8) See Reference 4, Chapter 8.

(9) Doctors, L. J., "An Application of the Finite Element Technique to Boundary Value Problems of Potential Flow," *IJNME*, Vol. 2, No. 2 (1970). Considers two-dimensional inviscid potential fluid flow. Handles flow conditions on a non-conducting boundary by introducing a fictitious node. Considers four possible boundary conditions: non-conducting, conducting, velocity-specified, and singularity-specified. Solves the problem of flow around a circular cylinder.

(10) Thompson, E. G., Mack, L. R., and Lin, F. S., "Finite-Element Method for Incompressible Slow Viscous Flow with a Free Surface," *Proc. 11th Midwestern Mech. Conf.*, Developments in Mechanics, Vol. 5, 1969.

(11) Oden, J. T., and Somogyi, D., "Finite-Element Applications in Fluid Dynamics," *Proc. ASCE, J. EM Dn*, Vol. 95, EM 3 (June 1969). Presents a finite element formulation for the flow of viscous, incompressible Newtonian fluids. Adopts fluid velocity as the unknown field variable. Derives the finite element equations from a functional expressed in terms of the viscosity and velocities. Both steady and unsteady flows are considered.

(12) See Reference 13, Chapter 13.

(13) Taylor, C., Patil, B. S., and Zienkiewicz, O. C., "Harbour Oscillation: A Numerical Treatment for Undamped Natural Modes," *Proc. Inst. of Civ. Engrs.*, Vol. 43 (June 1969).

(14) Arlett, P. L., Bahrani, A. K., and Zienkiewicz, O. C., "Application of Finite Elements to the Solution of Helmholz's Equation," *Proc. IEE*, Vol. 115, No. 12 (Dec. 1968). Solves the wave guide problem in which the Neumann and Dirich-

let boundary conditions are involved. Develops various three-dimensional elements such as a tetrahedron and 8-cornered element composed of five tetrahedra. Solves problems such as a dielectric-loaded wave guide and a lunar wave guide.

(15) Abel, J. F., Unpublished work conducted at the USAEWES, Vicksburg, Miss.

(16) Nielsen, H. B., "A Finite Element Method for Calculating a Two-dimensional Open Channel Flow," *Basic Research Progress Report No. 19*, Tech. Univ. of Denmark, Copenhagen (August 1969). Uses a finite element formulation for determining the fluid pressure as the unknown field variable. By using the computed pressures, heights, inclination of streamlines, and velocities are computed as element resultants. Since the basic momentum equation is nonlinear, an iterative scheme is used for the solution of the finite element equations.

(17) Leonard, J. W., "Compressible Flow by the Finite Element Method," Tech. Note TCTN-9500-920156, Bell Aerosystems Co., Niagara Falls, N. Y. 1969. Analysis of steady compressible flow of an isentropic perfect gas governed by the linearized equation. Uses Galerkin's residual method.

(18) Allik, H., and Hughes, T. J. R., "Finite Element Method for Piezoelectric Vibration," *IJNME*, Vol. 2, No. 2 (1970). On the basis of a variational principle, obtains the finite element equations for a tetrahedral element for analysis of piezoelectric or electrostatic effects. Draw a close analogy between various quantities involved in the finite element formulations for solid mechanics and for electrical fields.

(19) See Reference 11, Chapter 12.

(20) Bramlette, T. T., and Mallett, R. H., "A Finite Element Solution Technique for the Boltzmann Equation," *J. Fluid Mech.*, Vol. 42, Part 1 (1970). On the basis of Galerkin technique, develops a finite element formulation for the problem of dynamic behavior of gases at arbitrary Knudson numbers. The formulation is applied to the problem of one-dimensional linearized Couette flow.

(21) See Reference 7, Chapter 8.

(22) Kaufman, R. I., and Weaver, F. J., "Stability of Atchafalaya Levees," *Proc. ASCE, J. SM&F Dn*, Vol. 93, SM4 (July 1967).

(23) "Prediction of the Deformation of a Levee on a Soft Foundation," Res. Report R-69-18, MIT, Cambridge, to U.S. Army Corps of Engineers, New Orleans District, La. (Dec. 1968).

(24) See Reference 22, Chapter 10.

(25) "Soil Instrumentation for Interstate 95 Embankment, Saugus, Massachusetts," Research Report No. R-69-10, Soil Mech. Div., MIT, Cambridge, Mass. (1969).

(26) Nadai, A., *Theory of Flow and Fracture of Solids*," Vol. II, New York, McGraw-Hill Book Co., 1963.

FURTHER READING

Appa, K., "Kinematically Consistent Unsteady Aerodynamic Coefficients in Supersonic Flow," *IJNME*, Vol. 2, No. 4 (Oct–Dec. 1970).

Fischer, H. B., "A Lagrangian Method for Predicting Pollutant Dispersion in Bolinas Lagoon, California," Open File Report, U.S. Dept. of Int., Geological

Survey, Menlo Park, Calif. (December 1969). Broadly speaking, this is a finite element procedure for predicting the movement and dispersion of a pollutant in a tidal embayment. One-dimensional elements and the direct method of formulation are used.

Heitner, K. L., and Housner, G. W., "Numerical Model for Tsunami Run-Up," *Proc. ASCE, J. WW Dn.*, Vol. 96, WW3 (Aug. 1970). Obtains a finite element formulation for computation of wave run-up caused by tsunamis. The numerical results agreed well with experiments for run-up of solitary waves.

Kobayashi, A. S., and Woo, S. L. Y., "Analysis of Biological Structures," *Recent Advances.*

Leonard, J. W., and Bramlette, T. T., "Finite Element Solutions to Differential Equations," *Proc. ASCE, J. EM Dn.*, Vol. 96, EM6 (Dec. 1970). Uses Galerkin's method to derive formulation of a general class of field equations for combined initial-value and boundary-value problems in applied mathematics. Example given is Prandtl-Meyer steady compressible flow of an isentropic gas.

Lynch, F. De S., "A Finite Element Method of Viscoelastic Stress Analysis with Application to Rolling Contact Problem," *IJNME*, Vol. 1, No. 4 (1969).

McCorquodale, J. A., and Li, C. Y., "Finite Element Analysis of Sluice Gate Flow," *Trans, J. Engg. Inst. of Canada*, Vol. 54, No. 3 (Mar. 1971). Presents a procedure for computation of hydraulic properties of sluice gates with two-dimensional, irrotational flow.

Oden, J. T., "Finite-Element Analogue of Navier-Stokes Equation," *Proc. ASCE, J. EM Dn.*, Vol. 96, EM4 (Aug. 1970).

Orlob, G. T., Shubinski, R. P., and Feigner, K. D., "Mathematical Modelling of Water Quality in Estuarial Systems," Natl. Symp. on Estuarine Pollution, Stanford Univ., Aug. 1967. This network approach to two-dimensional problems is the homologue of the one-dimensional finite element method. Considers simultaneously the hydrodynamics and transport of conservative substance in a well mixed estuary.

Various authors, "Section 5: Analysis of Flow and Special Problems," *Recent Advances.*

15

CONCLUSION—ADVANTAGES AND LIMITATIONS OF THE METHOD

Undoubtedly several of the advantages and limitations of the finite element method have already become apparent to the reader. However, we shall conclude with a brief evaluation of the method, including a summary of the important advantages and limitations.

15-1 ADVANTAGES

Like all numerical approximations, the finite element method is based on the concept of discretization. Nevertheless, as either a variational or residual approach, the technique recognizes the multi-dimensional continuity of the body. Not only does the idealization portray the body as continuous, but it also requires no separate interpolation process to extend the approximate solution to every point within the continuum. Despite the fact that a solution is obtained at a finite number of discrete node points, the formulation of the field variable models inherently provides a solution at all other locations in the body. In contrast to other variational and residual approaches, the finite element method does not require trial solutions which must all apply to the entire multidimensional continuum. The use of separate subregions, or finite elements, for the separate trial solutions thus permits a greater flexibility in considering continua of complex shape.

Some of the most important advantages of the finite element method derive from the techniques of introducing boundary conditions. This is another area in which the method differs from other variational or residual approaches. Rather than requiring every trial solution to satisfy the boundary conditions, one prescribes the conditions after obtaining the algebraic equations for the assemblage. Since the boundary conditions do not enter into the equations for the individual finite elements, one can use the same field variable models for both internal and boundary elements. Moreover, the field variable models need not be changed when the boundary conditions change.

The introduction of boundary conditions into the assembled equations is a relatively easy process. It is simplified in that only the geometric boundary conditions need be specified in a variational approach because the natural conditions are implicitly satisfied. No special techniques or artificial devices are necessary, such as the non-centered difference equations or fictitious external points often employed in the finite difference method.

The finite element method not only accommodates complex geometry and boundary conditions, but it also has proven successful in representing various types of complicated material properties that are difficult to incorporate into other numerical methods. For example, formulations in solid mechanics have been devised for anisotropic, nonlinear, hysteretic, time-dependent, or temperature-dependent material behavior.

One of the most difficult problems encountered in applying numerical procedures of engineering analysis is the representation of nonhomogeneous continua. Nevertheless, the finite element method readily accounts for nonhomogeneity by the simple tactic of assigning different properties to different elements. If a refined representation of the variation of material characteristics is desired, it is even possible to vary the properties within an element according to a preselected polynomial pattern. For instance, it is possible to accommodate continuous or discontinuous variations of the constitutive parameters or of the thickness of a two-dimensional body.

The systematic generality of the finite element procedure makes it a powerful and versatile tool for a wide range of problems. As a result, flexible general-purpose computer programs can be constructed. Primary examples of these programs are the several structural analysis packages, which include a variety of element configurations and which can be applied to several categories of structural problems. Among these packages are ASKA,† STRUDL,‡ SAP,§ NASTRAN,|| and SAFE.¶ Another indicator of the

† Automatic System for Kinematic Analysis, Institute for Statics and Dynamics, Stuttgart, West Germany.

‡ STRUctural Design Language, Integrated Civil Engineering System (ICES), Massachusetts Institute of Technology.

generality of the method is that programs developed for one field of engineering have been applied successfully to problems in a different field with little or no modification.

Finally, an engineer may develop a concept of the finite element method at different levels. It is possible to interpret the method in physical terms. On the other hand, the method may be explained entirely in mathematical terms. The physical or intuitive nature of the procedure is particularly useful to the engineering student and practicing engineer. Nevertheless, it is significant that the method has mathematical foundations.

15-2 LIMITATIONS

One limitation of the finite element method is that a few complex phenomena are not accommodated adequately by the method at its current state of development. Some examples of such phenomena from the realm of solid mechanics are cracking and fracture behavior, contact problems, bond failures of composite materials, and nonlinear material behavior with work softening. Another example is transient, unconfined seepage problems. The numerical solution of propagation or transient problems is not satisfactory in all respects. Many of these phenomena are presently under research, and refinements of the method to accommodate these problems better can be expected.

The finite element method has reached a high level of development as a solution technique; however, the method yields realistic results only if the coefficients or material parameters which describe the basic phenomena are available. Material nonlinearity in solid mechanics is a notable example of a field in which our understanding of the material behavior has lagged behind the development of the analytical tool. In order to exploit fully the power of the finite element method, significant effort must be directed toward the development of suitable constitutive laws and the evaluation of realistic coefficients and material parameters.

Even the most efficient finite element computer codes require a relatively large amount of computer memory and time. Hence, use of the method is limited to those who have access to relatively large, high-speed computers. The advent of time-sharing, remote batch processing, and computer service bureaus or utilities has alleviated this restriction to some degree. In addition,

§ Structural Analysis Program, Professor E. L. Wilson, University of California, Berkeley.
|| NAsa STRuctural ANalysis, U.S. National Aeronautics and Space Administration.
¶ Structural Analysis by Finite Elements, Gulf Computer Sciences, Inc., and Gulf General Atomic, Inc.

the method can be applied indirectly to common engineering problems by utilizing tables, graphs, and other analysis aids that have been generated by finite element codes.

The most tedious aspects of the use of the finite element method are the basic processes of subdividing the continuum and of generating error-free input data for the computer. Although these processes may be automated to a degree, they have not been totally accomplished by computer because some engineering judgment must be employed in the discretization. Errors in the input data may go undetected and the erroneous results obtained therefrom may appear acceptable. Consequently, it is essential that the engineer/programmer provide checks to detect such errors. In addition to checks internal to the code, an auxiliary routine that reads the input data and generates a computer plot of the discretized continuum is desirable. This plot permits a rapid visual check of the input data.

Finally, as for any approximate numerical method, the results of a finite element analysis must be interpreted with care. We must be aware of the assumptions employed in the formulation, the possibility of numerical difficulties, and the limitations in the material characterizations used. A large volume of solution information is generated by a finite element routine, but this data is worthwhile only when its generation and interpretation are tempered by proper engineering judgment.

15-3 CONCLUSIONS

As of 1971, the finite element method has attained a plateau of maturity as an analysis technique. Future developments will probably consist of refinement of the techniques and a broader range of application, rather than major changes in the analysis procedure.

As evidenced in Parts B and C, a number of diverse element formulations were developed and applied during the 1960's. However, the trend appears to be toward a wider utilization of the isoparametric formulation. This approach has proven to be a simple, but elegant, generalization of the basic analysis of a single finite element.

FURTHER READING

Gallagher, R. H., "An Overview and Some Projections," *Recent Advances.*
See Reference 1, Chap. 8.
See Reference 8, Chap. 1.

APPENDIX I
ILLUSTRATIVE COMPUTER CODE

I-1 INTRODUCTION

This plane strain/stress computer code is a relatively simple one, written in FORTRAN IV to illustrate some of the basic concepts of Chapters 5 and 6. It will provide the beginner with an idea of how the important equations of the finite element method may be translated into computer instructions. Therefore, a number of comment cards have been inserted to provide cross-references to the appropriate sections and equations of the book. Moreover, the symbols used correspond as closely as possible to those in the text.

Because the code is an introductory one, it is not necessarily efficient for large scale problems. There is, however, a growing number of finite element codes being published, including the general purpose languages cited in Chapter 15 and in the Further Reading section of this appendix. The reader is referred to this body of literature for more advanced and efficient coding techniques.

The code is limited to linear, elastic, plane strain, or plane stress analysis of isotropic bodies. Only one loading case may be accommodated for each problem. The elements used are 4-CST quadrilaterals† (Example 6-2), and/or constant strain triangles. In its present form, the code has sufficient storage for 120 nodes, 100 elements, 20 surface tractions, and 10 different materials. It requires a machine with

† The 4-CST element is chosen to demonstrate the closed form integration in the element formulation and the condensation of an internal node. Consequently, this code does not include the numerical integration used with isoparametric elements.

a core storage of about 30 K. However, it is possible to change the capacity by modifying the COMMON, DIMENSION, and DATA statements. Since the number of elements is small, the code makes minimal use of peripheral storage; that is, it uses only one scratch tape. A large number of equations would require greater peripheral storage, as indicated by the partitioning schemes presented in Section 2-2.

One pitfall in employing a finite element code is data error. The present code contains a number of checks of the input data. When errors are located they are described, and execution is stopped. However, the checks in the code do not cover all possibilities. One method of checking input data more thoroughly is a special routine that generates a computer plot of the discretized structure and labels all nodes and elements in this plot. Such drawings will reveal most errors in the element and nodal data.

To further explain the computer code, most of the features of a user's manual are included in the subsequent sections. An explanation of the various subroutines is followed by a flow chart, a dictionary of variables, a guide for data input, a listing of the code, and sample input and output. Finally, possible modifications to the code are discussed.

I-2 EXPLANATION OF SUBROUTINES

Most of the computational steps are carried out in the various subroutines of the code. These subroutines are controlled by the main routine. The main routine also reads the title of a problem, computes the semi-band width, solves the overall equilibrium equations, and prints the displacements. Following is a list of subroutines with an explanation of their functions.

DATAIN	Reads and echo prints all input data. Performs checks for data.
ASEMBL	Initializes and assembles overall stiffness matrix and load vector. Temporarily stores data needed for later computation. Introduces geometric boundary conditions.
QUAD	Computes stress-strain matrix, stiffness matrix, body force vector, and strain-displacement matrix of either a 4-CST quadrilateral element or a triangular element.
CST	Computes strain-displacement matrix, stiffness matrix, and body force vector of a constant strain triangle (CST) element.
GEOMBC	Applies prescribed displacement boundary conditions at a single node.
BANSOL	Triangularizes the overall banded stiffness matrix by symmetric Gauss-Doolittle decomposition *or* solves for displacement vector corresponding to a particular load vector.
STRESS	Computes the strains, stresses, and principal stresses. Prints the stresses and principal stresses at element centroids.

I-3 FLOW CHART

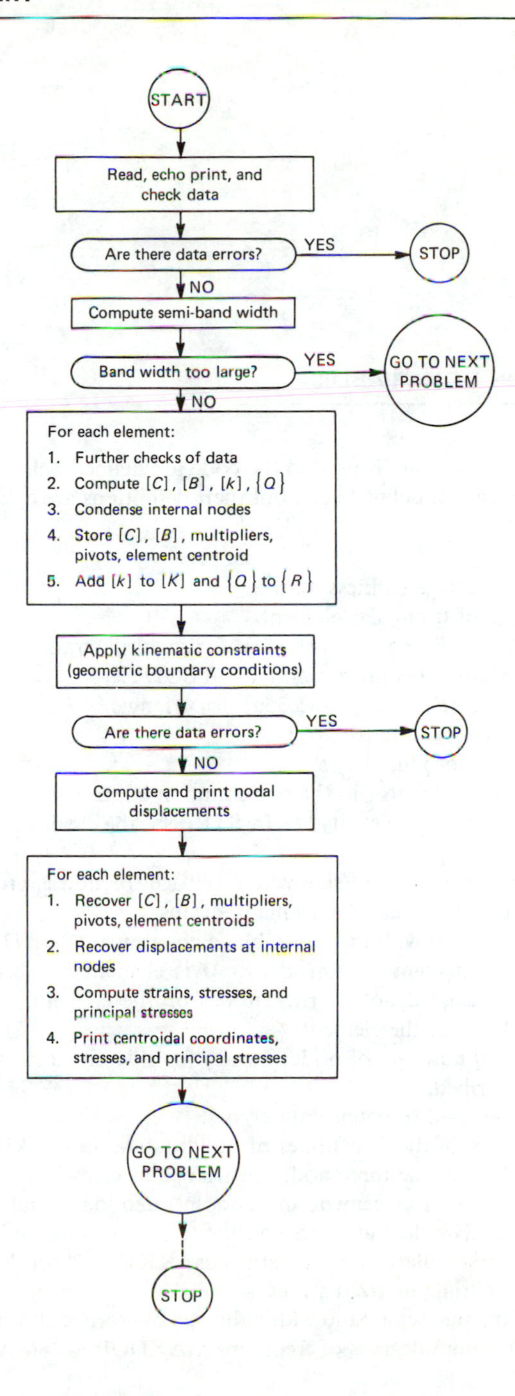
START
Read, echo print, and check data
Are there data errors?
YES
STOP
NO
Compute semi-band width
Band width too large?
YES
GO TO NEXT PROBLEM
NO
For each element:
1. Further checks of data
2. Compute [C], [B], [k], {Q}
3. Condense internal nodes
4. Store [C], [B], multipliers, pivots, element centroid
5. Add [k] to [K] and {Q} to {R}
Apply kinematic constraints (geometric boundary conditions)
Are there data errors?
YES
STOP
NO
Compute and print nodal displacements
For each element:
1. Recover [C], [B], multipliers, pivots, element centroids
2. Recover displacements at internal nodes
3. Compute strains, stresses, and principal stresses
4. Print centroidal coordinates, stresses, and principal stresses
GO TO NEXT PROBLEM
STOP

I-4 DEFINITION OF VARIABLES

The major arrays and symbols used in the code are defined below. Some temporary storage variables are not defined here, but their definitions are evident from the context.

AK	Assemblage stiffness matrix.
AREA	Area of triangular element.
B	Matrix $[B]$, equation (5.35a), for quadrilateral.
BODYF	Nodal load contribution from TBODY.
BT	Matrix $[B]$, equation (5.35a), for triangle.
C	Stress-strain matrix, $[C]$.
CB	Matrix product $[C][B]$.
CF	Commom factor in the computation of stress-strain matrix, $[C]$.
E	Modulus of elasticity, E. In STRESS, the strains at the element centroid.
EL	Length of element side where surface traction is prescribed.
FAC	Factor for averaging element strains.
IBAND	Semi-band width of assemblage equations, IBAND $\leq$ MAXBW.
IE(M,I)	Element identification array. M is element number, $1 \leq M \leq$ NEL. $I = 1,2,3,4$ denote corner nodes of the element, and $I = 5$ denotes MTYP for the element.
ISC,JSC	Nodal numbers of nodes i and j for side on which surface traction is prescribed.
ISTOP	Index used to count data errors.
I,J,K,L	Indices of the four nodes of quadrilateral in QUAD.
I,J,K	Indices of the three nodes of triangular element in CST.
KODE(I)	Index of displacement and concentrated load conditions at node I.
KKK	In BANSOL, index designating function to be performed. KKK = 1 for triangularization of stiffness. KKK = 2 for backward solution using triangularized stiffness.
MAXBW	Maximum semi-band width allowed by storage allocation declarations.
MAXDOF	Maximum degrees of freedom, MAXDOF = 2*MAXNP.

MAXEL	Maximum number of elements allowed by storage allocation declarations.
MAXMAT	Maximum number of materials allowed by storage allocation declarations.
MAXNP	Maximum number of nodes allowed by storage allocation declarations.
MAXSLC	Maximum number of surface traction cards allowed by storage allocation declarations.
MDIM	Maximum band width, MDIM = MAXBW.
MTYP	Material type number, 1 ≤ MTYP ≤ NMAT.
NBODY	Option for body force. NBODY = 0 for no weight, NBODY = 1 for weight force in the negative y direction.
NDIM	Maximum degrees of freedom, NDIM = MAXDOF.
NEL	Number of elements, NEL ≤ MAXEL.
NEQ	Total number of equations, NEQ = 2*NNP ≤ MAXDOF.
NMAT	Number of different materials, 1 ≤ NMAT ≤ MAXMAT.
NNP	Number of nodal points, NNP ≤ MAXNP.
NOLINE	In STRESS, index to limit output to 50 lines per page.
NOPT	Option for plane strain/stress. NOPT = 1 for plane strain, NOPT = 2 for plane stress.
NPROB	Problem number.
NSLC	Number of surface traction cards, NSLC ≤ MAXSLC.
PR	Poisson's ratio, ν.
PXI, PXJ	Nodal contributions of surface tractions in x-direction.
PYI, PYJ	Nodal contributions of surface traction in y-direction.
Q	Load vector of quadrilateral element. In STRESS, the element displacement vector.
QK	Stiffness matrix of quadrilateral element.
R	Assemblage load vector. Also computed displacements for the assemblage in MAIN, BANSOL and STRESS.
RO	Weight density of material.
SIG	Array for stresses.
SURTRX, SURTRY	x and y components of prescribed distributed tractions.
TBODY	Total weight of triangular element.
TH	Thickness, h.
TITLE	Array for title of the problem (72 alphanumeric characters).
TK	Stiffness matrix of triangular element.
TOTALA	Total area of quadrilateral element.
U	Prescribed displacement in x or y direction in GEOMBC.
ULX(I), VLY(I)	Concentrated load or displacement in x and y directions at node I.
X(I), Y(I)	x and y coordinates of node I.
XQ, YQ	Coordinates of the nodes of a quadrilateral or triangular element. (XQ(5) and YQ(5) are coordinates of centroid.)

I-5 GUIDE FOR DATA INPUT

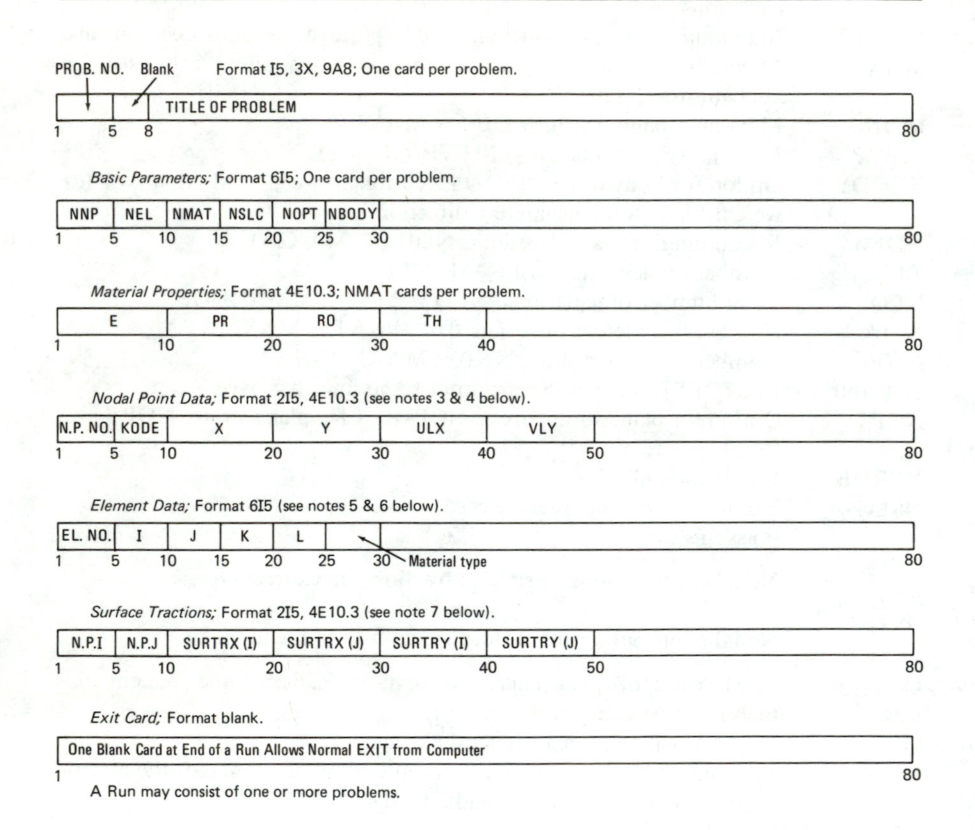

Notes on Input Data

1. Data cards must be in proper sequence.
2. Units must be consistent.
3. Usually one card is needed for each node. However, if some nodes fall on a straight line and are equidistant, data for only the first and the last points of this group are needed. Intermediate nodal point data are automatically generated by linear interpolation.

4. Forces and/or displacements prescribed at a node are identified by KODE as explained below:

KODE	*Force/Displacement Boundary Condition*
0	ULX = Prescribed Load in x direction
	VLY = Prescribed Load in y direction
1	ULX = Prescribed Disp in x direction
	VLY = Prescribed Load in y direction
2	ULX = Prescribed Load in x direction
	VLY = Prescribed Disp in y direction
3	ULX = Prescribed Disp in x direction
	VLY = Prescribed Disp in y direction

The sign of an applied force or displacement follows the sign of the coordinate directions. For instance, a force in the positive x direction is positive, and so on. For the nodes automatically generated as in Note 3, KODE = 0, ULX = 0 and VLY = 0 are assigned for the generated nodes.

5. IE(M,1), IE(M,2), IE(M,3), IE(M,4) denote four corner nodes, I,J,K,L, of a quadrilateral element, *M*. The program also permits use of triangular elements, which are indicated by repeating the third node; that is, IE(M,3) = IE(M,4), or K = L. For a right-handed coordinate system the nodes must be input counterclockwise around the element. IE(M,5) denotes the type of material in the element.

The maximum difference between numbers of any two nodes for a given element must be less than MAXBW/2.

6. Usually one card is needed for each element. However, if some elements are on a line in such a way that their corner node indices each increase by one compared to the previous element, only the data for the first element on the line need be input. However, note that data for the last element of the assemblage must be input. For example, in Exercise 6-3, data for only elements 1,4,7,10, and 12 are needed. The omitted element data is generated internally by the computer. The same material type as the previously input element is assigned to all generated elements.

7. Surface tractions must be specified between two adjacent nodes only. The three possible cases are shown in Fig. I-1. For case (a) only SURTRX(I) and (J) are input, and columns 31–50 are left blank. For case (b) only SURTRY(I) and (J) are input and columns 11–30 are left blank. For both tractions all columns from 1 to 50 are input. For case (c) the user may need to compute the components of tractions manually. Moreover, the user must multiply all surface intensities by the thickness of the element before the intensities are input in the computer.

The signs of tractions follow the directions of coordinate axes. A traction in the negative y direction is negative, and so on.

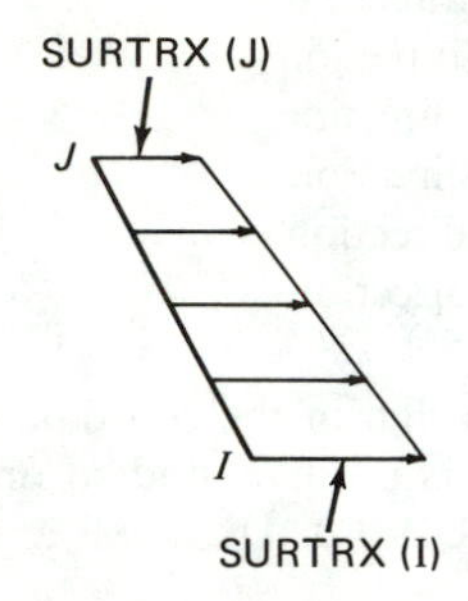

(a) Tractions in x direction only

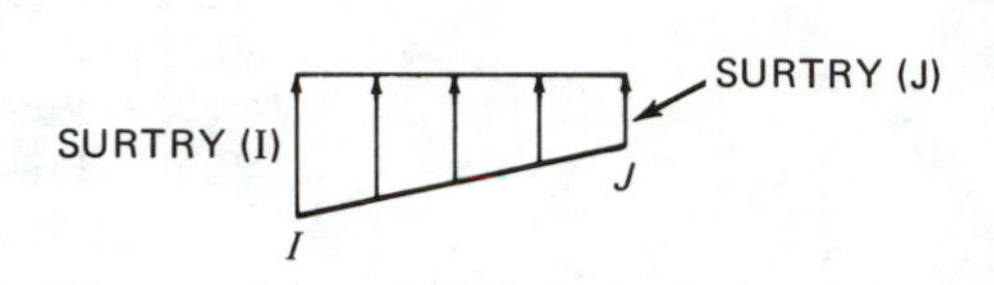

(b) Tractions in y direction only

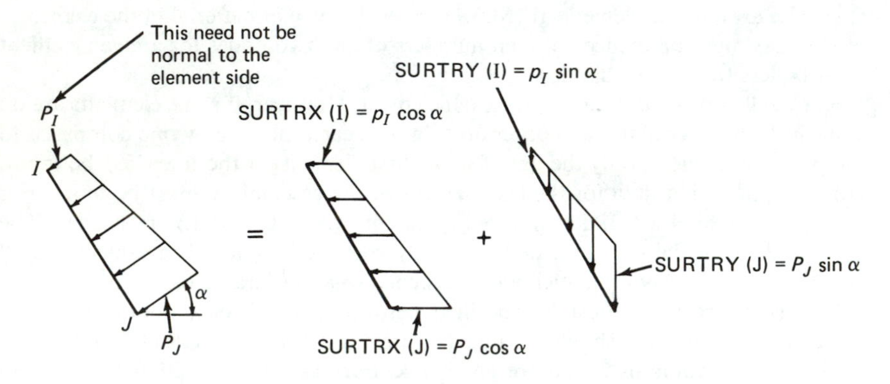

(c) Tractions in both x and y directions

Figure I-1 Three possible cases of surface tractions on element side I-J.

I-6 FORTRAN LISTING

```
C
C  INTRODUCTION TO THE FINITE ELEMENT METHOD - DESAI-ABEL
C  EXAMPLE CODE (PROGRAM) FOR PLANE STRAIN/STRESS
C
      DIMENSION  TITLE(9)
      COMMON NNP,NEL,NMAT,NSLC,NOPT,NBODY,MTYP,
     1E(10),PR(10),RO(10),TH(10),IE(100,5),
     2X(120),Y(120), ULX(120),VLY(120),KODE(120),ISC(20), JSC(20),
     3SURTRX(20,2),SURTRY(20,2)
      COMMON/ONE/ QK(10,10),Q(10),B(3,10),C(3,3),BT(3,6),XQ(5),YQ(5)
      COMMON/TWO/ IBAND, NEQ, R(240), AK(240,50)
      DATA MAXEL, MAXNP, MAXMAT, MAXBW, MAXSLC
     1    /  100,   120,     10,    50,     20/
C
C PROBLEM  IDENTIFICATION  AND  DESCRIPTION
C
 9999 READ 100,NPROB, (TITLE(I),I=1,9)
          IF(NPROB.LE.0) GO TO 999
 1020 PRINT 200,NPROB,(TITLE(I),I=1,9)
          CALL DATAIN (MAXEL,MAXNP,MAXMAT,MAXSLC,ISTOP)
               MAXDOF = 2*MAXNP
C
C COMPUTE  MAX. NODAL DIFFERENCE AND SEMI-BANDWIDTH, EQ. (6-1)
               MAXDIF = 0
          DO 1 I=1,NEL
          DO 1 J=1,4
          DO 1 K=1,4
               LL= IABS(IE(I,J)- IE(I,K))
          IF(LL.GT.MAXDIF) MAXDIF = LL
    1     CONTINUE
               IBAND = 2*(MAXDIF + 1)
               NEQ = 2*NNP
          IF(IBAND.GT.MAXBW) GO TO 900
          IF(ISTOP.GT.0) GO TO 999
          CALL ASEMBL(ISTOP)
          IF(ISTOP.GT.0) GO TO 999
C
C  TRIANGULARIZE STIFFNESS MATRIX, EQ. (2-2), KKK=1
          CALL BANSOL(1,AK,R,NEQ,IBAND,MAXDOF,MAXBW)
C
C  SOLVE FOR DISPLACEMENTS CORRESP. TO LOAD VECTOR R, EQ.(2-3), KKK=2
          CALL BANSOL(2,AK,R,NEQ,IBAND,MAXDOF,MAXBW)
      PRINT 300, (I, R(2*I-1),R(2*I),I=1,NNP)
C
          CALL STRESS
C
          GO TO 9999
  900 PRINT 901, IBAND, MAXBW
      GO TO 9999
  100 FORMAT(I5,3X,9A8)
  200 FORMAT(/8H1PROBLEM,I5,3H.. ,9A8/)
  300 FORMAT(37H1OUTPUT TABLE 1.. NODAL DISPLACEMENTS //
     1       13X,4HNODE, 9X, 11HU = X-DISP.,9X,11HV = Y-DISP./
     2       (5X,I12,2E20.8))
  901 FORMAT(///12H BANDWIDTH =,I4,25H EXCEEDS MAX. ALLOWABLE =,I4//
     1 30H GO ON TO NEXT PROBLEM          )
  999 STOP
      END
```

```
      SUBROUTINE DATAIN(MAXEL,MAXNP,MAXMAT,MAXSLC,ISTOP)
      COMMON NNP,NEL,NMAT,NSLC,NOPT,NBODY,MTYP,
     1E(10),PR(10),RO(10),TH(10),IE(100,5),
     2X(120),Y(120), ULX(120),VLY(120),KODE(120),ISC(20), JSC(20),
     3SURTRX(20,2),SURTRY(20,2)
C
      ISTOP = 0
      READ 1,NNP,NEL,NMAT,NSLC,NOPT,NBODY
C
      PRINT 100,NNP,NEL,NMAT,NSLC,NOPT,NBODY
C
C CHECKS TO BE SURE INPUT DATA DOES NOT EXCEED STORAGE CAPACITY
          IF(NNP.LE.MAXNP) GO TO 201
               ISTOP = ISTOP + 1
      PRINT 251, MAXNP
  201     IF(NEL.LE.MAXEL) GO TO 202
               ISTOP = ISTOP + 1
      PRINT 252, MAXEL
  202     IF(NMAT.LE.MAXMAT) GO TO 203
               ISTOP = ISTOP + 1
      PRINT 253, MAXMAT
  203     IF(NSLC.LE.MAXSLC) GO TO 204
               ISTOP = ISTOP + 1
      PRINT 254, MAXSLC
  204     IF(ISTOP.EQ.0) GO TO 205
      PRINT 255, ISTOP
      STOP
C
  205 READ 2, (E(I),PR(I),RO(I),TH(I),I=1,NMAT)
      PRINT 101
      PRINT 51, (I,E(I),PR(I),RO(I), TH(I),  I=1,NMAT)
C
C  READ AND PRINT NODAL DATA (REF. 1)
      PRINT 103
               N=1
    5 READ 3, M,KODE(M),X(M),Y(M),ULX(M),VLY(M)
         IF(M-N)4,6,7
    4 PRINT 105, M
      PRINT 52,M,KODE(M), X(M),Y(M),ULX(M),VLY(M)
               ISTOP= ISTOP +1
          GO TO 5
    7          DF = M + 1 - N
               RX=(X(M)-X(N-1))/DF
               RY=(Y(M)-Y(N-1))/DF
    8          KODE(N)=0
               X(N)=X(N-1)+RX
               Y(N)=Y(N-1)+RY
               ULX(N)=0.0
               VLY(N)=0.0
    6 PRINT 52,N,KODE(N),X(N),Y(N),ULX(N),VLY(N)
               N=N+1
          IF(M-N)9,6,8
    9     IF(N.LE.NNP) GO TO 5
C
C  READ AND PRINT ELEMENT PROPERTIES, TABLE 6-4
      PRINT 106
   13          L=0
   14 READ 15, M,(IE(M,I),I=1,5)
   16          L=L+1
          IF(M-L)117,17,18
```

```
  117 PRINT 118,M
      PRINT 53,M, (IE(M,I), I=1,5)
               ISTOP=ISTOP+1
          GO TO 14
   18          IE(L,1)= IE(L-1,1)+1
               IE(L,2)= IE(L-1,2)+1
               IE(L,3)=IE(L-1,3)+1
               IE(L,4)=IE(L-1,4)+1
               IE(L,5)=IE(L-1,5)
   17 PRINT 53, L,(IE(L,I),I=1,5)
          IF(M-L)20,20,16
   20     IF(NEL-L)21,21,14
   21     CONTINUE
C
C READ AND PRINT SURFACE LOADING(TRACTION) CARDS
          IF(NSLC.EQ.0) GO TO 31
   30 PRINT 108
          DO  40 L=1,NSLC
      READ 41,ISC(L),JSC(L),SURTRX(L,1),SURTRX(L,2),SURTRY(L,1),
     1 SURTRY(L,2)
   40 PRINT 42,ISC(L),JSC(L),SURTRX(L,1),SURTRX(L,2),SURTRY(L,1),
     1 SURTRY(L,2)
   31     IF(ISTOP.EQ.0) GO TO 999
      PRINT 900, ISTOP
C
    1 FORMAT(6I5)
  100 FORMAT(35H0INPUT TABLE 1.. BASIC PARAMETERS    //
    1        5X, 40H NUMBER OF NODAL POINTS. . . . . . . . . .,I5/
    2        5X, 40H NUMBER OF ELEMENTS. . . . . . . . . . . .,I5/
    3        5X, 40H NUMBER OF DIFFERENT MATERIALS . . . . . .,I5/
    4        5X, 40H NUMBER OF SURFACE LOAD CARDS. . . . . . .,I5/
    5        5X, 40H 1 = PLANE STRAIN, 2 = PLANE STRESS. . .,I5/
    6        5X, 40H BODY FORCES(1 = IN -Y DIREC., 0 = NONE),I5)
  251 FORMAT(////33H TOO MANY NODAL POINTS, MAXIMUM =, I5)
  252 FORMAT(////30H TOO MANY ELEMENTS, MAXIMUM = ,I5)
  253 FORMAT(////30H TOO MANY MATERIALS, MAXIMUM =,I5)
  254 FORMAT(////40H TOO MANY SURFACE LOAD CARDS, MAXIMUM = ,I5)
  255 FORMAT(////28H EXECUTION HALTED BECAUSE OF,I5,13H FATAL ERRORS/)
    2 FORMAT(4E10.3)
  101 FORMAT(36H0INPUT TABLE 2.. MATERIAL PROPERTIES //
    1        10H  MATERIAL,5X,10HMODULUS OF,6X,9HPOISSON'S,7X,
     28HMATERIAL,7X,  8HMATERIAL /
     34X,6HNUMBER,5X,10HELASTICITY,8X,7H  RATIO,8X,7HDENSITY,6X,
     49HTHICKNESS )
   51 FORMAT(I10,4E15.4)
  103 FORMAT(34H1INPUT TABLE 3.. NODAL POINT DATA //
    1        5X,5HNODAL,48X,7HX-DISP.,8X,7HY-DISP./
     25X,5HPOINT,6X,4HTYPE,14X,1HX,14X,1HY,8X,7HOR LOAD,8X,7HOR LOAD)
    3 FORMAT(2I5,4E10.3)
  105 FORMAT(5X,17HERROR IN CARD NO.,I5/)
   52 FORMAT(2I10,4E15.4)
  106 FORMAT(34H1INPUT TABLE 4.. ELEMENT DATA        //
    1        11X,31HGLOBAL INDICES OF ELEMENT NODES/3X,7HELEMENT,
     27X,1H1,7X,1H2,7X,1H3,7X,1H4,2X,8HMATERIAL)
  118 FORMAT(5X, 25HERROR IN ELEMENT CARD NO.,I5/)
   15 FORMAT(6I5)
   53 FORMAT(I10,4I8,I10)
  108 FORMAT(37H1INPUT TABLE 5.. SURFACE LOADING DATA   //
     117X, 33HSURFACE LOAD INTENSITIES AT NODES/
     24X,6HNODE I,4X,6HNODE J,10X,2HXI,10X,2HXJ,10X,2HYI,10X,2HYJ)
```

```
  41 FORMAT(2I5,4E10.3)
  42 FORMAT(2I10,4E12.4)
 900 FORMAT(///45H ASSEMBLY AND SOLUTION WILL NOT BE PERFORMED,,I5,
    121H  FATAL CARD ERRORS     )
 999 RETURN
     END



      SUBROUTINE  ASEMBL(ISTOP)
      COMMON NNP,NEL,NMAT,NSLC,NOPT,NBODY,MTYP,
     1E(10),PR(10),RO(10),TH(10),IE(100,5),
     2X(120),Y(120), ULX(120),VLY(120),KODE(120),ISC(20), JSC(20),
     3SURTRX(20,2),SURTRY(20,2)
      COMMON/ONE/ QK(10,10),Q(10),B(3,10),C(3,3),BT(3,6),XQ(5),YQ(5)
      COMMON/TWO/ IBAND, NEQ, R(240), AK(240,50)
      DIMENSION LP(8)
C
      REWIND 1
C  INITIALIZE
                  ISTOP = 0
C  INITIALIZE PARTS OF MATRICES C AND BT
                  BT(1,4) = 0.0
                  BT(1,5) = 0.0
                  BT(1,6) = 0.0
                  BT(2,1) = 0.0
                  BT(2,2) = 0.0
                  BT(2,3) = 0.0
                  C(1,3) = 0.0
                  C(2,3) = 0.0
                  C(3,1) = 0.0
                  C(3,2) = 0.0
C
C INITIALIZE OVERALL STIFFNESS MATRIX AK AND OVERALL LOAD VECTOR R
              DO  2 I=1,NEQ
                  R(I)=0.0
              DO  2 J =1,IBAND
    2             AK(I,J)=0.0
C
C COMPUTE  ELEMENT  STIFFNESSES AND LOADS ONE BY ONE
C
              DO 10 M=1,NEL
              IF(IE(M,5).GT.0) GO TO 11
                  ISTOP = ISTOP + 1
              GO TO 10
   11         CALL QUAD(M,AREA)
              IF(AREA.GT.0.0) GO TO 16
                  ISTOP = ISTOP + 1
         PRINT 20,M
C
C  CONDENSE ELEMENT STIFF. FROM 10X10 TO 8X8, EQ.(5-64), AND ELEMENT
C  LOADS FROM 10X1 TO 8X1, EQ.(5-64D).   (REF.2)
   16         IF(IE(M,3).EQ.IE(M,4)) GO TO 26
              DO 31 J = 1,2
                  IJ= 10-J
                  IK= IJ+1
                  PIVOT = QK(IK,IK)
              DO 32 K= 1,IJ
```

```
            F = QK(IK,K)/PIVOT
            QK(IK,K)=F
         DO 33 I=K,IJ
            QK(I,K)=QK(I,K)- F*QK(I,IK)
   33       QK(K,I) = QK(I,K)
   32       Q(K) =Q(K)-QK(IK,K)*Q(IK)
   31       Q(IK) =Q(IK)/PIVOT
C
C STORE MULTIPLIERS,PIVOTS,CONDENSED LOADS, STRAIN-DISP. AND STRESS-STRAIN
C MATRICES ON SCRATCH TAPE NO. 1 (TO BE USED LATER TO COMPUTE STRAINS AND
C STRESSES )
   26 WRITE (1) ((QK(I,J),J=1,10),I=9,10), Q(9), Q(10),
     1((B(I,J),J=1,10),I=1,3),((C(I,J),J=1,3),I=1,3),XQ(5),YQ(5)
C
C  ASSEMBLE STIFF.  AND  LOADS , DIRECT STIFF. METHOD, SEC, 6-5.
C
            LIM=8
         IF(IE(M,3).EQ.IE(M,4)) LIM = 6
         DO 40 I=2,LIM,2
            IJ = I/2
            LP(I-1) = 2*IE(M,IJ) - 1
   40       LP(I) = 2*IE(M,IJ)
         DO 50 LL=1,LIM
            I = LP(LL)
            R(I) = R(I) + Q(LL)
         DO 50 MM=1,LIM
            J = LP(MM) - I + 1
         IF(J.LE.0) GO TO  50
            AK(I,J)= AK(I,J)+ QK(LL,MM)
   50    CONTINUE
   10    CONTINUE
C
C ADD EXTERNALLY APPL. CONC. NODAL LOADS TO R
         DO 55 N=1,NNP
         IF(KODE(N).EQ.3) GO TO 55
            K=2*N
         IF( KODE(N).EQ.1) GO TO 57
            R(K-1) = R(K-1) + ULX(N)
         IF(KODE(N).NE.0) GO TO 55
   57       R(K) = R(K) + VLY(N)
   55    CONTINUE
C
C  CONVERT LINEARLY VARYING SURFACE TRACTIONS TO STATIC EQUIVALENTS,
C  AND ADD TO OVERALL LOAD VECTOR  R, EQ.(5-61A).
         IF(NSLC.EQ.0) GO TO 60
         DO 61 L = 1,NSLC
            I = ISC(L)
            J = JSC(L)
            II=2*I
            JJ=2*J
            DX = X(J) - X(I)
            DY = Y(J) - Y(I)
            EL = SQRT(DX*DX + DY*DY)
            PXI=SURTRX(L,1)*EL
            PXJ=SURTRX(L,2)*EL
            PYI=SURTRY(L,1)*EL
            PYJ=SURTRY(L,2)*EL
            R(II-1)=R(II-1)+PXI/3.0 + PXJ/6.0
            R(JJ-1)=R(JJ-1)+ PXI/6.0 + PXJ/3.0
```

```
              R(II)=R(II)+ PYI/3.0 + PYJ/6.0
              R(JJ)= R(JJ) + PYI/6.0 + PYJ/3.0
   61     CONTINUE
C
C INTRODUCE KINEMATIC CONSTRAINTS (GEOMETRIC BOUNDARY CONDITIONS),
C  EQ.(6-18),          REF. 1.
C
   60     DO 70 M=1,NNP
          IF(KODE(M).GE.0.AND.KODE(M).LE.3) GO TO 72
              ISTOP = ISTOP + 1
          GO TO 70
   72     IF(KODE(M).EQ.0) GO TO 70
          IF(KODE(M).EQ.2) GO TO 71
          CALL GEOMBC(ULX(M),2*M-1)
          IF(KODE(M).EQ.1) GO TO 70
   71     CALL GEOMBC(VLY(M),2*M)
   70     CONTINUE
      ENDFILE 1
          IF(ISTOP.EQ.0) GO TO 81
      PRINT 100, ISTOP
   20 FORMAT(/5X,17H AREA OF ELEMENT ,I5,14H  IS NEGATIVE  /)
  100 FORMAT(////42H SOLUTION WILL NOT BE PERFORMED BECAUSE OF  ,I5,
     1 15H   DATA ERRORS        /)
   81 RETURN
      END



      SUBROUTINE   QUAD(M,TOTALA)
      COMMON NNP,NEL,NMAT,NSLC,NOPT,NBODY,MTYP,
     1E(10),PR(10),RO(10),TH(10),IE(100,5),
     2X(120),Y(120), ULX(120),VLY(120),KODE(120),ISC(20), JSC(20),
     3SURTRX(20,2),SURTRY(20,2)
      COMMON/ONE/ QK(10,10),Q(10),B(3,10),C(3,3),BT(3,6),XQ(5),YQ(5)
      COMMON/TWO/ IBAND, NEQ, R(240), AK(240,50)
C
              I= IE(M,1)
              J= IE(M,2)
              K= IE(M,3)
              L= IE(M,4)
              MTYP = IE(M,5)
              TOTALA = 0.0
C
C  CONSTRUCT STRESS-STRAIN MATRIX C,EQ.(3-16C).  FOR PLANE STRAIN
C  NOPT=1, AND FOR PLANE STRESS  NOFT=2,  PRESENT  CODE  IS FOR
C  ISOTROPIC MATERIALS
          IF(NMAT.EQ.1.AND.M.GT.1) GO TO 5
          IF(NOPT.EQ.2) GO TO 2
              CF = E(MTYP)/((1.0+PR(MTYP))*(1.0-2.0*PR(MTYP)))
              C(1,1)=  CF*(1.0-PR(MTYP))
              C(1,2)=  CF* PR(MTYP)
              C(2,1)= C(1,2)
              C(2,2)= C(1,1)
              C(3,3)= CF*(1.0-2.0*PR(MTYP))/2.0
          GO TO 5
    2         CF = E(MTYP)/(1.0-PR(MTYP)*PR(MTYP))
              C(1,1)= CF
              C(1,2)= PR(MTYP)*CF
```

```
              C(2,1)= C(1,2)
              C(2,2)= CF
              C(3,3) = CF*(1.0-PR(MTYP))/2.0
    5         LIM = 4
          IF(K.EQ.L) LIM = 3
              XQ(5) = 0.0
              YQ(5) = 0.0
          DO 10 N=1,LIM
              NN = IE(M,N)
              XQ(N) = X(NN)
              YQ(N) = Y(NN)
              XQ(5) = XQ(5) + X(NN)/FLOAT(LIM)
   10         YQ(5) = YQ(5) + Y(NN)/FLOAT(LIM)
C
C INITIALIZE QUAD. STIFFNESS, LOAD VECTOR AND STRAIN-DISPLACEMENT VECTOR
          DO 13 II =1,10
              Q(II)=0.0
          DO 12 JJ = 1,10
   12         QK(II,JJ)=0.0
          DO 13 JJ=1,3
   13         B(JJ,II) = 0.0
          IF(K.NE.L) GO TO 15
      CALL CST(1,2,3,TOTALA)
          GO TO 999
   15 CALL CST(1,2,5,AREA)
              TOTALA = TOTALA + AREA
      CALL CST(2,3,5,AREA)
              TOTALA = TOTALA + AREA
      CALL CST(3,4,5,AREA)
              TOTALA = TOTALA + AREA
      CALL CST(4,1,5,AREA)
              TOTALA = TOTALA + AREA
  999 RETURN
      END


      SUBROUTINE   CST(I,J,K,AREA)
      COMMON NNP,NEL,NMAT,NSLC,NOPT,NBODY,MTYP,
     1E(10),PR(10),RO(10),TH(10),IE(100,5),
     2X(120),Y(120), ULX(120),VLY(120),KODE(120),ISC(20), JSC(20),
     3SURTRX(20,2),SURTRY(20,2)
      COMMON/ONE/ QK(10,10),Q(10),B(3,10),C(3,3),BT(3,6),XQ(5),YQ(5)
      COMMON/TWO/ IBAND, NEQ, R(240), AK(240,50)
      DIMENSION CB(3,6),LC(6),LT(3),TK(6,6)
C
              LT(1)= I
              LT(2)= J
              LT(3)= K
C
C     COMPUTE  STRAIN-DISPLACEMENT MATRIX B FOR TRIANGLE, EQ. (5-35A)
              BT(1,1)= YQ(J)-YQ(K)
              BT(1,2)= YQ(K)-YQ(I)
              BT(1,3) = YQ(I)-YQ(J)
              BT(2,4)=XQ(K)-XQ(J)
              BT(2,5) = XQ (I)-XQ(K)
              BT(2,6) = XQ(J) -XQ(I)
              BT(3,1)=BT(2,4)
```

```
               BT(3,2) = BT(2,5)
               BT(3,3) = BT(2,6)
               BT(3,4) = BT(1,1)
               BT(3,5)= BT(1,2)
               BT(3,6)= BT(1,3)
               AREA =(BT(2,4)*BT(1,3) - BT(2,6)*BT(1,1))/2.0
C
C  COMPUTE   C*B
          DO 10 II=1,3
          DO 10JJ = 1,6
               CB(II,JJ) = 0.0
          DO 10 KK = 1,3
   10          CB(II,JJ) = CB(II,JJ) + C (II,KK)*BT(KK,JJ)
C
C  COMPUTE (B**T)*C*B, EQ.(5-45A)
          DO 12 II = 1,6
          DO 12 JJ = 1,6
               TK(II,JJ)=0.0
          DO 12 KK=1,3
   12          TK(II,JJ)=   TK(II,JJ)+BT(KK,II)*CB(KK,JJ)
C
C  ADD TRIANGLE STIFNESS TO QUADRILATERAL STIFFNESS, EX.(6-2).
C  ADD TRIANGLE STRAIN-DISPLACEMENT MATRIX TO QUADRILATERAL  STRAIN-
C  DISPLACEMENT MATRIX
          DO 15 II=1,3
               LC(II) = 2*LT(II) - 1
   15          LC(II+3) = 2*LT(II)
          DO 30 II=1,6
               LL = LC(II)
               FK = 1.0/(4.0*AREA)
               FB = 2.0*FK
          DO 20 JJ=1,6
               MM = LC(JJ)
   20          QK(LL,MM) = QK(LL,MM) + TK(II,JJ)*TH(MTYP)*FK
          DO 30 JJ = 1,3
   30          B(JJ,LL) = B(JJ,LL) + BT(JJ,II)*FB
C
C  DEVELOP  BODY  FORCE  VECTOR,  EQ.(5-61B)
          IF(NBODY.EQ.0) GO TO 999
               TBODYF =  AREA* RO(MTYP)*  TH(MTYP)
               BODYF  =  -TBODYF/3.0
          DO 35 II=1,3
                JJ= 2* LT(II)
   35           Q(JJ)= Q(JJ)+ BODYF
  999 RETURN
      END.




      SUBROUTINE  STRESS
      COMMON NNP,NEL,NMAT,NSLC,NOPT,NBODY,MTYP,
     1E(10),PR(10),RO(10),TH(10),IE(100,5),
     2X(120),Y(120), ULX(120),VLY(120),KODE(120),ISC(20), JSC(20),
     3SURTRX(20,2),SURTRY(20,2)
      COMMON/ONE/ QK(10,10),Q(10),B(3,10),C(3,3),BT(3,6),XQ(5),YQ(5)
      COMMON/TWO/ IBAND, NEQ, R(240), AK(240,50)
      DIMENSION SIG(6)
```

```
C
      REWIND 1
      PRINT 300
                 NOLINE = 47
C
C   RETRIEVE MULTIPLIERS, PIVOTS, MATRICES B AND C, AND CENTROIDAL COORD.
C   FOR ELEMENT
            DO 5 M=1,NEL
      READ(1) ((QK(I,J),J=1,10),I=1,2),  Q(9), Q(10),
     1      ((B(I,J),J=1,10),I=1,3), ((C(I,J),J=1,3),I=1,3), XC,YC
C
C   SELECT NODAL DISPLACEMENTS FOR THE ELEMENT
                 LIM = 4
            IF(IE(M,3).EQ.IE(M,4)) LIM = 3
            DO 10 I=1,LIM
                 II = 2*I
                 JJ = 2*IE(M,I)
                 Q(II-1) = R(JJ-1)
   10            Q(II) = R(JJ)
C
C   RECOVER CONDENSED DISPLACEMENTS FOR THE QUADRILATERAL, EQ. (5-64G)
            IF(LIM.EQ.3) GO TO 16
            DO 15 K=1,2
                 JK = K + 8
                 IK = JK - 1
            DO 15 L=1,IK
   15            Q(JK) = Q(JK) - QK(K,L)*Q(L)
C
C   COMPUTE ELEMENT STRAINS, EQ.(5-35A)
                 LIM = 10
                 FAC = 0.25
            GO TO 17
   16            LIM = 6
                 FAC = 1.0
   17       DO 20 I=1,3
                 E(I) = 0.0
            DO 20 J=1, LIM
   20            E(I) = E(I) + B(I,J)*Q(J)*FAC
C
C   COMPUTE ELEMENT STRESSES   , EQ.(5-35B)
            DO 30 I=1,3
                 SIG(I) = 0.0
            DO 30 J=1,3
   30            SIG(I) = SIG(I) + C(I,J)*E(J)
C   COMPUTE PRINCIPAL STRESSES AND THE ANGLE WITH THE POSITIVE X AXIS
                 SP = (SIG(1)+SIG(2))/2.0
                 SM = (SIG(1)-SIG(2))/2.0
                 DS = SQRT(SM*SM+SIG(3)*SIG(3))
                 SIG(4) = SP + DS
                 SIG(5) = SP - DS
                 SIG(6) = 0.0
            IF(SIG(3).NE.0.0.AND.SM.NE.0.0) SIG(6) = 28.648*ATAN2(SIG(3),
     1                                                              SM)
C   PRINT STRESSES, 50 LINES PER PAGE
            IF(NOLINE.GT.0) GO TO 54
      PRINT 1000
                 NOLINE = 49
   54            NOLINE = NOLINE - 1
    5 PRINT 1010, M,XC,YC,(SIG(I),I=1,6)
      ENDFILE 1
```

```
 300 FORMAT(47H1OUTPUT TABLE 2.. STRESSES AT ELEMENT CENTROIDS //
    11X,7HELEMENT,9X,1HX,9X,1HY,4X,8HSIGMA(X),4X,8HSIGMA(Y),4X,
    28HTAU(X,Y),4X,8HSIGMA(1),4X,8HSIGMA(2), 7X,5HANGLE )
1000 FORMAT(1H1,    7HELEMENT,9X,1HX,9X,1HY,4X8HSIGMA(X),4X,8HSIGMA(Y),
    14X,8HTAU(X,Y),4X,8HSIGMA(1),4X,8HSIGMA(2), 7X,5HANGLE )
1010 FORMAT(I8, 2F10.2,1P6E12.4)
     RETURN
     END
```

```
      SUBROUTINE GEOMBC(U,N)
      COMMON/TWO/ IBAND, NEQ, R(240), AK(240,50)
C  THIS SUBROUTINE MODIFIES THE ASSEMBLAGE STIFFNESS AND LOADS FOR THE
C  PRESCRIBED DISPLACEMENT U AT DEGREE OF FREEDOM N, EQ.(6-18B). (REF.1)
          DO 100 M=2,IBAND
                K = N - M + 1
                IF(K.LE.0) GO TO 50
                R(K) = R(K) - AK(K,M)*U
                AK(K,M) = 0.0
   50           K = N + M - 1
                IF(K.GT.NEQ) GO TO 100
                R(K) = R(K) - AK(N,M)*U
                AK(N,M) = 0.0
  100     CONTINUE
                AK(N,1) = 1.0
                R(N) = U
      RETURN
      END
```

```
      SUBROUTINE BANSOL(KKK,AK,R,NEQ,IBAND,NDIM,MDIM)
C SYMMETRIC BAND MATRIX EQUATION SOLVER.  (REF. 2)
C
C KKK = 1 TRIANGULARIZES THE BAND MATRIX AK, EQ. (2-2)
C KKK = 2 SOLVES FOR RIGHT HAND SIDE R, SOLUTION RETURNS IN R, EQ.(2-3)
C
      DIMENSION AK(NDIM,MDIM), R(1)
              NRS = NEQ - 1
              NR = NEQ
          IF(KKK.EQ.2) GO TO 200
          DO 120  N= 1,NRS
              M= N-1
              MR = MIN0(IBAND,NR-M)
              PIVOT = AK(N,1)
          DO 120 L=2,MR
              CP= AK(N,L)/PIVOT
              I = M+L
              J = 0
          DO 110 K=L,MR
              J = J + 1
  110         AK(I,J)= AK(I,J) -CP*AK(N,K)
  120         AK(N,L) = CP
          GO TO 400
```

```
200       DO 220 N=1,NRS
                M= N-1
                MR = MIN0(IBAND,NR-M)
                CP= R(N)
                R(N)=CP/AK(N,1)
          DO 220  L=2,MR
                I = M + L
220             R(I)= R(I) - AK(N,L)*CP
                R(NR) = R(NR)/AK(NR,1)
          DO 320 I = 1,NRS
                N= NR- I
                M= N-1
                MR = MIN0(IBAND,NR-M)
          DO 320 K = 2,MR
                L = M+K
C  STORE COMPUTED DISPLACEMENTS IN LOAD VECTOR R
320             R(N)= R(N)- AK(N,K)*R(L)
400 RETURN
    END
```

I-7 SAMPLE INPUT/OUTPUT

Following is the raw data for a sample problem:

```
   1    PART OF THE PROBLEM ON REDUCIBLE NET, CHAPTER 6
   9    4    1    0    2    0
3,000E+07 3,000E-01 0,000E+00 1,000E+00
   1    3
   2    1 2,250E+00
   3    1 4,500E+00
   4    1 0,000E+00 1,500E+00
   6    0 4,500E+00 1,500E+00 7,500E+02
   7    1 0,000E+00 3,000E+00
   9    0 4,500E+00 3,000E+00 6,250E+02
   1    1    2    5    4    1
   3    4    5    8    7    1
   4    5    6    9    8    1
```

Following is a facsimile of the output for this problem:

PROBLEM 1.. PART OF THE PROBLEM ON REDUCIBLE NET, CHAPTER 6

INPUT TABLE 1.. BASIC PARAMETERS

NUMBER OF NODAL POINTS.	9
NUMBER OF ELEMENTS.	4
NUMBER OF DIFFERENT MATERIALS	1
NUMBER OF SURFACE LOAD CARDS.	0
1 = PLANE STRAIN, 2 = PLANE STRESS. . .	2
BODY FORCES(1 = IN -Y DIREC., 0 = NONE)	0

INPUT TABLE 2.. MATERIAL PROPERTIES

MATERIAL NUMBER	MODULUS OF ELASTICITY	POISSON'S RATIO	MATERIAL DENSITY	MATERIAL THICKNESS
1	0.3000E+08	0.3000E+00	0.0000E+00	0.1000E+01

INPUT TABLE 3.. NODAL POINT DATA

NODAL POINT	TYPE	X	Y	X-DISP, OR LOAD	Y-DISP, OR LOAD
1	3	0.0000E+00	0.0000E+00	0.0000E+00	0.0000E+00
2	1	0.2250E+01	0.0000E+00	0.0000E+00	0.0000E+00
3	1	0.4500E+01	0.0000E+00	0.0000E+00	0.0000E+00
4	1	0.0000E+00	0.1500E+01	0.0000E+00	0.0000E+00
5	0	0.2250E+01	0.1500E+01	0.0000E+00	0.0000E+00
6	0	0.4500E+01	0.1500E+01	0.7500E+03	0.0000E+00
7	1	0.0000E+00	0.3000E+01	0.0000E+00	0.0000E+00
8	0	0.2250E+01	0.3000E+01	0.0000E+00	0.0000E+00
9	0	0.4500E+01	0.3000E+01	0.6250E+03	0.0000E+00

INPUT TABLE 4.. ELEMENT DATA

	GLOBAL INDICES OF ELEMENT NODES				
ELEMENT	1	2	3	4	MATERIAL
1	1	2	5	4	1
2	2	3	6	5	1
3	4	5	8	7	1
4	5	6	9	8	1

OUTPUT TABLE 1.. NODAL DISPLACEMENTS

NODE	U = X-DISP.	V = Y-DISP.
1	0.00000000E+00	0.00000000E+00
2	0.00000000E+00	-.25496829E-04
3	0.00000000E+00	-.10327252E-03
4	0.00000000E+00	-.34237738E-05
5	0.34826087E-04	-.29368057E-04
6	0.71300991E-04	-.10636649E-03
7	0.00000000E+00	-.13794454E-04
8	0.69091631E-04	-.40046505E-04
9	0.13614441E-03	-.11612799E-03

continued

OUTPUT TABLE 2.. STRESSES AT ELEMENT CENTROIDS

ELEMENT	X	Y	SIGMA(X)	SIGMA(Y)	TAU(X,Y)	SIGMA(1)	SIGMA(2)	ANGLE
1	1.12	0.75	2.311E+02	-3.624E+00	2,046E+00	2,311E+02	-3.642E+00	4.995E-01
2	3.37	0.75	2.443E+02	3.624E+00	1,132E+01	2.448E+02	3,092E+00	2.689E+00
3	1.12	2.25	6.919E+02	-2.919E+00	-2,046E+00	6,919E+02	-2,925E+00	-1,687E-01
4	3.37	2,25	6.911E+02	2.919E+00	-1.132E+01	6,912E+02	2,732E+00	-9.425E-01

I-8 POSSIBLE MODIFICATIONS

It is possible to modify and generalize this code in several different ways. We indicate briefly here how some of these possible modifications could be accomplished.

Axisymmetric Case

The code could be modified to be applicable to axisymmetric problems (Example 5-6). A 4×4 stress-strain matrix and a 4×10 strain-displacement matrix would be necessary in QUAD. Corresponding changes in CST, ASEMBL, and STRESS would also be necessary. A routine for numerical integration would be required to compute the element stiffness.

Multiple Loading Cases

The present equations solver, BANSOL, is designed to accommodate multiple loadings. A new subroutine would be required to read the load data and compute the load vector for each case. The main routine could be modified to loop through this new subroutine, through the solution process, and through STRESS for each loading.

Anisotropy

Additional material parameters are required for anisotropic materials. Once these were included in the input data, only the matrix $[C]$ in QUAD would need to be modified. For example, an orthotropic stress-strain law is given in equation (3.13).

Material Nonlinearity

A more detailed description of constitutive behavior is necessary for a nonlinear code. One could use either a functional or digital representation of the material law. Once the data has been read and the semiband width computed, the remaining portion of the code must be an iterative, incremental, or mixed loop. The stress-strain matrix $[C]$ in QUAD would be modified according to the stresses computed in STRESS. The flow diagrams given in Chapter 7 indicate the necessary computational procedures for iterative or incremental methods.

REFERENCES

(1) Wilson, E. L., Program listing, Short Course on Finite Element Method, University of California, Berkeley, 1967. Also see Reference 14, Chapter 2 and References 14 and 20, Chapter 11.

(2) Felippa, C. A., See Reference 13, Chapter 2. Also, with Tocher, J. L., "Discussion: Efficient Solution of Load-Deflection Equations," *Proc. ASCE, J. ST Dn*, Vol 96, ST2 (Feb. 1970). Gives routine for triangularization of banded matrices.

(3) Desai, C. S., Code for nonlinear analysis of soils and rocks in "Solution of Stress-Deformation Problems in Soil and Rock Mechanics Using Finite

Element Methods," Ph. D. Dissertation, University of Texas at Austin, Aug. 1968.

(4) Abel, J. F., Codes for static and dynamic analysis of sandwich shells, included in "Static and Dynamic Analysis of Sandwich Structures with Viscoelastic Damping," Ph. D. Dissertation, University of California, Berkeley, 1968.

FURTHER READING

ICES-STRUDL II, Finite element program of the Integrated Civil Engineering System, (ICES), MIT, Cambridge. The initial system handles plane strain and stress, shallow shells, three-dimensional solids, plate bending and stretching, and plate bending. This is a highly versatile general system. Its use for a specialized problem may be expensive.

ASKA, Automatic System for Kinematic Analysis. Developed by J. H. Argyris, H. A. Kamel and others at the University of Stuttgart. A powerful general system which includes about 42 different finite elements. May be costly for a small specialized user.

Wilson, E. L. "A General Structural Analysis Program (SAP)," Rept. No. SESM 70-20, Department of Civil Engineering, University of California, Berkeley. Linear static and dynamic analysis of elastic structures. Includes elements for 3D frames, 3D beams, axisymmetric solids, 3D solids, plane stress and strain, plates, and shells. Element types may be combined or used separately.

Zienkiewicz, O. C., Programs such as in Reference 3, Chapter 9 and other programs developed at University of Wales, Swansea.

SAFE, Structural Analysis by Finite Elements. Developed by Gulf General Atomic, Inc. A library of programs which includes plane stress or strain, axisymmetric, three-dimensional, and shell problems. In addition to elastic solutions, capabilities for creep, cracking, and reinforcement yielding are available for reinforced and prestressed concrete. Programs for the digitizing of input data and for the graphical display of input and output are also part of the library.

NASTRAN, NAsa STRuctural ANalysis. Developed by U.S. National Aeronautics and Space Administration for elastic analyses of various structures. Capabilities include analyses for thermal expansion, dynamic response to transient loads and random excitation, computation of real and complex eigenvalues, and dynamic stability. Offers limited capability for nonlinear analysis. Adaptable to various computer systems. Includes a number of plotting options. Available from COSMIC.

SAMIS, Structural Analysis and Matrix Interpretive System. Developed by Jet Propulsion Laboratory, Langley Research Center, and Manned Spacecraft Center. Contains an element library including a general one-dimensional element and triangular elements for membrane and bending deformations. Available from COSMIC.

ELAS and ELAS8, General Purpose Computer Programs for the Equilibrium Problems of Linear Structures. Developed by the Jet Propulsion Laboratory. Elements incorporated include one-dimensional, triangular, quadrilateral, tetrahedral, hexahedral, conical frustra, and toroidal elements with quadrilateral and triangular cross sections. Available from COSMIC.

APPENDIX II
ABBREVIATIONS FOR REFERENCES

Abbreviations for some common references cited in this book are explained below.

Abbreviation	*Details*
Proc. ASCE	Proceedings of the American Society of Civil Engineers, New York.
J.EM Dn	Journal of the Engineering Mechanics Division.
J.ST Dn	Journal of the Structural Division.
J. SM&F Dn	Journal of the Soil Mechanics and Foundations Division.
J.HY Dn	Journal of the Hydraulics Division.
First Conf.	Proceedings of the First Conference (26–28 Oct. 1965) on Matrix Methods in Structural Mechanics, Wright-Patterson Air Force Base, Ohio, Nov. 1966. Available from NTIS, Springfield, Va, AD646300.
Second Conf.	Proceedings of the Second Conference (15–17 Oct. 1968) on Matrix Methods in Structural Mechanics, Wright-Patterson Air Force Base, Ohio, Dec. 1969. Available from NTIS, Springfield, Va, AFFDL-TR-68-150.
NTIS	National Technical Information Service, Springfield, Virginia 22151.
Symp. FEM	*Proceedings of the Symposium on Application of Finite Element Methods in Civil Engineering*, ASCE-Vanderbilt University, Nashville, Tenn., Nov. 1969.

IJNME	International Journal for Numerical Methods in Engineering, Wiley-Interscience, London.
USAEWES	U.S. Army Engineer Waterways Experiment Station, P.O. Box 631, Vicksburg, Mississippi 39180.
J. Appl. Mech.	Journal of Applied Mechanics (Proc. ASME, Series E)
HRB	Highway Research Board.
FEM Tapir	*Finite Element Methods in Stress Analysis*, Edited by Holland, I., and Bell, K. Published by TAPIR, The Technical University of Norway, Trondheim, 1969.
J. ACI	Journal of the American Concrete Institute.
AIAA J.	Journal of the American Institute of Aeronautics and Astronautics.
J. Aero. Sci.	Journal of Aerospace Sciences.
Recent Advances	*Recent Advances in Matrix Methods of Structural Analysis and Design* (Proc. of U.S.—Japan Seminar, Tokyo, Aug. 1969), edited by Gallagher, R. H., Yamada, Y., and Oden, J. T., University of Alabama Press, 1971.
COSMIC	Computer Software Management and Information Center, Barrow Hall, University of Georgia, Athens, Georgia 30601.

AUTHOR INDEX

SUBJECT INDEX